PLANT VACUOLES
Their Importance in Solute Compartmentation in Cells and Their Applications in Plant Biotechnology

NATO ASI Series

Advanced Science Institutes Series

A series presenting the results of activities sponsored by the NATO Science Committee, which aims at the dissemination of advanced scientific and technological knowledge, with a view to strengthening links between scientific communities.

The series is published by an international board of publishers in conjunction with the NATO Scientific Affairs Division

A	Life Sciences	Plenum Publishing Corporation
B	Physics	New York and London
C	Mathematical and Physical Sciences	D. Reidel Publishing Company Dordrecht, Boston, and Lancaster
D	Behavioral and Social Sciences	Martinus Nijhoff Publishers
E	Engineering and Materials Sciences	The Hague, Boston, Dordrecht, and Lancaster
F	Computer and Systems Sciences	Springer-Verlag
G	Ecological Sciences	Berlin, Heidelberg, New York, London,
H	Cell Biology	Paris, and Tokyo

Recent Volumes in this Series

Series A: Life Sciences

PLANT VACUOLES

Their Importance in Solute Compartmentation in Cells and Their Applications in Plant Biotechnology

Edited by

B.Marin

ORSTOM, Institut Français de Recherche Scientifique pour
le Dévelopment en Coopération
Paris, France

Springer Science+Business Media, LLC

Proceedings of a NATO Advanced Research Workshop on
Plant Vacuoles: Their Importance in Plant Cell Compartmentation
and Their Applications in Biotechnology
held July 6–11, 1986,
in Sophia-Antipolis, France

Library of Congress Cataloging in Publication Data

NATO Advanced Research Workshop on Plant Vacuoles: Their Importance in
    Plant Cell Compartmentation and Their Applications in Biotechnology
    (1986: Sophia-Antipolis, France)
    Plant vacuoles.

    (NATO ASI series. Series A, Life sciences; v. 134)
    "Proceedings of a NATO Advanced Research Workshop on Plant Vacuoles:
Their Importance in Plant Cell Compartmentation and Their Applications in
Biotechnology, held July 6–11, 1986, in Sophia-Antipolis, France"—T.p. verso.
    "Published in cooperation with NATO Scientific Affairs Division."
    Includes bibliographies and index.
    1. Plant vacuoles—Congresses. 2. Plant cell compartmentation—Congress-
es. 3. Plant cell culture—Congresses. 4. Plants—Metabolism—Congresses. 5.
Plant biotechnology—Congresses. I. Marin, Bernard P. II. North Atlantic Trea-
ty Organization. Scientific Affairs Division. III. Title. IV. Series. [DNLM: 1. Bio-
technology—congresses. 2. Cell Compartmentation—congresses. 3. Cells,
Cultured—congresses. 4. Organoids—metabolism—congresses. 5. Plants—
metabolism—congresses. QK 725 N279p 1986]
QK725.N37   1986   581.87′4   87-14121
ISBN 978-1-4684-5343-0   ISBN 978-1-4684-5341-6 (eBook)
DOI 10.1007/978-1-4684-5341-6

© 1987 Springer Science+Business Media New York
Originally published by Plenum Press New York in 1987
Softcover reprint of the hardcover 1st edition 1987

To Nicolas
and Michèle

These Proceedings of the First NATO/ORSTOM Advanced Research Workshop on Plant Vacuoles are dedicated to

Professor Philippe Matile

Through very active experimental research on lysosomes and vacuoles, Philippe Matile has made an impressive and extensive contributions to knowledge of the vacuolar compartment in fungi and plants.

During the past decades, he has played a vital role in the development of Biochemistry related to vacuoles. With his warm personality he has shared knowledge, technical expertise, and enthusiasm in a generous manner with students and colleagues all over the world.

PREFACE

The papers in this book, illustating the present status of knowledge related to the vacuolar compartment of fungi and plants, were presented at an Advanced Research Workshop entitled "Plant Vacuoles. Their Importance in Plant Cell Compartmentation and their Applications in Biotechnology" held in Sophia-Antipolis, France, on July 6-11, 1986.

The organizers were fortunate in being able to assemble representative leaders of all the above fields of research concerning this compartment. These scientists from all over the world were invited to present their latest results and to exchange views and plans for continued research in this highly exciting field, which is very important for the improving the industrial aspect of plant biotechnology related to the production of molecules with a high added value.

To ensure maximal flexibility and opportunity for dicussion, the chairmen of the various sessions were asked to introduce their respective topics, after which they were given the freedom of organizing their sessions in a more or less improvised fashion, with brief presentations of relevant new information by participants following a schedule fixed relatively shortly before the workshop. This procedure proved to be highly successful, giving rise to most productive and stimulating discussions among the representatives of the various fields.

These proceedings included the lectures by invited speakers, as well as poster presentations. The papers by W. Husemann and M. Roberts were not given during the workshop because these colleagues were unable to attend. However, they kindly provided their manuscripts for publication in this book. Certain other papers have been deleted, especially to limit too much overlapping . This is the case of the paper presented by X. Gidrol et al. on the existence of a third class of proton pump where it is possible to classify the tonoplast ATPase (see Proceedings of the XIth Yamada Conference on Energy Transduction through ATPase systems, May 26-27, 1985, Kobe, Japan). All the papers published in this volume provide a faithful image of the research carried out within this field, especially on the chemiosmotic proton circuits in tonoplast of fungi and higher plants. The papers have not been subjected to any cutting or editing, in order to preserve the individual nature of each contribution.

The systems examined here by the different authors cover most of the areas where very rapid and active research is being performed : Beta, sugarcane, Catharanthus, and Hevea. This list is not exhaustive. Spectacular progress has recently been made in the methods used to isolate the vacuoles. Another very important point which is not sufficiently considered is the nature of the tissue used and its physiological condition. Nevertheless, on the basis of the information presented here, the reader should, in the future, expect for new information in specific reviews.

This Workshop could not have taken place without the generous financial support of the Scientific Affairs Division, NATO, Brussels, Belgium, and DIVA, ORSTOM, Paris, France. Support from DIAA, Ministère de l'Agriculture, Paris, France, is also gratefully acknowledged. We would also like express our thank for the financial support of Conseil Régional de Provence et Côte-d'Azur as well as the support of the following firms : Elf-Biorecherches, Castanet-Tolosan, France; LKB-Instruments, Orsay, France; L'Oréal, Paris, France; Pharmacia, Saint-Quentinen-Yvelines, France; Roussel-UCLAF, Paris, France; and Wild-Leitz-France, Rueil-Malmaison, France. Particularly warm thanks are due to DGRCST, Ministère des Affaires Etrangères, Paris, France, for their travel fellowships.

We are deeply indebted to our colleagues in the staff of Department F of ORSTOM, Paris, France, for their enthusiastic help. We would like to express an especial word of thanks to Miss Michele Lanza-Marin, for her day-to-day help in all the administrative problems encountered during the Workshop. She rendered essential service in checking manuscripts, proofreading and redrawing figures. The book could not have ready in such a relatively short time without her invaluable help.

The authors have contributed with grace and a high standard of knowledge. They have updated their material to the time of publication of this volume. The book has been published so soon due to the efforts of all the contributors, and also to the cooperation and patience of the staff of Plenum Publishing Corporation.

Finally, we have pleasure in dedicating this workshop, organized under the auspices of the Minsitère de la Recherche et de la Technologie, Paris, France, the Scientific Affairs Division of NATO, Bruxelles, Belgium, and U.N.E.S.C.O., Paris, France, to Professor Philippe Matile.

Bernard Marin,
Montpellier, France

CONTENTS

PRODUCTION OF BIOCHEMICALS
(SECONDARY METABOLITES AND ENZYMES SEQUESTERATED
IN VACUOLES) BY PLANT CELL CULTURES

OVERVIEW OF RECENT DEVELOPMENTS

Bernard Marin

Research Unit "Biochemical and Physiological Mechanisms
of Plant Production", ORSTOM, 213, Rue LaFayette,
F-75.480-Paris-Cedex 10, France

Plant vacuoles are specific organelles in plant cells, just as much as chloroplasts, but possess very different properties, as has been shown by recent advances in knowledge of this compartment, which was for too long considered to be uninteristing.

What a lot of things can be said about them ! What a lot of abstruse quarrels, often resulting from observations made under different conditions. Thus, in meristem tissue, the vacuole is in the form of a polydispersed compartment with lysosomal properties. Then, in mature, differentiated tissue there is a single, central vacuole occupying up to 90% of the volume of the cell.

The vacuole was considered for a long time as a relatively inert compartment whose only function was to maintain turgor pressure, thus contributing to the growth and rigidity of plants. Later, it was thought that it was a compartment in which all that the cytoplasm or even other plant cell organelles did not want might accumulate: a place in which everything that might harm the functioning of the cell could collect. It had a detoxification function. However, as research on vacuoles progressed, it turned out that they handle several important functions with regard to the life of plants and their adaptation to the environment. The moment has come to draw up a balance of all the work that has been done on defining their functions.

The idea that the vacuole is a compartment that is incapable of effecting the least biosynthesis appears to be out of date. Vacuoles participate in a number of specific metabolic pathways. Thus, the final step of the synthesis of ethylene, the conversion of 1-amino-cyclo-propane-1-carboxylic acid takes place in the vacuole although the main site of ethylene action is not in this compartment. In addition, the first stage of the mobilisation of sugars which accumulate in vacuoles – such as sucrose and stachyose – is intravacuolar. Certain transformations of secondary metabolites such as that of o-sinapoyl-glucose into o-synapoyl-malate or those involving the various esters of hydroxycinnamoyl-malic acid are also intravacuolar.

Vacuoles are also concerned in degradation processes, where the type of involvement is varied. The way in which cytoplasm molecules are degraded in vacuoles is not understood and the mechanisms involved are the subject of much debate.

A large number of different molecules, which are often of economic interest (various alkaloids, saponins, steroid derivatives, coumaryl glycosides, various glycosides such as the glyco-conjugated derivative of gibberellic acid, etc.) are also accumulated and sequestered in vacuoles. In most cases vacuoles do not synthesize the molecules that they accumulate, but receive molecules generally synthesized in the cytoplasm or in certain cases in chloroplasts. The molecules are finally released. This is why vacuoles are not considered as classic cellular organelles but as an internal medium of the same functional nature nature as the external medium. The membranes surrounding them possess energization systems which are very similar but not identical. They have redox systems and proton pumping ATPases which contribute to the establishment of a difference in electrochemical proton potential providing the energy required for the functioning of most of the carriers described to date. It is therefore necessary now to analyse in detail the mechanisms involved in the transport of all these molecules, which generally takes place against a concentration gradient, and their accumulation in the intravacuolar medium.

The development of new fractionation techniques made it possible to isolate the various organelles in the plant cell without altering their natural properties and it has been possible to analyse the evolution of the contents of the various compartments in function of the physiological conditions encountered. It was then observed that vacuoles play a particularly important role in the compartmentation of solutes in plant cells. Most products of photosynthesis tend to accumulate in the vacuoles. This accumulation is reversible under normal conditions of cell functioning. In addition, the same process can be observed if these molecules are modified in the cytoplasm and if they are involved in a secondary metabolism. Nevertheless, accumulation may be in the form of sequestration.

In spite of considerable progress in organic synthesis and in pharmacy, plant cells reamin an important source of molecules used in most medicines. However, these molecules are often trapped in the vacuole. In certain cases, the conditions for releasing them into the cytoplasm are known. It will only be possible for them to be excreted as required into the external medium when details of the regulation mechanisms of the enzymes controlling vacuolar accumulation of secondary metabolites are known.

Most research on isolated plant cells and their ability to produce metabolites with a high added value aims, on a more or less long term basis, at using microbiological engineering techniques with comparable efficiency and scale. It is now necessary to identify the conditions required for the excretion into the external medium of molecules trapped in the vacuole. If it is accepted from recent results that an exchange of solutes and signals occurs between the cytoplasm and the external medium, it must be accepted that such exchange also exists between cytoplasm and vacuole, whence the importance of this type of research, which deserve further development and an inventory of knowledge drawn up for the still little-known field of plant vacuoles.

VACUOLES TODAY AND TWENTY YEARS AGO

Philippe Matile

Department of Plant Biology, University of Zurich,
Zurich, Switzerland

Nothing could demonstrate better the current awareness for the significance of plant vacuoles than the recent flood of reviews covering all the different aspects of vacuolar functions and reflecting the atonishing progress in the understanding of this compartment. Twenty years ago it was the search for the analogue in plant cells of animal lysosomes which provoked an attempt to isolate vacuoles and to study the subcellular compartmentation of hydrolytic enzymes. It appeared to be utterly impossible to isolate the large and fragile central vacuoles. However, it turned out that the small vacuoles from meristematic roots can be liberated by the chopping of plasmolysed tissue and enriched by appropriate centrifugation procedures (Fig. 1). The observation that several hydrolases had much higher specific activities in vacuolar fractions than in the homogenates represented an initial basis for postulating the lysosomal nature of plant vacuoles (Matile, 1966). The concept of the "lytic compartment of plant cells" which eventually emerged (Matile, 1975), was largely based on the localization of hydrolases in special types of vacuoles that could be isolated at that time, e.g. small vacuoles present in the latex of plants having articulated laticifers, protein bodies, or fungal vacuoles. Morphological evidence favouring the origin of the main vacuoles of mature cells from meristematic vacuoles was another basis of this concept. In addition, the lysosomal nature of

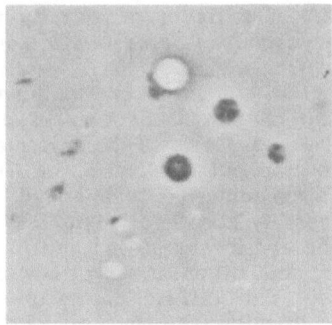

Figure 1

Vacuoles released upon the chopping
of plasmolysed corn root tips
and enriched by floatation
in a mannitol medium containing Urografin
(from Matile, 1966)

vacuoles was infered from electron micrographs showing the apparent decay of cytoplasmic structures suspended in the vacuolar fluid.

The breakthrough was achieved by Wagner (1975) who has thought us how large vacuoles can be isolated from protoplasts. Using this technique, Boller and Kende (1979) have succeeded in the unambiguous localization in vacuoles of a number of hydrolases. One of them, α-mannosidase, is now widely used as a useful marker enzyme of vacuoles. What is still missing now is not the knowledge about the lysosomal nature of vacuoles but rather the understanding of the lysosomal function. An important contribution with regard to the lysosomal function has recently been presented by Canut et al. (1985). They demonstrated that newly synthesized protein is broken down within the vacuoles of suspension cultured cells of Acer pseudoplatanus. This finding may represent a first step towards the elucidation of the role of vacuoles in the intracellular digestion of protein.

The present popularity of vacuoles does certainly not concern the lysosomal aspect but rather the various solutes accumulated in the cell saps and the transport processes across the tonoplast. The study of subcellular compartmentation of solutes in vacuoles is important because it is an indirect approach to the understanding of the cytoplasm, the truly living entity of the cell. The vacuole gradually emerges as the indispensable extraplasmatic compartment of plant cells which is employed by the cytoplasm for the purpose of maintaining proper metabolic conditions. A corresponding interpretation of vacuolar functions has first been proposed by Wiemken and Nurse (1973) who studied the subcellular compartmentation of amino-acids in baker's yeast.

One aspect of homeostasis is detoxification, i.e. the removal from the cytoplasm of substances that in one way or another could impair the finely regulated metabolic machinery. Some of the secondary metabolites are particularly interesting as they are only potentially toxic. Cyanogenic glycosides, o-coumarylglycoside, saponin glycosides, and glucosinolates are examples of such compounds which upon the rupture of tissues or cells are enzymatically converted to toxic products. It is obvious that the enzymes responsible for the production of toxic principles must be spatially separated from their substrates. The central vacuole has been identified as the storage compartment for the glycosides mentioned, and it has also been demonstrated that the glycosidases responsible for the release of toxic metabolites are located in extravacuolar compartments. Since toxic secondary metabolites are likely to represent an important part of the defense of plants against herbivores and biotrophic fungi, the vacuole must be regarded as an arsenal of chemical weapons. The subcellular organization of secondary metabolism which eventually leads to the safe deposition of toxic or potentially toxic compounds in the cell sap is still another fascinating aspect of detoxification (see Matile, 1984).

Small vacuoles present in the latex of Chelidonium majus could be isolated easily because they contain convenient markers in the form of several coloured alkaloids such as sanguinarine, berberine, chelerytrine and others (Matile et al., 1970). It was challenging to find out the cause for the retention of these solutes within the vacuoles. When sanguinarine was added to a suspension of isolated vacuoles, the alkaloid was readily absorbed (Fig. 2). It turned out that this process was neither dependent on metabolic energy nor did it appear to be due to carrier-mediated transport. These and other observations led to the hypothesis that uncharged lipophilic alkaloids can diffuse across membranes and are eventually trapped as cations in the acid milieu of the cell sap, or else they are complexed with vacuolar phenolics. This view has, however, recently been questionned upon the demonstration of highly specific carrier-mediated transport of alkaloids in vacuoles of several species (Deus-Neumann and Zenk, 1984 and 1986). Whereas the trap hypothesis predicts the unspecific accumulation of alkaloids depending on the pk of the lipophilic base, the vacuolar pH and the formation of non-diffusible complexes with vacuolar phenolics (see e.g. Renaudin and Guern, 1982), the results obtained in Zenk's laboratory suggest that the excess of alkaloids to cell saps is strictly dependent on the specific alkaloid carriers residing in the tonoplast. It is puzzling to see that on the one hand the accumulation of nicotine in vacuoles has been employed for the determination of vacuolar pH values in species that do not normally produce nicotine (e.g. Kurkdjian et al., 1981), whilst on the other hand in Zenk's laboratory, nicotine was taken up by vacuoles of Nicotiana but not by vacuoles of various species producing other alkaloids. It appears that under the experimental conditions employed in Zenk's laboratory the unspecific trapping of alkaloids in the acidic cell saps was abolished. Yet, it seems that these conditions were a prerequisite for the demonstration of specific carrier-mediated transport of alkaloids across the tonoplast which normally is masked by the unspecific diffusion and trapping. It is feasible that in vivo both mechanisms of alkaloid transport exist. In any case, it would be tempting to resume experimentation with alkaloid vacuoles of Chelidonium taking advantage of the enormous progress in the understanding of transport processes in tonoplasts which has been made in the past ten years.

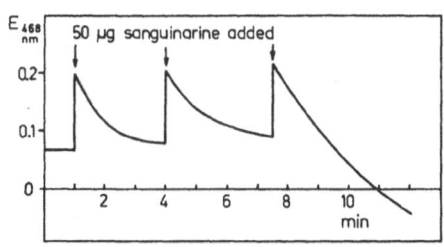

Figure 2

Absorption of the alkaloid sanguinarine
in vacuoles isolated from Chelidonium latex
(from Matile et al., 1970)

Both cuvettes of a double-beam spectrophotometer were filled with a suspension of vacuoles (turbidity nearly compensated). The changes in absorption at 450 nm are due to changes of sanguinarine concentration in the medium. Upon the repeated addition of alkaloid the vacuoles eventually burst and the turbidity of the reference is no longer compensated. Alkaloids were completely soluble at the end of the experiment.

REFERENCES

Boller, T., and Kende, H., 1979, Hydrolytic enzymes in the central vacuole of plant cells, Plant Physiol., 63:1123.

Canut, H., Alibert, G., and Boudet, A. M., 1985, Hydrolysis of intracellular proteins in vacuoles isolated from Acer pseudoplatanus L. cells, Plant Physiol., 79: 1090.

Deus-Neumann, B., and Zenk, M. H., 1984, A highly selective alkaloid uptake system in vacuoles of higher plants, Planta, 162:250.

Deus-Neumann, B., and Zenk, M. H., 1986, Accumulation of alkaloids in plant vacuoles does not involve an ion-trap mechanism, Planta, 167:44.

Kurkdjian, A., Morot-Gaudry, J. F., Wuilleme, S., Lamant, A., Jolivet, E., and Guern, J., 1981, Evidence for an action of fusicoccin on the vacuolar pH of Acer pseudoplatanus cells, Plant Science Letters, 23:233.

Matile, P., 1966, Enzyme der Vakuolen aus Wurzelzellen von Maiskeimlingen. Ein Beitrag zur funktionellen Bedeutung der Vakuole bei der intrazellulären Verdauung, Z. Naturforsch., 21b:871.

Matile, P., 1975, "The Lytic Compartment of Plant Cells", Cell Biology Monographs, Vol. 1, Springer-Verlag, Wien, New-York.

Matile, P., 1984, Das toxische Kompartment der Pflanzenzelle, Naturwissenschaften, 71:18.

Matile, P., Jans, B., and Rickenbacher, R., 1970, Vacuoles of Chelidonium latex : lysosomal property and accumulation of alkaloids, Biochem. Physiol. Pflanzen, 161:447.

Renaudin, J.-P., and Guern, J., 1982, Compartmentation mechanisms of indole alkaloids in cell suspension cultures of Catharanthus roseus, Physiol. Vég., 20:533.

Wiemken, A., and Nurse, P., 1973, The vacuole as a compartment of amino-acid pools in yeast, Proc. Third Int. Specialized Symp. on Yeasts, Part 2, pp 331-347.

METHODOLOGICAL AND OTHER ASPECTS OF INTACT

MATURE HIGHER PLANT CELL VACUOLES

George J. Wagner

Agronomy Department, University of Kentucky
Lexington, KY 40546-0091, USA

INTRODUCTION

I will, for the most part, restrict my discussion to methodological aspects of recent studies of intact, mature plant cell vacuoles. However, it is important to first review very briefly the history of discovery and study of plant vacuoles in general, since modern studies and methodologies are built upon that which was done before.

Observation of plant vacuoles began with the discovery of the mature cell vacuole by Meyen, Schleiden and others in the 1830's (Table 1) (Zirkle, 1937; Voeller, 1964; Wagner, 1983). Hartig recognized the occurrence of isolated protein bodies of seeds as early as 1856 (Pernollet, 1978). Later smaller vacuoles were observed in meristematic tissues and Went concluded that vacuoles were ubiquitous in plant tissues (Voeller, 1964).

During the 1880's and thereafter, utilizing the light microscope, selected plant tissues, endogenous pigments, vital stains, dissection, etc., various workers elucidated the basic functions of the mature plant cell vacuole. The English translation of Alexandre Guilliermond's papers entitled "The Cytoplasm of the Plant Cell", which was published in 1941, as well as books/reviews by Bailey and Kuster (Voeller, 1964), contain many fascinating discussions of the dynamic nature of plant vacuoles. During the period between about 1960 to 1975, considerable attention was focused on the origin of vacuoles. Ultrastructural studies by Buvat, Marty, Poux and others focused on the ontogeny of vacuoles using the cytochemistry of hydrolases as a tool (Marty et al., 1980).

In the late 1960's the era of isolation and biochemical characterization of plant vacuoles began with the description by Matile and colleagues of methods for isolating vacuoles from yeast and meristematic tissues of higher plants (Matile, 1973 and 1975). About the same time, the latex of <u>Hevea</u> was recognized as a rich source of hydrolase containing vacuoles, and these were isolated (Marin, 1985). Several basic approaches to isolating mature plant cell vacuoles were introduced in 1975 (Wagner, 1983 and 1985). Finally, pioneering work of Sze and Marin has spearheaded extensive studies of tonoplast transport which are rapidly making transport mechanisms of tonoplast among the best understood transport processes in plants (Marin, 1985; Sze, 1985).

In the past 10 years, about 100 papers have been published which have taken advantage of isolated mature plant cell vacuoles to study compartmentation and

Table 1

History of Plant Vacuoles

1830's	Discovery and observation	Meyen, Schleiden and others
1856	Isolation of protein bodies	Hartig
1880's	Recognition of basic functions	DeVries, Hof, Went, Pfeffer, Dangeard, Chambers, Guilliermond, and others
1960-1975	Studies of vacuole origin	Buvat, Marty, Poux, and others
1969	Isolation of immature higher plant vacuoles and isolation of vacuoles of Hevea latex	Matile, Pujarniscle, D'Auzac, Marin, and others
1967	Isolation of yeast and Neurospora	Matile, Wiemken, Cramer, Vaugh and Davis, and others
1975-1976	Isolation of mature higher plant vacuoles	Wagner and Siegelman, Leigh and Branton, Lorz et al. and others
1980	Pioneering studies of transport in plant membranes	Marin et al., Sze, Spanswick, Poole, Taiz, and collaborators

the biochemistry and transport properties of the tonoplast. Perharps such studies will help us explain some of the observations of vacuoles and vacuolar function made in situ by early cytologists.

I would like to take one additional diversion before discussing methodology. That is to point out the multifunctional nature of mature plant cell vacuoles (Table 2). Some of these functions will be discussed from a mechanistic point of view in papers to follow in this meeting while others are still outside the relm of mechanistic studies. But, all these functions, I believe, were recognized by early cytologists.

Osmotic functions of the mature vacuole are unique to higher and multicellular lower plants. In addition to providing the means for mechanical support and tropisms, and for cell expansion and stomatal movement, evolution of the vacuole - it can be argued - allowed plants to invade the terrestrial environment. The vacuole permits a small cytosol volume yet a large cell surface, allowing for increased interception of light for photosynthesis. It was long suspected from studies utilizing pulse chase analysis that the central vacuole was a major (slow turnover) reservoir for many inorganic ions. Some direct evidence now supports this conclusion. Similarly for organic metabolites, the metabolite-catabolite storage function of the vacuole is now well established, questions of what is a metabolite and what is a catabolite notwithstanding. Recent studies are consistent with a role for the vacuole in regulating cytosolic metabolism. Reports of calmodulin and calcium storage in vacuoles (Boudet et al., 1987; Poole and Blumwald, 1987) and the suggestion that tonoplast pyrophosphatase may serve to regulate cytosolic PP_i levels (Leigh and Pope, 1987) are tantalizing in this regard. Studies of the lytic role of the plant vacuole can now be focused to the level of asking questions about particular proteins using western blot methodology. However, a fundamental problem concerning studies of lytic functions where protoplasts are utilized is the question of how much perturbation of normal lytic events is brought about during protoplast isolation in preparation for recovery and analysis of vacuole constituents.

Early cytologists recognized that tissues were, in measure, a continuum, in that transvacuolar strands ans plasmodesmata interconnected the cytoplasm of tissue cells. The diagram of a leaf cross-section shown in Fig. 1 illustrates how well this was recognized when light microscopic observations of living tissue were the only means for making observations at the subcellular level. In contrast to this view, recall the picture of the plant cell projected by typical transmission light micrographs that occur in most modern textbooks. In such micrographs, the vacuole appears as an empty balloon - long after fixation, staining, post-fixation, etc. have disrupted the fragile transvacuolar strand system prevalent in the living cell. Perharps additional fragile ulstrastructural features are also disrupted and rendered unrecognizable.

Table 2

Known and Suggested Functions of Higher Plant Central Vacuoles

---

1) Osmotic - in concert with the cell wall and cytosol

   a) Mechanical support - tissue turgidity
   b) Tissue movements - nastic, geotropism phototropic
   c) Motive force for cell expansion
   d) Stomatal function
   e) Vacuole dynamics

2) Ion balance, storage and sequestration

3) Metabolite storage - primary and secondary metabolites

4) Sequestration - metabolite-catabolite, pollutant

5) Lytic - senescence, "normal" and "abnormal" turnover (?)

6) Intracellular and intercellular mixing - via transvacuolar strand system, a short transit distance from cell-to-cell

7) Intracellular organization

   a) Permit small cytosol volume yet large cell surface
   b) Provide for efficient distribution of chloroplasts
   c) Cytoplasmic cleavage in spore formation (?)
   d) Provide for protective positioning of nucleus (?)

8) Regulation of cytosolic metabolism ?

Finally, patterns of chloroplast distribution, nuclear location (often in center of cells, see Fig. 1), and the occurrence of vacuoles at spore cleavage planes, almost force one to speculate on the role of the vacuole - or perharps cytoskeletal positioning of the vacuole - in intracellular organization.

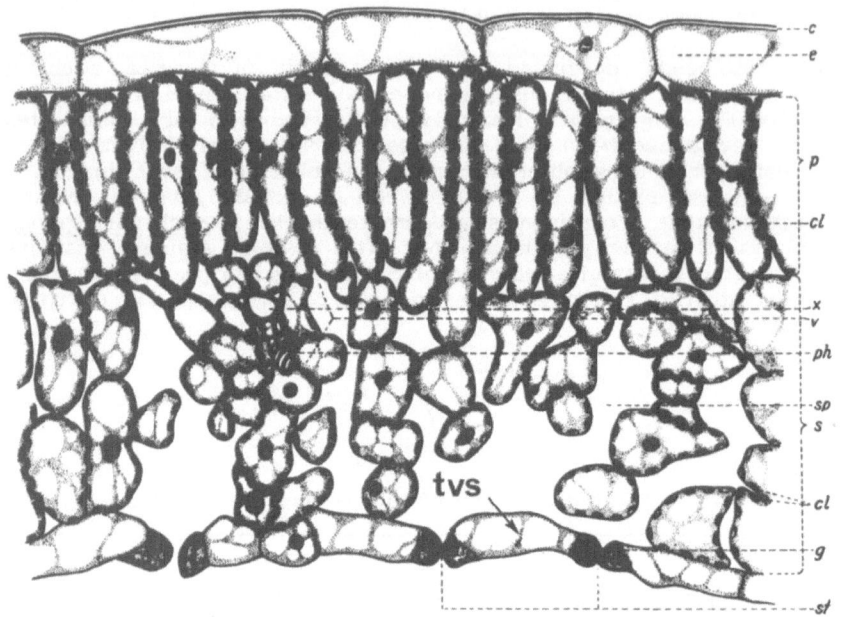

Figure 1

Depiction of the cross section of a leaf

Cross section of leaf of <u>Lonicera</u> <u>tatarica</u> illlustrating cuticle, c; epidermis, e; palisade mesophyll, p, with chloroplasts, cl; vein, v, with xylem, x, and phloem, ph; spongy mesophyll, s, with air spaces, sp, and chloroplasts, cl; lower epidermis with stomata, st, and guard cells, g. By permission from Botany, a textbook for Colleges, J. B. Hill, L. D. Overholts and H. W. Popp, McGraw-Hill Co., 1950. Note the transvacuolar strand network (tvas) and general position of nuclei in cells.

So, early cytologists such as De Vries, Chambers, Guilliermond, Dangeard and others did very well with the light microscope to discover the basic functions of mature plant vacuoles (Voeller, 1964; Guilliermond, 1941). Indeed this organelle is perharps the best example of how simple observations of living tissue can yield much information, even about a subcellular organelle. Of course, mature vacuoles - being extremely large, often filled with chromogenic or potentially chromogenic substances and usually acidic and therefore a site for concentration of weakly basic dyes such as neutral red - did and still do lend themselves to observational studies. However, while such observations of intact systems do relate to the real world and probably are of great importance to understanding relationships of vacuoles to broader questions such as agricultural productivity, they are nevertheless limited. When one wants to study membrane chemistry, enzymology, transport, etc., the object must be separated from the whole system, despite the dangers inherent in such separation. Since it is usually not possible to obtain the most pure and also most biochemically competent organelle membrane preparations using a single method, one needs to consider the goal of a study before adopting methods for isolation.

Currently, there are three main goals in studies which utilize isolated intact vacuoles (Table 3). If one's objective is to study compartmentation, purity, intactness (extent of leakiness to the solutes of concern), and the relationship of vacuolar solutes in vivo to that in isolated vacuoles are all of major concern. If transport studies are the goal, purity may be sacrificed (within limits) for the sake of retaining activity and preserving the cytosol/tonoplast interface - assuming this interface is important to the transport process being studied. Of course, these general arguments apply to the isolation of any subcellular or sub-tissue component which possesses both compartmentation and transport properties. For biochemical characterization of tonoplast, purity is essential, but loss of peripheral membrane components must also be controlled or understood.

Table 3

Current Studies Using Isolated, Intact Vacuoles

---

1) Vacuole/extravacuole compartmentation
    Main criteria:
    a) purity
    b) intactness
    c) relationship to tissue

2) Tonoplast transport
    Main criteria:
    a) retention of activity
    b) retention of tonoplast/cytosol interface

3) Biochemical characterization of tonoplast
    Main criteria:
    a) purity
    b) consideration of peripheral components

ISOLATION OF INTACT MATURE PLANT CELL VACUOLES

I would like to now discuss basic method for isolating mature plant cell vacuoles within the above context. There are three basic methods for releasing vacuoles from enzymatically released protoplasts : osmotic lysis, polybase induced lysis in isotonic medium, and lysis due to shear in isotonic medium (Wagner, 1986). A fourth basic method of vacuole isolation involves direct release of vacuoles from tissue into plasmolyzing medium. All of these methods generally, but not always, involve a pH adjustment to about pH 8.0. The protoplast based methods have been applied to many tissues including leaves, roots, stems, petals, cotyledons, etc., while the tissue slicing method appears to be restricted to root tissues which have certain characteristics. Yields of vacuoles using protoplast based procedures range from 5 to 75%, most commonly perharps 10 to 20%. The root slicing method yields in the range of 1 to 3%.

A recent survey of the literature showed that in about 80 papers describing isolation of vacuoles from protoplasts, less than 6% used a published method unchanged. In contrast, most investigators isolating vacuoles from tissue use essentially the original method (except that slicing is often done by hand instead of machine) and indeed the same tissue - red beet root.

Modification of basic procedures for vacuole isolation are often said to improve the yield, purity and functionality (stability, etc.) of vacuoles. To this author's knowledge, no study has compared basic procedures without modification of that previously used. Since slight modification in procedures (osmoticum level, density/composition of purification medium, etc.) are often needed with a single tissue, depending on physiological state of tissue, condition of protoplasts, it is difficult to interpret the general uselfulness of many modifications presented in the literature.

How, then, does one interpret the literature regarding the best method for performing objectives 1, 2 or 3 (Table 3) in an untested system or in attempting to improve a published protocol ? One approach is to examine the total literature for the most used methods and, as a first attempt, apply these.

Osmotic Lysis of Protoplasts

There are about 45 published, refereed papers describing vacuole isolation by osmotic lysis of protoplasts. This method has been applied to leaves, roots, stems, cotyledons, suspension cells, tubers, petals, and guard cells. The basic methods first described by Wagner and Siegelman (1975) and modified by Boller and Kende (1979) are often cited. This procedure involves lysis of protoplasts generally equilibrated in 0.6 to 0.7 M osmoticum, pH 5.5 by rapid procedure to 0.1 to 0.2 M KPO_4 buffer, pH 8 followed by washing and purification by various means. Generally, vacuoles isolated using this procedure are obtained in 10 to 20% yield and are relatively free from contamination by membraneous and soluble extravacuolar constituents. Numerous vacuole/extravacuole, microsolute compartmentation studies have utilized these methods successfully. Boller and Kende (1979) introduced a now widely employed procedure for purifying vacuoles obtained by osmotic lysis using step gradients of Ficoll. The above cited procedures fulfill reasonably well criteria required for successful vacuole/extravacuole compartmentation studies.

An alternative to lysis in KPO_4 buffer is to simply lower the level of osmoticum bathing protoplasts by 20 to 80%. In general, vacuoles isolated by lowering osmoticum appear to be contaminated to a greater degree than those prepared in KPO_4 buffer. However, KPO_4 buffer is known to equilibrate medium and sap pH, presumably due to K^+/H^+ exchange. Microsolutes other than H^+, K^+, and Na^+ appear to be largely retained in vacuoles prepared by osmotic lysis (Leigh, 1983; Wagner, 1986; Ryan and Walker-Simmons, 1983). The extent to which osmotic shock and relatively high ionic strength (in the case of KPO_4) may influence the vacuole/cytosol interface is unknown.

Polybase Induced Lysis of Protoplasts

Buser and Matile (1983) first adapted to plant protoplasts this method which was originally devised for preparation of vacuoles from yeast. It has been applied in at least 25 published, refereed publications for isolating vacuoles (under isotonic conditions) from protoplasts of leaves, petals, roots, suspension cells and other tissues. It appears to be gaining in popularity. The method described by Boudet et al. (1981) is widely cited. Here protoplasts are centrifuged at low speed through zones containing 2% Ficoll, 4 mg/ml DEAE dextran, 5% Ficoll, 4 mg/ml Dextran sulfate onto a 20% Ficoll cushion. All layers contain isotonic mannitol and are buffered at pH 7.7. Vacuoles accumulate at the lowest interface. Boudet's laboratory has recently modified this procedure to improve yield and purity (Canut et al., 1987). Reported yields vary from 3 to 50% and the level of contamination is reported to be comparable to that observed using osmotic shock in KPO_4. The extent to which reagents critical to this method (DEAE dextran, dextran sulfate) may influence the vacuole/cytosol interface is not known. This is also true for other reagents sometimes incorporated during vacuole isolation, such as EDTA, PVP, BSA, PEG, KCl, KPO_4, DTT, cysteine, and metabisulfite (Wagner, 1983). In this volume, Colombo et al. (1987) report proton pump activity in vacuoles isolated using the polybase procedure and also the osmotic shock procedure (lowered osmoticum). In the latter, but not the former, the pH gradient had to be reduced with NH_4^+ or by vesiculation (equilibration) in weak buffer before pump activity could be observed. This suggests that the polybase procedure results in loss of protons from the vacuole sap. It also indicates that vacuoles prepared by all four basic methods are protontransport competent, if they are poised to express pump activity. *Tulipa* vacuoles prepared by osmotic shock in KPO_4 were shown to be capable of proton transport earlier (Wagner, 1985).

Lysis of Protoplasts by Shear in Isotonic Medium

There are about 20 refereed publications from different laboratories describing the isolation of vacuoles from protoplasts by shear effected by centrifugation through viscous medium or by passage through a small orifice. In many of these procedures, adjustement of pH to about pH 8 is employed to destabilize protoplasts and stabilize vacuoles. The first described isotonic shear-lysis procedure was that of Lorz et al. (1976). Often cited modifications of this procedure are those by Guy et al. (1979) and Thom et al. (1982). In the Guy et al. procedure, protoplasts equilibrated at pH 5.8 are passed by centrifugation at high speed through an isotonic Ficoll containing step gradient adjusted to pH 7.7. Certain isotonic lysis procedures appear to yield more highly contaminated vacuole preparations, but in some cases, these preparations are shown to be "transport-competent" and it has been argued that purity is sacrificed for preservation of transport activity (Thom et al., 1982) and possibly the tonoplast/cytosol interface critical to transport functions. Proton and sugar transport have been successively studied using vacuoles isolated by this approach. Thom et al. (1982) report the occurrence of fragments of plasmalemma on vacuoles prepared in this manner. The isotonic, shear-lysis method of Martinoia et al. (see Wagner, 1985 and different contributions in this volume) appears to yield vacuoles having good purity, at least from barley leaf.

Isolation of Vacuoles Directly from Root Tissue

The basic method described by Leigh and Branton (1983) for slicing red beet root into plasmolyzing medium has been used extensively. The only major modification has been to employ hand chopping. Purity of vacuoles so prepared – after purification on metrizamide gradients – are as good as those obtained using osmotic lysis and polybase methods. The major drawbacks of the direct isolation approach are low yields (1%) and restriction to certain tissues. The major advantage of this approach is that compartmentation changes which may occur during protoplast preparation (minimum time required is ~1 hr in plasmolyzing medium) are largely avoided. Recently, Leigh and Deri Tomos (1984) recorded some loss of Na^+ and K^+ during the initial steps of the direct isolation procedure, but larger changes in at least ion compartmentation probably accompany protoplast based procedures.

CRITERIA/CONDITIONS FOR PURITY, STABILITY, INTEGRITY OF ISOLATED VACUOLES

Various membrane and soluble component markers (Wagner, 1986; Quail, 1979; Nagahashi, 1985) are useful for assessing purity of vacuole preparations to be used in compartmentation analysis and for biochemical characterization of isolated vacuoles (Table 4). Membrane markers are most crucial as evidenced by the presence of such activities recorded in many studies where soluble markers are reported to be absent. Presentation of data which suggest that all contamination occurs as intact residual protoplasts along with photographs which show debris as well as residual protoplasts suggests underestimation of contamination. Light micrographs of preparations containing few structures are less useful than those showing numerous ones. Presentation of useful light micrographs can provide for confidence in the state of purity concluded from marker enzyme assays.

Table 4

Criteria/Conditions for Purity, Stability, Integrity of Isolated Vacuoles

1) Compartmentation, Analysis and Biochemical Characterization of Tonoplast

Purity : Membrane and soluble contamination markers :

Mitochondria and inner membrane	- cytochrome c oxidase
Chloroplast and inner membrane	- chlorophyll
ER	- Antimycin A-insensitive NADH cyt c reductase is not appropriate
Cytosol	- glucose-6-P dehydrogenase,
	- malate dehydrogenase,
	- alcohol dehydrogenase,
Nuclear	- DNA, RNA
Golgi	- glucan synthase I
	- UDPase is not appropriate
Chloroplast envelope	- galactosyl transferase
Plasmalemma	- glucan synthase II
	- UDPG sterol transferase

Quantitation :
- counting structures
- correlative methods using pigments, secondary metabolites or enzymes
- use of ^3H_2O and ^{14}C-non-permeating polymer to measure vacuole pellet volume and extra-vacuole volume

Retention of Contents :

2) Transport Studies

Retention of activity and tonoplast/cytosol interface and stability
- pH 7.5 to 8.0, low temperature (4°C), rapid isolation
- stability/viability effectors Mg, Ca, or EDTA ? BSA, DTT, PVP, PEG, Ficoll, etc. ?

Cytochrome c oxidase and chlorophyll have been effectively used to monitor mitochondria and chloroplasts (or their inner membranes), respectively. An often used marker for endoplasmic reticulum - antimycin, a insensitive NADH cytochrome c reductase - is an invalid ER marker for vacuole preparations, since several investigators have shown the occurrence of this activity with tonoplast otherwise shown to be relatively free of membraneous contamination. Cinnamic acid-4-hydroxy-lase is probably also not appropriate as an ER marker since tonoplast may bear low levels of this (or another P-450) type hydroxylase (Wagner and Hrazdina, unpublished). It is often difficult to detect NADPH cytochrome c reductase in fractions of plant homogenates. Thus, an effective, generally applicable ER marker for vacuole preparations has not been identified. Several "cytosol" markers including glucose-6-phosphate dehydrogenase, malate dehydrogenase and alcohol dehydrogenase have been used effectively. Nuclear, Golgi, chloroplast envelope and plasmalemma markers are seldom utilized, though RNA and DNA, glucan synthase I, galactosyl transferase, and glucan synthetase II, UDPG sterol transferase are useful markers for these membranes, respectively. Since vacuoles preparations generally contain both acid phosphatase and ATPase, using UDPase as a Golgi marker may be problematic. Morre et al. (1987) has discussed the use of membrane thickness and PTA staining to distinguish some cellular membranes.

Accuracy in quantitation is crucial in compartmentation analysis where isolated structures are being compared with intact tissue or protoplasts derived from tissue. The most often cited procedure for comparing vacuolar and protoplast microsolute content is counting in the light microscope using a depth calibrated chamber and related contents on a per structure basis. Problems encountered in this method can easily lead to an error of perhaps 15 – 20% (Wagner, 1983; Leigh, 1983). Correlative methods based on the use of pigments, secondary metabolites, enzymes, where the component used for correlation is first shown to be restricted to or predominantly restricted to the vacuole have been widely used. Here -mannosidase is often used. Table 5 presents a summary of the reported occurrence of three acid hydrolase activities in sap of isolated vacuoles. Generally, but not always, α-mannosidase and acid protease are found to be predominantly vacuolar. While both α-mannosidase and acid protease are expected to occur in other cellular

Table 5

Experimentally Determined Vacuole/Extravacuole Distribution
of the Most-Studied Plant Acid Hydrolases[a]

Source of Vacuoles	Relative activity in vacuole (percent)[b]			Reference
	α-Mannosidase	Acid phosphatase	Acid protease	
Tomato leaf proto.	–	30	–	Walker-Simmons and Ryan (1977)
Castorbean endosperm proto.	–	29	77	Nishimura and Beevers (1978)
Tobacco suspension cell proto.	120–71	86	105	Boller and Kende (1979); Boller (1982)
Pineapple leaf proto.[c]	130	97	98–120	Boller and Kende (1979); Boller (1982)
Tulip petal proto.[d]	85	94	–	Boller and Kende (1979); Boller (1982)
Beet root tissue[d]	–	100	–	Leigh et al. (1979b)
Tobacco suspension proto.	–	100	–	Mettler and Leonard (1979)
Carrot suspension proto.[d]	–	55	–	Sasse et al. (1979)
Tobacco leaf proto.	–	93	–	Saunders (1979)
Horseradish root tissue	–	90–106	–	Grob and Matile (1979)
Bean cotyledon proto.	90	–	–	Van der Wilden and Chrispeels (1983)
Barley leaf proto.	109	59– 72	82– 87	Heck et al. (1981)
Wheat and corn leaf proto.	–	–	106–107	Lin and Wittenbach (1981)
Wheat leaf proto.	–	–	112–103	Wittenbach et al. (1982)
Barley leaf proto.[e]	100	–	85	Martinoa et al. (1981)
Wheat leaf proto.[e]	–	–	87	Wagner et al. (1981)
Hippeastrum petal proto.[e]	–	–	15– 29	Wagner et al. (1981)
Melilotus leaf proto.	116	50	–	Boudet et al. (1981)
Sugar cane suspension proto.	–	96	48– 49	Thom et al. (1982a)
Barley leaf proto.	98	–	–	Kaiser et al. (1982)
Gentiana root proto.	97	123	–	Keller and Wiemken (1982)
Barley leaf proto.	84–89	82	–	Grandstedt and Huffaker (1982)
Wheat leaf proto.	47	52	100–102	Waters et al. (1982)
Barley leaf proto.	112	77	–	Wagner et al. (1983)
Barley leaf proto.	85	–	95	Thayer and Huffaker (1984)
Buckwheat suspension proto.	100	–	–	Hollander–Czytko and Amrhein (1983)[34]
Tobacco leaf proto.	90	–	–	Saunders and Gillespie (1984)[35]
Pea leaf proto.	95	103	–	Guy and Kende (1984)[36]
Digitalis suspension proto.	100	–	–	Kreis and Reinhard (1985)[37]
Raphanus cotyledon proto.	no act.	60	–	Sharma and Strack (1985)[38]
Parsley suspension proto.	low	95	–	Matern et al. (1986)[39]

[a]Modified from (9)
[b]Unless otherwise noted, quantitation was made by using the error prone method of counting numbers of protoplasts and vacuoles (estimated error ±15 to 20%)
[c]Assumed α-mannosidase was 100% vacuolar
[d]Quantitation by correlation method using vacuolar pigment
[e]Quantitation by total sap volume determination

15

organelles, the first in Golgi (Szumilo et al., 1986) and the second in chloroplasts (Nettleton et al., 1985) and perharps elsewhere, the bulk of total cellular activity of these enzymes is expected to be contained within the vacuole. In contrast, acid phosphatase is less reliable as a vacuole sap marker (Table 5).

Various secondary metabolites including endogenous pigments, alkaloids, ascorbic acid, etc., have been used in compartmentation analysis. Again, this method is most often applied after determining the degree to which the compound to be used for correlation is vacuolar, and that determination is generally made by counting structures (Wagner, 1986). Therefore, accuracy of quantitation is relative to accuracy of the initial determination. A third, but not often used, approach to quantitation is to measure total preparation volume and extra-vacuole volume of a vacuole preparation after centrifugation through a hydrophobic liquid using a ^{14}C-labeled nonpermeating component and $^{3}H_2O$ (Wagner, 1985).

Two further considerations which must be made in attempting to relate results of compartmentation analysis to in vivo subcellular distribution of metabolites are 1) changes in compartment contents which can occur during protoplast isolation and vacuole release from protoplast or tissue, and 2) the difference in size of isolated vacuoles in hypertonic medium verses their size in vivo. These aspects have been most thoroughly discussed by Leigh and Deri Tomos (1984).

For studies of transport utilizing isolated vacuoles, one is principally interested in retaining stability and activity of the isolated structures for the period of assay. It seems clear that isolated vacuoles for most plants prefer pH 7.5 - 8.0, low temperature storage and mechanical disturbance. While numerous additives have been suggested as having stabilizing effects, results are contradictory. The incorporation of EDTA in isolation medium appears to be growing in popularity. This chelator is said by some to improve vacuole release from protoplasts and also to stabilize vacuoles. Others have not seen dramatic effects of EDTA in other systems, but the question has only been studied systematically in one case (Komor et al., 1982).

The question of the importance of the vacuole/cytosol interface has been speculated upon (Wagner, 1986; Boudet et al., 1985). There is no experimental evidence to describe the condition of the tonoplast face after vacuole isolation by various methods, but vacuoles prepared by shear induced protoplast lysis are competent in proton and sugar transport - albeit not with tissue level efficiency (Komor et al., 1982). Vacuoles recovered by slicing of red beet root are also competent in relatively efficient sugar and proton transport (see Wagner, 1986). The Hevea lutoid system continues to yield new information regarding primary and secondary transport pathways (Marin, 1985).

Various laboratories are beginning to address the question of how secondary metabolites which are accumulated in the vacuole are transported to and retained in the vacuole. These studies are in their infancy, however, compound specific uptake has been demonstrated in several cases using isolated vacuoles (Rataboul et al., 1985; Werner and Matile, 1985; Deus-Neumann and Zenk, 1984; Matern et al., 1986). It has been speculated that specific transporters may exist in the tonoplast for secondary metabolites. There is currently much debate concerning the specific porter versus the ion-trap models for alkaloid transport (see Guern and et al., 1987; Renaudin and Guern, 1987). How the cytosol might tolerate the occurrence of potentially reactive secondary metabolites en route to the vacuole is little discussed. An alternative route involving secondary metabolites packaged in vesicles will be discussed in a later paper (Wagner, 1987). There is ultrastructural evidence to support a vesicular transport pathway in the case of the highly reactive tannins (Parham and Kaustinen, 1977; Wagner, 1981) and terpenoids (Carde et al., 1980). There may be several different modes of transporting secondary products through the cytosol.

SUMMARY

The large size, sap occurrence of chromogenic or potentially chromogenic substances, extreme sap pH, and perharps tonoplast extensibility were all factors which allowed early cytologists to learn so much about mature plant vacuoles through observations made of living cells. These factors now allow us to isolate vacuoles intact (particularly tonoplast extensibility) and study transport and compartmentation properties of this organelle. The trend in the development of methodology for isolating and manipulating intact vacuoles is to use basic approaches and modify these in an attempt to optimize the system being tested. There is still a need to devise new methods for isolating vacuoles directly from tissues other than roots, because problems inherent in the use of protoplasts (as useful as these may be) are not likely to be easily circumvented.

There is now a great need to devise efficient methods for separating cell types prior to vacuole isolation so that vacuoles of a particular cell type can be studied. When vesicles from homogenates are used to study transport, the extent of inside-out versus outside-in vesicles must be determined, and these separated. It is now possible to determine the proportion of sealed vesicles in a population (Poole and Blumwald, 1987). Also, more use should be made of agents which selectively permeabilize plasma membrane in intact tissue/cells Cu^{2+} (Anraku, 1987), DMSO, etc. and NMR (Pfeffer et al., 1987) to study less disrupted systems. Finally, evacuolated protoplasts (Steingraber and Hampp, 1987), capable of regenerating a vacuole could be very useful for studying vacuole ontogeny and function.

REFERENCES

Anraku, Y., 1987, Active transport of amino-acids and calcium ions in fungal vacuoles, in : "Plant Vacuoles. Their Importance in Solute Compartmentation and Their Applications in Biotechnology", B. Marin, ed., Plenum Publishing Corporation, New-York.

Boller, T., and Kende, H., 1979, Hydrolytic enzymes in the central vacuole of plant cells, Plant Physiol. 63:1123.

Boudet, A. M., Alibert, G., and Marigo, G., 1985, Vacuoles and tonoplast in regulation of cellular metabolism, Physiol. Veg.,

Boudet, A. M., Alibert, G., Marigo, G., and Ranjeva, R., 1987, The vacuole : Possible role in signal transduction versus cytoplasmic homeostasis ?, in : "Plant vacuoles. Their importance in Solute Compartmentation and Their Applications in Biotechnology", B. Marin, ed., Plenum Publishing Corporation, New-York.

Boudet, A. M., Canut, H., and Alibert, G., 1981, Characterization of vacuoles from Melilotus alba mesophyll protoplasts, Plant Physiol. 68:1354.

Canut, H., Alibert, G., Carasco, A., and Boudet, A. M., 1987, Protein degradation in vacuoles from Acer pseudoplatanus L. cells, in : "Plant Vacuoles. Their Importance in Solute Compartmentation and their Applications in Biotechnology", B. Marin, ed., Plenum Publishing Corporation, New-York.

Carde, J. P., Bernard-Dagan, C., and Gleizes, M., 1980, Membrane systems involved involved in the synthesis and transport of monoterpene hydrocarbons in pine leaves, in : "Biogenesis and Function of Plant Lipids", P. Mazliak, P. Benveniste, C. Costes, and R. Douce, eds., p. 441, Elsevier Biomedical Biomedical Press, Amsterdam.

Colombo, R., Cerana, R., and Lado, P., 1987, Mg,ATP-dependent H^+ transport in vacuoles and tonoplast vesicles from Acer pseudoplatanus cells, in : "Plant Vacuoles. Their Importance in Solute Compartmentation and Their Applications in Biotechnology", B. Marin, ed., Plenum Publishing Corporation, New-York.

Deus-Neumann, B., and Zenk, M. H., 1984, A highly selective alkaloid uptake system in vacuoles of higher plants, Planta, 162:250.

Guern, J., Renaudin, J. P., and Barbier-Brygoo, H., 1987, Accumulation of organic solutes in plant vacuoles : The interpretation of data is not so easy, in : "Plant Vacuoles. Their Importance in Solute Compartmentation and Their Applications in Biotechnology", B. Marin, ed., Plenum Publishing Corporation, New-York.

Guilliermond, A., 1941, "The Cytoplasm of the Plant Cell", Chronica Botanica, Waltham, MA.

Guy, M., and Kende, H., 1984, Conversion of 1-amino-cyclopropane-1-carboxylic acid to ethylene by isolated vacuoles of Pisum sativum L., Planta, 160:281.

Guy, M., Reinhold, L., and Michaeli, D., 1979, Direct evidence for a sugar transport mechanism in isolated vacuoles, Plant Physiol., 64:61.

Hollander-Czytko, H., and Amrhein, N., 1983, Subcellular compartmentation of shikimic acid and phenylalanine in buckwheat cell suspension cultures grown in the presence of shikimate pathway inhibitors, Plant Science Letters, 29:89.

Knuth, M. E., Keith, B., Clark, C., Garcia-Martinez, J. L., and Rappaport, L., 1983, Stabilization of transport capacity of cowpea and barley vacuoles, Plant Cell Physiol., 24:423.

Komor, E., Thom, M., and Maretzki, A., 1982, Vacuoles from sugarcane cells. III. Proton-motive potential difference, Plant Physiol., 69:1326.

Kreis, W., and Reinhard, E., 1985, Rapid isolation of vacuoles from suspension-cultured Digitalis lanata cells, J. Plant Physiol., 12:385.

Leigh, R. A., 1983, Methods, progress and potential for the use of isolated vacuoles in studies of solute transport in higher plants, Physiol. Plant., 57:390.

Leigh, R. A., and Deri Tomos, A., 1984, An attempt to use isolated vacuoles to determine the distribution of sodium and potassium in cells of storage roots of red beet, Planta, 159:469.

Leigh, R. A., and Pope, A. J., 1987, Understanding tonoplast function : Some emerging problems, in : "Plant Vacuoles. Their Importance in Solute Compartmentation and Their Applications in Biotechnology", B. Marin, ed., Plenum Publishing Corporation, New-York.

Lorz, H., Harms, C. T., and Potrykus, I., 1976, Isolation of vacuoplasts from protoplasts of higher plants, Biochem. Physiol. Pflanzen, 169:617.

Marin, B. P., ed., 1985, "Biochemistry and Function of Vacuolar Adenosine-triphosphatase in Fungi and Plants", Springer-Verlag, Berlin, Heidelberg, New-York, and Tokyo.

Marty, F., Branton, D., and Leigh, R.A., 1980, in: "The Biochemistry of Plants", P. Stumpf and E. Conn, eds., Vol. 1, Academic Press, New-York.

Matern, V., Reichenbach, C., and Heller, W., 1986, Efficient uptake of flavonoids into parsley vacuoles requires acetylated glycosides, Planta, 167:183.

Matile, P., 1975, "The Lytic Compartment of Plant Cells", Cell Biology Monographs, Vol. 1, Springer-Verlag, Berlin and New-York (1975), and Annu. Rev. Plant Physiol., 29:193.

Morre, D. J., Brightman, A., Scherer, G., Von Dorp, B., Penel, C., Auderset, G., Sandelius, A. S., and Greppin, H., 1987, Highly purified tonoplast fractions by preparative free-flow electrophoresis, in : "Plant Vacuoles. Their Importance in Solute Compartmentation and Their Applications in Biotechnology", B. Marin, ed., Plenum Publishing Corporation, New-York.

Nagahashi, G., 1985, in : "Modern Methods of Plant Analysis", H. F. Linskens and J. F. Jackson, eds., Vol. 1, Springer-Verlag, Berlin., Heidelberg, New-York, Tokyo.

Nettleton, A. M., Bhalla, P. L., and Dalling, M. J., 1985, Characterization of peptide hydrolase activity associated with thylakoids of the primary leaves of wheat, J. Plant Physiol., 119:35.

Parham, R. A., and Kaustinen, H. M., 1977, On the site of tannin synthesis in plant cells, Bot. Gaz., 138:465.

Pernollet, J. C., 1978, Protein bodies of seeds, Phytochemistry, 17:1473.

Pfeffer, P. R., Tu, S. I., Gerasimowicz, W. V., and Boswell, R. T., 1987, Role of the vacuole in metal ion trapping as studied by in vivo ^{31}P-NMR spectroscopy, in : "Plant Vacuoles. Their Importance in Solute Compartmentation and Their Applications in Biotechnology", B. Marin, ed., Plenum Publishing Corporation, New-York.

Poole, R. A., and Blumwald, E., 1987, Transport of inorganic ions in tonoplast vesicles, in : "Plant Vacuoles. Their Importance in Solute Compartmentation and Their Applications in Biotechnology", B. Marin, ed., Plenum Publishing Corporation, New-York.

Quail, P. H., 1979, Plant cell fractionation, Annu. Rev. Plant Physiol., 30:425.

Rataboul, P., Alibert, G., Boller, T., and Boudet, A. M., 1985, Intracellular transport and vacuolar accumulation of o-coumaric acid glucoside in Melilotus alba mesophyll cell protoplasts, Biochim. Biophys. Acta, 816:25.

Renaudin, J. P., and Guern, J., 1987, Ajmalicine transport into vacuoles isolated from Catharanthus roseus cells, in : "Plant Vacuoles. Their Importance in Solute Compartmentation and Their Applications in Biotechnology", B. Marin, ed., Plenum Publishing Corporation, New-York.

Ryan, C. A., and Walker-Simmons, M., 1983, in : Methods in Enzymol., 96:580.

Saunders, J. A., and Gillepsie, J. M., 1984, Localization and substrate specificity of glycosidases in vacuoles of Nicotiana rustica, Plant Physiol., 76:885.

Sharma, V., and Strack, D., 1985, Vacuolar location of 1-sinapoyl-glucose:L-malate-sinapoyl-transferase in protoplasts from cotyledons of Raphanus sativus, Planta, 163:563.

Steingraber, M., and Hampp, R., 1987, Vacuolar and cytosolic metabolite pools by comparative fractionation of vacuolate and evacuolate protoplasts, in : "Plant Vacuoles. Their Importance in Solute Compartmentation and Their Applications in Biotechnology", B. Marin, ed., Plenum Publishing Corporation, New-York.

Sze, H., 1985, H^+-translocating ATPases: Advances using membrane vesicles, Annu. Rev. Plant Physiol., 36:175.

Szumilo, T., Kaushal, G. P., Hori, H., and Elbein, A. D., 1986, Purification and properties of a glycoprotein processing α-mannosidase from mung bean seedlings, Plant Physiol., 81:383.

Thom, M., Maretzki, A., and Komor, E., 1982, Vacuoles from sugarcane suspension cultures. Isolation and partial characterization, Plant Physiol., 69:1315.

Voeller, B. R., 1964, in : "The Plant Cell", Vol. 6, J. Brachet and A. E. Mirsky, eds., Academic Press, New-York.

Wagner, G. J., 1981, Compartmentation in plant cells : The role of the vacuole, Recent Advan. Phytochemistry, 16:1.

Wagner, G. J., 1983, in : "Isolation of Membranes and Organelles from Plant Cells", J. L. Hall, and A. L. Moore, eds., Academic Press, London, New-York, Paris, San-Diego, San-Francisco, Sau Paulo, Sydney, Tokyo, Toronto.

Wagner, G. J., 1985, in : "Modern Methods of Plant Analysis", H. F. Liskins, and J. F. Jackson, eds., Vol. 1, Springer-Verlag, Berlin, Heidelberg, New-York, Tokyo.

Wagner, G. J., 1987, "Isolation of Higher Plant, Mature Vacuoles: General Principles Criteria for Purity and Integrity", in : Methods in Enzymol., Academic Press, New-York (in press).

Wagner, G. J., 1987, The possible role of endoplasmic reticulum in the biosynthesis and transport of anthocyanin pigments, in : "Plant Vacuoles. Their Importance in Solute Compartmentation and Their Applications in Biotechnogy", B. Marin, ed., Plenum Publishing Corporation, New-York.

Werner, C., and Matile, P., 1985, Accumulation of coumaryl-glucosides in vacuoles of barley mesophyll protoplasts, J. Plant Physiol., 118:237.

Zirkle, C., 1937, The plant vacuole, Bot. Rev., 3:1.

PROPERTIES OF VACUOLES

AS A FUNCTION OF THE ISOLATION PROCEDURE

Hélène Barbier-Brygoo, Jean-Pierre Renaudin, Pierre Manigault, Yves Mathieu, Armen Kurkdjian, and Jean Guern

Laboratoire de Physiologie Cellulaire Végétale, C.N.R.S.-I.N.R.A. Boite Postale No 1, F-91190-Gif-sur-Yvette, France

INTRODUCTION

A more and more detailed picture of the properties of vacuoles is emerging from the literature of the past few years. The vacuoles are major storage compartments for acid hydrolases, sucrose, organic acids such as malate and citrate, basic amino-acids and many different secondary metabolites. Aside this general and coherent picture, large discrepancies exist in the literature concerning the intensity of the electrochemical potential difference of protons ($\Delta\overline{\mu}_H{}^+$) across the tonoplast of isolated vacuoles. The electrical potential difference (E_m) across the tonoplast of isolated vacuoles was reported to be negative when calculated from the equilibrium distribution of permeant lipophilic cations, whereas positive E_m were measured with microelectrodes (for a review, see Leigh, 1983, and Gibrat et al., 1985 a). As to the transtonoplast pH gradient, the only agreement is on the fact that the vacuoles are acidic relative to the cytoplasm. But very few measurements of the transtonoplast ΔpH in cells or isolated vacuoles have been performed. Furthermore, reports of the dissipation of this pH gradient during the isolation of vacuoles and their subsequent manipulation (Schmitt and Sandermann, 1982; Matern et al., 1986) contrast with other results demonstrating the stability of the ΔpH.

Such large uncertainties in the values of ΔpH and E_m across the tonoplast limit considerably our knowledge of the intensity and regulation mechanisms of the driving force for transport processes coupled to $\Delta\overline{\mu}_H{}^+$. Possible origins of these discrepancies could be due to : i) Differences in the biological materials from which the vacuoles were isolated; ii) Important bias introduced by the different techniques used to measure vacuolar parameters. The fact that E_m values were shown to differ markedly when measured on the same vacuolar suspension either with the lipophilic cation TPP$^+$ (E_m = - 60 mV) or with microelectrodes (E_m = + 13 mV) established clearly that the contradiction lies in the methods themselves (Barbier-Brygoo et al., 1985). It was shown that in fact TPP$^+$ binding to the tonoplast was responsible for an apparent intravacuolar accumulation and simulated the existence of a negative transmembrane potential (Gibrat et al., 1985 b). Furthermore, the modulation of the apparent negative E_m by classical effectors (KCl, FCCP, MgATP) was demonstrated to result from variations of probe binding. It was thus concluded that lipophilic cationic probes were unsuitable for measuring E_m on vacuolar suspensions and iii) Modifications of the properties of the vacuoles induced by the isolation procedure or selection by this procedure of subclasses of vacuoles from a population with a large dispersion of properties.

This paper will concentrate mainly on this third possible origin of the discrepancies encountered in the literature. Results will be presented about the properties (E_m, ΔpH) of vacuoles isolated in highly concentrated saline conditions – e.g. NaCl 500 mM (Deus-Neumann and Zenk, 1984) or KCl 800 mM (Korzun et al., 1984) – compared to those of vacuoles isolated in neutral osmotica such as mannitol or sorbitol. The variability of pH in populations, and the possible selection of subclasses by the isolation procedure will be then considered.

ISOLATION OF VACUOLES FROM CATHARANTHUS ROSEUS CELLS IN NEUTRAL OR SALINE OSMOTICA

The procedure to isolate vacuoles from suspension–cultured cells of Catharanthus roseus has been described elsewhere (Renaudin et al., 1986). Protoplasts were prepared from the cells by enzymic digestion of the cell walls in a medium containing 550 mM sorbitol during 2 h. They were then washed thrice to eliminate the cell-wall degrading enzymes. The yield of protoplast recovery was ca 80%. Dramatic losses of solutes have been shown to occur during protoplast preparation. Isolated protoplasts contained only 13% NO_3^-, near 50% malate, citrate, ajmalicine, K^+ and Ca^{2+}, 80% serpentine and Mg^{2+} and 100% sugars relative to cells (Renaudin et al., 1986). This important and differential loss was related both to the removal of the cell wall, a site of Ca^{2+} accumulation, and to the large shrinkage induced by the initial plasmolysis of the cells.

To prepare vacuoles, protoplasts were ruptured osmotically by incubating for 30 min in 200 mM sorbitol, 5 mM EDTA-Na$_2$ (pH 7.2). Isolated vacuoles were purified in a single-step gradient made of 13 ml sample phase containing ruptured protoplasts, 360 mM sorbitol, 5 mM EDTA-Na$_2$, 6.7% (w/v) Nycodenz (a neutral triiodinated derivative of benzoic acid, Nyegaard & Co), pH 7.3, above which 1 ml upper phase comprising 550 mM sorbitol, 10 mM Hepes-KOH pH 7.3 was layered. The vacuoles were recovered in the upper 0.7 ml after centrifugation for 3 min at 160 g. The yield of purified vacuoles relative to the initial protoplasts was 20% based on α-mannosidase, a vacuolar marker, with approximately 2.0 vacuoles deriving from one protoplast. From the measurement of the contamination by extravacuolar marker enzymes the purity of the vacuolar preparation was rather good. The content and concentration of solutes in vacuoles (organic acids, mineral ions, alkaloids) have been estimated and compared to those of the protoplasts (Renaudin et al., 1986).

In order to isolate vacuoles in NaCl or KCl, a fraction of the protoplasts was resuspended, during the washings, in 500 mM NaCl (or KCl), 10 mM Hepes-KOH pH 7.3. The osmotic shock was done in the same way as for sorbitol, at a final concentration of 200 mM NaCl (or KCl). The lower phase of the gradient was made of 330 mM NaCl (or KCl), 5 mM EDTA-Na$_2$, 10 mM Hepes-KOH pH 7.3, 10% (w/v) Nycodenz. The upper phase was 500 mM NaCl (or KCl), 10 mM Hepes-KOH pH 7.3 and 5% (w/v) Nycodenz (in order to compensate for the lower density of the salts relative to sorbitol). Vacuoles isolated in salts had a yield, based on number, lower than those isolated in sorbitol, since they were ca 1.0 10^6/ml in salts instead of 2.5 10^6/ml in sorbitol. The two kinds of vacuoles were relatively stable since 75% (on a number basis) remained after 5 h.

INFLUENCE OF THE ISOLATION PROCEDURE ON POTENTIAL DIFFERENCE AND pH GRADIENT ACROSS THE TONOPLAST

Transtonoplast Potential Difference

The electrical potential difference (E_m) was measured with microelectrodes as already described (Barbier and Guern, 1982). Briefly, glass micropipettes filled with 1 M KCl and connected to Ag/AgCl electrodes were used. An external macro-reference electrode (Ag-AgCl) was connected to the suspension droplet through an agar-KCl bridge. The vacuoles were held under microscope with a microholder and individually impaled. The time constant and the electrical resistance of the tonoplast were determined using a single electrode (Rona et al., 1980). Diameters were measured on each vacuole impaled and the membrane surface area was calculated. From these data, membrane specific resistance and capacitance were then calculated.

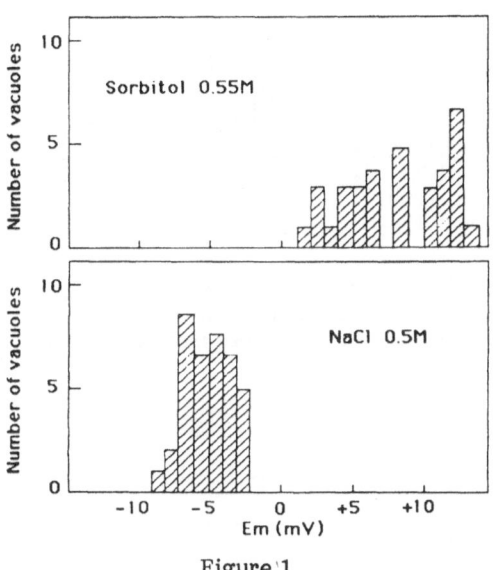

Figure 1

Distribution of E_m values
measured with microelectrodes
on isolated <u>Catharanthus</u> <u>roseus</u> vacuoles

The medium comprised 10 mM Hepes-KOH pH 7.4 and either 0.55 M sorbitol or 50 mM NaCl.

The distributions of E_m values for vacuoles suspended in either sorbitol or NaCl appeared markedly different (Fig. 1). The first ones were positive inside, as already reported for <u>Catharanthus</u> <u>roseus</u> vacuoles isolated in sorbitol medium (Barbier-Drygoo et al., 1985), with a mean value (+ standard error for 35 vacuoles)

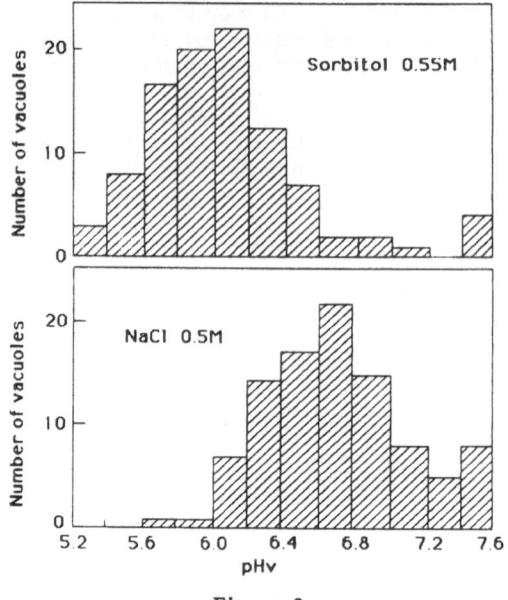

Figure 2

Distribution of vacuolar pH values (pH$_v$)
calculated from the accumulation ratio of 9-amino-acridine
in Catharanthus roseus isolated vacuoles

The medium comprised 10 mM Hepes-KOH pH 7.4 and either 0.55 M sorbitol or 0.50 M NaCl.

of + 8.0 $\pm$ 0.6 mV. The second ones were negative inside, with a mean value of 5.0 $\pm$ 0.2 mV (39 vacuoles). This comparison was completed by measuring E_m, specific resistance and capacitance of vacuoles isolated from the same protoplast suspension in media containing either neutral (sorbitol) or saline (NaCl or KCl) osmotica. The results showed (Table 1) that E_m was negative in saline buffer irrespective of the salt used (NaCl or KCl). They indicated also that specific resistance and capacitance of the tonoplast were strongly reduced (up to 50%) in saline buffer with regard to sorbitol buffer.

These results bring a significant contribution to the controversial problem of the sign of the transtonoplast potential difference. For the first time, it is shown on the same biological material and with the same technique that vacuoles can be either positive or negative inside according to the isolation medium. This calls for further comparative investigations concerning cases where negative E_m were reported for vacuoles isolated in special ionized osmotica such as betain (John and Miller, 1986).

Table 1

Influence of the Osmoticum on the Electrical Characteristics of the Tonoplast

Medium	Em $\pm$ SE[a] (mV)	R $\pm$ SE[a] (kΩ. cm^2)	C $\pm$ SE[a] (μF . cm^{-2})
Sorbitol 550 mM	+ 8.8 $\pm$ 0.8	1.7 $\pm$ 0.3	1.3 $\pm$ 0.2
NaCl 500 mM	- 4.4 $\pm$ 0.3	0.9 $\pm$ 0.1	0.6 $\pm$ 0.1
KCl 500 mM	- 2.8 $\pm$ 0.1	0.7 $\pm$ 0.2	0.8 $\pm$ 0.1

[a]The potential difference (E_m), specific resistance (R) and capacitance (C) were individually on 20 vacuoles in each condition, and mean values were calculated.

Vacuolar pH

The vacuolar pH measured first with the fluorescent probe 9-aminoacridine according the microfluorimetric technique already described (Manigault et al., 1983). The histograms (Fig. 2) gathering individual pH measurements on vacuole populations isolated in different osmotica confirm the large pH dispersion already described (Brown et al., 1984; Kurkdjian et al., 1984). The characteristics of this variability will be discussed later on. The most interesting result was that the pH distribution of vacuoles prepared in NaCl was strongly shifted towards higher pH values when compared to the pH distribution in sorbitol. The mean pH values calculated from the mean ratio of internal to external 9-aminoacridine concentrations were 5.90 and 6.55 for vacuoles isolated in sorbitol and NaCl, respectively.

The same investigation was conducted with a different technique of vacuolar pH measurement, namely the ^{31}P-NMR technique. This technique previously described for cell suspensions of Acer pseudoplatanus and Catharanthus roseus (Martin et al., 1982) was adapted to vacuolar preparations of Catharanthus roseus according to a procedure which will be detailed elsewhere. Briefly, to obtain the intravacuolar P_i signal in a reasonable time (usually 2-10 min) cells were loaded with P_i 12 hours before the preparation of vacuoles. Vacuoles were suspended in a 20 mm NMR tube (1 - 2 10^6 vacuoles . ml^{-1}) and a methylene diphosphonic acid (MDP) solution in a capillary tube was used as the reference signal. Fig. 3 shows a typical spectrum revealing aside the MDP signal at 16.2 ppm two P_i peaks, the extravacuolar one corresponding to pH 7.08 and the vacuolar one corresponding to pH 5.43. Thus, such spectra allowed the measurement of the transtonoplast pH (about 1.7 pH unit here) and the checking of its stability with time. When tested in a suspension medium without Na$^+$, the transtonoplast pH appeared reasonably stable for several hours (less than 0.1 pH unit decrease per hour). This stability contrasts with already reported results describing a rapid alkalinization of vacuoles isolated from Glycine max (Schmitt and Sandermann, 1982) or Petroselinum hortense (Matern et al., 1986).

Table 2

Influence of the Osmoticum on the Vacuolar pH

Medium	Vacuolar pH $\pm$ SE
Sorbitol 0.55 M	5.21 $\pm$ 0.04
NaCl 0.50 M	6.35 $\pm$ 0.02
KCl 0.50 M	6.04 $\pm$ 0.04

The spectrum corresponds to the accumulation of 512 scans (1.2 s each). Twelve ml of vacuolar suspension (2 10^6 vacuoles . ml^{-1} in 0.55 M sorbitol, 10 mM Hepes-KOH pH 7.4) were used. The extravacuolar phosphate peak and the vacuolar one were used to monitor the external pH (pH_o) and the intravacuolar one (pH_v). MDP was used as internal reference.

The table reports mean vacuolar pH values measured on vacuole suspensions prepared in different osmotica (sorbitol, NaCl or KCl).

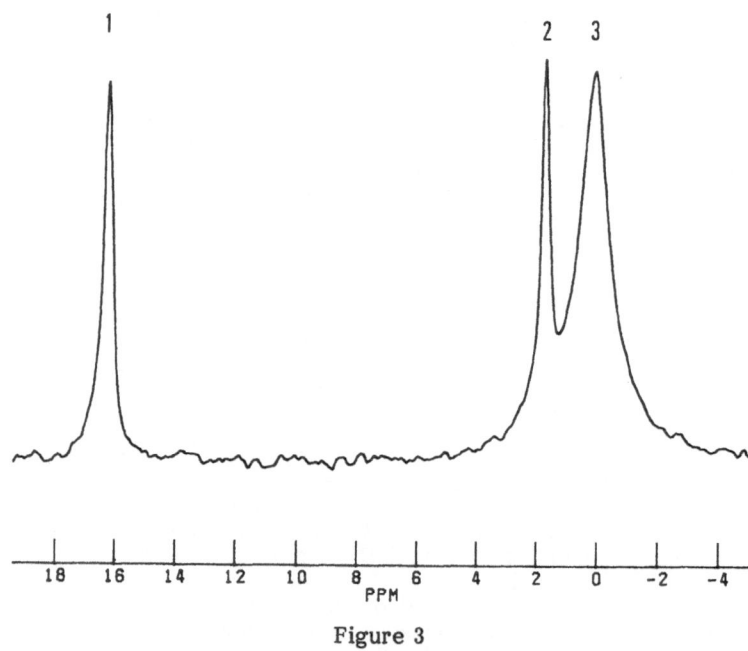

Figure 3

Measurement of vacuolar pH with the ^{31}P-NMR technique on Catharanthus roseus isolated vacuoles

Comparative measurements were made on vacuole populations prepared from the same cell suspension with different osmotica in the isolation medium. Fig. 3 shows that the mean vacuolar pH of the vacuole suspensions isolated in NaCl was markedly higher than the one corresponding to sorbitol vacuoles, in good agreement with what was observed for individual measurements with 9-aminoacridine fluorimetry (for a discussion of the discrepancy between the pH values measured with the two techniques, see Kurkdjian et al., 1985). The internal pH of vacuoles prepared with KCl was also shifted to higher values but less extensively than with NaCl, as already observed for the electrical potential difference.

Possible Origins of the Salt-induced Modifications of Vacuolar Properties

As regards the electrical potential difference, an effect of the salts on the tonoplast surface potential could be responsible for a part of the measured E_m decrease. As already described on isolated vacuoles, external cations such as Na^+ or K^+ induce a depolarization of the surface potential (Gibrat et al., 1985 b). Such a depolarization could modify the transmembrane potential in two ways, as the surface potential contributes significantly to E_m (Barbier-Brygoo et al., 1984), and controls the ionic concentrations at the membrane surface. In other respects, the variations observed in membrane resistance and capacitance probably reflect salt-induced modifications of the membrane ionic conductance and dielectric properties. The transtonoplast E_m of the non-energized isolated vacuoles was shown to be a diffusion potential in which anions, especially citrate, play a major role (Gibrat et al., 1985 a). Modifications of the tonoplast specific permeabilities to the vacuolar ions and/or modifications of the ionic composition of the vacuoles during their isolation could be at the origin of E_m changes.

As to the mechanisms of the salt-induced vacuolar alkalinization, one can hypothesize that the Na^+/H^+ antiport at the tonoplast described by Blumwald and Poole (1985) could be involved, at least in part. As a matter of fact, Catharanthus roseus vacuoles isolated in sorbitol and treated with 10 mM NaCl increased their internal pH by 0.6 pH unit in two hours (unpublished data).

Another point which deserves further investigation concerns the possible relationship between the salt-induced tonoplast E_m modification and the vacuolar alkalinization. These two events can be either independent consequences of tonoplast modifications or directly related (e.g. an important loss of organic acids during isolation in NaCl would alkalinize the vacuoles and render E_m more negative).

VARIATION OF VACUOLAR PARAMETERS AND POSSIBLE SELECTION DURING THE ISOLATION OF VACUOLES

The dispersion of several parameters inside populations of vacuoles is of common evidence from the simple observation of size, absorption or fluorescence of secondary metabolites (Brown et al., 1984; Chaprin and Ellis, 1984), neutral red staining, or individual measurements of the vacuolar pH (Kurkdjian et al., 1984; Kurkdjian et al., 1985). In term of vacuolar pH, Fig. 2 and previous results (Kurkdjian et al., 1984) show that the dispersion ranges up to 2.0 pH units. It is induced neither by the isolation step, since a similar dispersion is observed on the corresponding cell or protoplast populations (Kurkdjian et al., 1984), nor by the technique of pH measurement (9-aminoacridine microfluorimetry, H^+-sensitive microelectrodes, or NMR, for which the pH dispersion is revealed in Fig. 3 by the broadness of the vacuolar phosphate peak). Various treatments which induce modifications of the mean internal pH of a vacuole population (e.g. NaCl, see Fig. 2) do not modify the amplitude of the dispersion of pH values.

The variability of the vacuolar pH has been shown to originate likely from the properties of the vacuolar buffer (mainly malate, citrate, and K^+). Small variations (+12%) in the concentration of one buffer component can induce variations in the 100-fold range in the concentration of protons, i.e. 2 pH unit changes (Kurkdjian et al., 1985). As the yield of vacuoles with regard to the initial number of protoplasts is usually rather low from 3% to 50% (Wagner, 1983), one can wonder whether the purified vacuoles are fully representative of the whole vacuole population in cells, and not in fact a special subclass selected by the isolation procedure. Such a selection appears as one simple explanation when some metabolites are found in greater amounts in isolated vacuoles than in the corresponding protoplasts, as it was the case for serpentine in isolated Catharanthus roseus vacuoles (Renaudin et al., 1986; Renaudin and Guern, 1986). In this latter case, a positive correlation between serpentine concentration in protoplasts and vacuolar acidity had been previously demonstrated (Brown et al., 1984). The possibility that the differences observed between the vacuoles isolated in neutral or saline osmotica could originate from the selection of a peculiar class of vacuoles in salts could not be discarded. Against it was the reported alkalinization of sorbitol vacuoles by 10 mM NaCl, and the absence of difference in size, stability, and extent of the pH dispersion for the two kinds of vacuoles.

CONCLUSION

The results presented here demonstrate that dramatic changes in the electrochemical gradient of H^+ across the tonoplast are induced by the isolation of vacuoles in saline media. Saline osmotica (especially NaCl) appeared to reduce both the positive E_m and the pH gradient. These combined effects decreased the $\Delta\overline{\mu}H^+$ values from 9 - 13 kJ . mol^{-1} for sorbitol vacuoles to 4 - 5 kJ . mol^{-1} for NaCl vacuoles. Thus the capacity of vacuoles to mediate solute uptake coupled to the $\Delta\overline{\mu}H^+$ appears drastically dependent upon the isolation procedure. This calls for a more generalized evaluation of the transtonoplast E_m and pH as major characteristics of vacuolar preparations, especially in the case of uptake studies.

This study also brings further data demonstrating the large variability of the transtonoplast ΔpH which is a constant feature of cell, protoplast and vacuole populations. As a consequence other vacuolar properties (H^+-coupled uptake capacity) should likely be submitted to the same variability. Whereas no evidence was obtained to demonstrate a selection of vacuoles during their isolation, such a possibility calls for further attention.

REFERENCES

Barbier, H., and Guern, J., 1982, Transmembrane potential of isolated vacuoles and sucrose accumulation by Beta vulgaris roots, in : "Plasmalemma and Tonoplast : Their Function in the Plant Cell", D. Marmé, E. Marré, and R. Hertel, eds., Elsevier Biomedical Press, Amsterdam.
Barbier-Brygoo, H., Gibrat, R., Renaudin, J. P., Brown, S. C., Pradier, J. M., Grignon, C., and Guern, J., 1985, Membrane potential difference of isolated plant vacuoles : positive or negative ? II. Comparison of measurements with microelectrodes and cationic probes, Biochim. Biophys. Acta, 819:215.
Barbier-Brygoo, H., Romieu, C., Grouzis, J. P., Gibrat, R., Grignon, C., and Guern, J., 1984, Evidence for the contribution of surface potential to the transtonoplast potential difference measured on isolated vacuoles with microelectrodes, Z. Pflanzenphysiol., 114:215.
Blumwald, E., and Poole, R. J., 1985, Na$^+$/H$^+$ antiport in isolated tonoplast vesicles from storage tissue of Beta vulgaris, Plant Physiol., 78:163.
Brown, S. C., Renaudin, J. P., Prévot, C., and Guern, J., 1984, Flow cytometry and sorting of plant protoplasts : Technical problems and physiological results from a study of pH and alkaloids in Catharanthus roseus, Physiol. Vég., 22:541.
Chaprin, N., and Ellis, B. E., 1984, Microspectrophotometric evaluation of rosmarinic acid accumulation in single cultured plant cells, Can. J. Bot., 62:2278.

Deus-Neumann, B., and Zenk, M. H., 1984, A highly selective alkaloid uptake system in vacuoles of higher plants, Planta, 162:250.

Gibrat, R., Barbier-Brygoo, H., Guern, J., and Grignon, C., 1985 a, Transtonoplast potential difference and surface potential of isolated vacuoles, in : "Biochemistry and Function of Vacuolar ATPase in Fungi and Plants", B. P. Marin, ed., Springer-Verlag, Berlin, Heidelberg, New-York, Tokyo.

Gibrat, R., Barbier-Brygoo, H., Guern, J., and Grignon, C., 1985 b, Membrane potential difference of isolated plant vacuoles : Positive and negative ? I. Evidence for membrane binding of cationic probes, Biochim. Biophys. Acta, 819:206.

John, P., and Miller, A. J., 1986, Electrogenic proton translocation by the adenosine triphosphatase of intact vacuoles isolated from beet (Beta vulgaris L.), J. Plant Physiol., 122:1.

Korzun, A. M., Salyev, R. K., and Kuzevanov, V. Ya., 1984, Peculiarities of electrophysiological investigations of the vacuolar membrane in cells of higher plants, Soviet. Plant Physiol., 31:156.

Kurkdjian, A. C., Barbier-Brygoo, H., Manigault, J., and Manigault, P., 1984, Distribution of vacuolar pH values within populations of cells, protoplasts and vacuoles isolated from suspension cultures and plant tissues, Physiol. Veg., 22:193.

Kurkdjian, A. C., Quiquampoix, H., Barbier-Brygoo, H., Péan, M., Manigault, P., and Guern, J., 1985, Critical evaluation of methods for estimating the vacuolar pH of plant cells, in : "Biochemistry and Function of Vacuolar Adenosine-triphosphatase in Fungi and Plants", B. P. Marin, ed., Springer-Verlag, Berlin, Heidelberg, New-York, Tokyo.

Leigh, R., 1983, Methods, progress and potential for the use of isolated vacuoles in studies of solute transport in higher plant cells, Physiol. Plant., 57:390.

Manigault, P., Manigault, J., and Kurkdjian, A. C., 1983, A microfluorimetric method for vacuolar pH measurement in plant cells using 9-aminoacridine, Physiol. Vég., 21:129.

Martin, J. B., Bligny, R., Rebeille, F., Douce, R., Leguay, J. J., Mathieu, Y., and Guern, J., 1982, A ^{31}P nuclear magnetic resonance study of intracellular pH of plant cells cultivated in liquid medium, Plant Physiol., 70:1156.

Matern, U., Reichenbach, C., and Heller, W., 1986, Efficient uptake of flavonoids into parsley (Petroselinum hortense) vacuoles requires acylated glycosides, Planta, 167:183.

Renaudin, J. P., Brown, S. C., Barbier-Brygoo, H., and Guern, J., 1986, Quantative characterization of protoplasts and vacuoles from suspension cultured cells of Catharanthus roseus, Physiol. Plant., in press.

Renaudin, J. P., and Guern, J., 1986, Ajmalicine transport in vacuoles isolated from Catharanthus roseus cells, in : "Plant Vacuoles. Their Importance in Solute Compartmentation and Their Applications in Biotechnology", B. Marin, ed., Plenum Publishing Corporation, New-York.

Rona, J. P., Van De Sype, G., Cornel, D., Grignon, C., and Heller, R., 1980, Plasmolysis effect on electrical characteristics of free cells and protoplasts of Acer pseudoplatanus L., J. Bioelectrochem. Bioenerg., 7:377.

Schmitt, R., and Sandermann Jr., H., 1982, Specific localization of β-glucoside conjugates of 2,4-dichlorophenoxyacetic acid in soybean vacuoles, Z. Naturforsch., 37:772.

Wagner, G. J., 1983, Higher plant vacuoles and tonoplast, in : "Isolation of Membranes and Organelles from Plant Cells", J. L. Hall, ed., Academic Press, London.

HIGHLY PURIFIED TONOPLAST FRACTIONS

BY PREPARATIVE FREE-FLOW ELECTROPHORESIS

D. James Morre, A. Brightman, G. Scherer[*], B. Vom Dorp[*],
C. Penel[**], G. Auderset[**], A. S. Sandelius[***],
and H. Greppin[**]

Department of Medicinal Chemistry, Purdue University, West
Lafayette, IN 47907, USA; [*]University of Bonn, Bonn, Federal
Republic of Germany; [**]University of Geneva, Geneva, Switzer-
land; and [***]University of Göteborg, Göteborg, Sweden

INTRODUCTION

For the preparation of tonoplasts, the different approaches followed include isolation from homogenates by density gradient fractionation or from vacuoles prepared from protoplasts by osmotic lysis (Wagner, 1983). For our studies, we have used preparative free-flow electrophoresis, a procedure whereby highly purified fractions of both plasma membrane and of tonoplast are obtained from the same homogenate (Sandelius et al., 1986; Auderset et al., in press). In this technique, based on the methodology of Hannig and coworkers (Hannig and Heidrich, 1977), a mixture of components to be separated is introduced into a separation buffer moving perpendicular to the flux lines of an electric field (Fig. 1). Membranes bearing different electrical charge densities will migrate different distances across the separation chamber and thus may be resolved.

According to our criteria for tonoplast and plasma membrane identification, fractions nearest the point of sample injection (farthest from the anode) are enriched in plasma membranes while fractions farthest from the point of sample injection are enriched in tonoplast. Other membranous cell components (dictyosomes, endoplasmic reticulum, plastids and mitochondria) are located in fractions intermediate between the plasma membrane and the tonoplast. The identification of the various cell components is based both upon electron microscope morphology and upon the biochemical estimation of various characteristic marker enzyme constituents (Sandelius et al., 1986).

MATERIALS AND METHODS

Plant Material

Segments were cut just below the cotyledons of dark grown (4 - 5 days) soybean (Glycine max) or zucchini (Cucurbita pepo) hypocotyls. Spinach (Spinacia oleracea) was grown for 4 weeks in a chamber with short-day illumination.

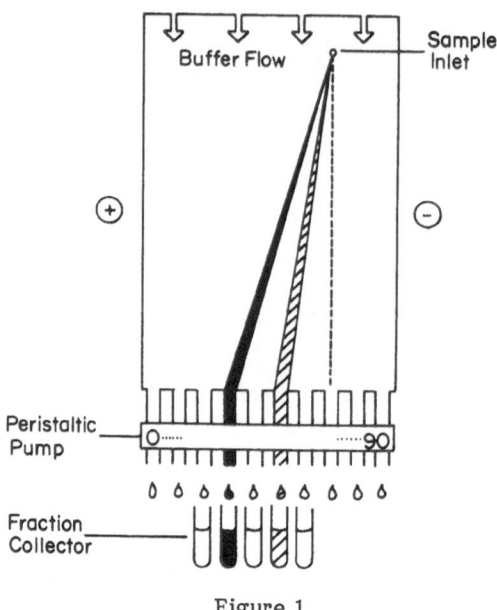

Figure 1

Principles of Free-Flow Electrophoresis

Membranes

Membranes were prepared as described by Sandelius et al. (1986) for soybean and by Scherer and Fischer (1985) for zucchini. Homogenates were by razor blade chopping.

Free-flow electrophoresis

The electrophoresis medium (chamber buffer) contained 0.25 M sucrose, 2 mM KCl, 10 M $CaCl_2$, 10 mM triethanolamine, and 10 mM acetic acid, pH 7.5. The equipment was a VaP-5 or VaP-21 continuous free-flow electrophoresis unit (Bender and Hobein, Munich, FRG). The conditions for the separation were constant voltage of 800 V/9.2 cm field, 165 mA, buffer flow 1.7 ml/fraction/h, sample injection 2.7 ml/h and constant temperature of 6°C.

Analyses

Determination of sealed vesicles was on 10% dextran gradients as described by Sze and Churchill (1981). Assays of proton transport by fluorescent quenching of quinacrine was as adapted from Schuldiner et al. (1972) and by acridine orange according to Glickman et al. (1983). Details of fraction composition and marker enzyme analyses are provided by Sandelius et al. (1986) and by Auderset et al. (in press).

TONOPLAST IDENTIFICATION IN ISOLATED FRACTIONS

To identify tonoplast vesicles when isolated in the absence of the usual positional relationships as found in the intact cell, we have used membrane thickness and differential staining with phosphotungstic acid at low pH. This approach incorporates information determined from in situ conditions in which tonoplast and plasma membranes are undisturbed and most easily identified. With plant cells, reaction of the plasma membranes with either phosphotungstic acid or silicotungstic acid at low pH serves both to identify plasma membrane and to estimate the purity of plasma membrane fractions when used in combination with electron microscope morphometry (Roland. et al., 1972). This method has the advantage of reacting with plasma membrane both in the intact cell where positional relationships adjacent to the cell wall are maintained and in broken cell preparations or in isolated fractions where these positional relationships are lost. Binding of the auxin transport inhibitor, N-1-naphtylphtalamic acid (NPA), also is sufficiently specific for plasma membranes in plants to serve as a marker (Lembi et al., 1971). Unfortunately, there are no enzyme activities that unequivocally mark the plasma membrane in plants. A vanadate-inhibited, K^+-stimulated, Mg^{2+}-ATPase is present but may be more widespread in its distribution (Sandelius et al., 1986; Hodges and Mills, 1986). Among the best enzymatic markers is the so-called glucan synthetase II (Ray, 1977), a glycosyltransferase exhibiting a high K_m for UDP-glucose (Van Der Woude et al., 1974).

Table 1

Membrane Dimensions in Leaves of Spinach
Following a 24 h Period of Continuous Fluorescent Light (10 mmol m^{-2} s^{-1})
Sufficient to induce Flowering

Cell Component	Membrane Thickness (nm $\pm$ S.D.)
Plasma membrane	10.5 $\pm$ 0.3
Tonoplast	8.1 $\pm$ 0.9
Endoplasmic reticulum	6.3 $\pm$ 0.8
Nuclear envelope	
Outer membrane leaflet	6.8 $\pm$ 0.6
Inner membrane leaflet	5.8 $\pm$ 0.7
Mitochondria	
Outer membrane	6.4 $\pm$ 0.8
Inner membrane	6.0 $\pm$ 0.9
Chloroplasts	
Outer envelope membrane	5.1 $\pm$ 0.8
Inner envelope membrane	6.3 $\pm$ 0.8
Stroma thylakoids	6.1 $\pm$ 0.7
Grana thylakoids (both appressed membranes together)	13.0 $\pm$ 1.1
Peroxisomes	7.0 $\pm$ 0.4
Golgi apparatus	6.6 $\pm$ 1.8

Values are based on measurements from 30-300 individual membranes from several different fixations. From Auderset et al. (in press).

Tonoplast and plasma membranes, as well as some membranes from the mature Golgi apparatus face, are thick (7 - 10 nm) as determined visually (Fig. 2) and from measurements from electron micrographs (Table 1). However, these different thick membranes may be differentiated readily on the basis of their abilities to react with phosphotungstic acid at low pH. The plasma membrane is reactive whereas the tonoplast is not. All other membranes are both thin (5 - 6 nm) and unreactive with phosphotungstic acid at low pH.

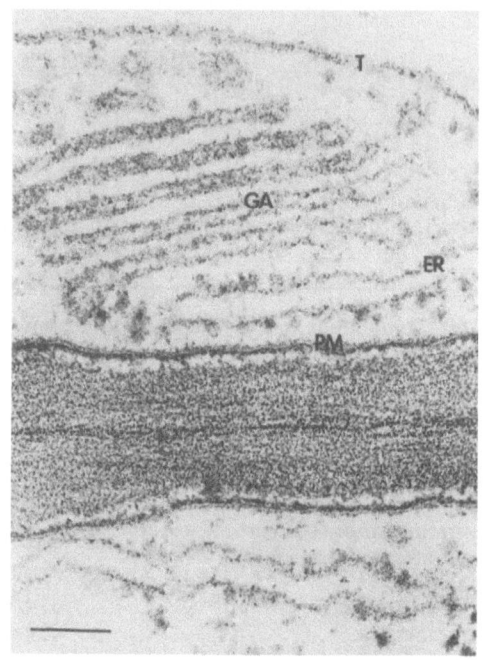

Figure 2

High magnification electron micrograph
of a portion of a photoinduced spinach leaf[1]

---

[1] illustrating the "thick" appearance of the plasma membrane (PM) and tonoplast (T) as well as some mature membranes of the Golgi apparatus (GA). Other membranes of the cell as exemplified by endoplasmic reticulum (ER) appear thin. Scale bar = $0.1\,\mu$m.

By projecting electron microscope negative at magnifications approaching 1,000,000 times, these different classes of membranes were identified based on thickness measurements (Table 2). The distribution of membranes by quantitative morphometry from a spinach homogenate in which tonoplast and plasma membranes were first partially enriched by differential centrifugation and then resolved by preparative free-flow electrophoresis (e.g. Fig. 3) indicated a marked enrichment of tonoplast in fraction "A" whereas plasma membranes are enriched at the opposite end of the separation in fraction "E". There is perharps only 3% or less of cross contamination of tonoplast by plasma membrane and vice versa comparing these fractions.

Table 2

Composition based on Morphometry of Free-Flow Electrophoresis Fractions Prepared from Leaves of Photo-induced Spinach Plants

Free-Flow Fraction	Percentage of Membranes in Fraction				
	11-12 nm	9-11 nm	7-9 nm	5-7 nm	PTA-stained
Fraction A	2 ± 1	2 ± 1	93 ± 2	2 ± 1	2 ± 1
Fraction B	8 ± 5	9 ± 2	57 ± 8	25 ± 4	7 ± 2
Fraction C	16 ± 5	21 ± 3	37 ± 5	26 ± 4	11 ± 1
Fraction D	10 ± 3	33 ± 6	29 ± 5	28 ± 2	29 ± 6
Fraction E	3 ± 1	87 ± 1	2 ± 2	8 ± 3	89 ± 3

Each value is based on a minimum of 300 measurements. From Auderset et al. (in press).

BIOCHEMICAL MARKER FOR TONOPLAST

The search for a membrane-associated biochemical marker for tonoplast vesicles was aided greatly by the assignment of an anion-stimulated, nitrate-inhibited Mg^{2+}-APase to the tonoplast (Marin et al., 1982; Marin, 1985). The free-flow electrophoresis fractions identified from measurements of membrane thickness do contain an anion-stimulated, nitrate-inhibited Mg^{2+}-ATPase when prepared from soybean hypocotyls as expected. Additionally, they retain ATP-dependent proton transport activity as determined using a fluorescence quench assay with quinacrine as described (Schuldiner et al., 1972). Tonoplast vesicles appear to lack glucan synthetase II activity (Sandelius et al., 1986) but appear to contain sterol glycoside synthetase in hypocotyls of soybean (Sandelius et al., 1986) but not in those of pumpkin (Scherer, 1984).

TONOPLAST VESICLES PREPARED BY PREPARATIVE FREE-FLOW ELECTROPHORESIS ARE SEALED

Based on assay by centrifugation on 10% dextran gradients (Sze and Churchill, 1981) more than 90% of the tonoplast vesicles from fraction "A" by preparative free-flow electrophoresis appear to be sealed (Table 3).

The numbers of sealed vesicles are increased by homogeneization using the razor blade chopping as compared to the Polytron homogenizer so that for studies involved with proton transport, razor blade homogenization was used.

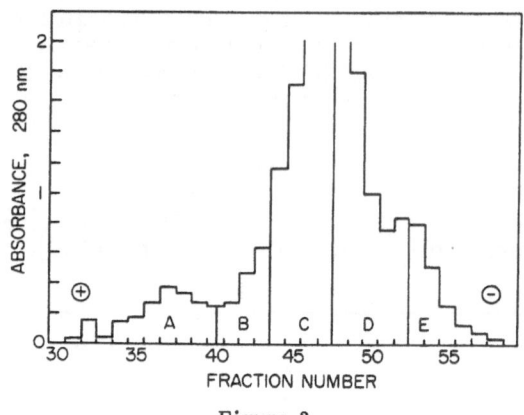

Figure 3

The absorbance at 280 nm of fractions
from a representative preparative free-flow electrophoresis separation
from leaves of photo-induced spinach plants

A-E indicate pooled fractions. The composition of these fractions would be similar to that summarized in Table 2.

Table 3

Assay For Integrity of Membrane Vesicles
From Preparative Free-Electrophoresis

		% Total Protein Recovered	
Fraction Loaded	Fraction Recovered	Razor Blade Chopped	Polytron Homogenized
A (Tonoplast)	Interface	92	68
	Pellet	8	32
E (Plasma Membrane)	Interface	66	57
	Pellet	34	43

Assay is based on centrifugation of soybean membranes on 10% dextran gradients according Sze and Churchill (1981).

TONOPLAST VESICLES (AS WELL AS PLASMA MEMBRANE VESICLES) APPEAR TO BE RESOLVED BY FREE-FLOW ELECTROPHORESIS INTO TWO DISTINCT SUBPOPULATIONS

As summarized in Figure 4, when different plasma membrane markers from soybean membranes are expressed as total activity, there are clearly two peaks seen in the electrophoretic separations.

Preliminary observations from cultured millet (Paspalum) cells provided through the courtesy of Prof. N. Carpita, Purdue University, and studied in collaboration with Prof. C. E. Bracker, Purdue University, suggest that the membranes of fraction "E" (the least electronegative fraction of plasma membrane) are cytoplasmic side out while cytoplasmic side in vesicles are found in fraction "C". This conclusion is based on the appearance of vesicles formed from plasma membrane in this system which are asymmetric and exhibit a dark outer leaflet and a less dark inner leaflet making possible assignment of absolute orientations to the vesicles obtained. Additionally, the external surface of soybean plasma

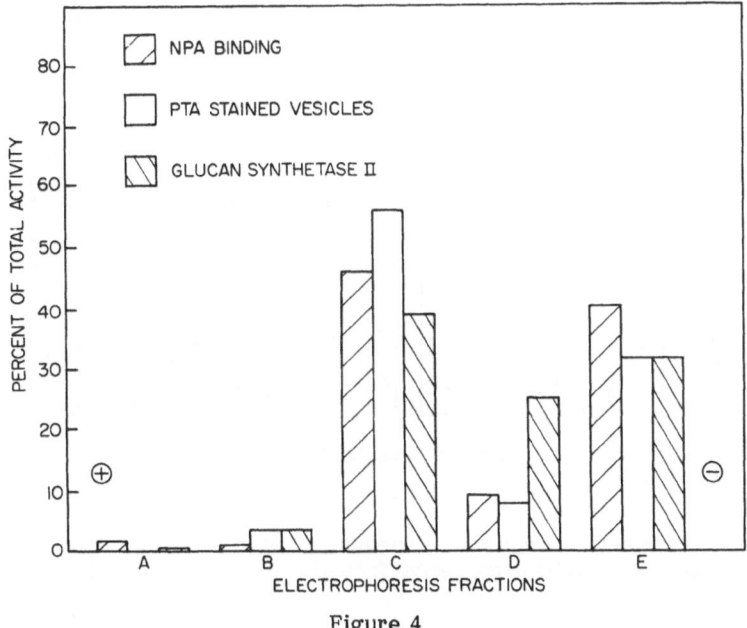

Figure 4

Expression of plasma membrane markers
as percent of total activity recovered
in each of the different pooled free-flow electrophoresis from soybean

---

Evidence is for two populations of plasma membrane vesicles of opposite sidedness. One is primarily in fraction "E" and the other is in fraction "C".

membranes reacts more strongly with the lectin concanavalin A than does the cytoplasmic surface. The differential reactivity with the lectin also is consistent with the above assignment of absolute conformation (A. Brightman, Purdue University, results unpublished).

When a similar analysis is done for tonoplast markers, at least two fractions again are observed (Fig. 5). Both proton pumping and nitrate-inhibited ATPase are concentrated in fraction "A" but are present as well in fractions "B" and "C". It appears that the tonoplast vesiculated into at least two fractions perharps having opposite orientations.

The question of the occurrence of a vanadate-inhibited ATPase activity in fractions "A" and "B" of the electrophoretic separations cannot be explained on the basis of simple plasma contamination since the specific plasma membrame markers of PTA staining and NPA binding are absent from these fractions. It may be that this activity is endogenous under some conditions to the tonoplast or that our assay conditions are not sufficiently rigourous to discriminate between tonoplast and plasma membrane ATPases with respect to vanadate sensitivity. The rather large activity in fraction "C" is probably the result of the presence of the activity in endomembranes and/or organelles other than surface membranes. Fraction "C" contained the bulk of the marker enzyme activities for endoplasmic reticulum (NADH- and NADPH-cytochrome c reductase) and for mitochondria (succinate

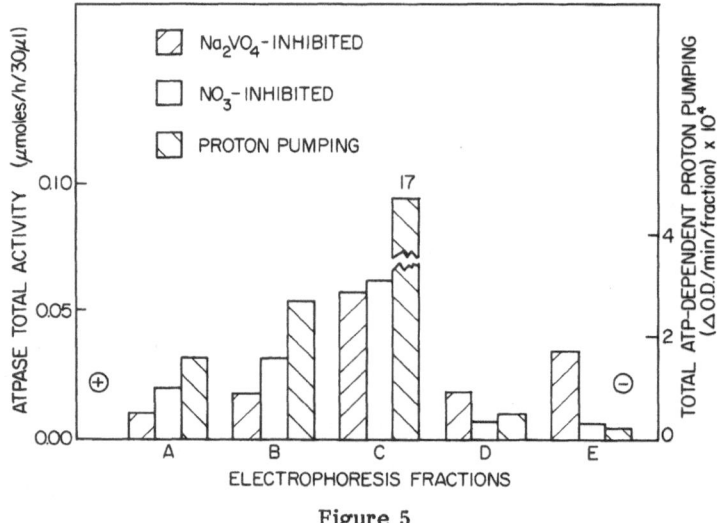

Figure 5

ATPase activity and ATP-dependent proton pumping measured in fractions obtained by preparative free-flow electrophoresis from soybean

---

Results are expressed as total activity recovered in each of the different pooled fractions. Nitrate-inhibited ATPase as well as ATP-dependent proton pumping were found not only in fractions "A" and "B" but also in fraction "C". Fraction "C" contained plasma membrane and tonoplast in addition to the bulk of the endoplasmic reticulum and mitochondria applied to the separation. The vanadate-inhibited ATPase was found in fractions "C" and "E" corresponding to plasma membrane but activity located with fractions "A" and "B" was not ascribed readily to plasma membrane contamination.

dehydrogenase and cytochrome oxidase) whereas fraction "B" was the dominant location of the Golgi apparatus markers latent IDPase and fucosyl-transferase (Sandelius et al., 1986).

RESOLUTION OF PLASMA MEMBRANE AND TONOPLAST FRACTIONS BY CONSECUTIVE SUCROSE AND GLYCEROL GRADIENT CENTRIFUGATIONS

In an attempt to resolve the plasma membrane and tonoplast activities of fractions "B" and "C" observed from the free-flow electrophoretic separations, tonoplast and plasma membrane-enriched fractions were obtained first by the preparative procedure detailed by Scherer and Fisher (1985) that uses two consecutive centrifugations (Fig. 6). First, three fractions were separated on a sucrose step gradient. Fraction "A-2" when recentrifuged on an isopycnic glycerol gradient (Gradient "B") yielded additional fractions of which fraction "B-1" was taken for further analysis. Fraction "A-2" from the sucrose step gradient applied to the second isopycnic glycerol gradient yielded plasma membrane fractions "B-5" and "B-6" also selected for analysis.

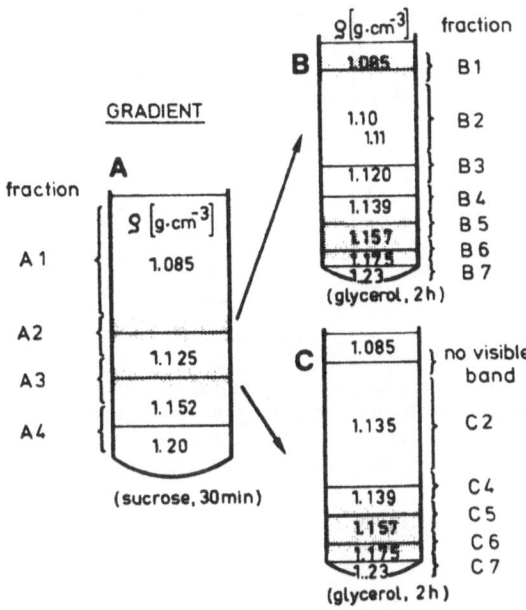

Figure 6

Sequential sucrose and glycerol gradient procedure
for isolation of tonoplast- and plasma membrane-enriched fractions
from plant stems
(from Scherer and Fischer, 1985)

As illustrated in Fig. 7, fractions "B-1" (tonoplast-enriched) and "B-5,6" (plasma membrane-enriched) were resolved by the free-flow electrophoresis technique although a substantially different pattern was observed compared with that observed when crude microsome fractions were the starting material (Sandelius et al., 1986). The electrophoretic mobilities corresponded most closely to those given by the inner fractions of tonoplast and plasma membrane (electrophoresis fractions "D" and "C") as depicted in Figures 4 and 5. The results given here are from zucchini hypocotyls although similar results were obtained using the same procedures with soybean hypocotyls and with cress (Lepidium sativum) roots. While the observations appear sound and reproducible, the interpretation must be regarded as preliminary. Nevertheless, the combined techniques of density centrifugation and free-flow electrophoresis offer the exciting possibility now of obtaining in relatively pure form with minimal cross contamination both right side out and inside out tonoplast and plasma membrane vesicles from the same starting homogenates. While the resolution of the two inner fractions remains tentative (e.g. Fig. 7), the outer two fractions ("A" = tonoplast of unknown orientation and "E" the plasma membrane of cytoplasmic side out orientation) can be achieved by preparative free-flow electrophoresis alone.

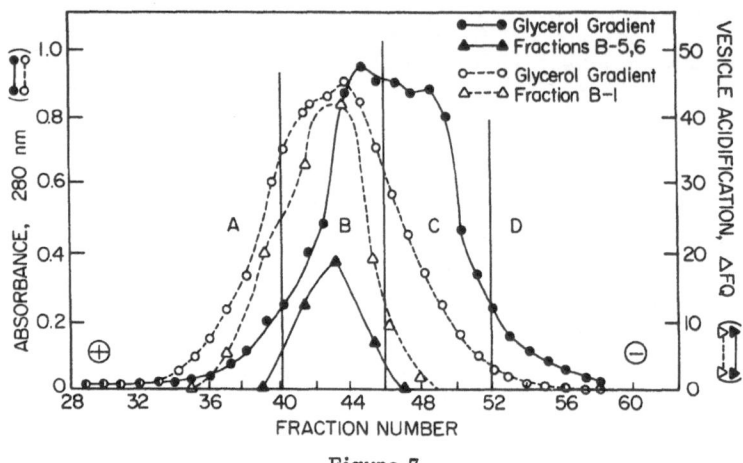

Figure 7

Free-flow electrophoresis separation of plasma membrane-enriched (solid symbols) or tonoplast-enriched (open symbols) fractions from zucchini hypocotyls obtained initially using the combined glycerol and sucrose gradients diagrammed in Figure 6

Absorbance at 280 nm, an approximation of the total amount of material present in each of the fractions, is given by the circles and vesicle acidification as measured by fluorescence quenching of quinacrine is given by the triangles. Free-flow electrophoresis resolved each fraction into a more electronegative region with good proton pumping activity which dominated in the tonoplast-enriched fractions with evidence of a major (fraction "B") and a minor (fraction "A") subpopulation. Plasma membrane from the glycerol gradient fractions "B-5" and "B-6" also resolved into a fraction corresponding to free-flow electrophoresis fraction "C" (= plasma membrane with a cytoplasmic side in orientation) and which did not pump protons as well as a fraction corresponding to fraction "B" which did pump protons and may represent, in part, contaminating tonoplast or Golgi apparatus that coisolated with the glycerol fractions "B-5" and "B-6". A fraction clearly coincident with plasma membrane fraction "E" by free-flow electrophoresis was not obvious from the glycerol gradient fractions "B-5" and "B-6".

TONOPLAST VESICLES ISOLATED
BY PREPARATIVE FREE-FLOW ELECTROPHORESIS
ARE ACTIVE IN PROTON PUMPING

As illustrated in Figures 5 and 7, tonoplast vesicles isolated by preparative free-flow electrophoresis are active in inward pumping of protons as determined by quenching of fluorescence of quinacrine contained within the vesicle interiors. The tonoplast activity was sensitive to nitrate.

The apparent absence of proton pumping activity in the highly purified plasma membranes obtained both by free-flow electrophoresis apparently independent of orientation or by aqueous two-phase partition (not illustrated) remains problematic. Sometimes weak proton pumping is given by free-flow fraction "E" which should have the correct final orientation to be active in this regard but no activity is seen with fractions "B-5" and "B-6" after free-flow electrophoresis except that which may be ascribed to tonoplast or Golgi apparatus contamination (Fig. 7). This fraction and the fraction obtained by aqueous two phase partitioning may have the incorrect orientation to pump protons in agreement with the interpretation of Kjellbom and Larsson (1984) based on measurements of ATPase latency.

SUMMARY

1. Highly purified fractions of tonoplast and plasma membrane vesicles were obtained from the same homogenates of both etiolated stems and of green leaves by the technique of preparative free-flow electrophoresis.

2. From a starting material of 35 - 50 g fresh weight of tissue, 0.8 to 1 mg of purified tonoplast (fraction "A") or of plasma membrane (fraction "E") were obtained.

3. Within the tonoplast fraction, about 90% of the vesicles had membranes with thicknesses in the range 7 - 9 nm that did not stain with phosphotungstic acid at low pH suggesting an origin from the tonoplast. The remaining 15% of membranes were 6 nm in thickness or thinner. The absence of plasma membrane contamination in the tonoplast fraction was verified by the absence of binding activity of the inhibitor of polar auxin transport, N-1-naphtyl-phtalamic acid, and very low specific activity of the plasma membrane marker enzyme glucan synthetase II.

4. A chloride-stimulated, nitrate-inhibited ATPase activity was associated dominantly with the putative tonoplast vesicles.

5. The highly purified tonoplast but not the plasma membrane vesicles showed ATP-driven proton-pumping activity (both from soybean as well as from zucchini).

6. By combining sucrose and glycerol gradient centrifugation in conjunction with preparative free-flow electrophoresis, it may be possible to obtain a total of four fractions from the same homogenate consisting of highly purified isolates of both inside out and right side out plasma membrane and tonoplast vesicles without significant (perharps less than 15%) cross contamination of one by the other.

ACKNOWLEDGEMENTS

Work supported in part by a grant from the National Science Foundation (to DJM) PCM 8206222 and from the Fonds National Suisse de la Recherche Scientifique. We thank Keri Safranski and Dorothy Werderitsh for technical assistance and Prof. Charles E. Bracker for access to electron microscope facilities. Collaboration with Prof. C. E. Bracker and Prof. N. Carpita, Department of Botany and Plant Pathology, Purdue University, on the studies with vesicle fractions of millet is gratefully acknowledged.

REFERENCES

Auderset, G., Sandelius, A. S., Penel, C., Brightman, A., Greppin, H., and Morré, D. J., 1986, Isolation of plasma membrane and tonoplast fractions from spinach leaves by preparative free-flow electrophoresis and effect of photoinduction, Plant Physiol., in press.

Glickman, J., Croen, K., Kelly, S., and Al-Awquati, Q., 1983, Golgi membranes contain an electrogenic H^+ pump in parallel to a chloride conductance, J. Cell Biol., 97:1303.

Hannig, K., and Heidrich, H. G., 1977, Continuous free-flow electrophoresis and its application in biology, in : "Cell Separation Methods. Part IV", H. Bloemendal, ed., p. 93, Elsevier/North-Holland Biomedical Press, New-York.

Hodges, T. K., and Mills, D., 1986, Isolation of the plasma membrane, Methods In Enzymology, 118:41.

Kjellbom, P., and Larsson, C., 1984, Preparation and polypeptide composition of chlorophyll-free plasma membranes from leaves of light-grown spinach and barley, Physiol. Plant., 62:501.

Lembi, C. A., Morré, D. J., St. Thompson, K., and Hertel, R., 1971, N-1-naphthylphthalamic acid-binding activity of a plasma membrane-rich fraction from maize coleoptiles, Planta, 99:37.

Marin, B., 1985, "The Biochemistry and Physiology of Vacuolar Adenosine-triphosphatase from Fungi and Higher Plants", Springer-Verlag, Berlin, Heidelberg, New-York, Tokyo.

Marin, B., Crétin, H., and D'Auzac, J., 1982, Energization of solute transport and accumulation at the tonoplast in Hevea latex, Physiol. Vég., 20:333.

Ray, P. M., 1977, Auxin-binding sites of maize coleoptiles are localized on membranes of endoplasmic reticulum, Plant Physiol., 59:594.

Roland, J. C., Lembi, C. A., and Morré, D. J., 1972, Phosphotungstic acid - chromic acid as a selective electron-dense stain for plasma membrane of plant cells, Stain Technol., 47:195.

Sandelius, A. S., Penel, C., Auderset, G., Brightman, A., Millard, M., and Morré, D. J., 1986, Isolation of highly purified fractions of plasma membrane and tonoplast from the same homogenate of soybean by free-flow electrophoresis, Plant Physiol., 81:177.

Scherer, G. F. E., 1984, Stimulation of ATPase activity by auxin is dependent on ATP concentration, Planta, 161:394.

Scherer, G. F. E., and Fischer, G., 1985, Separation of tonoplast and plasma membrane H^+-ATPase from zucchini hypocotyls by consecutive sucrose and glycerol gradient centrifugation, Protoplasma, 129:109.

Schuldiner, S., Rottenberg, H., and Avron, M., 1972, Determination of pH in chloroplasts. 2. Fluorescent amines as a probe for the determination of pH in chloroplasts, Eur. J. Biochem., 25:64.

Sze, H., and Churchill, K. A., 1981, Mg^{2+}/KCl-ATPase of plant plasma membranes is an electrogenic pump, Proc. Natl. Acad. Sci. U.S.A., 78:5578.

Van Der Woude, W. J., Lembi, C. A., Morré, D. J., Kidinger, J. A., and Ordin, L., 1974, Beta-glucan synthetases of plasma membrane and Golgi apparatus from onion stem, Plant Physiol., 54:333.

Wagner, G. J., 1983, Higher plant vacuoles and tonoplasts, in : "Isolation of Membranes and Organelles from Plant Cells", J. L. Hall and A. L. Moore, eds., Academic Press, New-York.

ANALYSIS OF VACUOLAR POLYPEPTIDES FROM BARLEY MESOPHYLL

BY IMMUNOBLOTTING

Jürgen M. Schmitt, Enrico Martinoia,
Dirk K. Hincha, and Georg Kaiser

Botanisches Institut der Universität
Mittlerer Dallenbergweg 64
D-8700-Würzburg, Federal Republic of Germany

INTRODUCTION

Mesophyll vacuoles perform an important metabolic role as an intermediate storage compartment for photosynthetic products like sucrose, malate or citrate (Kaiser et al., 1982). Futhermore, salt-tolerant plants like barley take up salts and accumulate them in the vacuole when exposed to high salt concentrations (Yeo, 1983), thereby maintaining turgor without expending energy for the synthesis of organic osmotica. We have intensely studied the physiology of vacuolar transport of sucrose, malate, citrate, and chloride during the last years (Kaiser and Heber, 1984; Martinoia et al., 1985 and 1986). From kinetic experiments it became clear that the tonoplast contains several translocating proteins which are capable of catalysing fast and specific exchange between vacuolar and cytosolic compartments. At least one type of proton ATPase drives some of these transport processes.

Very little is known about the tonoplast membrane proteins catalysing the translocation of metabolites. This lack of data has at least in part been caused by the difficulties in isolating mesophyll vacuoles in pure form and in sufficient amounts for biochemical studies. Mesophyll vacuoles comprise approximately 80% of the volume of the cell but less than 4% of the total cellular protein (Kaiser et al., 1986). A contamination of the vacuoles which corresponds to only a minor fraction of the 96% non-vacuole protein of the cell represents a heavy contamination if it is expressed as a fraction of the isolated vacuole protein.

In a recent publication (Kaiser et al., 1986) we have characterized the proteins of barley mesophyll vacuoles using gel electrophoretic techniques. Approximately 35 polypeptides of different electrophoretic mobilities in SDS gels have been found by sensitive silver staining. Of these, 15 were found in the soluble fraction and approximately 20 were associated with the membrane fraction. The polypeptide patterns of thylakoids and vacuoles show differences but also several similarities in electrophoretic mobilities, although contamination of the vacuolar fraction by constituents of other cellular compartments, as checked by marker enzymes, was below 1% compared to the corresponding activities in the protoplast.

In this paper, we determine the purity of isolated vacuoles by immunoblotting with monospecific antisera against the most likely contaminants from the stroma (RuBPcase) and the thylakoids (CF_1) of chloroplasts. In addition we analyze the antigenic properties of vacuole proteins using a polyspecific rabbit antiserum.

MATERIALS AND METHODS

Isolation of protoplasts, vacuoles, cytoplasts, and vacuoplasts

Mesophyll protoplasts were isolated from 7 to 9 days old barley and wheat primary leaves (Hordeum vulgare L. cv. Gerbel and Triticum aestivum cv. Kanzler) as described by Kaiser et al. (1982). Vacuoles were liberated from the protoplasts and purified following the method of Martinoia et al. (1985). Tonoplast membranes and soluble vacuolar proteins were separated after rupture of the vacuoles by freezing as described by Kaiser et al. (1986). Vacuoplasts (Lörz et al., 1976) and cytoplasts were liberated from protoplasts using the evacuolation technique of Griesbach and Sink (1983).

Antiserum against vacuole proteins

Approximately 600 μg of the tonoplast membrane fraction was emulsified in Freund's incomplete adjuvant and injected subcutaneously into a rabbit named Vakju. Booster injections using the same amount of protein in 150 mM NaCl were given at days 30 and 50 after the immunization. Blood was collected at day 45 (Vakju 1) and at day 57 (Vakju 2). The serum was stored frozen at - 20° C.

Antiserum against chloroplast proteins

Chloroplast coupling factor was isolated as described by Wolter et al. (1984). The rabbit antiserum was raised as described by Hincha et al. (1985). RuBPcase was purified from leaves of Oenothera hookeri strain albicans Johannsen (kindly provided by Prof. W. Stubbe) as described by Höfner et al. (1986).

Immunoblotting

Proteins separated on SDS-gels were transferred to nitrocellulose membranes (Sartorius SM 26) in a buffer comprising 20 mM Tris, 150 mM glycine, 20% (v/v) methanol at 0.5 A for 3 h at 4°C, using a LKB transphor cell (Towbin et al., 1979). Filters were blocked in 1% gelatin (Bio-Rad) in TBS (TBS : 20 mM Tris, pH 7.5, 150 mM NaCl), incubated with antiserum (diluted as given in the figure legends in 1% BSA, TBS) for 1 h, washed in TBS-0.1% Tween 20, incubated for 1 h with goat anti-rabbit antibody coupled to horseradish-peroxidase (Bio-Rad), washed and developed with 4-chloronaphtol/H_2O_2 as suggested by the manufacturer. Standard proteins (phosphorylase B, 97 kDa; bovine serum albumin, 66 kDa; ovalbumin, 45 kDa; carboanhydrase, 29 kDa; myoglobin, 17 kDa) were labeled with fluorescein isothiocyanate using the labeling conditions of Nowotny (1979). The standard migrated marginally slower in comparaison with unlabeled controls. Myoglobin yielded a series of bands. The position of the fastest band, migrating close to the unstained myoglobin, is given.

Other Techniques

All other techniques have been described in Kaiser et al. (1986).

RESULTS

Chloroplasts contain the major part of the cellular protein. Chloroplast proteins are there the most likely protein contaminants of isolated vacuoles. We have used monospecific antisera against two chloroplast proteins to check, whether they are present in isolated vacuole preparations. RuBPcase is a soluble stromal protein and comprises up to 50% of the total soluble leaf protein. The antiserum against coupling factor reacted weakly against the alpha, gamma, delta, and epsilon subunits and strongly against the beta subunit (Fig. 1, lane 3). The antiserum against RuBPcase reacted strongly with the large subunit of the enzyme, but only weakly against the small subunit (Fig. 1, lane 5).

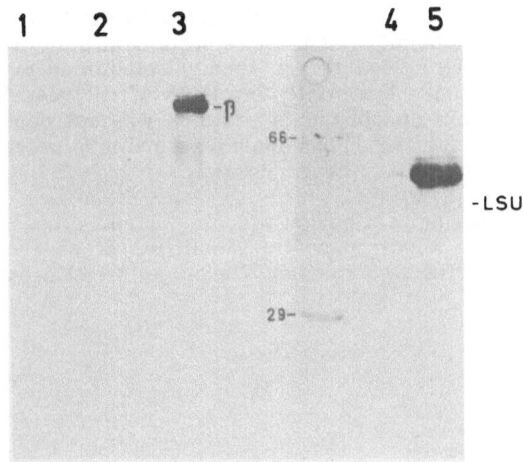

Figure 1

The purity of isolated vacuoles and vacuoplasts
as estimated by immunoblotting

Lane 1 : 3.2 10^5 isolated vacuoles; lane 2 : 3.2 10^5 vacuoplasts; lane 3 : isolated thylakoid membranes, corresponding to 4 μg chorophyll; lane 4 : 3 10^5 isolated vacuoles; lane 5 : crude homogenate from barley leaves. Filter A was probed with a 1:1000 dilution of anti-spinach chloroplast coupling factor. Filter B was probed with a 1:100 dilution of anti-RuBPcase. Beta : a large subunit of coupling factor. LSU : the large subunit of RuBPcase.

A small amount of RuBPcase large subunit protein was detected in the isolated vacuoles at a position corresponding to the authentic subunit (Fig. 1, lane 4). No material reacting with antiserum against CF_1 was detected in both isolated vacuoles (Fig. 1, lane 1) and vacuoplasts (Fig. 1, lane 2).

A polyspecific rabbit antiserum raised against an enriched tonoplast fraction from highly purified barley vacuoles recognized approximately 17 polypeptides of different apparent molecular weights in lanes loaded with purified vacuoles (Fig. 2-A, lane 1). The strongest reaction was against a polypeptide of approximately 24 kDa.

Protoplasts can be evacuolated by ultracentrifugation, yielding two plasmalemma-bound subfractions : cytoplasts which contain cytosol and organelles, and vacuoplasts which contain cytosol and vacuole (Lörz et al., 1976). Isolated vacuoplasts loaded at the same number of individuals as vacuoles produced an immunopattern virtually identical to that of the vacuoles (Fig. 2-A, lane 2).

Two polypeptides in the thylakoid pattern (Fig. 2-A, lane 3) were labeled by the anti-vacuole serum. Both had no counterpart of identical electrophoretic mobility in the vacuole and vacuoplast lanes, respectively. The staining intensities of the two thylakoid bands were found to be somewhat variable between experiments. A control using pre-immune serum showed that the immune reactions were specific (Fig. 2-B).

Vacuolar proteins have been distinguished by their solubility properties after freeze-thaw treatment in different media (Kaiser et al., 1986). Fig. 3 shows that most of the immunoreactive polypeptides from vacuoles can be solubilized by freeze-thawing in either water or 300 mM NaCl. Three prominent polypeptides, most notably a part of the heavy band of 24 kDa are readily solubilized by freeze-thawing in the presence of NaCl but are essentially insoluble after freeze-thawing in water. The fractionation of soluble and insoluble proteins was not complete, because low volumes (400 μl) were used for the freeze-thaw treatments in order to avoid excessive dilution. This resulted in some cross-contamination.

The antiserum raised against barley vacuole proteins cross-reacts with proteins from wheat (Fig. 4). The electrophoretic mobilities of several polypeptides are clearly different between the two species. The major 24 kDa polypeptide migrates identically.

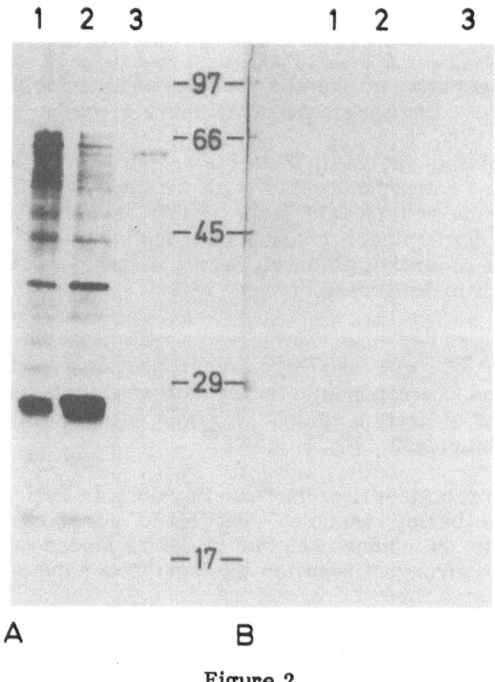

Figure 2

Immunoblot of subcellular fractions and isolated vacuoplasts

Lanes 1 : 3.2 10^5 isolated vacuoles; lanes 2 : 3.2 10^5 vacuoplasts; lanes 3 : isolated thylakoid membranes, corresponding to 4 μg chlorophyll. Filter A : The filter was probed with a 1:100 dilution of anti-vacuole serum Vakju 2. Filter B : Pre-immune serum Vakju 0. Identical samples were loaded on the lanes of Filters A and B and on the lanes of Filter A in Fig. 1.

DISCUSSION

The analysis of possibly contaminating proteins in vacuolar preparations by immunoblotting represents a method independent of the enzymatic activity of the contaminant. It permits the sensitive analysis of proteins, even if they are denatured or have a high K_m like RuBPcase. A slight contamination of isolated vacuoles with RuBPcase was inferred on the basis of gel electrophoresis in the presence of SDS after silver staining (Kaiser et al., 1986). The electrophoretic identification was correct as shown in Fig. 1. RuBPcase is known to be notoriously "sticky" and has been found also in preparations of envelope proteins from chloroplasts (Joyard et al., 1982). No contamination of vacuoles with chloroplast coupling factor could be found (Fig. 1). Conversely, this means that no major vacuolar protein was cross-reactive with the beta subunit of chloroplast coupling factor.

A polyspecific rabbit antiserum raised against partially purified tonoplast membranes derived from highly purified vacuoles recognized approximately 17 polypeptides in lanes loaded with isolated vacuoles (approximately 14 μg of protein) and two proteins of different molecular mass in lanes loaded with thylakoid membranes (approximately 20 μg of protein; Figs. 2 and 3).

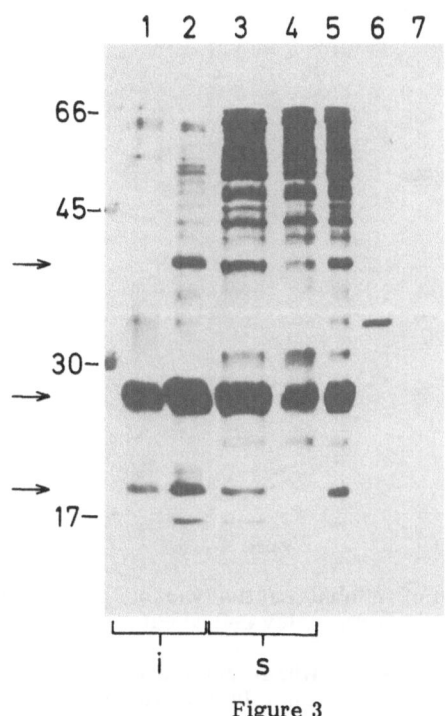

Figure 3

The solubility of vacuolar antigens after freeze-thawing
as estimated by immunoblotting

The filter was probed with 1:500 dilution of anti-vacuole serum Vakju 2. Vacuoles were freeze-thawed and centrifuged to separate the insoluble (i) proteins appearing in the pellet from the soluble (s) proteins in the supernatant. Lanes 1,3 : Freeze-thawed in the presence of 300 mM NaCl; lanes 2,4 : Freeze-thawed in the presence of water; lane 5 : Untreated vacuoles; lane 6 : Thylakoid membranes corresponding to 4 g chlorophyll; lane 7 : Protoplasts corresponding to approximately 40 μg of protein. Arrows indicate the positions of polypeptides insoluble in water but soluble in 300 mM NaCl.

It is unlikely that the bands decorated by the antiserum represent a series of proteolytic breakdown products, since the pattern was reproducible between experiments and virtually identical between vacuoles and vacuoplasts (Fig. 2). Most of these immunogenic vacuole proteins could be solubilized by freeze-thaw treatment and represent thus peripheral membrane proteins or soluble proteins proper (Fig. 3).

The major antigen of approximately 24 kDa (corresponding to protein 9 of Kaiser et al., 1986) has the same electrophoretic mobility as the apoprotein of the light harvesting complex from thylakoid membranes. This protein is one of the most prominent proteins in barley thylakoid gel patterns (Machold et al., 1977). Fig. 2-A clearly shows that the 24 kDa vacuole polypeptide bears no immunological relationship to the thylakoid protein.

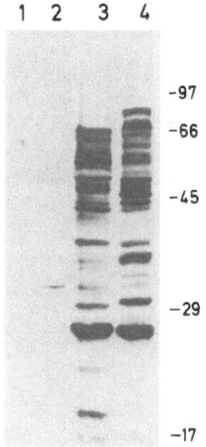

Figure 4

Comparison of immunoreactive vacuole polypeptides
from barley and wheat

Lane 1 : Barley cytoplasts; lane 2 : Wheat cytoplasts; lane 3 : Barley vacuoplasts; lane 4 : Wheat vacuoplasts. Approximately 2.5 10^5 individuals were loaded per lane. The filters were probed with a 1:500 dilution of Vakju 2 serum.

Immunological detection of vacuolar proteins is highly sensitive. Only approximately 14 μg of total vacuole protein has been loaded per lane in Figs. 1 – 3. Contaminating proteins, like plasmalemma and cytosolic proteins in the vacuoplast lanes in Fig. 2 do not impair sensitivity. Vacuolar proteins can even be detected at low staining intensity in lanes loaded with total protoplast protein.

Figure 4 shows that vacuolar polypeptides which are heavily stained in the vacuoplast lanes cannot be detected in the cytoplast lanes. This supports the conclusion of Griesbach and Sink (1983) that evacuolation of protoplasts by ultracentrifugation is a very efficient process.

Peroxidase reactions of varying intensities between experiments in lanes loaded with thylakoid proteins can be readily distinguished from the reactions of immunodecorated vacuolar antigens by their electrophoretic mobilities. It is unclear at present, whether this color dvelopment is caused by an immunological recognition of the polypeptides in question, or whether thylakoid haem proteins (Høyer-Hansen, 1980) catalyze the reaction directly.

The differences in electrophoretic mobilities and stoichiometry between several vacuolar proteins from two species of the family Poaceaea (Fig. 4) suggests that the vacuole is not a "conservative" compartment in evolutionary terms.

The availability of an antiserum against vacuole proteins is expected to facilitate the analysis of these proteins by enabling the monitoring of the purification of proteins of yet unknown function or of proteins which are non-functional when released from the membranes like translocators. The serum will furthermore allow to probe for sequences coding for vacuolar proteins using cDNA libraries cloned into expression vectors. The detection of a vacuole specific clone during a preliminary screening of a Mesembryanthemum leaf cDNA library (Schmitt and Bohnert, unpublished results) constructed in lambda gt11 shows that this approach is possible. Also, the amount of individual vacuolar proteins under specific physiological situations like salt stress can be determined.

REFERENCES

Griesbach, R., and Sink, K., 1983, Evacuolation of mesophyll protoplasts, Plant Science Letters, 30:297.
Hincha, D. K., Heber, U., and Schmitt, J. M., 1985, Antibodies against individual thylakoid membrane proteins as molecular probes to study chemical and mechanical freezing damage in vitro, Biochim. Biophys. Acta, 809:337.
Höfner, R., Vazquez-Moreno, L., Winter, K., Bohnert, H. J., and Schmitt, J. M., 1986, Induction of crassulacean acid metabolism in Mesembryanthemum crystallinum by high salinity : Mass increase and de novo synthesis of PEP-carboxylase, submitted for publication.
Høer-Hansen, G., 1980, Identification of haem-proteins in thylakoid polypeptide patterns of barley, Carlsberg Res. Commun., 45:167.
Joyard, J., Grossman, A., Bartlett, S. G., Douce, R., and Chua, N. H., 1982, Characrcterization of envelope membrane polypeptides from spinach chloroplasts, J. Biol. Chem., 257:1095.
Kaiser, G., and Heber, U., 1984, Sucrose transport into vacuoles isolated from barley mesophyll protoplasts, Planta, 161:562.
Kaiser, G., Martinoia, E., Schmitt, J. M., Hincha, D. K., and Heber, U., 1986, Polypeptide pattern and enzymic character of vacuoles isolated from barley mesophyll protoplasts, Planta, in press.
Kaiser, G., Martinoia, E., and Wiemken, A., 1982, Rapid appeearence of photosynthetic products in the vacuoles isolated from barley mesophyll protoplasts by a new fast method, Z. Pflanzenphysiol., 107:103.
Lörz, H., Harms, C. T., and Potrykus, I., 1976, Isolation of "vacuoplasts" from protoplasts of higher plants, Biochem. Physiol. Pflanzen, 169:617.
Machold, O., Simpson, D. J., and Hoyer-Hansen, G., 1977, Correlation between the freeze-fracture appearance and polypeptide composition of thylakoid membranes in barley, Carlsberg Res. Commun., 42:499.

Martinoia, E., Flügge, U. I., Kaiser, G., Heber, U. and Heldt, H. W., 1985, Energy-dependent uptake of malate into vacuoles isolated from barley mesophyll protoplasts, Biochim. Biophys. Acta, 806:311.

Martinoia, E., Schramm, M. J., Kaiser, G., Kaiser, W., and Heber, U., 1986, Transport of anions in isolated barley vacuoles, Plant Physiol., 80:895.

Nowotny, A., 1979, "Basic Exercises in Immunochemistry", 2nd Edition, Springer-Verlag, Berlin.

Towbin, H., Staehelin, T., and Gordon, J., 1979, Electrophoretic transfer of proteins from polyacrylamide gels to nitrocellulose sheets : Procedure and some applications, Proc. Natl. Acad. Sci. U.S.A., 76:4350.

Wolter, F. P., Schmitt, J. M., Bohnert, H. J., and Tsugita, A., 1984, Simultaneous isolation of three peripheral proteins - a 32 kDa protein, ferredoxin NADP$^+$ reductase and coupling factor - from spinach thylakoids and partial characterization of a 32 kDa protein, Plant Science Letters, 34:323.

Yeo, A. R., 1983, Salinity resistance : Physiologies and prices, Physiol. Plant., 58:214.

A HYPERPOLARIZATION-ACTIVATED K+ CURRENT IN ISOLATED VACUOLES

OF ACER PSEUDOPLATANUS

Roberta Colombo[*], Piera Lado[*] and Antonio Peres[**]

[*]Dipartimento di Biologia, Centro di Studio del C.N.R. per la Biologia Cellulare e Molecolare delle Piante and [**]Dipartimento di Fisiologia e Biochimica Generali dell'Università di Milano Via Celoria, 26, I-20133-Milano, Italy

INTRODUCTION

Transport processes across the vacuolar membrane of higher plants are very important for cytoplasmic homeostasis and in general in the plant cell physiology. Tonoplast transport properties are difficult to study because of the intracellular localization of the vacuolar membrane.

Many advancements have been made since the use of membrane vesicles to study transport systems was developed. Using this approach in tonoplast vesicles derived both from tonoplast enriched microsomal fractions (see Sze, 1985 for references) and from isolated vacuoles (Mandala and Taiz, 1985; Poole et al., 1984; Jochem et al., 1984; Thom and Komor, 1985), the operation of an electrogenic H^+-ATPase on the tonoplast has been established and some data suggesting the presence of antiports (Na^+/H^+, Ca^{2+}/H^+) or anion transports driven by the H^+ electrochemical gradient have been obtained (Blumwald and Poole, 1985 a; Schumaker and Sze, 1985; Lüttge et al., 1981; Blumwald and Poole, 1985 b). To date little is known about the existence of ionic channels on vacuolar membrane of higher plants as electrophysiological measurements have been performed mainly on giant algal cells (Lunewsky et al., 1983).

The improvement of the technique of vacuole isolation on one hand, and the application of the patch–clamp technique on the other, have opened new possibilities to study ionic channels on vacuolar membranes (Takeda et al., 1985; Kado et al., 1986; Kolb et al., 1986; Hedrich et al., 1986; Bentrup et al., 1986).

In this work we used the patch–clamp technique in the "whole–cell" (whole-vacuole) recording mode (Hamill et al., 1980; Sakmann and Neher, 1983) to investigate ionic currents in the entire membrane of isolated vacuoles from Acer pseudoplatanus protoplasts.

MATERIALS AND METHODS

Protoplasts were isolated from cultured Acer pseudoplatanus cells by enzymatic digestion (Kurkdjian and Barbier-Brygoo, 1984). Experiments were performed on the vacuoles spontaneously released from protoplasts. Only vacuoles with diameters greater than $30\mu m$ were used for the experiments.

Pipettes had resistances between 2 and 6 MOhm. The clamp amplifier was built in our electronic shop. It included facilities to switch between a 100 MOhm and a 10 GOhm feedback resistors and to compensate up to 70% of the series resistance. These were used depending on the current amplitudes. Signals were filtered at 2 KHz and recorded on FM Tape.

The standard external solution (pH 7.2) included (in mM) : 10 Hepes, 153.4 K^+, 1 Ca^{2+}, 2 Mg^{2+}, 154 Cl^-, 1 SO_4^{2-}. External solutions with reduced K^+ concentrations were made by substituting K^+ with equimolar TMA^+ (tetramethylammonium). The pipette was filled with (in mM) : 10 malate, 150 K^+, 134 Cl^- (pH 5). The osmolarity of both solutions was adjusted to 0.6 M with mannitol. The temperature during the experiments was 22 – 25°C.

RESULTS AND DISCUSSION

The formation of gigaseals between pipette tip and tonoplast was very fast and seals with resistance greater than 10 GOhm were easily obtained. As already reported (Takeda et al., 1985; Kado et al., 1986; Kolb et al., 1986) discrete channel openings were sometimes observed when in "cell-attached" conditions. The establishment of the "whole-cell" mode was, on the contrary, rather difficult to obtain due to the extreme fragility of the tonoplast (Korzun et al., 1985) and to the tendency of the ruptured patch to reseal. We have partially overcome this problem by using an electronically-controlled electrovalve to apply negative pressure to the pipette. In this way short (down to 100 msec) suction pulses could be applied, which led to a higher rate of success in rupturing the patch without destroying the vacuole. In symmetrical KCl the vacuole resting potential was slightly positive (+ 10.3 ± 4.0 mV SD, n = 6), in agreement with the data obtained in isolated vacuoles from other materials by microelectrode measurements (Gibrat et al., 1985). When voltage pulses were applied from a 0 mV holding potential records such those of Fig. 1-A and B were observed. Currents flowing in response to various 500 msec depolarizations are shown in Fig. 1-A, while the currents elicited by hyperpolarizations in the same vacuole are shown in Fig. 1-B (note the different calibration). Hyperpolarizations clearly induce the development of a voltage- and time-dependent inward current of a few nA with a small inward tail upon returning to 0 mV. The instantaneously-developing, fastly-decaying, outward current seen upon depolarization appears to be due to the voltage-dependent closure of some channels which were open at 0 mV. An evidence to support this statement is given in Fig. 1-C where several hyperpolarizations of different durations to - 80 mV were followed by an instantaneous jump to + 40 mV. It is clear that the amplitude of each outward tail is proportional to the amplitude reached by the inward current at the end of the corresponding hyperpolarizing pulse, showing that the tails currents flow through the same channels which carry the inward current.

Fig. 1-D shows the relation between steady-state current (at the end of the 500 msec pulses) and tonoplast voltage. The tonoplast is thus strongly inward-rectifying, showing an extremely high resistance in the positive range of potentials. We have obtained similar results in 10 other vacuoles.

These results are very similar to those recently reported on vacuoles isolated from barley mesophyll protoplasts (Hedrich et al., 1986).

The reversal potential (E_{rev}) of this current is at a slightly positive value (+ 7.4 ± 2.2 mV SD, n = 5) as it is shown in Fig. 2-A where postpulses to various voltages were given after a conditioning hyperpolarization. To study the ionic nature of the hyperpolarization-activated current we have performed other experiments changing the composition of the external solution. The fact that the E_{rev} is close to 0 mV in symmetrical KCl suggests that these ions are good candidates to carry

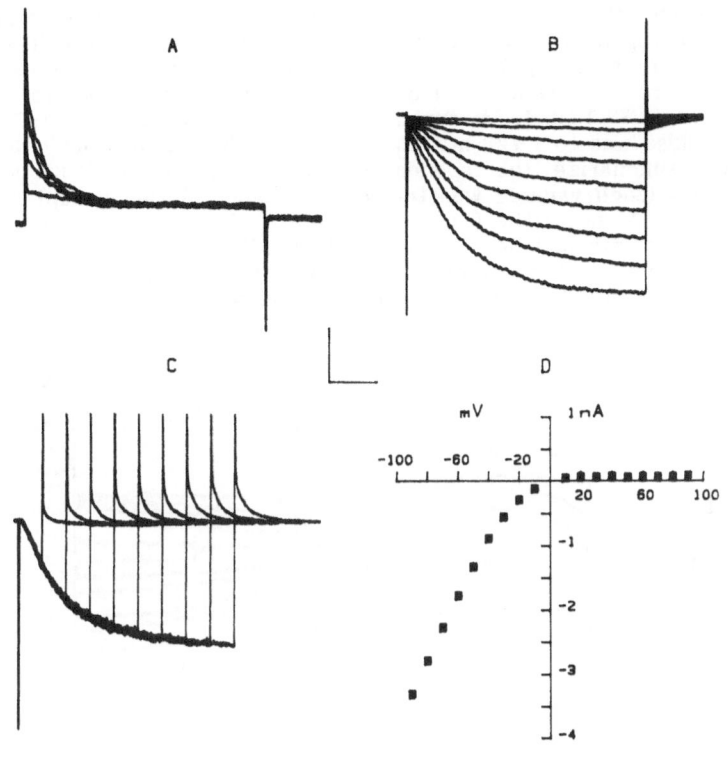

Figure 1

Typical recordings of membrane currents
elicited by various command potentials applied with the whole–cell technique
to a vacuole $45\,\mu m$ in diameter

Holding potential 0 mV throughout this figure. A. Rapidly decaying outward currents elicited by depolarizations to + 20, + 40, + 60, + 80 and + 100 mV. Calibration bars correspond to 100 pA and 100 msec. B. The responses to nine hyperpolarizations from – 10 to – 90 mV (in 10 mV steps) are shown. Calibrations bars represent 1 nA and 100 msec. C. Conditioning pulses to – 80 mV were applied for various durations; at the end of each pulse the potential was stepped back to + 40 mV to elicit outward tail currents. The amplitudes of the tails are clearly proportional to the amplitudes of the inward currents at the end of each corresponding hyperpolarization. Calibration bars represent 1 nA and 200 msec. D. The current amplitudes at the end of the pulses in A and B are shown against the membrane potential. The threshold for the appearance of the current is around 0 mV. For positive potentials the steady–state membrane conductance is extremely low.

the current (rather than the other ionic components which are asymmetrically distributed).

We have therefore changed the external K^+ concentration by substituting it with equimolar amounts of TMA^+ (a well-known impermeant ion for animal membranes (Almers et al., 1984; Fukushima and Hagiwara, 1985)), leaving the external Cl^- unaltered. Fig. 2-B shows a set of current traces elicited by various hyperpolarizations in low external K^+ (8 mM). For moderate hyperpolarizations a time-dependent decline of the inward current is visible, while for stronger hyperpolarizations the inward current increases with time. A substantially flat trace (easily recognizable by the presence of single channel transitions on it) is visible in response to the pulse to − 60 mV (E_{rev}). Upon returning to 0 mv large outward tails are observed. We have also tested an intermediate external K^+ concentration of 33.3 mM. A typical result is shown in Fig. 2-C where the conditioning pulse was to − 120 mV. E_{rev} has been shifted to about − 30 mV in this case. Fig. 2-D summarizes the dependence of E_{rev} on the external K^+ concentration for 10 vacuoles (each vacuole was tested in a single solution except one that was tested in the 153.4 and in the 8 mM K^+ solutions). The continuous line is the best fit through the points and has a slope of 49.3 mV per decade; the dotted line is the theoretical K^+ equilibrium potential under the assumption that the vacuolar content had been completely substituted by the pipette solution.

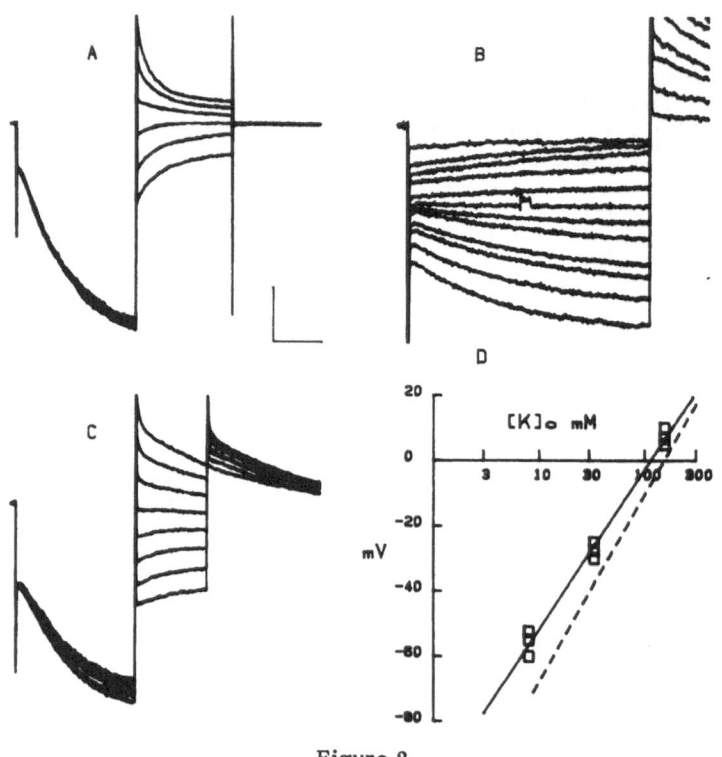

Figure 2

Reversal potential of the inward current
and its dependence on the external potassium concentration

The holding potential was 0 mV throughout. A. In 153.4 mM external K^+ the tails current reverted their direction between 0 and + 10 mV. The conditioning pulse to − 70 mV was followed by postpulses from − 20 to + 30 mV, in 10 mV steps. Calibration bars represent 1 nA and 200 msec; vacuole diameter 60 μm. B. 8 mM external K^+ (E_{rev} − 60 mV in this case). Currents in response to voltage steps from − 10 to − 120 mV (in 10 mV steps) are shown. Calibration bars represent 100 pA and 100 msec; vacuole diameter 50 μm. Notice the single channel transition. C. In 33.3 mM external K^+ E_{rev} was between − 25 and − 30 mV. In the case shown the conditioning pulses were to − 120 with postpulses from − 70, to 0 mV in 10 mV steps. Calibration bars represent 200 pA, 200 msec; vacuole diameter 35 μm. D. Plot of E_{rev} measured in 10 vacuoles in three different external potassium concentrations.

These results indicate that the current is substantially carried by K^+, and hence that tonoplast possesses hyperpolarization-activated K^+ channels. This is in agreement with the findings of Hedrich et al. (1986), although these authors also suggest a malate contribution to the current.

The presence of inward rectifier channels may therefore be a general feature of higher plant vacuoles. The understanding of their physiological role requires further work to define their selectivity and to ascertain the existence of other channels. However the fact that these channels are voltage–dependent suggests that their activity is likely to be associated to already known electrogenic transport processes.

REFERENCES

Almers, W., McCleskey, E. W., and Palade, P. T., 1984, A non-selective cation conductance in frog muscle membrane blocked by micromolar external calcium ions, J. Physiol., 353:565.

Bentrup, F.-W., Gogarten-Boekels, M., Hoffmann, B., Gogarten, J. P., and Baumann, C., 1986, ATP-dependent acidification and tonoplast hyperpolarization in isolated vacuoles from green suspension cells of Chenopodium rubrum L., Proc. Natl. Acad. Sci. U.S.A, 83:2431.

Blumwald, E., and Poole, R. J., 1985 a, Na^+/H^+ antiport in isolated tonoplast vesicles from storage tissue of Beta vulgaris, Plant Physiol., 78:163.

Blumwald, E., and Poole, R. J., 1985 b, Nitrate storage and retrieval in Beta vulgaris L. : Effects of nitrate and chloride on proton gradients in tonoplast vesicles, Proc. Natl. Acad. Sci. U.S.A., 82:3683.

Fukushima, Y., and Hagiwara, S., 1985, Currents carried by monovalent cations through calcium channels in mouse neoplastic B lymphocytes, J. Physiol., 358:255.

Gibrat, R., Barbier-Brygoo, H., Guern, J., and Grignon, C., 1985, Transtonoplast potential difference and surface potential of isolated vacuoles, in : "Biochemistry and Function of Vacuolar Adenosine-triphosphatase in Fungi and Plants", B. P. Marin, ed., Springer-Verlag, Berlin, Heidelberg, New-York, Tokyo.

Hamill, O. P., Marty, A., Neher, E., Sakmann, B., and Sigworth, F. J., 1980, Improved patch–clamp techniques for high-resolution current recording from cells and cell-free membrane patches, Pflüg. Arch. Ges. Physiol., 391:85.

Hedrich, R., Flügge, U. I., and Fernandez, J. M., 1986, Patch–clamp studies of ion transport in isolated plant vacuoles, FEBS Letters, 204:228.

Jochem, P., Rona, J. P., Smith, J. A. C., and Lüttge, U., 1984, Anion-sensitive ATPase activity and proton transport in isolated vacuoles of species of the CAM genus Kalanchoë, Physiol. Plant., 62:410.

Kado, R. T., Kurkdjian, A., and Takeda, K., 1986, Transport mechanisms in plant cell membranes : An application for the patch clamp technique, Physiol. Vég., 24:227.

Kolb, H. A., Kohler, K., and Martinoia, E., 1986, Potassium channels in membranes of isolated mesophyll barley vacuoles, Plant Physiol., 80 S:137.

Korzun, A. M., Kuzevanov, V. Ya., and Salyaev, R. K., 1985, Peculiarities of electrophysiological investigations of the vacuolar membrane in cells of higher plants, Soviet Plant Physiol., 31:925.

Kurkdjian, A. C., and Barbier-Brygoo, H., 1984, A hydrogen ion-selective liquid-membrane microelectrode for measurements of the vacuolar pH of plant cells in suspension culture, Anal. Biochem., 132:96.

Lunevsky, V. Z., Zherelova, O. M., Vostrikov, I. Y., and Berestovsky, G. N., 1983, Excitation of Characeae cell membranes as a result of activation of calcium and chloride channels, J. Membrane Biol., 72:43.

Lüttge, U., Smith, J. A. C., Marigo, G., and Osmond, C. B., 1981, Energetics of malate accumulation in the vacuoles of Kalanchoë tubiflora cells, FEBS Letters, 126:81.

Mandala, S., and Taiz, L., 1985, Proton transport in isolated vacuoles° from corn coleoptiles, <u>Plant Physiol.</u>, 78:104.

Poole, R. J., Briskin, D. P., Krátky, Z., and Johnstone, R. M., 1984, Density gradient localization of plasma membrane and tonoplast from storage tissue of growing and dormant red beet, <u>Plant Physiol.</u>, 74:549.

Sakmann, B., and Neher, E., 1983, "Single Channel Recording", Plenum Publish. Corp., New-York.

Schumaker, S. K., and Sze, H., 1985, A Ca^{2+}/H^+ antiport system driven by the proton electrochemical gradient of a tonoplast H^+-ATPase from oat roots, <u>Plant Physiol.</u>, 79:1111.

Sze, H., 1985, H^+-translocating ATPases : Advances using membrane vesicles, <u>Annu. Rev. Plant. Physiol.</u>, 36:175.

Takeda, K., Kurkdjian, A. C., and Kado, R. T., 1985, Single-channel currents from the tonoplast membrane of isolated <u>Catharanthus roseus</u> vacuoles, <u>Plant Physiol.</u>, 77 S:88.

Thom, M., and Komor, E., 1985, Electrogenic proton translocation by the ATPase of sugarcane vacuoles, <u>Plant Physiol.</u>, 77:329.

A MICROCOMPUTER-BASED SYSTEM FOR THE RECORDING AND ANALYSIS

OF FLUORESCENCE TRANSIENTS FROM MEMBRANE VESICLES

Ian R. Jennings, Philip A. Rea and Dale Sanders

Department of Biology, University of York
Heslington, York Y01-5DD, England, U. K.

INTRODUCTION

Until a few years ago, the problem of studying the kinetics of transport at the tonoplast seemed almost insurmountable. The only technique available was that of compartmental analysis, which, as a prerequisite for its successful application, necessitated that the cell be at steady state with respect of solute concentration. Thus, although unidirectional fluxes could be derived from the analysis of isotopic efflux, dynamic studies were precluded.

The availability of well-characterized preparations of tonoplast vesicles has changed this picture considerably. Transport can now be studied in vitro using defined and simple media on both sides of the membrane, and intracellular regulators, which can severely compromise the interpretation of data from whole cells, may be excluded. Increased experimental access to the tonoplast has however demanded the development of new techniques which take account of the small size of the vesicles and exploit the possibility of non-steady state kinetic studies.

The transmembrane electrochemical gradient for protons ($\Delta\overline{\mu}_H{}^+$) is a central factor which couples primary and secondary transport processes at the tonoplast. Since straightforward electrophysiological techniques are not yet available for measurement of $\Delta\overline{\mu}_H{}^+$ in membrane vesicles, fluorescent probes —reporting either the chemical (ΔpH) or electrical ($\Delta\Psi$) components of $\Delta\overline{\mu}_H{}^+$ —are commonly used. For example, in the case of activation of an inwardly-directed electrogenic proton pump, fluorescence emission at 540 nm from the monoamine dye acridine orange is quenched as intravesicular pH falls, and emission at 640 nm from the dye Oxonol V is quenched as an inside-positive $\Delta\Psi$ is generated. If the dyes are appropriately calibrated, the absolute magnitudes and rates of formation of ΔpH and $\Delta\Psi$ can be computed from the extent and rate of change of fluorescence. Similarly, kinetics of dissipation of $\Delta\overline{\mu}_H{}^+$ can be assessed after sudden cessation of H$^+$ pumping or after the initiation of H$^+$/solute antiport (Blumwald et al., 1987).

Output from the fluorescence spectrometer is normally to a chart recorder. Thus, the extent to which the data yield reliable information on transport **kinetics** is ultimately dictated by the accuracy with which the investigator can draw a tangent to the fluorescence trace. Besides the possibility of imprecision, manual procedures are laborious.

We have therefore developed software for one-line display and recording by an IBM microcomputer of fluorescence data from a Perkin-Elmer LS-5 fluorescence spectrometer. Storage of the data in digitalized form enables editing and subsequent analysis by a non-linear least squares routine. Consequently, it is possible to determine precisely the magnitude of the components and rates of transients in the fluorescence signal.

MATERIALS

Hardware

The essential hardware is as follows : an LS-5 (or LS-3) fluorescence spectrometer and an LS-X Communications Interface, both supplied by Perkin-Elmer; an IBM-PC/XT system unit (with a Hercules graphics card and a minimum of 256K of memory), keyboard, monochrome screen and asynchronous communications adapter; a dot matrix printer (driven in parallel to the IBM) such as the Epson FX or LX series. Curve-fitting makes use of the 8087 math coprocessor. It is also useful (though not essential) to have the GP-100 printer from Perkin-Elmer (see below).

Software

The only commercial software required is DOS 2.1, or the above, for the IBM. Potential users can contact the authors to obtain more information on the programs outlined below.

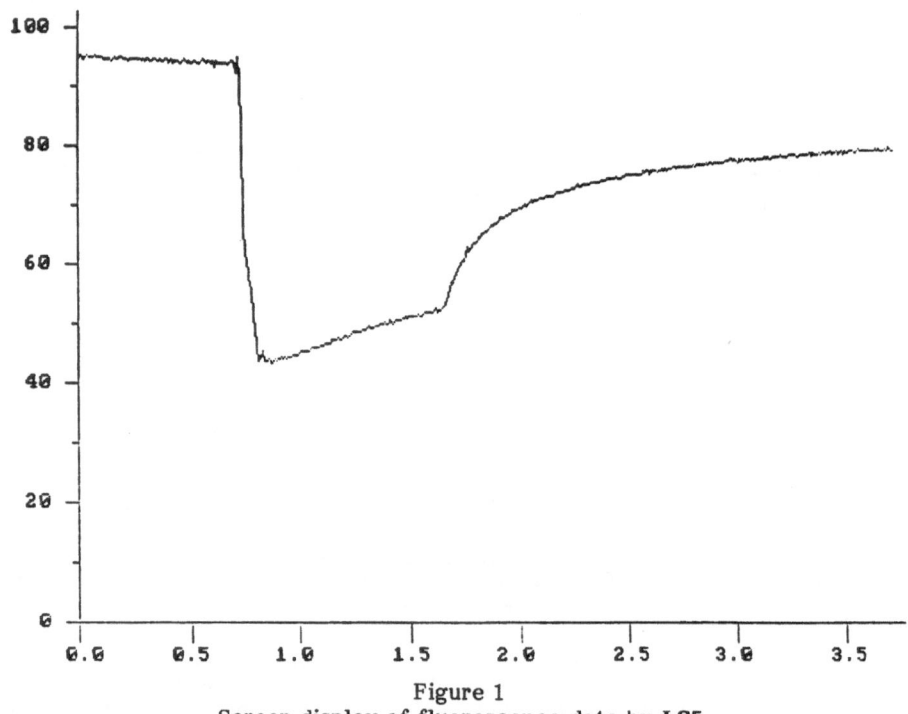

Figure 1

Screen display of fluorescence data by **LS5**

Ordinate = fluorescence (%); abscissa = time (min). Tonoplast vesicles from red beet were subjected to a "pH jump" from pH 6 to 8 in a medium containing 5 μM acridine orange 0.73 min after the start of the recording. The slow relaxation represents the passive efflux of protons down their electrochemical gradient. At 1.65 min, Na^+/H^+ antiport was initiated by the addition of 30 mM Na^+, and rapid recovery of the fluorescence ensued.

METHODS AND RESULTS

Data Collection From The Fluorescence Spectrometer

Data collection is performed by the **LS5** program. The major features of LS5 are as follows :

- LS5 is initialized for future reference by recording the details of the experimental conditions.

- Fluorescence signals are sent from the spectrometer every 100 ms and are then stored in memory.

- As data arrive, they are displayed on the video display unit in the format "relative fluorescence = f(time)". The axes can be scaled automatically or set manually. The screen is therefore acting as a chart recorder (Fig. 1).

- The maximum time for a single recording is 40 min.

- When the experimental run is complete, the program is terminated, and data are written to disk.

Editing The Data

Once collected, the data can be edited for review or curve-fitting using the **LS5EDIT** program. The major features of this program are summarized below.

LS5EDIT allows specific sections of the experiment to be reviewed on screen in detail. Thus, after selection of the segment for detailed analysis, the interesting

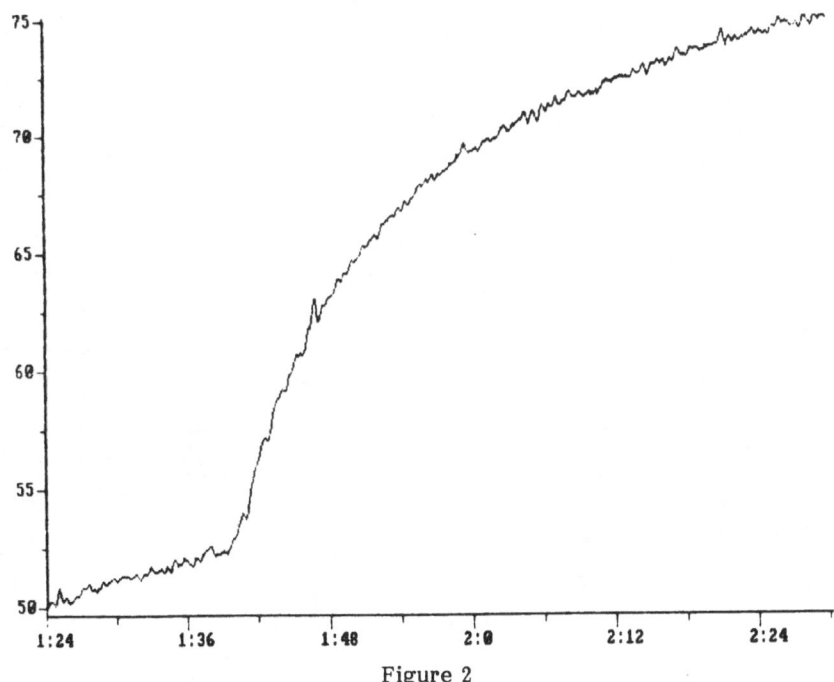

Figure 2

Scale expansion of selected data from Fig. 1
using **LS5EDIT** in review mode

Abscissa units: min:s. Figure focuses on the period of fluorescence relaxation after addition of Na^+, to enable more detailed visual analysis.

59

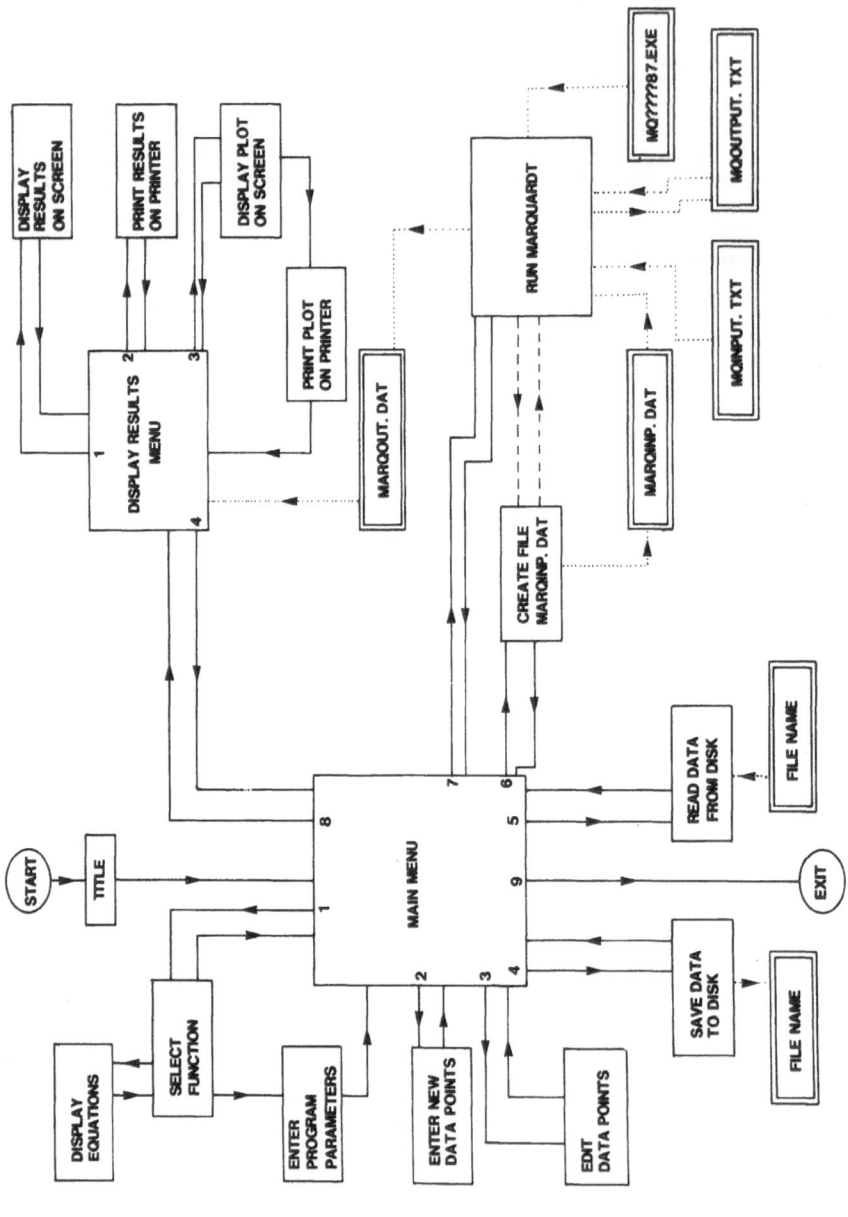

<u>Figure 3.</u> Flow diagram summarizing the program structure of **MARQDAT.** Single-lined boxes and the solid lines connecting them indicate user-interactive pathways; double-lined boxes represent files created by the program, and the dotted lines show pathways followed by the data.

portion can be expanded on the ordinate or abscissa (Fig. 2). While in this review mode, a hard copy of the analog display unit can be obtained on the dot matrix printer.

The data selection mode of **LS5EDIT** enables specific segments of the data to be written into new text files, in a format suitable for subsequent statistical analysis (see below). Time can then be re-defined as starting from the point at which the treatment begins. The sampling frequency can also be adjusted at this point : for example, not all 1000 data points in a 100 s interval have to be used.

LS5EDIT also allows a complete analog print-out of the experimental data on the GP-100. This facility can be useful if the experiment lasts for longer than the time encompassed by the video display unit (operational maximum : 25 min).

Non-linear Least Squares Fitting

Curve fitting of stored data is executed by the program **MARQDAT** which is based on the Marquardt algorithm (Marquardt, 1963). A flow diagram summarizing the structure of MARQDAT is given in Fig. 3. The program offers a range of functions but, in practice, the most convenient for $\Delta\mu_H{}^+$-related fluorescence changes is an exponential of the form :

$$F = p1 \left[p2 - \frac{p3}{\exp\left[p4\,(t - p5)\right]} \right]$$

in which F is relative fluorescence, t is time and p1 trough p5 are constants derived by curve-fitting. Normally, p1 and p5 are constrained to 1 and 0, respectively. Results of fitting the data of Fig. 1 to this equation are displayed in Fig. 4 and Fig. 5.

In this example, the initial rate of Na^+-dependent fluorescence quenching can easily be calculated as the difference between the derivatives of the two functions at t = 1.65 min real time (i.e. t = 0.76 min and t = 0 min in Fig. 4 and

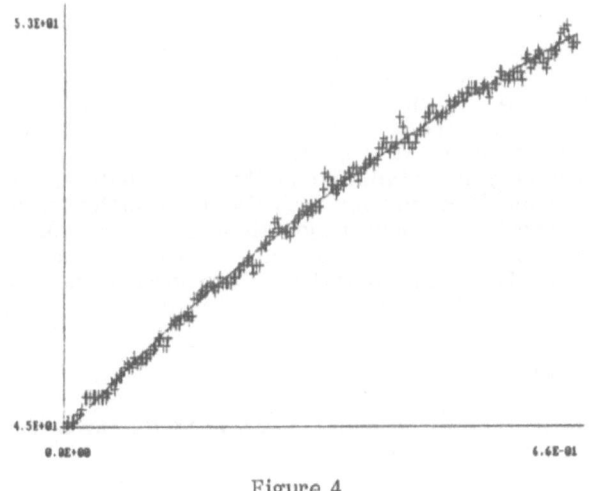

Figure 4

Non-linear least squares fit (solid line) using MARQDAT

Fluorescence data (Fig. 1) for the period of relaxation due to passive efflux of protons (0.89 - 1.65 min) were fitted according to Eq. 1, with time redefined to zero at the start of the period. Parameter estimates $\pm$ standard errors: p2, 70.3 $\pm$ 2.4 % min^{-1}; p3, 27.1 $\pm$ 2.41 % min^{-1}; p4, 0.547 $\pm$ 0.060 min^{-1}.

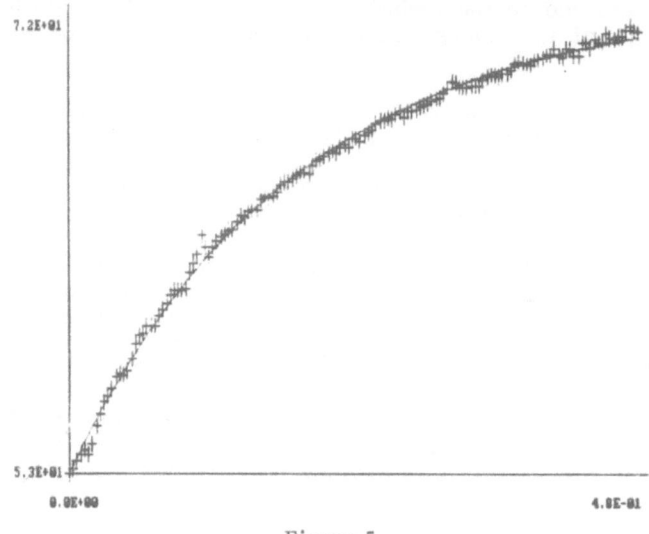

Figure 5

As in Fig. 4, but for period of relaxation following Na^+ addition to medium (1.65 – 2.31 min real time). Time redefined to zero at point of Na^+ addition. Parameter estimates $\pm$ standard errors: p2, 74.1 $\pm$ 0.1 % min^{-1}; p3, 22.4 $\pm$ 0.1 % min^{-1}; p4, 4.51 $\pm$ 0.07 min^{-1}.

Fig. 5, respectively). Thus, the relaxation due to passive H^+ efflux at 1.65 min is 9.8 % min^{-1}, while after the addition of Na^+ this value increases to 101.0 % min^{-1}. The rate of Na^+-dependent fluorescence quenching is therefore 91.2 % min^{-1}.

DISCUSSION

Until now, analysis of fluorescence data from membrane vesicles has generally been a rather arbitrary and laborious affair, with few attempts to quantify rigorously non-steady state events. The software described here enables the casual user to undertake a more detailed analysis which is not subject to the imprecision of subjectively-drawn tangents. Although the case of the ΔpH probe acridine orange has been dealt with here in detail, the method is equally applicable to analysis of the fluorescence from $\Delta\Psi$-reporting dyes. One noteworthy feature of the system we describe is its cost; the only major item of expenditure apart from the fluorescence spectrometer is the microcomputer. The software is simple to use, and, with practice, enables results to be derived considerably quicker than by analysis of analog data.

ACKNOWLEDGEMENTS

We are grateful to the Agricultural and Food Research Council (U. K.) and the University of York for financial support.

REFERENCES

Blumwald, E., Rea, P. A., and Poole, R. J., 1987, Preparation of tonoplast vesicles : Applications to H^+-coupled secondary transport in plant vacuoles, Methods In Enzymology, in press.

Marquardt, D., 1963, An algorithm for least squares estimation of non-linear parameters, J. Soc. Industr. Appl. Math., 11:431.

BENEFITS AND COSTS OF VACUOLATION

John A. Raven

Department of Biological Sciences
University of Dundee
Dundee, DD 1 4 HN, United Kingdom

INTRODUCTION

This paper outlines the benefits and costs of vacuolation in the cells of photo-trophs. It elaborates on the material in Chapter 8 of Raven (1984); a detailed exposition may be found in Raven (1987).

BENEFITS OF VACUOLATION

Preamble

The possible benefits of the presence of a vacuole occupying a large fraction of the volume of the protoplast can be divided into those which relate solely to the increased volume and surface area of the protoplast per unit of cytoplasm, independent of the nature of the vacuolar contents; and those which depend on the nature of the solutes in the vacuole.

Benefits Independent of the Nature of Vacuolar Contents

Increasing the volume of the protoplast per unit volume of cytoplasm increases (for a given shape of cell, e.g. spherical) the total surface area of the cell per unit of cytoplasmic volume. This area amplification has possible benefits in terms of the rate of resource acquisition under resource-limiting conditions. At a given chromophore concentration in the cytoplasm, dispersal of a given volume of cytoplasm at the periphery of a vacuole generally increase the efficiency of photon absorption of an "average" chromophore molecule relative to an otherwise comparable non-vacuolate cell. No increased efficiency is found if the chromophore concentration is low and/or the cell is very small (no package effect) or the total chromophore per unit projected area is such that almost all of the incident light is absorbed, regardless of vacuolation. Area amplification resulting from vacuolation also increases the plasmalemma area per unit cytoplasmic volume through which "lipid solution", or mediated, transport of chemical resources can occur on a cytoplasmic volume basis. Vacuolation of a cell of a given shape also decreases the impediment to resource transfer from bulk aqueous phase to the cell surface resulting from boundary layer effects, again on a cytoplasmic volume basis. The increased capacity for uptake of chemical resources from low external concentrations due to vacuolation can increase overall solute acquisition per unit cytoplasmic volume, except in cells of picoplankton size. Other remedies for a restricted capacity for chemical nutrient uptake per unit cytoplasmic volume include evaginations of the cell surface.

The discussion here has, implicitly, largely concerned unicells; macrophytes can also benefit directly from vacuolation with respect to the rate of resource acquisition.

Benefits Dependent On The Nature Of Vacuolar Contents

Putative benefits of vacuolation which depend on the nature of vacuolar contents include increases in the rate of acquisition of resources, and an improved capacity to manipulate, transform and protect resources which have already been aquired. Effects of vacuolation which involved essentially irreversible deposition (within the life of the cell) of specific solutes in the vacuole embrace the acquisition, manipulation and protection of resources, but not true ("reversible") storage. deposition-related increases in resource (phosphate and iron) acquisition capacity can, in attached rhizophytic macrophytes, result from rhizosphere acidification involving accumulation of cation salts of organic acid anions in the vacuole. Related to resource (CO_2, H_2O) acquisition in terrestrial halophytes is the disposal of "excess" salt entering the roots by deposition in hypertrophied vacuoles. Resource manipulation involving deposition of salts of organic acid anions occurs as a means of OH^- disposal in the reduction of NO_3^- in the shoot, a process which can, under some conditions, increase the photon and water use efficiency of N assimilation relative to other locations of NO_3^- reduction and modes of OH^- disposal. Resource protection in plants often involves the deposition of antibiophage solutes in the vacuoles. The possible osmotic engendered by cation organate deposition related to phosphate and iron acquisition, or NO_3^- reduction, can be ameliorated by production of insoluble calcium oxalate rather than soluble potassium malate.

Reversible accumulation of solutes in vacuoles is involved in CAM,which can increase the water (and nitrogen ?) use efficiency of carbon assimilation, and may be involved in the storage of fermentation products during temporary anoxia. The storage of soluble compounds in vacuoles during CAM cannot, apparently, be replaced by that of insoluble material. There is scope for storage of insoluble waxes during fermentation; this is only known to occur in Euglena. Storage of energy and carbon (as reduced carbon), phosphorus and nitrogen often involves soluble low-molecular weight materials in large vacuoles, but can also involve polymers in smaller vacuoles or in other parts of the cytoplasm.

Other potentially beneficial effects of vacuolation include a (mechanistically unexplained) stimulation of the velocity of cytoplasmic streaming in large, vacuolate cells, and a role in buffering the cytoplasm from rapid, large changes in volume in cells (e.g. stomata) undergoing large changes in protoplast volume.

COSTS OF VACUOLATION

Against any putative benefits must be set, in terms of natural selection, costs of vacuolation. The costs include those (energy, solutes) of producing, and (energy) of maintaining, the vacuoles; and (energy, carbon) of synthesising extra wall material. The additional energy and carbon costs (all on a unit cytoplasm basis) can be of a similar magnitude to the benefits of additional energy and carbon acquisition resulting from vacuolation. Additional costs may be incurred in relation to the energetics of large cells maintaining a constant turgor pressure in environments (e.g. estuaries) with large and frequent changes in external osmolarity, and of habitat opportunities lost if vacuolation reduces desiccation tolerance.

CONCLUSIONS

The analysis conducted by Raven (1987) suggests that the potential maximum specific growth rate of a cell of a given cytoplasmic volume under optimal growth conditions is decreased if it is vacuolated. However, the cell's resource acquisition rate under conditions of low resource availability is frequently (but not invariably) enhanced by vacuolation even when the resource costs of vacuolation are taken into consideration. Although the cost-benefit analysis of resource acquisition may not always favour vacuolation, the vacuolation may still lead to an <u>increase</u> in inclusive fitness when benefits to the storage, manipulation and protection of resources are also considered. Clearly most species of phototrophs <u>are</u> vacuolate, major exception being many planktophytes (where the lack of vacuoles may be rationalised, at least for smaller cells) and essentially all microalga-invertebrate symbioses.

ACKNOWLEDGEMENT

Dr. B. Marin has provided essential catalysis for the production of the analysis outlined in this paper.

REFERENCES

Raven, J. A., 1984, "Energetics and Transport in Aquatic Plants", A. R. Liss, New-York.
Raven, J. A., 1987, The role of vacuoles, New Phytologist, in press.

THE PHOTOSYNTHETIC POTENTIAL AND GROWTH

OF PHOTOAUTOTHROPHIC CELL SUSPENSION CULTURES

FROM CHENOPODIUM RUBRUM: A CASE STUDY

Wolfgang Hüsemann

Institute of Plant Biochemistry, University of Münster
D-4400-Münster, Federal Republic of Germany

INTRODUCTION

Chlorophyllous cell cultures from different plant species, mostly from dico-tyledonous angiosperms are used in several laboratories for studies on the nutritional and cultural conditions for chlorophyll formation and chloroplast differentiation as well as for the induction of high photosynthetic CO_2 assimilation rates as recently reviewed (Dalton and Peel, 1983).

Sustained photoautotrophic growth of in vitro cultured plant cells in the absence of an exogenous sugar but in the presence of CO_2 enriched air (1-2% CO_2; v/v) has been achieved now for a variety of different plant species. Recent progress and still existing problems in selecting cell cultures with a high photosynthetic potential capable of sustained photoautotrophic growth, has been reviewed by Yamada and Sato (1984), Horn and Dalton (1984), Horn and Widholm (1984), Hüsemann (1984 and 1985) and Yamada (1985).

The culture technique has advanced to a stage that is possible now to propagate the cells under photoautotrophic conditions in small vials (Yamada and Sato, 1978; Katoh, 1979), in two-tier culture flasks (Hüsemann and Barz, 1977) as well as in various fermenter and continuous culture systems (Dalton, 1980; Bender et al., 1981; Yamada et al., 1981; Peel, 1982; Hüsemann, 1982 and 1983). However, the unresolved difficulty of the induction of chlorophyll formation and chloroplast differentiation in morphologically unorganized cell cultures from monocotyledonous plants remains, especially in the case of grasses. Another problem concerns the demand of cultured plant cells for high CO_2 concentrations far above ambient air level for photoautotrophic growth.

The present report summarizes some essential experimental results on the photosynthetic potential and growth of photoautotrophic cell suspensions from Chenopodium rubrum.

PHOTOAUTOTROPHIC BATCH-CULTURE GROWTH

Photoautotrophic cell suspensions from Chenopodium rubrum have been established using the two-tier culture vessel method. The cells are propagated in a sugar-free mineral salt medium with 2% CO_2 as the sole carbon source. The cells are cultivated under continuous white light of 8,000 lux at 25°C on a gyrotory shaker (Hüsemann and Barz, 1977 ; Hüsemann, 1984). The photoautotrophic, vitamin and phytohormone independent cell suspension cultures from Chenopodium rubrum consist mainly of small cell aggregates (10-20 cells) and many single cells with a high chlorophyll content (400-500 μg chlorophyll/g fresh weight ; 20-25 μg chlorophyll/10^6 cells). Starting with 1 g inoculum of cells (approximately 15 million cells), the increase in fresh weight and cell number is 5- to 6-fold during a 14 days' growth period.

Summarizing the experimental results from growth kinetic studies, the transfer of stationary growth phase cells into fresh culture medium results immediately in a high protein formation, followed by an exponential phase of cell division after a 2 days' lag, whereas the onset of rapid chlorophyll formation is delayed for about 3 days. During exponential culture growth there is no net accumulation of sugar and starch. When the cells enter stationary growth phase, there is a steady increase in chlorophyll, sugar and starch content of the cells. This correlates with changes in the metabolic activities of the cells as documented by in vitro activities of enzymes related to different metabolic pathways as well as by the pattern of photosynthetic product formation (Table 1-4). Cytoplasmic and mitochondrial activities dominate during the phase of rapid cell division, while chloroplast-associated activities increase after the transition to stationary growth phase. These observations confirm the proposal of a sequential development of cytoplasmic and chloroplastic activities during the growth cycle of photosynthetically cultured plant cells (Nato et al., 1985).

Such fluctuations in the fine structure of the cells, in the accumulation of cell components (chlorophyll, protein, carbohydrates) as well as in the pattern of photosynthetic product formation and in vitro activities of enzymes related to different metabolic pathways indicate unbalanced growth of batch-propagated cells.

CONTINUOUS CULTURE GROWTH

Continuous growth of cultured cells under photoautotrophic conditions was first achieved with spinach (Dalton, 1980) and asparagus (Peel, 1982) cells. Photoautotrophic cell suspensions from Chenopodium rubrum have successfully been grown in continuous culture using an airlift culture system with a working volume of 1.5 l (Hüsemann, 1983). Establishing a dilution rate of 0.16 day^{-1}, the mean generation time of the cells during steady-state growth was about 100 hours. Consistency was achieved in the levels of nutrient uptake, cell division activity, cellular content of chlorophyll, protein and starch, photosynthetic CO_2 assimilation and activities of some enzymes related to different metabolic pathways. These findings clearly show, that an inverse relationship between cell growth and cytodifferentiation with respect to chloroplast differentiation cannot be applied to those cells that exclusively meet their energy and carbon demand by photosynthesis. This interpretation is supported further by the fine structure of the cells from continuous photoautotrophic cell cultures from Chenopodium rubrum, possessing well developed mitochondria and chloroplasts during steady-state of continuous culture growth (Hüsemann, 1985).

Finally, some distinctive features of batch propagated and continuously grown photoautotrophic cell suspensions from Chenopodium rubrum are summarized for comparison (Table 1).

STABILITY OF PHOTOAUTOTROPHISM

Photoautotrophism in cell suspension cultures of Chenopodium rubrum is a rather stable physiological modification. The chlorophyllous cells retain their

Table 1

Some Physiological Characteristics
of Batch-propagated and Steady-state Continuous Photoautotrophic Cell Cultures
of Chenopodium rubrum

	Batch-culture		Continuous culture (steady-state)
	day 5	day 14	
Cell doubling time (h)	50	120	100
Protein ($\mu g/10^6$ cells)	380	440	240
Chlorophyll ($\mu g/10^6$ cells)	17	28	12
Starch ($\mu g/10^6$ cells)	25	80	5
Photosynthetic activity (μmol CO_2/ mg chlorophyll/h)	80	65	100
Enzyme activity :			
– RuBP-carboxylase	105	190	72
– PEP-carboxylase	85	65	82

capacity for sustained photoautotrophic growth even after the cells have been propagated for 3 months under conditions of counter-selection in the presence of 2% sucrose in the culture medium. Though the heterotrophic mode of nutrition leads to a substancial decrease in the chlorophyll content and CO_2 assimilation rate of the cells as well as to changes in the pattern of photosynthetic products (Hüsemann et al., 1979; Herzbeck, 1985), the cells rapidly took up photoautotrophic growth again.

PHOTOSYNTHETIC STUDIES

The potential of light-dependent CO_2 assimilation and photosynthetic product in cultured plant cells has been studied extensively with photoautotrophic cell suspensions from Chenopodium rubrum (Hüsemann et al., 1979 and 1984; Herzbeck and Hüsemann, 1985; Herzbeck, 1985).

Rapidly dividing photoautotrophic cell suspensions assimilate about $80 \mu mol$ CO_2/mg chlorophyll/h (1.2 μmol $CO_2/10^6$ cells/h), whereas in stationary growth phase cells the photosynthetic activity accounts for about $60 \mu mol$ CO_2/mg chlorophyll/h (1.8 μmol $CO_2/10^6$ cells/h). This discrepancy in growth-phase dependent changes in the photosynthetic CO_2 assimilation rates (increase: when calculated on a cell basis; decrease: when calculated on a chlorophyll basis) are obviously due to a disproportional relationship between the increase in the chlorophyll content (+ 53%) and the stimulation of the CO_2 assimilation capacity of the cells (+ 28%), showing unbalanced growth. The question of an increase in the chlorophyll content of the cells is not followed by a stimulation of the rate of light-dependent CO_2 assimilation has still to be examined.

As shown by the ^{14}C-labelling pattern of photosynthetic intermediates after short-term [14]CO_2-photosynthesis (10 sec), the light-dependent CO_2 fixation under standard culture conditions (pH-value and composition of the culture medium, 2% (v/v) CO_2 in the gaseous atmosphere) predominantly occurs by a $C_1 \rightarrow C_5$-carboxylation using the ribulose–bis–phosphate–carboxylase (RuBP–carboxylase) following the Calvin cycle. Interestingly, ^{14}C–label in malate and aspartate – representing primary products of a $C_1 \rightarrow C_3$–carboxylation reaction mediated by the phosphoenol-pyruvate–carboxylase (PEP–carboxylase) – accounts for about 10% to 5% of total ^{14}C–radioactivity incorporated by photoautotrophic cell suspension cultures of Chenopodium rubrum during exponential and stationary growth phase, respectively (Hüsemann et al., 1979; Hüsemann, 1981). This pattern of $^{14}CO_2$ fixation is accompanied by a corresponding change in the in vitro activities of the RuBP–carboxylase (Table 5).

Higher amounts of $^{14}CO_2$ fixation into C_4–acids, accounting for 20% to 40%, have been found with photosynthetic cell cultures from Arachis hypogaea (Bender et al., 1985) and Nicotiana tabacum (Nishida et al., 1980). Because these authors did not perform their $^{14}CO_2$ fixation experiments under normal culture conditions (i.e. the cells were incubated in buffer solutions with pH–values around 7) the experimental results may not necessarily reflect the actual pathway of CO_2 fixation of the cells which have been maintained under standard culture conditions.

As found by my own studies with photoautotrophic cell suspension cultures of Chenopodium rubrum, changes in the pH–value of the incubation medium of the cells during photosynthesis, severely influence the pathway of primary CO_2–fixation ($C_1 \rightarrow C_5$–carboxylation; $C_1 \rightarrow C_3$–carboxylation). Raising the pH of the culture medium from initially 4.5 (normally established in the phase of rapid cell division) to 5.5 greatly stimulates the C_4–pathway of CO_2 assimilation by enhanced incorporation of CO_2 into malate and aspartate at the expense of phosphoglyceric acid and sugar–phosphates after short–term (10 sec) $^{14}CO_2$–photosynthesis. There is only a small increase in the total content of inorganic carbon (CO_2 + HCO_3^-) in the incubation medium due to the increased formation of HCO_3^-. Studies on the potential mechanisms (CO_2/HCO_3^-–uptake and intracellular accumulation, intracellular pH–variation) possibly responsible for the observed pH–dependency of photosynthesis in photoautotrophic cell cultures from Chenopodium rubrum have to be performed.

Table 2

pH Dependency of Changes
in the Pattern of ^{14}C–incorporation into Photosynthetic Intermediates
in Photoautotrophic Cell Suspensions of Chenopodium rubrum

	pH of the culture medium	
	4.5	5.5
^{14}C–incorporation (%) :		
Malate	15.4 + 0.4	50.7 + 1.0
Aspartate	1.7 + 0.2	3.6 + 1.0
Phosphoglyceric acid	18.7 + 2.8	13.3 + 1.5
Sugar–phosphates	60.2 + 3.0	25.2 + 5.0
Phosphoenolpyruvate Citrate, Glutamate	4.0 + 0.5	7.2 + 1.6

Cells from exponential growth were used. The pH of the culture medium were adjusted to 4.5 and 5.5 in the presence of 50 mM Mes buffer. A 10 sec $^{14}CO_2$ fixation in the presence of 2% (v/v) CO_2 was started by injecting 377 KBq of a NaH$^{14}CO_3$ solution directly into the cell suspension.

Table 3

^{14}C-incorporation into Photosynthetic Products
After 60 min $^{14}CO_2$ Photosynthesis by Photoautotrophic Cell Cultures
from Chenopodium rubrum during different phases of batch-growth

	$^{14}CO_2$ incorporation (%)		
	day 5	day 14	day 21
Chloroform fraction	2.3 ± 0.2	2.5 ± 0.3	1.2 ± 0.1
Ion-exchange separation			
Basic fraction	13.5 ± 1.1	8.2 ± 1.3	2.8 ± 0.4
Acidic fraction	9.8 ± 0.6	5.8 ± 0.6	4.7 ± 0.4
Neutral fraction	42.7 ± 4.0	52.0 ± 5.2	38.0 ± 4.5
Starch fraction	21.2 ± 3.0	26.2 ± 4.2	49.0 ± 6.2
Protein fraction	5.7 ± 0.4	2.6 ± 0.3	2.3 ± 0.3
Insoluble residue	4.8 ± 0.6	2.7 ± 0.4	2.0 ± 0.1

$^{14}CO_2$ photosynthesis was performed under standard culture conditions (i.e. pH and composition of the culture medium, 2% (v/v) CO_2 in the presence of 180 KBq $^{14}CO_2$.

Table 4

Growth-phase Dependent Changes
in the ^{14}C-label Distribution in Photosynthetic Metabolites
in Photoautotrophic Cell Suspensions
from Chenopodium rubrum

Chemical fractions	Changes in ^{14}C-label after a 5 h $^{12}CO_2$ chase Bq $(10^6 \text{ cells})^{-1}$		
	day 5	day 14	day 21
Chloroform fraction	(+) 400	(+) 1100	(+) 170
Ion-exchange separation			
Basic fraction	(−) 840	(−) 530	(−) 160
Acidic fraction	(−) 1080	(−) 580	(−) 760
Neutral fraction	(−) 4000	(−) 3270	(−) 3520
Starch fraction	(+) 790	(+) 1720	(+) 2140
Protein fraction	(+) 1850	(+) 1610	(+) 1140
Insoluble residue	(+) 310	(+) 1910	(+) 790

The intensity of carbon flow has been calculated by the loss (increase) of radio-activity in individual metabolites or chemical fractions after a 5 h $^{12}CO_2$ chase following a preceding 1 h $^{14}CO_2$ photosynthesis under standard culture conditions.

Meanwhile, there is a good deal of evidence for growth-phase correlated changes in the photosynthetic product formation and in the flow of carbon from photosynthetic intermediates into different cell metabolites as demonstrated by photosynthetic studies with photoautotrophic cell cultures from Chenopodium rubrum (Hüsemann et al., 1984; Herzbeck and Hüsemann, 1985).

Irrespective of the growth phase, most $^{14}CO_2$ is assimilated into sugars (glucose, fructose, sucrose) and starch, thus representing primary "sinks" for photosynthetically fixed carbon. Rapidly dividing cells (day 5) incorporate considerably higher amounts of photosynthetically assimilated carbon into amino-acids, organic acids, sugar-phosphates and protein compared to the merely non-dividing cells from stationary growth phase (day 14 and 21), where the ^{14}C-label in starch reaches maximum values (Table 3).

Table 5

In vitro Activities of Enzymes related to Different Metabolic Pathways
in Photoautotrophic Cell Suspensions from Chenopodium rubrum
at Different Phases of Batch Culture Growth

Enzyme	nmol substrate (mg protein)$^{-1}$ min^{-1}		
	growth phase		
	day 5	day 14	day 21
RuBP-carboxylase	107	193	225
PEP-carboxylase	86	64	55
NADP$^+$-glyceraldehyde--3-phosphate-dehydrogenase	221	398	363
NAD$^+$-glyceraldehyde--3-phosphate-dehydrogenase	980	690	680
NADP$^+$-malate-dehydrogenase	100	194	179
NAD$^+$-malate-dehydrogenase	5900	5300	3260
NADP$^+$-malic enzyme	69	35	28
NAD$^+$-malic enzyme	68	38	37
Pyruvate-orthophosphate-dikinase	8	22	23
Pyruvate-kinase	123	108	42
Cytochrome c-oxidase	34	9	5

Metabolic carbon flow of photosynthetically fixed carbon from photosynthetic intermediates into different cell components has been studied by $^{14}CO_2$-pulse/$^{12}CO_2$-chase photosynthesis. The experimental data are summarized in Table 4 and show a strict correlation of the intensity and the pattern of photosynthetic carbon flow and the actual growth phase of the cells.

The basic fraction (amino-acids), the acidic fraction (organic acids and sugar-phosphates) and the neutral fraction (sugars) always function as "carbon sources" to provide with carbon skeletons for divers biosynthetic pathways as indicated by the loss of ^{14}C-label during a 5 hours $^{12}CO_2$ chase. Higher proportions of photosynthetically assimilated carbon are channelled into protein, lipids and structural components, for example cell walls (insoluble residue) in actively dividing cells (day 5) compared to the merely non-dividing cells from stationary growth phase (day 14 and 21), where photosynthetically assimilated carbon is preferentially accumulated as starch.

These growth-phase dependent fluctuations in the metabolic flow of photosynthetically assimilated carbon into different metabolites corroborate with corresponding changes in the in vitro activities of glycolytic, citric acid cycle and chloroplastic enzymes in photoautotrophic cell suspensions of Chenopodium rubrum during batch growth (Table 5).

These observations further confirm data on photosynthesis and carbon metabolism in plant leaves of different development stages, which also indicate a strict correlation between the pattern of photosynthetic carbon metabolism and leaf ages (Dickson and Larson, 1981; Giaquinta, 1978; Herold, 1980).

Fluctuations in the in vitro activities of carboxylation reactions mediated by the RuBP-carboxylase and PEP-carboxylase in photosynthetic cell cultures are of special interest as concerns the possible many-faceted function of the

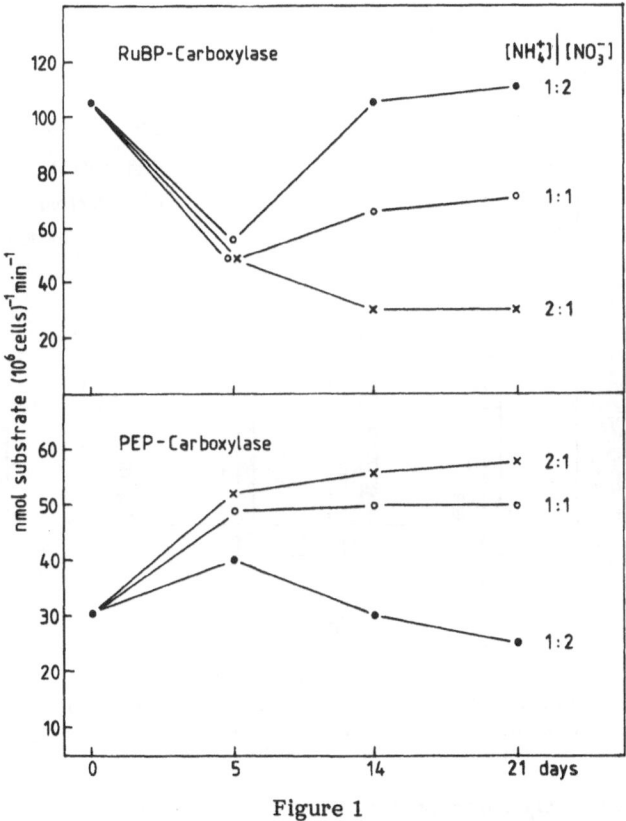

Figure 1

The effect of different concentrations of ammonia and nitrate ions on the in vitro activities of the RuBP-carboxylase and PEP-carboxylase in photoautotrophic cell cultures from Chenopodium rubrum

PEP-carboxylase. In agreement with some other workers (Yamada et al., 1982; Nato et al., 1985; Sato et al., 1986; Neumann, 1986) one may assume that the PEP-carboxylase opens some anaplerotic pathways to provide additional organic acids to the tricarboxylic acid cycle to maintain its synthetic function.

Furthermore, a regulatory relationship exists between the molar ratio of ammonia to nitrate nitrogen supply and the in vitro activities of the RuBP- and PEP-carboxylase in photoautotrophic cell suspensions of Chenopodium rubrum (Barz and Hüsemann, 1982) as shown in Fig. 1.

An increase in the ammonia concentration (20.6 mM, 30.6 mM, 40.6 mM) and a corresponding reduction in the nitrate content (39.4 mM, 29.4 mM, 19.4 mM) of the culture medium brought about a substantial reduction in the in vitro activities of the RuBP-carboxylase, but maintained high activities of the PEP-carboxylase irrespectively of the growth phase (day 5 : exponential growth; day 14 and 21 : stationary growth) as compared to cells grown under standard culture conditions (20.6 mM ammonia, 39.4 mM nitrate).

Preliminary results with photoautotrophic cell suspensions from Chenopodium rubrum indicate, that ammonia obviously exerts a regulatory effect on diverting photosynthetically assimilated carbon into different metabolic pathways, with preference into the metabolic sequence of organic acids, amino-acids and proteins at the expense of carbohydrate accumulation (Herzbeck, 1985). Similar results

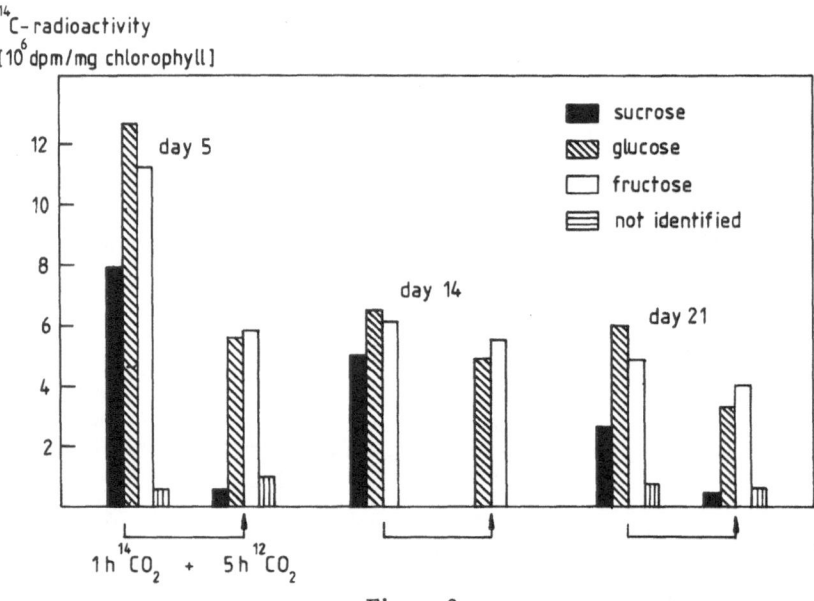

Figure 2

Metabolic ^{14}C turnover
from photosynthetically formed malate, citrate and sugar-phosphates
in photoautotrophic cell cultures from Chenopodium rubrum
at different phases of batch growth.

The $^{14}CO_2$-pulse/$^{12}CO_2$-chase photosynthetic measurements were performed under standard culture conditions.

were obtained with isolated cells from spinach (Woo and Canvin, 1980) and Papaver somniferum (Hammel, 1979). More extended research on this area is necessary for a better understanding of the molecular mechanisms of the profound ammonia effect on enzyme activity, CO_2 assimilation and carbon partitioning. Photoautotrophic cell cultures should be well suited for further analyzing the mechanisms by which higher plant cells controll the metabolic flow of photosynthetically assimilated carbon to biosynthetic reactions in response to physiological conditions.

As already mentioned, protein formation in batch-propagated photoautotrophic cell suspension cultures from Chenopodium rubrum is highest in rapidly dividing cells. Because sustained amino-acid synthesis depends on the supply with keto acids, mostly intermediates of the citric acid cycle, the metabolism of malate was studied.

As found by $^{14}CO_2$-pulse/$^{12}CO_2$-chase photosynthesis, 75% to 90% of ^{14}C-label in malate formed during a 1 h $^{14}CO_2$ photosynthesis is turned over during a 5 h $^{12}CO_2$-chase in cells from stationary and exponential growth phase, respectively, as shown in Fig. 2. This corroborates well with maximum values for aminoacid and protein synthesis as indicated by its ^{14}C-labelling (Tables 3 and 4).

In order to avoid any interference of carbon flow from malate into different biosynthetic pathways by the metabolism of other ^{14}C-labeled photosynthetic intermediates, the metabolism of exogenously supplied uniformly ^{14}C-labeled malate was measured. Photoautotrophic cell suspensions of Chenopodium rubrum were incubated under standard culture conditions in the presence of 10 mM (U-^{14}C-)-malate as described previously (Herzbeck and Hüsemann, 1985). The uptake of malate by the cells is linear with time reaching 70%-80% within 30 min. 50%-60% of the total malate taken up by the cells is further metabolized (Table 8).

Table 6

Metabolism of Exogenously Supplied (U-^{14}C)-malate
in Photoautotrophic Cell Cultures from Chenopodium rubrum

| | Growth phase | | |
| | exponential | stationary | |
	day 5	day 14	day 21
^{14}C-malate metabolized :			
(% of ^{14}C-malate uptake)	63	52	51
^{14}C-distribution pattern :			
(% of ^{14}C-malate metabolized)			
Amino-acids	62	32	31
Citrate, succinate	9	30	21
Protein	10	8	8
Lipids	6	5	10
Starch	5	7	14
Sugars	4	14	10
Not identified	4	4	6

The metabolism of exogenously supplied ^{14}C-labeled malate was measured under standard culture conditions of the cells.

Irrespectively of the growth phase, maximum proportions of ^{14}C-label from malate have been recovered in amino-acids, organic acids and protein. During transition from exponential (day 5) to stationary growth (day 14 and 21), the transfer of ^{14}C-label from malate into amino-acids and proteins is reduced, while at the same time increasing amounts of ^{14}C from malate are channelled into organic acids and surprisingly into sugars and starch as indicated in Table 6. This suggests that malate serves as a source of carbon for diverse biosynthetic pathways which imply the operation of glycolytic and citric acid cycle reactions in photosynthesizing cells in the light.

The formation of ^{14}C-alanine from (U-^{14}C)-malate requires decarboxylation and transamination reactions mediated by the malic enzyme and by the glutamate-pyruvate-transaminase. Both enzymes are present with substantial activities in the cell cultures.

The formation of ^{14}C-labeled sugars and starch from (U-^{14}C)-malate may be accomplished by a refixation of $^{14}CO_2$ evolved by decarboxylation and/or by reassimilation of respiratory $^{14}CO_2$ mediated by the RuBP-carboxylase.

Besides the malic enzyme, the pyruvate-orthophosphate-dikinase (PPDK), a key enzyme involved in C_4-photosynthesis, is present in photoautotrophic cell suspension cultures of Chenopodium rubrum, which have been proven to exhibit a C_3-type of photosynthesis. Recently, pyruvate-orthophosphate-dikinase activity has also been reported for photoautotrophic cell cultures from Nicotiana tabacum, a C_3-plant (Sato et al., 1986), as well as in leaves of C_3-plants like wheat (Aoyagi and Bassam, 1983 and 1986) and spinach (Aoyagi and Bassham, 1985). It may be expected that the malic enzyme in C_3-cells has a limited role in conversion of C_4-acids (malate) to C_3-carbon skeletons (pyruvate), but the role of the pyruvate-orthophosphate-dikinase in C_3-photosynthesis still remains dubious. Yet it should not be ruled out that there might exist a pathway for C_4-intracellular carbon transport and CO_2 concentration mechanism in the chloroplast of a C_3 plant cell, depending on the presence of the malic enzyme, the pyruvate-orthophosphate-dikinase, the PEP-carboxylase and the malate dehydrogenase (NADP-specific), as recently discussed by Aoyagi and Bassham (1986). This would mean that the major difference in C_3- and C_4-photosynthesis might be a significative rather than a qualitative difference in enzyme activity.

Photoautotrophic cell suspension cultures from Chenopodium rubrum should represent an appropriate experimental system for further research on this area of cellular photosynthesis.

ACNOWLEDGEMENTS

The author's studies on photoautotrophic cell cultures have been supported by the Deutsche Forschungsgemeinschaft.

REFERENCES

Aoyagi, K., and Bassham, J. A., 1983, Pyruvate-orthophosphate-dikinase in wheat leaves, Plant Physiol., 73:853.

Aoyagi, K., and Bassham, J. A., 1984, Pyruvate-orthophosphate-dikinase mRNA organ specificity in wheat and maize, Plant Physiol., 76:278.

Aoyagi, K., and Bassham, J. A., 1986, Appearance and accumulation of C_4 carbon pathway enzymes in developing wheat leaves, Plant Physiol., 80:334.

Barz, W., and Hüsemann, W., 1982, Aspects of photoautotrophic cell suspension cultures, in : "Plant Tissue Culture 1982", A. Fujiwara, ed., Mazuren, Tokyo.

Bender, L., Kumar, A., and Neumann, K. H., 1981, Photoautotrophe pflanzliche Gewebekulturen in Laborfermenten, in : "Fermentation", R. M. Lafferty, ed., Springer-Verlag, Berlin.

Bender, L., Kumar, A., and Neumann, K. H., 1985, On the photosynthetic system and assimilate metabolism of Daucus and Arachis cell cultures, in : "Primary and Secondary Metabolism of Plant Cell Cultures", K. H. Neumann, W. Barz, and E. Reinhard, eds., Springer-Verlag, Berlin, Heidelberg.

Dalton, C. C., 1980, Photoautotrophy of spinach cells in continuous culture : Photosynthetic development and sustained photoautotrophic growth, J. Exptl. Bot., 31:791.

Dalton, C. C., and Peel, E., 1983, Product formation and plant cell specialization : A case study of photosynthetic development in plant cell cultures, Progr. Ind. Microbiol., 17:109.

Dickson, R. E., and Larson, P. R., 1981, ^{14}C fixation, metabolic patterns, and translocation profiles during leaf development in Populus deltoides, Planta, 152: 461.

Giaquinta, R., 1978, Source and sink leaf metabolism in relation to phloem translocation, Plant Physiol., 61:380.

Hammel, K. E., Cornwell, K. L., and Bassham, J. A., 1979, Stimulation of dark CO_2 fixation by ammonia in isolated mesophyll cells of Papaver somniferum, Plant Cell Physiol., 20:1523.

Herold, A., 1980, Regulation of photosynthesis by sink activity – the missing link, New Phytol., 86:131.

Herzbeck, H., 1985, Untersuchungen zum Kohlenstoffmetabolismus in photoautotrophen Zellsuspensionskulturen von Chenopodium rubrum, Doctoral Thesis, University of Münster.

Herzbeck, H., and Hüsemann, W., 1985, Photosynthetic carbon metabolism in photoautotrophic cell suspension cultures of Chenopodium rubrum, in : "Primary and Secondary Metabolism of Plant Cell Cultures", K. H. Neumann, W. Barz and E. Reinhard, eds., Springer-Verlag, Berlin, Heidelberg.

Horn, M. E., and Dalton, C. C., 1984, Photosynthetic cell cultures and their biotechnological applications, in : IAPTC New Letter, University of Illinois, IAPTC Secretary, Urbana, U.S.A.

Horn, M. E., and Widholm, J. M., 1984, Aspects of photosynthetic plant tissue cultures, in : "Application of Genetic Engineering to Crop Improvement", G. B. Collins and J. F. Petolino, eds., M. Nijhoff/Dr. W. Junk Publish. Co.

Hüsemann, W., 1981, Growth characteristics of hormone and vitamin independent photoautotrophic cell suspension cultures from Chenopodium rubrum, Protoplasma, 109:415.

Hüsemann, W., 1982, Photoautotrophic growth of cell suspension cultures from Chenopodium rubrum in an airlift fermenter, Protoplasma, 113:214.

Hüsemann, W., 1983, Continuous culture growth of photoautotrophic cell suspensions from Chenopodium rubrum, Plant Cell Reports, 2:59.

Hüsemann, W., 1984, Photoautotrophic cell cultures, in : "Cell Culture and Somatic Cell Genetics of Plants", I. K. Vasil, ed., Vol. 1 : Laboratory Techniques, Academic Press, New-York.

Hüsemann, W., 1985, Photoautotrophic growth of cells in culture, in : "Cell Culture and Somatic Cell Genetics of Plants", I. K. Vasil, ed., Vol. II : Cell Growth, Nutrition, Cytodifferentiation, and Cryopreservation, Academic Press, New-York.

Hüsemann, W., Plohr, A., and Barz, W., 1979, Photosynthetic characteristics of photomixotrophic and photoautotrophic cell suspension cultures of Chenopodium rubrum, Protoplasma, 100:101.

Hüsemann, W., Herzbeck, H., and Robenek, H., 1984, Photosynthesis and carbon metabolism in photoautotrophic cell suspensions of Chenopodium rubrum from different phases of batch growth, Physiol. Plant., 62:349.

Katoh, K., Ohta, Y., Hirose, Y., and Iwamura, T., 1979, Photoautotrophic growth of Marchantia polymorpha L. cells in suspension culture, Planta, 144:509.

Nato, A., Hoarau, J., Brangeon, J., Hirel, B., and Suzuki, A., 1985, Regulation of carbon and nitrogen assimilation pathways in tobacco cell suspension cultures in relation with ultrastructural and biochemical development of the photosynthetic apparatus, in : "Primary and Secondary Metabolism of Plant Cell Cultures", K. H. Neumann, W. Barz, and E. Reinhard, eds., Springer-Verlag, Berlin, Heidelberg.

Neumann, K. H., 1986, Photosynthesis in cell and tissue cultures of higher plants, in : "Abstracts of VI International Congress of Plant Tissue and Cell Culture", D. A. Somers, B. G. Gegenbach, D. D. Biesboer, W. P. Hackett and C. E. Green, eds., University of Minnesota, Minneapolis, U.S.A.

Nishida, K., Sato, F., and Yamada, Y., 1980, Photosynthetic carbon metabolism in photoautotrophically and photomixotrophically cultured plant cells, Plant Cell Physiol., 21:47.

Peel, E., 1982, Photoautotrophic growth of suspension cultures of Asparagus officinalis L. cells in turbidostats, Plant Science Letters, 24:147.

Sato, F., Koizumi, N., Takeda, S., and Yamada, Y., 1986, Photoautotrophic culture of tobacco cells and their character : Carboxylation enzyme and herbicide resistance of photoautotrophic cells, in : Abstracts of VI International Congress of Plant Tissue and Cell Culture", University of Minnesota, Minneapolis, U.S.A.

Yamada, Y., 1985, Photosynthetic potential of plant cell cultures, in : "Advances in Biochemical Engineering/Biotechnology, Plant Cell Culture", A. Fiechter, ed., Springer-Verlag, Berlin, Heidelberg.

Yamada, Y., and Sato, F., 1978, The photoautotrophic culture of chlorophyllous cells, Plant Cell Physiol., 19:691.

Yamada, Y., and Sato, F., 1984, Selection of photoautotrophic cells, in : "Handbook of Plant Cell Cultures", D. A. Evans, W. R. Sharp, P. V. Ammirato, and Y. Yamada, eds., Vol. 1 : Techniques for propagation and breeding, McMillan, New-York.

Yamada, Y., Sato, F., and Watanabe, K., 1982, Photosynthetic carbon metabolism in cultured photoautotrophic cells, in : "Plant Tissue Culture 1982", A. Fujiwara, ed., Mazuren, Tokyo.

VACUOLAR ACCUMULATION OF ORGANIC ACIDS AND THEIR ANIONS
IN CAM PLANTS

J. Andrew Smith

Department of Botany, University of Edinburgh
The Kings' Buildings, Mayfield Road, Edinburgh EH9-3JH
Scotland, U. K.

INTRODUCTION

In plants showing crassulacean acid metabolism (CAM), assimilation of atmospheric CO_2 occurs mainly or exclusively at night. This is the time at which CO_2 uptake incurs the minimum cost in transpirational water loss in the semi-arid habitats characteristic of CAM plants. Associated with nocturnal CO_2 fixation, there is a gradual accumulation of malic acid in the vacuoles of the assimilatory tissue. During the subsequent day, while the stomata are closed, this malic acid is decarboxylated and the CO_2 thereby released is refixed through the C_3 photosynthetic carbon reduction cycle (Winter, 1985).

Because the biochemical interconversions involving malic acid in the cytoplasm are spatially separated from the site of its storage in the vacuole, great importance accrues to the mechanism and control of the processes mediating malic-acid transport across the tonoplast. Like the other vacuolar solutes, the nocturnally synthesized malic acid is also osmotically active, so its accumulation has consequences for the water relations of the assimilatory tissue and the whole plant. Thus, as well as describing the day-night fluxes of malic acid at the membrane level, we should try to discern how these transport processes contribute to what we regard as the "ecological adaptation" shown by CAM plants and their relative success in semi-arid enrironments.

ORGANIC ACIDS AND THEIR ANIONS AS VACUOLAR OSMOTICA

A wide spectrum of carboxylates is found in higher plants as major vacuolar osmotica (at concentrations of the order of 10 to 100 mol . m^{-3} on a tissue-volume basis). The most common are aconitate, citrate, iso-citrate, fumarate, malate, oxalate and tartrate (Kinzel, 1982). But we must make the distinction (ultimately a quantitative one) between the accumulation of organic acids (i.e. anions balanced by protons) and their salts (i.e. anions balanced by inorganic cations). At one extreme, the large day-night changes in malic-acid content characteristic of CAM involve solely the free acid, with 2 mol titratable H^+ accumulated per mol malate; bulksap pH may reach values as low as 3.3 towards the end of the night (Lüttge et al., 1981 and 1982). At the other, organic-acid anion accumulation may be stoichiometrically balanced by inorganic cations. A good example is accumulation of K_2-malate in guard cells of some species during stomatal opening (McRobbie, 1981; Fitzsimons and Weyers, 1986), which results in an _increase_ in vacuolar pH (see Willmer, 1983). More commonly, charge balance is probably provided by a mixed complement of protons and inorganic cations. Further empirical studies of this relationship would improve our understanding of the importance of the vacuole in acid-base regulation in plant cells (Kurkdjian et al., 1985; Marigo et al., 1986).

Apart from the day-night rhythm in malic-acid content associated with dark CO_2, CAM plants also have a background level of organic-acid anions as their major vacuolar osmotica. In _Kalanchoë_ species, for example, the principal anions are citrate and iso-citrate (together with some malate that does not participate in the CAM rhytm); these are largely balanced by, and to some extent complexed with, calcium and magnesium (Phillips and Jennings, 1976; Phillips, 1980; Kinzel, 1982, Kenyon et al., 1985). It is also characteristic of CAM plants that bulk-cell-sap osmotic pressure (π) tends to be very low, so the malic acid involved in the CAM rhythm can have a relatively large impact on the osmotic relations of the assimilatory tissue. In well-watered plants of _Kalanchoe daigremontiana_, leaf-cell-sap π typically increases from 0.32 MPa at the start of the night to 0.64 MPa at the end (Smith and Lüttge, 1985). Furthermore, the malate participating in the CAM rhythm apparently behaves as an ideal osmoticum, having an osmotic coefficient of unity (Lüttge and Nobel, 1984; Smith and Lüttge, 1985; Smith et al., 1986). Its osmotic effectiveness implies an important role for the nocturnally accumulated malic acid in tugor maintenance in the assimilatory tissue, above and beyond its role as a temporary storage form for CO_2. Despite the increasingly negative leaf water potentials associated with nocturnal transpiration, leaf turgor pressure (P) stays constant or even increases slightly in the course of the night. In _Kalanchoe daigremontiana_, leaf P then reaches its maximum value of about 0.20 MPa towards midday, at the time when transpiration rate is at a minimum (Smith and Lüttge, 1985).

These major changes in plant water relations during the day-night cycle are thus directly dependent on the processes transferring malic acid into and out of the vacuole. Given the magnitude and regulatory of these malic-acid fluxes, work with CAM plants should help to clarify the importance of solute transport at the tonoplast in cell water relations and bioenergetics.

MALIC-ACID TRANSPORT AT THE TONOPLAST

What then do we know of the mechanism of malic-acid transport into and out of the vacuole in CAM plants ? The importance of the vacuole as a storage site for malic acid is apparent on anatomical grounds alone. In the succulent tissues characteristic of CAM plants, the vacuole can constitute about 97 % of the volume of assimilatory cells in mature organs (Steudle et al., 1980; Lüttge et al., 1982). As the volume-fraction of the cell wall and cytoplasmic organelles is so low, the vacuole provides the only sink of quantitative importance for storage of the nocturnally synthesized malic acid. But it is also clear that transport at the tonoplast must be finely geared to malic-acid metabolism, not least to preserve the acid-base balance of the cytoplasm. In the following sections, I shall briefly summarize what we know of the properties of the transport systems in CAM plants. Further details of the biochemical and physiological context of these events can be found in Lüttge and Smith (1985 and 1987).

Transport at the Tonoplast and Control fo Cytosolic Composition

Apart from its biochemical importance in storage of the nocturnally assimilated CO_2, removal of malic acid from its site of synthesis into the vacuole is essential if control of cytosolic composition is to be maintained. By the same argument, decarboxylation and further metabolism of malic acid in the light period must match its rate of efflux from the vacuole during this time.

The potential magnitude of the "acid load" on the cytosol can be appreciated by considering an increase in malic-acid concentration of 150 mol m^{-3} (on a cell-sap volume basis) during the course of a 12-h dark period. This is approximately the maximal nocturnal acidification observed in the chlorenchyma tissue of the leaf-succulent Kalanchoë daigremontiana (Lüttge et al., 1981; Lüttge and Smith, 1984). With the cytosol occupying about 1 % of the cell volume (Steudle et al., 1980; Lüttge et al., 1982), and considering that the malic acid, with two mol dissociable H$^+$ per mol, is produced from neutral carbohydrate precursors, this corresponds to an average rate of H$^+$ production of 694 mmol H$^+$ per m^3 of cytosol per second. With a buffer capacity of 20 mol H$^+$ m^{-3} per pH unit across the "permissible" range of cytosolic pH values of 6.5 to 8.0 (Raven, 1985), this rate of H$^+$ production would lead to a decrease in cytosolic pH of 0.0347 units s^{-1}. In other words, in the absence of net H$^+$ transport into the vacuole, cytosolic pH would decrease from a typical value of 7.4 (Raven, 1985) to the minimum "tolerable" value of 6.5 in only 26 s.

The position with respect to malate is analogous. The cytosolic enzyme phosphoenolpyruvate (PEP) carboxylase, which catalyzes the penultimate reaction in the biosynthetic pathway, is particularly sensitive to feedback inhibition by malate, with a K_i of the order of 1 mol m^{-3} (see Buchanan-Bollig et al., 1984). Even taking account of the likely concentrations in vivo of effectors such as glucose-6-phosphate (Nott and Osmond, 1982), and of the fact that the maximum catalytic activity of PEP carboxylase in vitro can be an order of magnitude higher than that required to explain the observed rates of nocturnal malic-acid synthesis (Osmond, 1978), it is essential that malate is transferred into the vacuole if PEP carboxylase is to remain active during the night. Unfortunately, direct measurements of cytosolic malate concentration have not yet been possible. But in the absence of malate transport into the vacuole, the rate of malic-acid synthesis considered above would increase the cytosolic concentration of malate by 10 mol m^{-3} every 29 s.

Driving Forces for Malic-acid Transport at the Tonoplast

Despite the lack of direct measurements of cytosolic pH or malate concentration in CAM plants, it is nevertheless possible to make reasonable predictions of the driving forces involved in malic-acid transport at the tonoplast. Bulk-tissue values for sap pH and malate concentration give a fairly accurate indication of the vacuolar contents in view of the large volume-fraction of this compartment. The pK values for malic-acid dissociation have also been measured to be pK_1 = 3.2 and pK_2 = 4.3 in the leaf-cell sap, at an ionic strength of approximately 0.2 (Lüttge and Smith, 1984). Thus, over the range of vacuolar pH values found in Kalanchoe daigremontiana (viz. pH 6.0 to 3.3), malate will exist as a mixture of the anionic forms, mal^{2-} and Hmal$^-$, and the undissociated acid, H$_2$mal. At a typical cytosolic pH of 7.4, more than 99.9 % of the total malate is in the form of mal^{2-}.

Together with the transtonoplast electrical potential difference of + 25 mV (inside positive) measured on intact cells (Rona et al., 1980), the Nernst criterion can be applied to the distribution of the different ionic species between cytosol and vacuole. This reveals that both H$^+$ and Hmal$^-$ must be actively accumulated against their electrochemical potential difference, whereas mal^{2-} appears to be passively distributed across the tonoplast (Lüttge and Ball, 1979). In fact, the observed range of vacuolar malate concentrations and pH values, together with the assumption of an equilibrium distribution of mal^{2-}, leads to the prediction of cytosolic concentrations between 2.0 and 6.9 mol m^{-3}.

In contrast with the accumulation process, malic-acid efflux from the vacuole during the light period is thermodynamically downhill. Consistent with this, malic-acid efflux shows a low Q_{10} and is unaffected by a variety of metabolic inhibitors (see Lüttge and Ball, 1979).

Energetics of Malic-acid Transport

The simplest hypothesis to explain malic-acid accumulation in the vacuole invokes primary active transport of protons at the tonoplast, with malate^{2-} ions following passively to maintain electroneutrality (Lüttge and Ball, 1979; Lüttge et al., 1982). Before considering the experimental evidence for such a mechanism, two aspects of its feasibility can be examined from a bioenergetic standpoint.

First, if proton transport was to be driven by a tonoplast-bound ATPase, the free energy available for ATP hydrolysis (ΔG_{ATP}) in the cytosol must exceed the electrochemical potential difference for protons ($\Delta \bar{\mu}_H$+) across the tonoplast. Based on measurements of adenine-nucleotide levels in the bulk tissue of Kalanchoe tubiflora and some assumptions about their compartmentation, ΔG_{ATP} was calculated to be between - 49 and - 54 kJ mol^{-1}, similar to values in other systems (Smith et al., 1982; cf. Raven, 1984). $\Delta \bar{\mu}_H$+ is estimated to increase from 16 kJ mol^{-1} to, at most, 26 kJ mol^{-1} in the course of the dark period (Lüttge et al., 1981). Towards the end of the night, therefore, the "proton pump" would either be working as a 1 H$^+$ATPase under strong kinetic control, or as a 2 H$^+$-ATPase close to thermodynamic equilibrium. At equilibrium, of course, the ATPase reaction is poised against the $\Delta \bar{\mu}_H$+, and no net H$^+$ transport occurs.

Second, the rate of ATP supply during the night seems just adequate to support the operation of a 2 H$^+$-ATPase, but insufficient for a 1 H$^+$-ATPase. Measurements of respiration rate show that oxidative phosphorylation would provide half the required ATP, the other half being generated by substrate-level phosphorylation during the phosphorolytic breakdown of starch (or glucans) to malate (Lüttge et al., 1981; Lüttge and Smith, 1987). The overall stoichiometry would then be 2 H$^+$: 1 mal^{2-} transferred into the vacuole per ATP hydrolyzed.

Hence, even though a 1 H$^+$-ATPase could bring about a greater accumulation of malic acid in the vacuole, and at a faster rate for a given $\Delta \bar{\mu}_H$+ across the tonoplast, it seems that only a 2 H$^+$-ATPase could be supported by the rate at which ATP is generated. No instance is yet known for a CAM plant in which $\Delta \bar{\mu}_H$+ exceeds 26 kJ mol$^-$ (which would invalidate the 2 H$^+$-ATPase mechanism). Malate concentrations of up to 320 mol m^{-3} have been observed in some CAM bromeliads (Smith et al., 1986), but these have not yet been analyzed in terms of $\Delta \bar{\mu}_H$+ at the tonoplast. For example, vacuolar pH may remain at higher values than it does in Kalanchoë spp. Moreover, Bennett and Spanswick (1984) have shown by careful measurements of ATP hydrolysis and pH changes that the tonoplast ATPase in Beta vulgaris functions as a 2 H$^+$-ATPase would be approaching thermodynamic equilibrium towards the end of the night could be one reason for the characteristic reduction in the rate of malic-acid accumulation seen at this time in CAM plants.

Vacuolar ATPase Activity

What experimental evidence is there for the transport scheme outlined above ? The existence of a vacuolar ATPase in the assimilatory cells of CAM plants was first demonstrated in Kalanchoe daigremontiana. Vacuoles were isolated from mesophyll-cell protoplasts using either polybase-induced lysis by DEAE-dextran (Smith et al., 1984 b), or by osmotic shock in the presence of EGTA (Aoki and Nishida, 1984). Assay of marker enzymes showed that the substrate-specific ATPase activity could not attributed to cytoplasmic contaminants in the isolated-vacuole fraction. Further, the vacuolar ATPase activity was insensitive to vanadate, azide and phlorizin, inhibitors that block the plasmalemma, mitochondrial and thylakoid ATPases, respectively. However, the vacuolar ATPase activity was very sensitive to N,N'dicyclohexylcarbodiimide (DCCD) and organotin compounds, which are potent inhibitors of H$^+$-translocating ATPases (Smith et al., 1984 a).

Amongst its other characteristics, the hydrolytic activity of the vacuolar ATPase of Kalanchoe daigremontiana has a relatively high pH optimum (pH 8.0), an apparent $K_m(MgATP^{2-})$ of 0.31 mol m^{-3} can function as well with Mn^{2+} as with Mg^{2+} but is inhibited by Cu^{2+}, and is insensitive to cations but sensitive to anions. In particular, both Cl$^-$ and malate^{2-} at 50 mol m^{-3} stimulate the ATPase activity by about 40% in desalted extracts, whereas NO$_3^-$ at 50 mol m^{-3} inhibits the activity by 40% (Jochem et al., 1984). These general characteristics have now been shown to be diagnostic of the vacuolar ATPase from a wide range of plant sources (Marin, 1985; Sze, 1985), and have been demonstrated in two facultative CAM species, Mesembryanthenum crystallinum and Kalanchoë blossfeldiana (Struve et al., 1985).

ATP-dependent H$^+$ Transport at the Tonoplast

Some evidence that the vacuolar ATPase might drive proton transport in CAM plants was obtained by Jochem et al. (1984) from microelectrode measurements on intact vacuoles of Kalanchoe tubiflora. Vacuoles incubated in the presence of MgATP showed a significantly lower internal pH and more positive transtonoplast membrane potential than the controls.

More detailed studies have been possible with sealed membrane vesicles using fluorescent probes such as quinacrine to follow pH-gradient formation. Using a tonoplast-enriched preparation of sealed membrane vesicles obtained by differential centrifugation of protoplast homogenates, we were able to demonstrate highly substrate-specific proton accumulation in Kalanchoe daigremontiana (Smith et al., 1987). The plasmalemma and mitochondrial ATPases contributed 20% to 25% towards proton transport, which was blocked almost completely by relatively low concentrations of diethylstilbestrol, DCCD and NO$_3^-$. To explore the characteristics of H$^+$ transport driven by the tonoplast ATPase alone, quinacrinefluorescence quenching was also studied in the presence of both vanadate and azide. As for the ATPase activity, this revealed that H$^+$ accumulation was unaffected by inorganic cations but was highly dependent on the nature of the accompanying anion. Unlike the ATPase activity, however, malate^{2-} was more than three times as effective as Cl$^-$ in supporting H$^+$ transport.

One interpretation of these results is that, while the tonoplast ATPase clearly has regulatory sites affected by anions (Walker and Leigh, 1981; Mandala and Taiz, 1985; Marin et al., 1985; Sze, 1985; Griffith et al., 1986; Randall and Sze, 1986), the stimulation of H$^+$ accumulation reflect the effective permeability of the tonoplast to the various anions. Thus, the high rate of H$^+$ accumulation in the presence of malate^{2-} could indicate the presence of a specific membrane carrier for this anion in CAM cells. This would also explain the observation that the anion-channel blocker 4,4'-diisothiocyanatostilbene-2,2'-disulphonate (DIDS) inhibits ATP-dependent H$^+$ accumulation but not the hydrolytic activity of the ATPase (Smith et al., 1987). In the facultative CAM plant Mesembryanthemum crystallinum, however, ATP-dependent H$^+$ transport occurs at higher rates in the presence of Cl$^-$ than malate^{2-} (Struve and Lüttge, 1987).

The results for K. daigremontiana suggest that malate^{2-} anions themselves, synthesized in the cytosol as a result of CO$_2$ fixation by PEP carboxylase, might be involved in activating net H$^+$ influx into the vacuole at night. Indeed, even when no further acidification is occurring, continued operation of the tonoplast ATPase is needed to maintain the steady-state ΔpH : When the ATPase is blocked, ΔpH is observed to dissipate relatively rapidly (Smith et al., 1987). More information is now needed on the characteristics of malate uptake into the vacuoles of CAM plants. Buser-Suter et al. (1982) demonstrated that malate uptake is carrier-mediated, and Nishida and Tominaga (this volume) have observed ATP stimulation of malate uptake. But the exact nature of the flux coupling between H$^+$ and malate transport remains unclear. Specifically, we need to test the hypothesis that malate accumulation is directly related to the magnitude of the proton-motive force generated by the tonoplast H$^+$-ATPase. The tonoplast of Kalanchoe daigremontiana has also been shown to contain an H$^+$-translocating inorganic pyrophophatase (Jochem, 1986), with properties similar to those described in several other tissues

(Chanson et al., 1985; Rea and Poole, 1985; Wang et al., 1986). The extent to which this enzyme contributes to the generation and maintenance of H^+ across the tonoplast in vivo urgently requires assessment.

Malic-acid Efflux from the Vacuole

Whereas malic-acid accumulation is an energy-requiring process, efflux from the vacuole during the light period is thermodynamically downhill. Unfortunately, details of the efflux mechanism are still unclear. It appears, however, that at the start of the day, when vacuolar pH is still near its lowest values, a considerable part of the malic acid leaves the vacuole in the form of the undissociated acid, H_2mal. Two arguments support this view (Lüttge and Smith, 1984). First, the membrane permeability coefficient for H_2mal required permit efflux (or malate decarboxylation) at the rates observed in the intact tissue, viz. $2.2 \ 10^{-8} \ m \ s^{-1}$, is close to the value estimated from Collander plots ($4 \ 10^{-8} \ m \ s^{-1}$) based on the partition coefficient of malic acid between ether and water. And second, if the concentration-dependence of malic-acid efflux from leaf slices in solution (Lüttge and Ball, 1977) is related to the amounts of H_2mal, $Hmal^-$ and mal^{2-} in the tissue, there is found to be an approximately linear relation between efflux and the concentration of H_2mal; moreover, the slope of a log/log plot of these two variables is 1.16 at the highest malic-acid concentrations, which is close to the value of 1.00 predicted for passive diffusion.

The conclusion is that at high malic-acid concentrations, efflux of malate from the vacuole occurs predominantly in the form of the undissociated acid, H_2mal, by passive diffusion, i.e. by a "lipid-solution" mechanism (Lüttge and Smith, 1984). At lower malic-acid contents, when vacuolar pH is higher and H_2mal is not present in significant amounts, efflux must be in the form of $Hmal^-$ and/or mal^{2-}. Efflux of the anionic forms will also be passive, but is presumably carrier-mediated. At any rate, the back-flux of H_2mal into the cytosol represents a "leak" against which the proton pump must work during the dark period to bring about malic-acid accumulation. In addition to the thermodynamic restrictions on the tonoplast ATPase discussed above, this back-flux of H_2mal may limit the maximum malic-acid concentrations that can attained towards the end of the dark period.

Quantitative Significance of Malic-acid Transport at the Tonoplast

Because malic-acid accumulation associated with the CAM rhythm occurs in discrete phases, it is relatively straightforward to relate the rate of acidification to net fluxes of protons and malate across the tonoplast. For example, the maximum nocturnal accumulation of 150 mol m^{-3} malic acid observed in Kalanchoe daigremontiana converts to a net influx of H^+ (2 H^+ per $malate^{2-}$) across the tonoplast of 124 nmol $m^{-2} \ s^{-1}$. The mean ATPase activity of isolated vacuoles of 14.5 pmol h^{-1} vacuole^{-1} can also be expressed on a membrane surface-area basis : it corresponds to an ATPase activity of 46 nmol $m^{-2} \ s^{-1}$ at the equivalent night temperature (Smith et al., 1984 a). If the proton pump functioned as a 2 H^+-ATPase, the resulting H^+ flux of 92 nmol $m^{-2} \ s^{-1}$ would then at least be of the same order as the maximum observed rates of malic-acid accumulation.

As well as considering fluxes averaged over the entire dark period, the high activity of the tonoplast ATPase in CAM plants can also be appreciated from the fact that maximum short-term fluxes of the order of 300 nmol $H^+ \ m^{-2} \ s^{-1}$ can occur. This is about three-fold higher than the maximum rates estimated for stomatal guard cells. Indeed, in terms of a typical maximum reaction velocity for H^+-ATPases of around 100 mol ATP (mol ATPase)$^{-1} \ s^{-1}$, a transmembrane flux of 300 nmol $H^+ \ m^{-2} \ s^{-1}$ would represent an area density of 1.5 nmol ATPase m^{-2} for a 2 H^+-ATPase. This is within the range of area densities observed for H^+-ATPases in coupling membranes (Raven, 1984), but indicates that an upper limit to the rates of malic-acid accumulation in CAM plants may also be set by the molecular architecture of the tonoplast. The high activity of this transport system may be an aid to further progress in our understanding of solute movement across the tonoplast.

REFERENCES

Aoki, K., and Nishida, K., 1984, Activity associated with vacuoles and tonoplast vesicles isolated from the CAM plant, Kalanchoë daigremontiana, Physiol. Plant., 60:21.

Bennett, A. B., and Spanswick, R. M., 1984, H^+-ATPase activity from storage tissue of Beta vulgaris. II. H^+/ATP stoichiometry of an anion-sensitive H^+-ATPase, Plant Physiol., 74:545.

Buchanan-Bollig, I. C., Kluge, M., and Müller, D., 1984, Kinetic changes with temperature of phosphoenolpyruvate carboxylase from a CAM plant, Plant Cell Environ., 7:63.

Buser-Suter, C., Wiemken, A., and Matile, P., 1982, A malic acid permease in isolated vacuoles of a crassulacean acid metabolism plant, Plant Physiol., 69:456.

Chanson, A., Fichmann, J., Spear, D., and Taiz, L., 1985, Pyrophosphate-driven proton transport by microsomal membranes of corn coleoptiles, Plant Physiol., 79:159.

Fitzsimons, P. J., and Weyers, J. D. B., 1986, Potassium ion uptake by swelling Commelina communis guard cell protoplasts, Physiol. Plant., 66:469.

Griffith, C. J., Rea, P. A., Blumwald, E., and Poole, R. J., 1986, Mechanism of stimulation and inhibition of tonoplast H^+-ATPase of Beta vulgaris by chloride and nitrate, Plant Physiol., 81:120.

Jochem, P., 1986, Dr. rer.-nat.-Dissertation, Technische Hochschule Darmstadt.

Jochem, P., Rona, J.-P., Smith, J. A. C., and Lüttge, U., 1984, Anion-sensitive ATPase activity and proton transport in isolated vacuoles of species of the CAM genus Kalanchoë, Physiol. Plant., 62:410.

Kinzel, H., 1982, "Pflanzenökologie und Mineralstoffwechsel", Eugen Ulmer, Stuttgart.

Kurkdjian, A., Quiquampoix, H., Barbier-Brygoo, H., Pean, M., Manigault, P., and Guern, J., 1985, Critical evaluation of methods for estimating the vacuolar pH of plant cells, in : "Biochemistry and Function of Vacuolar Adenosine-Triphosphatase in Fungi and Plants", B. P. Marin, ed., Springer-Verlag, Berlin, Heidelberg, New-York, Tokyo.

Lüttge, U., and Ball, E., 1977, Concentration and pH dependence of malate efflux and influx in leaf slices of CAM plants, Z. Pflanzenphysiol., 83:43.

Lüttge, U., and Ball, E., 1979, Electrochemical investigation of active malic acid transport at the tonoplast into the vacuoles of the CAM plant Kalanchoë daigremontiana, J. Membrane Biol., 47:401.

Lüttge, U., Smith, J. A. C., and Marigo, G., 1982, Membrane transport, osmoregulation, and the control of CAM, in : "Crassulacean Acid Metabolism", I. P. Ting and M. Gibbs, eds., American Society of Plant Physiologists, Rockville, Maryland.

Lüttge, U., Smith, J. A. C., Marigo, G., and Osmond, C. B., 1981, Energetics of malate accumulation in the vacuoles of Kalanchoë uniflora cells, FEBS Letters, 121:81.

Lüttge, U., and Nobel, P. S., 1984, Day-night variations in malate concentration, osmotic pressure, and hydrostatic pressure in Cereus validus, Plant Physiol., 75:804.

Lüttge, U., and Smith, J. A. C., 1984, Mechanism of passive malic-acid efflux from vacuoles of the CAM plant Kalanchoë daigremontiana, J. Membrane Biol., 81:149.

Lüttge, U., and Smith, J. A. C., 1985, Transport of malic acid in cells of CAM plants, in : "Biochemistry and Function of Vacuolar Adenosine-Triphosphatase in Fungi and Plants", B. P. Marin, ed., Springer-Verlag, Berlin, Heidelberg, New-York, Tokyo.

Lüttge, U., and Smith, J. A. C., 1987, CAM plants, in : "Solute Transport in Plant Cells and Tissues", J. L. Hall and D. A. Baker, eds., Longman, London, in press.

McRobbie, E. A. C., 1981, Ionic relations of stomatal guard cells, in : "Stomatal Physiology", P. G. Jarvis and T. A. Mansfield, eds., Cambridge University Press, Cambridge.

Mandala, S., and Taiz, L., 1985, Partial purification of a tonoplast ATPase from corn coleoptiles, Plant Physiol., 78:327.

Marigo, G., Bouyssou, H., and Boudet, A. M., 1986, Accumulation des ions nitrate et malate dans des cellules de Catharanthus roseus et incidence sur le pH vacuolaire, Physiol. Vég., 24:15.

Marin, B. P., 1985, "Biochemistry and Function of Vacuolar Adenosine-Triphosphatase in Fungi and Plants", Springer-Verlag, Berlin, Heidelberg, New-York, Tokyo.

Marin, B., Preisser, J., and Komor, E., 1985, Solubilization and purification of the ATPase from the tonoplast of Hevea, Eur. J. Biochem., 151:131.

Nott, D. L., and Osmond, C. B., 1982, Purification and properties of phosphoenol-pyruvate carboxylase from plants with crassulacean acid metabolism, Aust. J. Plant Physiol., 9:409.

Osmond, C. B., 1978, Crassulacean acid metabolism : A curiosity in context, Annu. Rev. Plant Physiol., 29:379.

Phillips, R. D., 1980, Deacidification in a plant with crassulacean acid metabolism associated with anion-cation balance, Nature, 287:727.

Phillips, R. D., and Jennings, D. H., 1976, Succulence, cations and organic acids in leaves of Kalanchoë daigremontiana grown in long and short days in soil and water culture, New Phytol., 77:599.

Randall, S. K., and Sze, H., 1986, Properties of the partially purified tonoplast H^+-pumping ATPase from oat roots, J. Biol. Chem., 261:1364.

Raven, J. A., 1984, "Energetics and Transport in Aquatic Plants", A. R. Liss, New-York.

Raven, J. A., 1985, pH regulation in plants, Sci. Prog. (Oxford), 69:495.

Rea, P. A., and Poole, R. J., 1985, Proton-translocating inorganic pyrophosphatase in red beet (Beta vulgaris L.) tonoplast vesicles, Plant Physiol., 77:46.

Rona, J.-P., Pitman, M. G., Lüttge, U., and Ball, E., 1980, Electrochemical data on compartmentation into cell wall, cytoplasm, and vacuole of leaf cells in the CAM genus Kalanchoë, J. Membrane Biol., 57:25.

Smith, J. A. C., Griffiths, H., Lüttge, U., Crook, C. E., Griffiths, N. M., and Stimmel, K.-H., 1986, Comparative ecophysiology of CAM and C_3 bromeliads. IV. Plant water relationships, Plant Cell Environ., 9:395.

Smith, J. A. C., Jochem, P., and Lüttge, U., 1987, Anion-sensitive, ATP-dependent proton accumulation in membrane vesicles from the CAM plant Kalanchoë daigremontiana, submitted for publication.

Smith, J. A. C., and Lüttge, U., 1985, Day-night changes in leaf water relations associated with the rhythm of crassulacean acid metabolism in Kalanchoë daigremontiana, Planta, 163:272.

Smith, J. A. C., Marigo, G., Lüttge, U., and Ball, E., 1982, Adenine-nucleotide levels during crassulacean acid metabolism and the energetics of malate accumulation in Kalanchoë tubiflora, Plant Science Letters, 26:13.

Smith, J. A. C., Uribe, E. G., Ball, E., Heuer, S., and Lüttge, U., 1984 a, Characterization of the vacuolar ATPase activity of the crassulacean acid metabolism plant Kalanchoë daigremontiana, Eur. J. Biochem., 141:415.

Smith, J. A. C., Uribe, E. G., Ball, E., and Lüttge, U., 1984 b, ATPase activity associated with isolated vacuoles of the crassulacean acid metabolism plant Kalanchoë daigremontiana, Planta, 162:299.

Steudle, E., Smith, J. A. C., and Lüttge, U., 1980, Water-relation parameters of individual mesophyll cells of the crassulacean acid metabolism plant Kalanchoë daigremontiana, Plant Physiol., 66:1155.

Struve, I., Weber, A., Lüttge, U., Ball, E., and Smith, J. A. C., 1985, Increased vacuolar ATPase activity correlated with CAM induction in Mesembryanthemum crystallinum and Kalanchoë blossfeldiana cv. Tom Thumb, J. Plant Physiol., 117:451.

Struve, I., and Lüttge, U., 1987, MgATP^{2-}-dependent electrogenic H$^+$-transport in tonoplast vesicles of the facultative CAM plant Mesembryanthemum crystallinum L., Planta, in press.

Sze, H., 1985, H$^+$-translocating ATPases : Advances using membrane vesicles, Annu. Rev. Plant Physiol., 36:175.

Walker, R. R., and Leigh, R. A., 1981, Characterization of a salt-stimulated ATPase activity associated with vacuoles isolated from storage roots of red beet (Beta vulgaris L.), Planta, 153:140.

Wang, Y., Leigh, R. A., Kaestner, K. H., and Sze, H., 1986, Electrogenic H$^+$-pumping pyrophosphatase in tonoplast of oat roots, Plant Physiol., 81:497.

Willner, C. M., 1983, "Stomata", Longman, London.

Winter, K., 1985, Crassulacean acid metabolism, in : "Topics in Photosynthesis, Vol. 6, Photosynthetic Mechanisms and the Development", J. Barber and N. R. Baker, eds., Elsevier Publishers B. V., Amsterdam.

OSMOTIC FACTORS AFFECTING THE MOBILISATION OF SUCROSE
FROM VACUOLES OF RED BEET STORAGE ROOT TISSUE

Caroline A. Perry[*],[**], Roger A. Leigh[*], A. Deri Tomos[***]
and J.L. Hall[**]

[*]Rothamsted Experimental Station,
Harpenden, Hertfordshire
[**]Department of Biology, University of Southampton,
Southampton; and [***]Adran Biocemeg a Gwyddor Pridd,
Coleg Prifysgol Gogledd Cymru, Bangor, Gwynedd, U. K.

INTRODUCTION

The accumulation of solutes in the vacuoles of plant cells is a reversible process and vacuolar solutes can be made available both for metabolism and to buffer cytoplasmic solute concentrations. However, some of the storage compounds in the vacuole may also make important contributions to sap osmotic pressure and hence turgor pressure. This raises the question of what happens to turgor pressure when these solutes are exported from the vacuole. To test this, the maintenance of sap osmotic pressure and turgor pressure during the mobilisation of sucrose from red beet storage root vacuoles was studied. Sucrose is localised in the vacuoles of this tissue (Leigh et al., 1979) and contributes about 40% of the sap osmotic pressure (Tomos et al., 1984). Thus if it was completely mobilised without a compensatory accumulation of other solutes, there would be a large decrease in turgor pressure. Mobilisation of sucrose can be induced by washing disks of beet tissue in aerated solutions for several days. This induces a vacuolar acid invertase and the concentration of sucrose in the vacuole declines (Leigh et al., 1979). In addition, washing causes an increase in the rate of salt uptake by the tissue (Van Steveninck, 1975), therefore the effect of salts on turgor opressure and solute content were also studied. The capacity of any turgor pressure regulation mechanism was tested by washing disks at three different external osmotic pressures.

MATERIALS AND METHODS

Disks measuring 8 mm diameter by 1 mm thick were cut from storage roots of red beet (Beta vulgaris L.), rinsed briefly in 0.5 mol m^{-3} CaSO$_4$ to remove solutes released from broken cells and transferred to the aerated experimental solutions. The solutions initially contained 0, 200 or 350 mol m^{-3} mannitol and 0.5 mol m^{-3} CaSO$_4$ but after 118 h 5 mol m^{-3} KCl and 5 mol m^{-3} NaCl were also present. To inhibit the growth of bacteria, solutions were changed four times during the first 24 h and twice daily thereafter. Samples of disks were taken at the times indicated. Those used for turgor pressure measurements were kept in an aliquot of experimental solution but those used for sap extraction were rinsed briefly in 0.5 mol m^{-3} CaSO$_4$. Cell turgor pressure was measured using a pressure probe and sap was extracted and analysed as described previously (Tomos et al., 1984).

RESULTS

The experimental period was divided into two parts : the initial 118 h when NaCl and KCl were absent from the experimental solutions, and the subsequent phase when they were present. The first period was characterised by a decrease in the concentration of sucrose in the disks and the second by an accumulation of salts. The changes occurring during each of these periods will be described separately.

Sucrose Mobilisation Phase

Washing disks in aerated solutions caused their sucrose content to decline but the rate of decrease depended on the external osmotic pressure (Fig. 1). The decline was fastest in disks in 0 mol m^{-3} mannitol and slowest in disks in 350 mol m^{-3} mannitol. Sucrose had disppeared from the disks in 0 mol m^{-3} mannitol by 46 h and from all disks by 118 h.

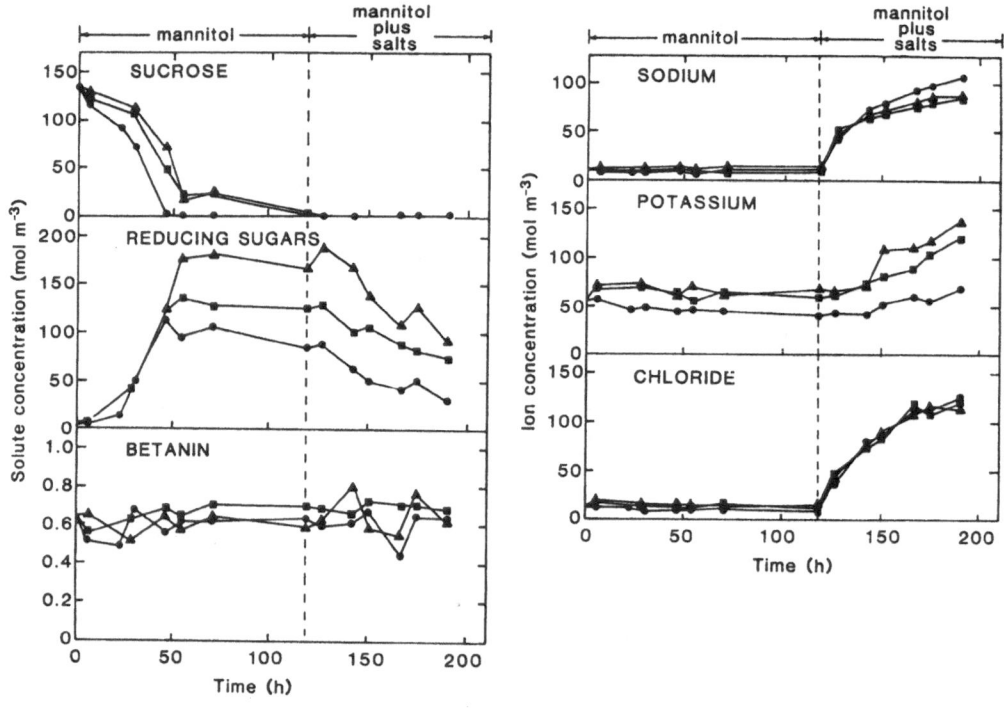

Figure 1

Changes in solute concentrations
in beet disks washed for 118 h
in 0 (●), 200 (■) or 350 (▲) mol m^{-3} mannitol
and then transferred to solutions
containing 5 mol m^{-3} KCl and 5 mol m^{-3} NaCl
in addition to mannitol

The decline in sucrose was accompanied by an increase in reducing sugar concentrations which began after a lag of 24 h (Fig. 1). This accumulation stopped after about 50 h and reasonably steady reducing sugar concentrations were maintained thereafter. However, the final steady concentration depended on the mannitol treatment. The accumulation was highest in the disks in 350 mol m^{-3} mannitol and lowest in the disks in 0 mol m^{-3} mannitol.

During the first 118 h there were no changes in the concentrations of Na$^+$, K$^+$, Cl$^-$, or betanin in the disks (Fig. 1) although concentrations of K$^+$ were lower in disks in the 0 mol m^{-3} mannitol treatment (Fig. 1). Thus the only substantial changes in solute concentrations accompanying the loss of sucrose from the vacuole was an accumulation of reducing sugars.

Coincident with the changes in sugar content, there was an decrease in the osmotic pressure of the sap from the disks in 0 mol.m^{-3} mannitol, a more or less constant value in the disks in 200 mol m^{-3} mannitol and an increase in those in 350 mol m^{-3} mannitol (Fig. 2). For the disks in 0 mol m^{-3} mannitol the decrease in osmotic pressure began immediately upon transfer to the experimental solutions and a new steady value was reached by 30 h. In contrast, the increase in osmotic pressure in disks in 350 mol m^{-3} mannitol began only after a lag of about 24 h and a steady value was not achieved until after 50 h.

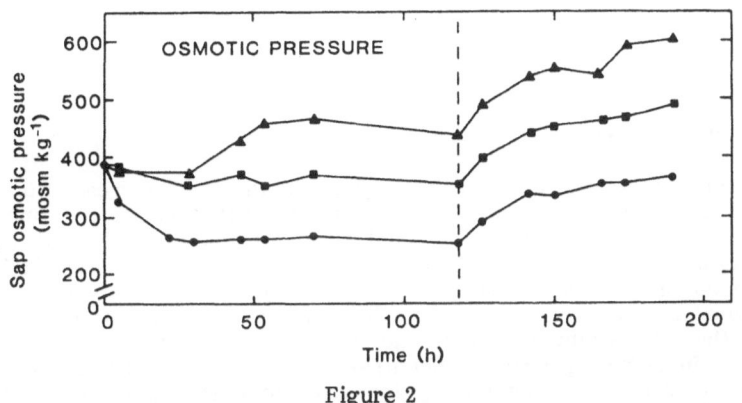

Figure 2

Changes in the osmotic pressure of sap from disks
washed for 118 h
in 0 (●), 200 (■) or 350 (▲) mol m^{-3} mannitol
and then transferred (at dashed line)
to solutions containing 5 mol m^{-3} NaCl and 5 mol m^{-3} KCl
in addition to the mannitol

Changes in turgor pressure were as expected from the changes in osmotic pressure. Disks in 0 mol m^{-3} mannitol initially had a turgor pressure of 0.88 MPa which declined to a new steady value of 0.7 MPa by the second day of washing (Fig. 3). This decrease in turgor pressure occurred without a lag period. The turgor pressure of disks in 200 mol m^{-3} mannitol remained constant at about 0.5 MPa, while that in disks in 350 mol m^{-3} mannitol was initially 0.12 MPa but increased to a new steady value of 0.31 MPa after a lag of about 1 d.

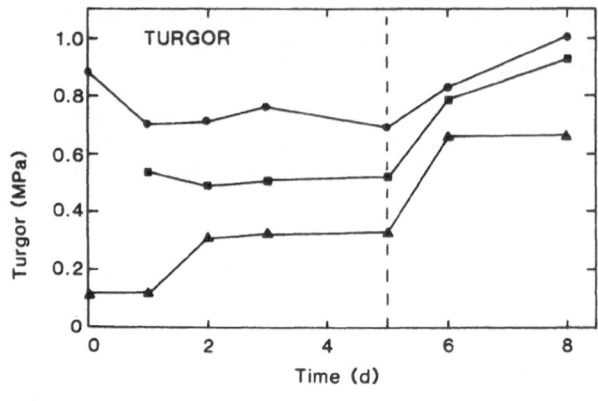

Figure 3

Changes in cell turgor pressure in disks
washed for 118 h
in 0 (●), 200 (■) or 350 (▲) mol m^{-3} mannitol
and then transferred to solutions
containing 5 mol m^{-3} NaCl and 5 mol m^{-3} KCl
in addition to the mannitol

The dashed line indicates the time of transfer.

Salt Accumulation Phase

The concentrations of Na$^+$, K$^+$ and Cl$^-$ in the disks increased upon transfer
to solutions containing NaCl and KCl (Fig. 1). In all treatments the amount of Na$^+$
+ K$^+$ absorbed was approximately equal to the amount of Cl$^-$ but the Na$^+$/K$^+$
discrimination was dependent on the mannitol concentration. In the 72 h following
transfer to the salt-containing solutions, the disks in 350 mol m^{-3} mannitol absorbed
equal amounts of Na$^+$ and K$^+$ but those in 0 mol m^{-3} mannitol took up three times
more Na$^+$ than K$^+$. Coincident with the increase in salt concentrations there was
an decrease in reducing sugar concentrations (Fig. 1). These changes in solute
concentrations were accompanied by increases in both sap osmotic pressure and
cell turgor pressure (Figs. 2 and 3).

DISCUSSION

The results show that both internal osmotic pressure and turgor pressure
were regulated during the mobilisation of sucrose from vacuoles of red beet strorage
root. However, only disks in 200 mol m^{-3} mannitol maintained these parameters
at constant values. The osmotic pressure of 200 mol m^{-3} mannitol is about 0.5
MPa and this is similar to the extracellular water potential in intact storage roots
of red beet (Leigh and Tomos, 1983). The observation that, at this external osmotic
pressure, turgor pressure is maintained constant strongly suggest that the cells
recognize this as being equivalent to their natural external water potential.

In the absence of salts, regulation of turgor pressure was achieved by the
accumulation of reducing sugars. The use of hexoses for osmotic purposes thus

limited the extent to which sucrose–derived C was released from the vacuole. The retention of sugars in the vacuole to maintain turgor pressure was given priority over any metabolic processes that required them. However, the accumulated hexoses were utilised when salts were supplied. The salts presumably provided an alternative vacuolar osmoticum that could be used to maintain sap osmotic pressure and so overcome the need to use sugars for this purpose. The uptake of salts and export of sugars were accompanied by an increase in osmotic pressure and turgor pressure. Sutcliffe (1954) also observed an increase in osmotic pressure during salt absorption by washed beet disks. It is unclear whether this is because the cells naturally develop a higher turgor pressure when salts are the main vacuolar osmotica or whether it is the result of changes induced by the washing procedure.

For the disks in 200 and 350 mol m^{-3} mannitol, the turgor pressure changes that occurred during sucrose mobilisation could be accounted for by the accumulation of reducing sugars. The lag in the onset of the increase in turgor pressure in the 350 mol m^{-3} mannitol treatment was probably due to the time needed for the induction of vacuolar acid invertase activity. However, for the disks in 0 mol m^{-3} mannitol, the decrease in turgor pressure was larger than the decrease in total sugar concentrations and occurred without a lag. Other experiments (Perry et al., 1986) have shown that this decrease is achieved by the leakage of several solutes from the tissue. The solutes were lost in proportion to their concentration in the tissue. However, betanin concentration did not decrease significantly (Fig. 1) indicating that the losses were not due to cell death. This turgor-driven leakage probably explains the lower K$^+$ concentrations in the disks washed in 0 mol m^{-3} mannitol.

Although the sap osmotic pressure and cell turgor pressure of disks in 0 or 350 mol m^{-3} mannitol did not return to the values maintained in disks in 200 mol m^{-3} mannitol, this was not due to any obvious limitation imposed by internal solute concentrations. In the disks in 350 mol m^{-3} mannitol, the initial sucrose concentration was sufficient to have provided, upon hydrolysis, an increase in turgor pressure of almost twice that observe. Similarly, solutes were still available for leakage from the disks in 0 mol m^{-3} mannitol when turgor pressure stabilised. The results seem to indicate that turgor pressure is regulated so that it falls within a particular range of values. Once perturbed it does not appear to be adjusted back to exactly the original value. Such a strategy has the advantage of preventing excessive leakage of solutes when turgor pressure is being reduced and of limiting the amount of solutes, particularly organic compounds, that need to be diverted into an osmotic role when turgor pressure is being increased.

Although the beet disk system has proved a useful "model" with which to study turgor pressure regulation during utilisation of vacuolar solutes, it is clearly an artificial system. However, the ideas it has generated should be applicable to other tissues and it will be interesting to see whether similar constraints limit the mobilisation of solutes in other tissues.

REFERENCES

Leigh, R. A., ap Rees, T., Fuller, W. A., and Banfield, J., 1979, The location of acid invertase activity and sucrose in vacuoles of storage roots of beet root (Beta vulgaris L.), Biochem. J., 178:539.
Leigh, R. A., and Tomos, A. D., 1983, An attempt to use isolated vacuoles to determine the distribution of sodium and potassium in cells of storage roots of roots of red beet (Beta vulgaris L.), Planta, 159:469.
Perry, C. A., Leigh, R. A., Tomos, A. D., Wyse, R. E., and Hall, J. L., 1986, The regulation of turgor pressure during sucrose mobilisation and salt accumulation by excised storage root tissue, Planta, in press.
Sutcliffe, J. F., 1954, The absorption of potassium by plasmolysed cells, J. Exp. Bot., 5:215.

Tomos, A. D., Leigh, R. A., Shaw, C. A., and Wyn-Jones, R. G., 1984, A comparison of methods for measuring turgor pressures and osmotic pressures of cells of red beet storage root tissue, J. Exp. Bot., 35:1675.

Van Steveninck, R. F. M., 1975, The "washing" or "aging" phenomenon in plant tissues, Annu. Rev. Plant Physiol., 26:237.

DISTRIBUTION OF POTASSIUM BETWEEN VACUOLE AND CYTOPLASM

IN RESPONSE TO POTASSIUM DEFICIENCY

Richard Storey[*] and Roger A. Leigh[**]

[*]C.S.I.R.O., Division of Horticultural Research,
Merbein, Victoria, Australia; and [**]Rothamsted
Experimental Station, Harpenden, Hertfordshire, U. K.

INTRODUCTION

Differential changes in the distribution of potassium between cytoplasm and vacuole may be an important response of plants to K^+ deficiency. Leigh and Wyn-Jones (1984) suggested that as tissue K^+ concentration declines, the K^+ concentration in the cytoplasm will remain constant while that in the vacuole will decrease. They proposed that decreases in cytoplasmic K^+ concentration would only be observed at low tissue K^+ concentrations and that growth would decline in response to these decreases. Evidence for such differential behaviour of the cytoplasmic and vacuolar K^+ pools in barley leaf cells has now been obtained using X-ray microanalysis.

MATERIALS AND METHODS

Barley (Hordeum vulgare L., cv. Clipper) was grown in nutrient solution containing 0.02 or 2.5 mol m^{-3} K^+ (as K_2SO_4), 1 mol m^{-3} $MgSO_4$, 1 mol m^{-3} $NH_4H_2PO_4$, 5 mol m^{-3} $Ca(NO_3)_2$ and micronutrients at the concentrations given by Sutcliffe and Baker (1974) except that Fe was added as FeNaEDTA. Four seedlings were grown in 2.5 dm^3 of solution in a controlled environment cabinet (250 μmol m^{-2} s^{-1} PAR, 12 h day, and day/night temperatures of 23/17°C). After 13 d, three plants from each pot were harvested, weighed, then cations were extracted with HCl and measured by atomic absorption. The fourth plant was used for X-ray microanalysis.

All X-ray microanalytical observations were made on bulk-frozen, fully-hydrated tissue from the first fully expanded leaf. The procedures used to prepare the tissue are described elsewhere (Leigh et al., 1986). Briefly, a section of leaf was frozen in melting N_2, fractured in vacuo at - 100°C, etched at - 80°C to reveal cell detail, recooled and evaporatively coated with Cr. X-ray spectra were recorded using an EDAX energy-dispersive X-ray analyzer fitted to a Philips 500 scanning electron microscope. Spectra from 0 to 5 keV were recorded using an accelerating voltage of 12 kV, an emission current of 40 μA, a spot diameter of 0.4 μm, and a data collection time of 100 live s. The beam was focussed in the "reduced area" mode.

RESULTS

After 13 d, the plants grown in 0.02 mol m^{-3} K$^+$ yielded only 25% of the fresh weight of those grown in 2.5 mol m^{-3} K$^+$ and their shoot K$^+$ concentrations were 66% lower (Table 1). To compensate for lack of K$^+$ they accumulated more Na$^+$, Ca^{2+} and Mg^{2+}.

Table 1

Yield and Cation Concentrations in Barley Plants
after 13 d growth in nutrient solution
containing 0.02 or 2.5 mol m^{-3} K

External K concentration (mol m^{-3})	Fresh weight	Tissue cation concentration (mol m^{-3})			
		K	Ca	Mg	Na
0.02	0.25	94	52	40	47
2.5	1.0	281	19	10	2

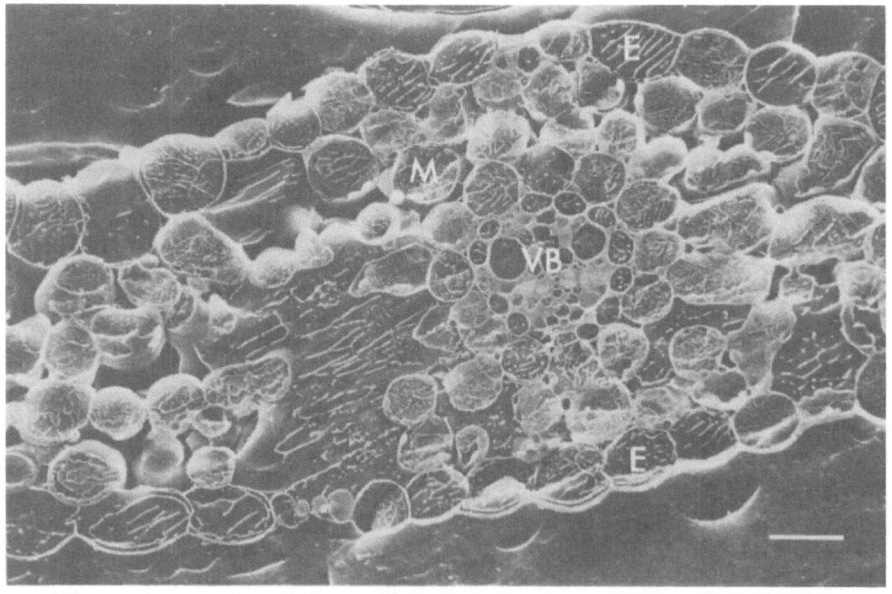

Figure 1

Appearance of cells in a barley Leaf
after preparation for X-ray microanalysis

The leaf is seen in transverse section with epidermal cells (E), mesophyll cells (M) and a vascular bundle (VB) clearly visible. Bar is approximately 25 μm.

After preparation of tissue for X-ray microanalysis all major cell types in the leaves of barley could be identified (Fig. 1). In mesophyll cells, it was possible to distinguish vacuole, cytoplasm and chloroplasts but in epidermal cells only the vacuole could be identified.

Potassium was the major cation in the vacuole and cytoplasm of mesophyll cells of plants grown in 2.5 mol m^{-3} K$^+$ (Fig. 2). Other cations were at or below the limits of their detection (probably about 20 mol m^{-3}). Although Ca^{2+} was not detectable in the mesophyll, significant concentrations were present in vacuoles of epidermal cells.

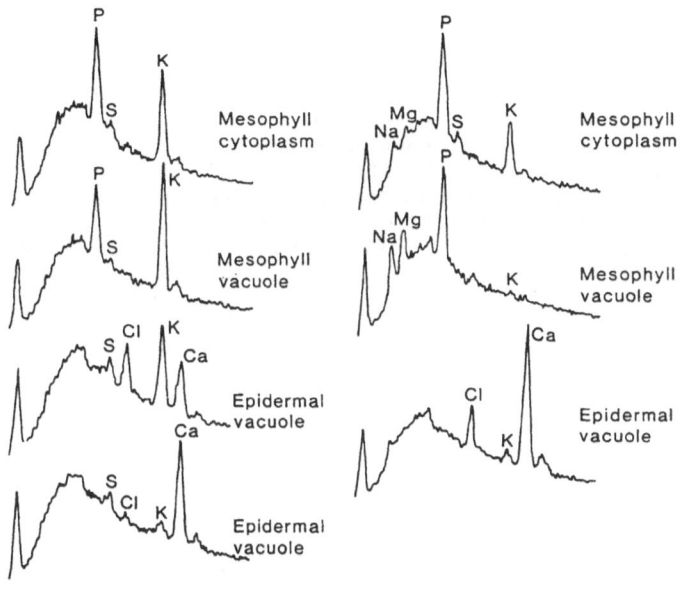

Figure 2

X-ray microanalysis spectra for cell compartments
in leaves from barley plants
grown in 2.5 mol m^{-3} K (left) or 0.02 mol m^{-3} K (right)

In plants grown in 0.02 mol m^{-3} K$^+$, there was no detectable K$^+$ in the vacuoles of mesophyll cells but these compartments contained Na$^+$ and Mg^{2+} instead (Fig. 2). In contrast, K$^+$ was still detectable in the cytoplasm of these cells. Despite the increased levels of Ca^{2+} in these K$^+$-deficient plants (Table 1) this element was not detectable in mesophyll cells. As in the plants from the high K$^+$ treatment, Ca^{2+} appeared to be preferentially located in epidermal vacuoles.

DISCUSSION

These results confirm the suggestion of Leigh and Wyn-Jones (1984) that K$^+$ concentrations in the vacuole and cytoplasm respond differently to changes in tissue K$^+$ concentrations. Thus while K$^+$ was undetectable in the vacuole of mesophyll cells of plants grown in 0.02 mol m^{-3} K$^+$, it was still present in the cytoplasm of these cells. The results indicate the importance of the tonoplast in controlling the distribution of K$^+$ between vacuole and cytoplasm. When K$^+$ supply is good, transport systems at this membrane allow the accumulation of this ion in the vacuole but, when supply is inadequate, they maintain the cytoplasmic K$^+$

concentration at the expense of vacuolar K^+ pool. There is now a need to understand how the tonoplast is able to control K^+ distribution in this way.

In addition to highlighting the importance of compartmentation at the subcellular level, the results also indicate that cations may be differentially distributed between different cell types within barley leaves. Calcium appears to be preferentially located in the epidermis whereas Na^+ and magnesium accumulate in the mesophyll. It is unclear how this distribution is achieved but such intercellular compartmentation means that changes at the tissue level cannot be interpreted in terms of the "average" cell. Instead account must be taken of the way in which individual cell types are responding.

REFERENCES

Leigh, R. A., Chater, M., Storey, R., and Johnston, A. E., 1986, Accumulation and subcellular distribution of cations in relation to the growth of potassium-deficient barley, Plant Cell Environ., in press.

Leigh, R. A., and Wyn-Jones, R. G., 1984, A hypothesis relating critical potassium concentrations for growth to the distribution and functions of this ion in the plant cell, New Phytol., 97:1.

Sutcliffe, J. F., and Baker, D. A., 1974, "Plants and Mineral Salts", 2nd Edition, Edward Arnold, London.

VACUOLAR pH LEVEL IS RELATED TO PHOTOSYNTHESIS

IN ISOLATED MESOPHYLL PROTOPLASTS OF <u>MELILOTUS ALBA</u>

Claudine Nef[*], Gilbert Alibert[**]
and Alain M. Boudet[**]

[*]Laboratoire de Physiologie Végétale
Centre ORSTOM d'Adiopodoumé, Abidjan, Côte-d'Ivoire
and [**]Centre de Physiologie Végétale
Université Paul Sabatier, F-31.062-Toulouse-Cedex
France

Freshly isolated protoplasts of <u>Melilotus alba</u> were kept in dark conditions during two hours, then transferred under light and the vacuolar pH change of the protoplasts was monitored during the next hours using benzylamine as a radiochemical probe.

The vacuolar pH decreases steadily of about 0.5 pH unit during the first 90 min of illumination and then stabilizes at this value. Returning back the protoplasts in dark induces an increase of the vacuolar pH until the initial value was recovered (this took about 60 min). Moreover :

- the vacuolar acidification is related to the light intensity,

- vacuolar pH and $^{14}CO_2$ fixation by the protoplasts follow the same pattern of variation when the protoplasts are transferred from dark to light,

- DCMU and cyanid inhibit both photosynthesis and light-dependent acidification of the vacuole,

- photosynthesis and pH variations show the same dependence upon the external pH medium.

Thus, the light-induced variations of the vacuolar pH appears to be linked to photosynthesis.

Additional works indicated that during photosynthesis, anionic compounds are synthesized from CO_2 and rapidly transferred into the vacuole. However these compounds are not released from the vacuoles in dark while the vacuolar pH is modified as stated before. Thus the vacuolar acidification cannot be attributed to the transfer of photosynthetats into the organelle but may be rather found in an increase in the disponibility of photosynthetic ATP, fueling a tonoplastic proton pump. Works are in progress to test this hypothesis.

UNDERSTANDING TONOPLAST FUNCTION :

SOME EMERGING PROBLEMS

Roger A. Leigh and Andrew J. Pope

Rothamsted Experimental Station
Harpenden, Hertfordshire, U. K.

INTRODUCTION

Although it is only 10 years since methods for the large-scale isolation of intact vacuoles from mature, higher plant cells were first reported (Wagner and Siegelman, 1975; Leigh and Branton, 1976), progress has been rapid and an understanding of the primary and secondary transport systems at the tonoplast is beginning to emerge (Fig. 1; see Marin, 1985; Sze, 1985 for reviews).

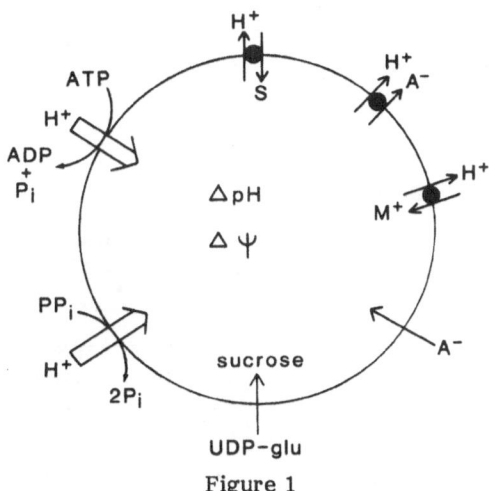

Figure 1

Transport systems at the tonoplast

It is now established that the primary transport process at the tonoplast is inwardly-directed, electrogenic proton translocation which creates gradients of both pH (ΔpH, inside acid) and membrane potential ($\Delta\Psi$, inside positive). There are two enzymes capable of catalysing this transport, an anion-sensitive ATPase and a K^+-stimulated inorganic pyrophosphatase (PPase). The ΔpH   generated   by these activities can be used to drive the uptake of both cations and neutral organic solutes through a variety of solute/H$^+$ antiports whereas anion uptake is thought to result from electrophoretic movement of anions in response to the $\Delta\Psi$. In addition to these systems, there is now evidence that sucrose accumulation in the vacuole is mediated by a group translocator that simultaneously synthesises and transports sucrose using UDP-glucose as the sole substrate (Thom and Maretzki, 1985 and 1987; Thom et al., 1986).

To a large extent the progress that has been made in understanding tonoplast transport systems is due to work <u>in vitro</u> with intact vacuoles (Leigh, 1983) or sealed tonoplast vesicles (Sze, 1985). The intention here, however, is not to review this work but instead to point out some interesting problems that are emerging and to examine the extent to which these can be answered using <u>in vitro</u> techniques.

THE PHYSIOLOGICAL ROLE OF THE H$^+$-TRANSLOCATING PPase

Walker and Leigh (1981 a and b) discovered that isolated beet vacuoles possess both membrane-bound ATPase and PPase activities and speculated that both might function as proton pumps. Convincing evidence that both catalyse inwardly directed, electrogenic H$^+$-transport at the tonoplast has now been obtained (e.g. Bennett

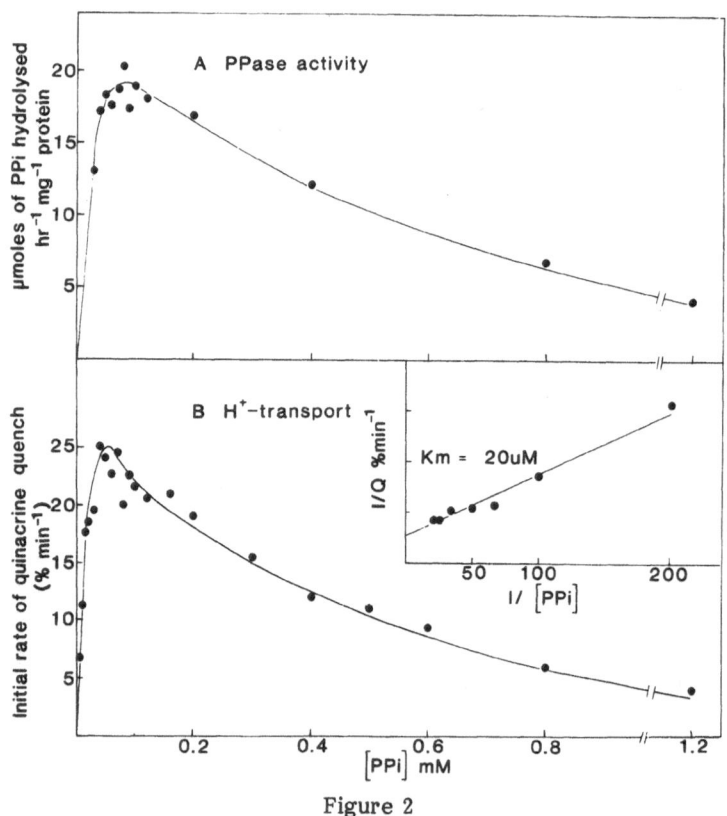

Figure 2

The response of PPase activity (A) and PP$_i$-dependent H$^+$-transport (B)
to changes in PP$_i$ concentration

The Mg^{2+} concentration was 5 mol m^{-3}.

and Spanswick, 1983; Churchill and Sze, 1983; Bennett et al., 1984; Poole et al., 1984; Chanson et al., 1985; Rea and Poole, 1985; Wang et al., 1986). Thus, the question which now needs to be answered is : What is the physiological function of the PPase ?

In the work of Rea and Poole (1985) the rates of H^+-transport with 3 mol m^{-3} PP_i were very much lower than those with 3 mol m^{-3} ATP which seemed to indicate the PP_i-dependent H^+-pump was quantitatively less important than the ATPase. However, more recent work with oat root tonoplast has shown that both PP_i hydrolysis and PP_i-dependent H^+-transport are very sensitive to the concentrations of PP_i and Mg^{2+} (Figs. 2 and 3). With 5 mol m^{-3} Mg^{2+}, maximum rate of PP_i hydrolysis and H^+-transport are obtained with 60 - 90 mmol m^{-3} PP_i, with progressive inhibition at higher PP_i concentrations (Fig. 2). At low PP_i concentrations the response to Mg^{2+} more or less conforms to Michaelis-Menten kinetics, but at higher PP_i concentrations a distinct Mg^{2+} optimum is observed (Fig. 3). These, and other experiments, indicate that in vitro maximum rates of H^+-transport are obtained with 1.5 mol m^{-3} Mg^{2+} and 60 - 90 mmol m^{-3} PP_i. The fact that both PP_i hydrolysis and H^+-transport show the same behaviour indicates that the observed responses are due to direct alteration of the enzyme activity and not to any biophysical constraints on H^+-transport. If H^+-pumping by the PPase and the ATPase are compared with each operating at optimal substrate and Mg^{2+}

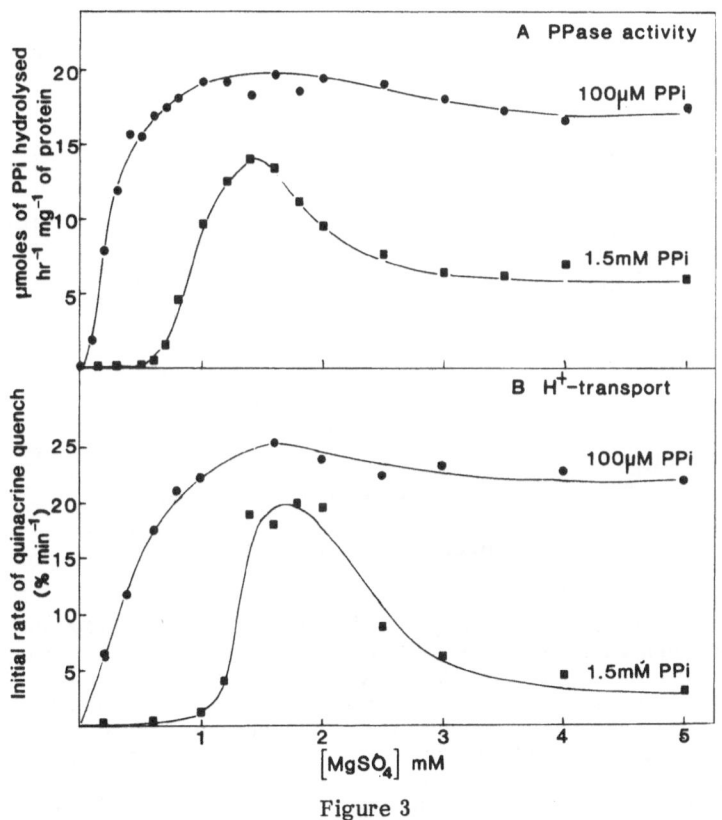

Figure 3

The response of PPase activity (A) and PP_i-dependent H^+-transport (B)
to changes in Mg^{2+} concentration
at 0.1 (●) and 1.5 (■) mol m^{-3} PP_i

concentrations, both the rate and extent of pH gradient formation are similar (Fig. 4). Thus the PPase could be a quantitatively important H^+-pump in vivo.

Whether the optimal conditions for the operation of the PP_i-dependent pump are likely to exist in vivo is unclear because it is difficult to make reliable measurements of cytoplasmic PP_i and Mg^{2+} concentrations. However, both can be estimated. For instance, studies of protein synthesis in vitro indicate that this process requires about 5 mol m^{-3} Mg^{2+} (Gibson et al., 1984). Since the ionic requirements of protein synthesis are thought to be major determinants of the ionic composition of the cytoplasm (Wyn-Jones et al., 1979), this probably approximates to the concentration in the cytoplasm in vivo. Measurements of the PP_i concentration in tissues fall within the range 5 - 39 nmol/g fresh weight (Edwards et al., 1984; Smyth and Black, 1984). If this is all assumed to be in the cytoplasm, which occupies about 10% of the cell volume, then the cytoplasmic concentrations of PP_i will be in the range 50 - 400 mmol m^{-3}. These values would suggest that the PPase could be a quantitatively important H^+-pump in vivo. Although the PPase requires K^+ for maximal activity (Walker and Leigh, 1981 b; Rea and Poole, 1985), this requirement saturates at about 20 mol m^{-3} (Wang et al., 1986). Since the cytoplasmic K^+ concentration is about 100 - 150 mol m^{-3} (Leigh and Wyn-Jones, 1984), K^+ is unlikely to limit the rate of H^+-pumping in vivo.

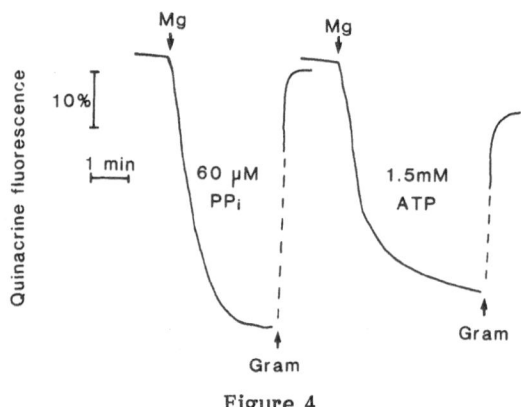

Figure 4

PP_i and ATP-dependent H^+-transport
(measured by quinacrine fluorescence quenching)
in oat tonoplast vesicles

Finally, although it is usually assumed that both the ATPase and PPase reside on the same membrane, this has still to be established. As pointed out by Rea et al. (1987 a), the fact that both ATPase and PPase activities are found in the same preparations of vacuoles or vesicles does not rule out the possibility that they are associated with different subpopulations of membranes. However, if this is the case, the generation of an ATP-dependent ΔpH in one subpopulation should not affect the generation of a PP_i-dependent ΔpH in the other. In other words, if the two reside on different membranes, the effects of ATP and PP_i on ΔpH formation would be additive. If non-additivity is found then it is likely that both pumps reside on the same membrane. Preliminary experiments with oat root tonoplast vesicles have shown that while PP_i can cause a slight stimulation of ΔpH formation when added after a steady-state H^+-gradient has been formed with ATP, the total gradient is no larger than that with PP_i alone (Pope and Leigh, unpublished results). Addition of ATP after formation of a steady-state H^+-gradient with PP_i slightly decreased the gradient. These experiments suggest that the effects of ATP and PP_i are not additive and support the idea that both pumps reside on the same membrane. Kaestner and Sze (1986) have reported similar results, also with oat tonoplast vesicles. Other evidence that the two pumps are associated with the same membrane is discussed by Rea et al. (1987 a).

Thus the available evidence suggests that both the ATP- and PP_i-dependent H^+-pumps reside on the tonoplast and that both are capable of generating quantitatively similar pH gradients. The question which remains unresolved at this time is why the tonoplast possesses two H^+-pumps. The additivity experiments suggest that the size of the gradient formed is no greater with both substrates than with either alone so the answer must lie elsewhere. One possibility is that though both are present they function at different times during the cell's development and so the proportion of the $\Delta\mu_{H^+}$ contributed by each pump changes with time. Alternatively, the PPase might be a back-up system for the ATPase, maintaining the $\Delta\mu_{H^+}$ when ATP levels fall. This could be studied using ^{31}P-NMR (Roberts, 1984) to measure the response of the ΔpH across the tonoplast as ATP and PP_i levels alter. However, highly sensitive NMR systems will be needed to detect PP_i if it is present at the levels indicated above. A third possibility is that the PPase is poised at equilibrium and, depending on conditions, can either hydrolyse PP_i to pump H^+ into the vacuole or utilise the pH gradient to synthetise PP_i. Its function would be to maintain reasonably steady levels of PP_i in the cytoplasm. The likelihood that the PPase is a reversible H^+-pump is discussed in detail by Rea et al. (1987 a). Their calculations show that if the stoichiometry is 2 H/PP_i or greater, then reversal is possible if the pH is about 2 units. There would now seem to be a need to determine this stoichiometry using isolated tonoplast vesicles, as was done for the ATPase (Bennett and Spanswick, 1984).

CONTROL OF CYTOPLASMIC ION CONCENTRATIONS

It has long been assumed that the vacuole plays an important role in the control of cytoplasmic pH (Smith and Raven, 1979) but more recently it has become obvious that it also helps to regulate the concentrations of other ions in the cytoplasm. Two studies which indicate a role for the tonoplast in this process are those by Lee and Ratcliffe (1983) and Leigh et al. (1987). Using ^{31}P-NMR, Lee and Ratcliffe (1983) showed that in pea roots P_i was located both in the vacuole and cytoplasm. When the plants were starved of phosphorus, the P_i concentration in the cytoplasm was maintained constant while that in the vacuole fell. This suggests that cytoplasmic P_i concentration is maintained at the expense of that in the vacuole. Similar behavior was observed for K^+ in mesophyll cells of barley leaves (Storey and Leigh, 1987; Leigh et al., 1987). In plants grown in nutrient solution with 2.5 mol m^{-3} K^+, X-ray microanalysis showed that this cation was present in both the vacuole and the cytoplasm of the mesophyll cells. However, in plants grown in 0.02 mol m^{-3} K^+ the concentration in the vacuole dropped to undetectable levels even though K^+ was still present at significant concentrations in the cytoplasm.

Both of these studies indicate important features of the transport mechanisms at the tonoplast. In cells with adequate supplies of nutrients the transport systems load the excess into the vacuole for storage. When the rate of supply drops, however, mechanisms exist for reversing this process to maintain cytoplasmic concentrations, at the expense of those in the vacuole. This presumably indicates that the transport systems at the tonoplast are sensitive to the concentrations of nutrient ions in the cytoplasm and the overall direction of transport is governed by this parameter. It should be possible, using in vitro techniques, to study the sensitivity of the direction of transport to changes in the concentrations of ions on each side of the membrane and so test to the above suggestion.

One problem with in vitro studies of transport is that the membrane may be changed by the isolation procedure so that inherent control systems are lost. Recent work by Blumwald and Poole (1985) provides an example of a futile transport cycle across the tonoplast in vitro that is unlikely to exist in vivo. The experiments of Blumwald and Poole (1985) provided evidence for the existence of a NO_3^-/H^+ symport on the tonoplast of beet cells and it was suggested that this symport was responsible for the export of NO_3^- from the vacuole. However, there was no obvious control of this system. In the isolated tonoplast vesicles, NO_3^- entering the vesicles in response to $\Delta\Psi$ was immediately exported via the symport. Yet, in intact tissues NO_3^- may be temporarily accumulated in vacuoles to concentrations as high as 100 molm^{-3} before being utilised at later stages of growth when its concentration declines to zero (Leigh and Wyn-Jones, 1986). Thus it would appear that in vivo the activity of the symport is controlled to allow the accumulation of NO_3^- in the vacuole at certain times (presumably when N supply is plentiful) and its release at others (when N supply is poor). The apparent absence of regulation in the experiments of Blumwald and Poole (1985) could be a tissue-specific effect because beet storage roots do not normally accumulate NO_3^- to significant concentrations (Greenwood and Hunt, 1986). However it could also indicate the loss on inherent control systems during isolation. This may indicate a limitation in the use of isolated membranes to study control of transport mechanisms and it may be worthwhile studying methods of isolation more intensively to see whether inherent control systems can be retained in vitro. One particular practise that should be investigated is the washing of membranes with KI to remove acid phosphatase (Poole et al., 1984). This procedure may also remove other proteins (see Rea et al., 1987 b) and some of these could have important regulatory functions.

CONTROL OF TONOPLAST TRANSPORT SYSTEMS BY TURGOR PRESSURE

The accumulation of solutes in vacuoles is a reversible process and vacuolar solutes can be made available both for metabolism and to buffer cytoplasmic solute concentrations (Leigh and Wyn-Jones, 1986). However, some of the storage compounds in the vacuole may also make important contributions to sap osmotic pressure and hence turgor pressure. This raises the question of whether turgor pressure can control the rate or extent to which the solutes are released from the vacuole.

To test this, changes in osmotic pressure accompanying the mobilisation of sucrose in beet storage root were studied (see also Perry et al., 1987). Sucrose is located in the vacuoles of this tissue and contributes about 40% of the sap osmotic pressure (Leigh et al., 1979; Tomos et al., 1984). Washing disks in aerated solutions induces an acid invertase activity in the vacuole which hydrolyses the sucrose and is the first step in its mobilisation (Leigh, 1984). For the experiment in Fig. 5, disks of beet storage root were washed in aerated 200 mol m^{-3} mannitol for 118 h and transferred to a solution containing 5 mol m^{-3} KCl and 5 mol m^{-3} NaCl in addition to the mannitol. The washing solution contained 200 mol m^{-3} mannitol because this has an osmotic pressure similar to the water potential in the extracellular space of intact beet storage roots (Leigh and Tomos, 1983).

Disappearance of sucrose from the tissue was accompanied by an increase in the concentration of reducing sugars that kept total sugar (sucrose + reducing sugars) constant (Fig. 5). As a result, the sap osmotic pressure did not change as sucrose was lost from the vacuole. Since the hydrolysis of sucrose produced two molecules of hexose, the results imply that only half of the reducing sugars were made available to the cell, the remainder staying in the vacuole to maintain turgor pressure. The need to prevent a large change in turgor pressure as sucrose disappeared thus limited the extent to which sucrose-derived C was lost from the vacuole.

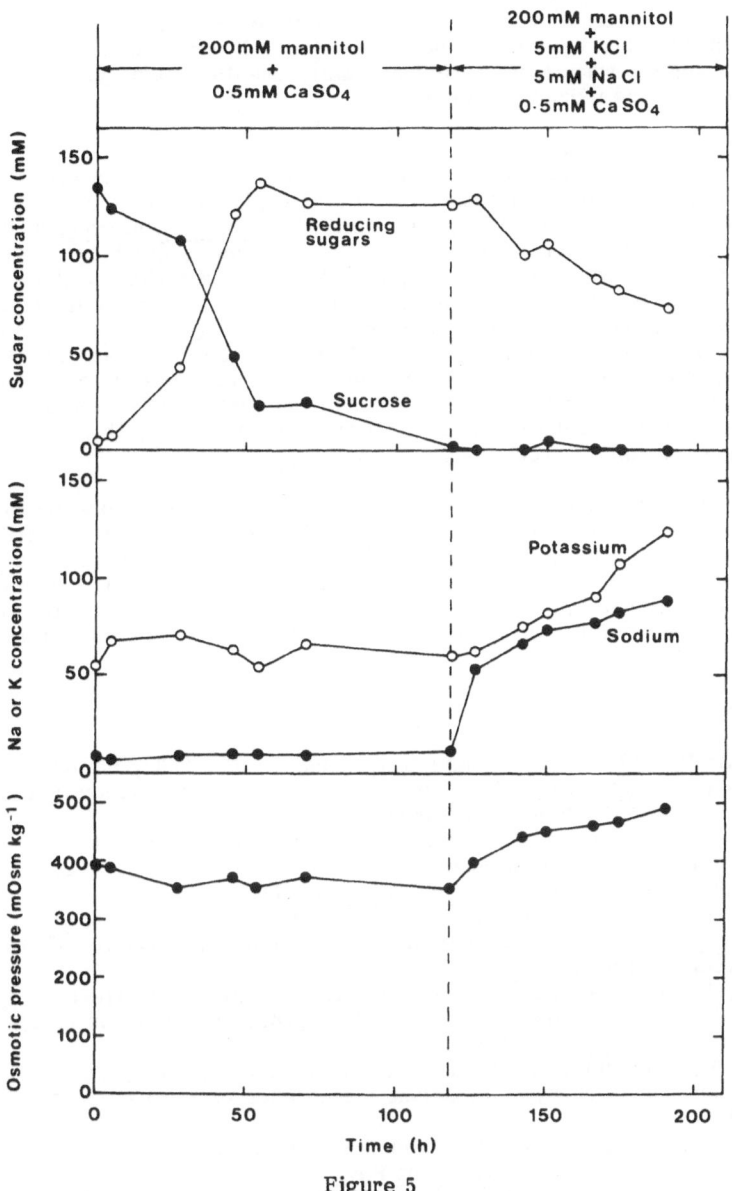

Figure 5

Changes in the solute concentrations and osmotic pressure
of beet disks washed initially in 200 mol m^{-3} mannitol
and then transferred (at the point indicated by the dashed line)
to solutions containing 5 mol m^{-3} NaCl and 5 mol m^{-3} KCl
in addition to the mannitol

Addition of salts to the external solution caused a decline in the concentration of reducing sugars and an increase in the salt concentration in the disks (Fig. 5). This is interpreted as indicating that the salts provided an alternative vacuolar osmoticum that could be used to replace the reducing sugars in the vacuole. This replacement of sugars by salts was accompanied by an increase in osmotic pressure. The reason for this increase is not understood. It could indicate that the tissue develops a higher turgor pressure when salts are the major vacuolar osmotica or it could be a change induced by the washing procedure.

This work clearly suggests that the need to maintain turgor pressure can limit the extent to which solutes are made available from the vacuole and it will be interesting to see how far these conclusions apply to other solutes. In both spinach leaves (Steingröver et al., 1986) and cultured Catharanthus roseus cells (Marigo et al., 1986), utilisation of vacuolar NO_3^- is accompanied by increases in the concentration of organic acids or sugars. One interpretation of these observations is that these solutes are accumulated in the vacuole to prevent changes inturgor pressure as NO_3^- is utilised. The need to maintain turgor pressure during utilisation of stored vacuolar solutes has important implications because the rate at which the solutes are exported from the vacuole could be limited or controlled by the rate at which replacement solutes are made available. This would be particularly important for NO_3^- because this could limit the rate of growth when vacuolar NO_3^- is providing the main source of N for growth.

Studying the mechanism by which turgor pressure controls the rate of export from the vacuole is unlikely to be possible using currently available in vitro techniques. Generation of turgor pressure requires the presence of a cell wall or some other rigid structure around the membrane and thus it is not developed in isolated membranes or vacuoles. New methods will have to be found if the relationships between turgor pressure and tonoplast are to be studied further.

REFERENCES

Bennett, A. B., O'Neill, S. D., and Spanswick, R. M., 1984, H^+-ATPase of storage tissue of Beta vulgaris. I. Identification and characterization of an anion-sensitive H^+-ATPase, Plant Physiol., 74:538.

Bennett, A. B., and Spanswick, R. M., 1983, Optimal measurement of ΔpH and $\Delta \Psi$ in corn root membrane vesicles : Kinetic analysis of Cl^- effects on a proton-translocating ATPase, J. Membrane Biol., 71:95.

Bennett, A. B., and Spanswick, R. M., 1984, H^+-ATPase of storage tissue of Beta vulgaris. II. H^+/ATP stoichiometry of an anion-sensitive H^+-ATPase, Plant Physiol., 74: 545.

Blumwald, E., and Poole, R. J., 1985, Nitrate storage and retrieval in Beta vulgaris L. : Effects of nitrate and chloride on proton gradients in tonoplast vesicles, Proc. Natl. Acad. Sci. U.S.A., 82:3683.

Chanson, A., Fichmann, J., Spear, D., and Taiz, L., 1985, Pyrophosphate-driven proton transport by microsomal membranes of corn coleoptiles, Plant Physiol., 79:159.

Churchill, K. A., and Sze, H., 1983, Anion-sensitive, H^+-pumping ATPase in membrane vesicles from oat roots, Plant Physiol., 71:610.

Edwards, J., ap Rees, T., Wilson, P. W., and Morell, S., 1984, Measurement of the inorganic pyrophosphate in tissues of Pisum sativum L., Planta, 162:188.

Gibson, T. S., Speirs, J., and Brady, C. J., 1984, Salt tolerance in plants. II. In vitro translation of m-RNAs from salt-tolerant and salt-sensitive plants on wheat germ ribosomes. Responses to ions and compatible organic solutes, Plant Cell Environ., 7:579.

Greenwood, D. J., and Hunt, J., 1986, Effect of nitrogen fertiliser on the nitrate contents of field vegetables grown in Britain, J. Sci. Food Agric., 37:373.

Kaestner, K. H., and Sze, H., 1986, Potential-dependent anion transport across tonoplast vesicles from oat roots, Plant Physiol., 80 S:81.

Lee, R. B., and Ratcliffe, R. G., 1983, Phosphorus nutrition and the intracellular distribution of inorganic phosphate in pea root tips : A quantitative study using ^{31}P-NMR, J. Exp. Bot., 34:1222.

Leigh, R. A., 1983, Methods, progress and potential for the use of isolated vacuoles in studies of solute transport in higher plant cells, Physiol. Plant., 57:390.

Leigh, R. A., 1984, The role of the vacuole in the accumulation and mobilization of sucrose, Plant Growth Regulation, 2:339.

Leigh, R. A., ap Rees, T., Fuller, W. A., and Banfield, J., 1979, The location of acid invertase activity and sucrose in vacuoles of storage roots of beet root (Beta vulgaris L.), Biochem. J., 178:539.

Leigh, R. A., and Branton, D., 1986, Isolation of vacuoles from root storage tissue of Beta vulgaris L., Plant Physiol., 58:656.

Leigh, R. A., Chater, M., Storey, R., and Johnston, A. E., 1987, Accumulation and subcellular distribution of cations in relation to the growth of potassium-deficient barley, Plant Cell Environ., in press.

Leigh, R. A., and Tomos, A. D., 1983, An attempt to use isolated vacuoles to determine the distribution of sodium and potassium in cells of storage roots of red beet (Beta vulgaris L.), Planta, 159:469.

Leigh, R. A., and Wyn-Jones, R. G., 1984, A hypothesis relating critical potassium concentrations for growth to the distribution and functions of this ion in the plant cell, New Phytol., 97:1.

Leigh, R. A., and Wyn-Jones, R.G., 1986, Cellular compartmentation in plant nutrition : The selective cytoplasm and the promiscuous vacuole, in : :"Advances in Plant Nutrition", Vol. 2, P. B. Tinker and A. Lauchli, eds., Praeger, New-York.

Marigo, J., Bouyssou, H., and Boudet, A. M., 1986, Accumulation des ions nitrate et malate dans les cellules de Catharanthus roseus et incidence sur le pH vacuolaire, Physiol. Vég., 24:15.

Marin, B. P., 1985, "Biochemistry and Function of Vacuolar Adenosine-triphosphatase in Fungi and Plants", Springer-Verlag, Berlin, Heidelberg, New-York, and Tokyo.

Perry, C. A., Leigh, R. A., Tomos, A. D., and Hall, J. L., 1987, Osmotic factors affecting the mobilisation of sucrose from vacuoles of red beet storage root tissue, in : "Plant Vacuoles. Their Importance in Solute Compartmentation and Their Applications in Biotechnology", B. Marin, ed., Plenum Publishing Corporation, New-York.

Poole, R. J., Briskin, D. P., Ktratky, Z., and Johnstone, R. M., 1984, Density gradient localization of plasma membrane and tonoplast from storage tissue of growing and dormant red beet. Characterization of proton-transport and ATPase in tonoplast vesicles, Plant Physiol., 74:549.

Rea, P. A., and Poole, R. J., 1985, Proton-translocating inorganic pyrophosphatase in red beet (Beta vulgaris L.) tonoplast vesicles, Plant Physiol., 77:46.

Rea, P. A., Griffith, C. J., and Sanders, D., 1987 a, A third-category and fourth-category H^{+}-phosphohydrolase at the tonoplast, in : "Plant Vacuoles. Theit Importance in Solute Compartmentation and Their Applications in Biotechnology", B. Marin, ed., Plenum Publishing Corporation, New-York.

Rea, P. A., Griffith, C. J., and Sanders, D., 1987 b, Differential susceptibilities of tonoplast ATPase and PPase to irreversible inhibition by chaotropic anions, in : "Plant Vacuoles. Their Importance in Solute Compartmentation and Their Applications in Biotechnology", B. Marin, ed., Plenum Publishing Corporation, New-York.

Roberts, J. K., 1984, Study of plant metabolism in vivo using NMR spectroscopy, Annu. Rev. Plant Physiol., 35:375.

Smith, F. A., and Raven, J. A., 1979, Intracellular pH and its regulation, Annu. Rev. Plant Physiol., 30:289.

Smyth, D. A., and Black, C. C., 1984, Measurement of the pyrophosphate content of plant tissues, Plant Physiol., 75:862.

Steingröver, E., Ratering, P., and Siesling, J., 1986, Daily changes in uptake, reduction and storage of nitrate in spinach grown at low light intensity, Physiol. Plant., 66:550.

Sze, H., 1985, H^{+}-translocating ATPases : Advances using membrane vesicles, Annu. Rev. Plant Physiol., 36:175.

Thom, M., Leigh, R. A., and Maretzki, A., 1986, Evidence for the involvement of a UDP-glucose-dependent group translocator in sucrose uptake into vacuoles of storage root of red beet, Planta, 167:410.

Thom, M., and Maretzki, A., 1985, Group translocation as a mechanism for sucrose transfer into vacuoles from sugarcane cells. Proc. Natl. Acad. Sci. U.S.A., 82:4697.

Tomos, A. D., Leigh, R. A., Shaw, C. A., and Wyn-Jones, R. G., 1984, A comparison of methods for measuring turgor pressures and osmotic pressures of cells of red beet storage root tissue, J. Expt. Bot., 35:1675.

Wagner, G. J., and Siegelman, H. W., 1975, Large-scale isolation of intact vacuoles and chloroplasts from protoplasts of mature plant tissues, Science, 190:1298.

Walker, R. R., and Leigh, R. A., 1981 a, Characterization of a salt-stimulated ATPase activity associated with vacuoles from storage roots of red beet (Beta vulgaris L.), Planta, 153:140.

Walker, R. R., and Leigh, R. A., 1981 b, Mg^{2+}-dependent, cation-stimulated inorganic pyrophosphatase associated with vacuoles from storage roots of red beet (Beta vulgaris L.), Planta, 153:150.

Wang, Y., Leigh, R. A., Kaestner, K. H., and Sze, H., 1986, Electrogenic H^{+}-pumping pyrophosphatase in tonoplast of oat roots, Plant Physiol., 81:497.

Wyn-Jones, R. G., Brady, C. J., and Speirs, J., 1979, Ionic and osmotic relations in plant cells, in : "Recent Advances in the Biochemistry of Cereals", D. L. Laidman and R. G. Wyn-Jones, eds., Academic Press, London..

ANALYSIS OF THE ELECTRICAL FLUCTUATIONS

IN RED BEET (BETA VULGARIS L.) VACUOLES

Jean-Paul Lassalles, Joël Alexandre
and Michel Thellier

Laboratoire sur les Echanges Cellulaires, Unité Associée CNRS
No. 203, Faculté des Sciences, Boite Postale No. 67, F-76.130
Mont-Saint-Aignan, France

INTRODUCTION

Like animal cells, plant cells possess ionic pumps, such as ATPases, and ionic channels. The plant ATPases have been mainly studied by use of biochemical methods. Membrane fractions enriched with plasmalemma or with tonoplast have been prepared, and vesicles reconstituted from them. ATPases activities have been described, both on the plasmalemma and on the tonoplast. These activities are increased by MgATP. In the higher plants, the plasmalemma pump is strongly inhibited by vanadate, while the proton pump of the tonoplast is inhibited by NO_3^- (Leigh, 1983; Marre and Ballarin-Denti, 1985). Besides the specific ionic effects, the ATPase activities are sensitive to the ionic strength of the bathing medium, which affects the surface electrostatic interactions in the membrane (Thibaud et al., 1984; Gibrat et al., 1985). In contrast with the active transports, the biochemical study of the passive ionic transports of the plant cells has practically not begun.

Electrophysiological methods can help the study of the transport properties of plant cells. Recently, the "patch clamp" technique has revealed the existence of K^+ channels in guard-cell protoplasts of Vicia faba (Schroeder et al., 1984). However, patch clamping implies perfusing the interior of the cell with an artificial solution. To study the response of a cell under more natural conditions, "fluctuation analysis" thus may be better adapted. We have used this technique first with entire Acer cells (Alexandre et al., 1985), then with isolated vacuoles of red beet (Alexandre et al., 1986).

MEMBRANE NOISE ANALYSIS : BASIC PRINCIPLES

Statement of the Problem

Consider a model membrane bearing a single type of ionic channel (a K^+ channel, for instance) with a specific conductance, γ. One can study the ionic intensity, I(t), as a function of the time, t, through decreasing values of the surface area of the membrane : S, S/2, S/4, etc.. (Fig. 1). If the ionic channels are distributed at random, it is clear that, globally, I(t) will be proportional to the surface area of membrane under study. However, in the details, the phenomenon will depend on the properties of the channels. If the channels are always open (Fig. 1-a), and

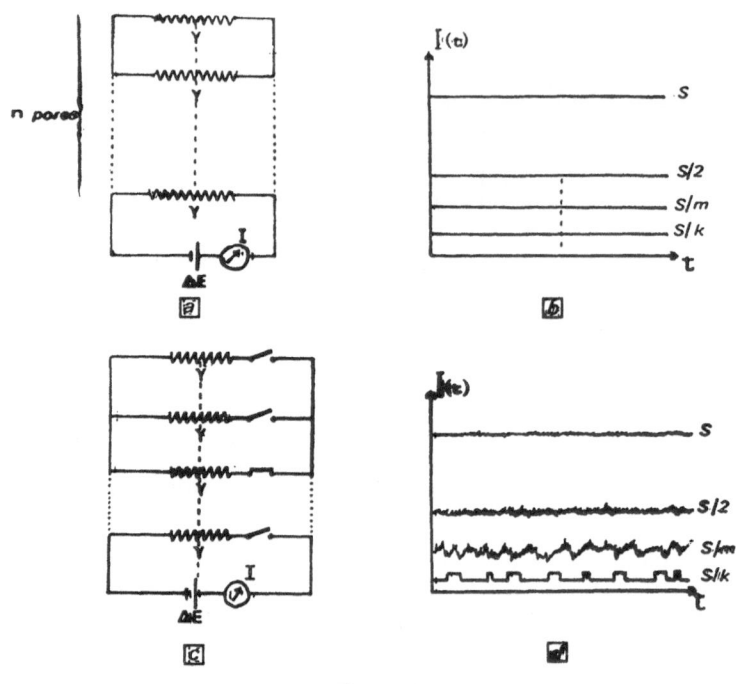

Figure 1

Theoretical behavior of a model membrane with ionic channels :

(a) electrical model of a membrane with permanently open channels, and (b) time-course of the intensity through decreasing surface areas of this membrane; (c) electrical model of a membrane with transient ionic channels (1 channel has been represented open, and 3 closed, on this graph), and (d) time-course of the intensity through decreasing areas of this membrane. Symbols : t : time, ΔE : overall force acting on the ions, I(t) : electric intensity corresponding to the ionic flux, γ : conductance of a single ionic channel, S, S/2, .. S/k : decreasing values of the surface area of membrane.

if the force (i.e. the electrochemical potential difference), ΔE, acting on the ion movements, is constant, then the representative curves, I(t), will be horizontal straight lines as long as the smallest surface area of membrane under study, S/k, contains at least 1 channel (Fig. 1-b). Now, if the channels open and close at random (Fig. 1-c). For each value of the surface area of membrane, I(t) will fluctuate around a mean value. Moreover, the smaller the surface area of membrane under study, the greater will be the relative importance of the fluctuations with regard to the mean. For a membrane area small enough to contain only 1 channel, one can observe only small square signals corresponding to the channel being open or closed (Fig. 1-d).

The "patch-clamp" techniques apply to situations when only one channel, or, at the most, a few channels are considered at a time. "Patch-clamp" thus provides direct information on the characteristics of individual channels. However, i) the technical difficulties for achieving "patch-clamp" with plant systems are still very cumbersome, and ii) the devices necessary for performing "patch-clamp", especially the computing ones, are quite elaborate and expensive. The "fluctuation analysis" endeavours to decode the fluctuations of I(t), as obtained when large membrane areas, containing many ion channels, are studied, using ordinary microelectrodes:

$$I(t) = n(t)\, \mathcal{V}\, \Delta E \qquad (1)$$

where n(t) is the number of channels open at time t. Studying the passive permeabilities of a membrane by use of " fluctuation analysis" is straightforward only when there is a single type of ionic channel in this membrane, which is generally not the case with real biological membranes. In most cases, "patch-clamp" will then be better adapted than "fluctuation analysis" to study the passive transports through a membrane.

On the contrary, "fluctuation analysis" will generally be more useful than "patch-clamp" to study the properties of active transport of a membrane. The reason is that an active pump (e.g. an ATPase) exchanges only a very small number of ions at a time. The corresponding fluctuation is thus much smaller than that corresponding to the opening of a channel (which results in exchanging hundreds or thousands of ions at a time). The individual signals corresponding to the pump functioning are thus far below the sensitivity of the "patch-clamp" techniques. However, the value of ΔE will depend on the activity of the pump. Hence according to eq. (1), one can link the intensity of the fluctuations to the activity of the pump. This link can be either direct (ATPase function) or indirect (ionic fluxes in channels), or both.

Principle of Using Fourier's Transform

When a great number of channels are involved, I(t) is very complex and the direct observation of the corresponding graph is practically not interpretable. Fourier's transform associates to a wide class of functions of t, x(t), an image function, X(f), of the frequency, f :

$$X(f) = \int_{-\infty}^{+\infty} x(t)\, e^{-j2\pi f t}\, dt \qquad (2)$$

where j is the basis of the complex numbers. For instance, when x(t) is a sine function, $x(t) = a \cos 2\pi f_0 t$, X(f) is a function which is equal to zero everywhere except at two points, $f = \pm f_0$, where it is represented by a δ-function of Dirac. When x(t) is a deterministic finite function, eq. (2) is reduced to the well known series of Fourier : x(t) can then be represented by a sum of sine functions, as follows $x(t) = \sum a_n \sin n\, wt$, and the spectrum of X(f), i.e. the a_n's, is shown to be discrete (n = 1, 2, 3, etc .). A square signal (⎍⎍⎍) can be discribed in this way by the superposition of a basic sine function and of its harmonics. Using eq. (2), one can also represent "infinite" functions. A special case of the latter is that of the function of Dirac, $\delta_t(t)$ (⌐⌐), where δ is zero everywhere, except at time $\bar{t}$ when it is infinite : in this case, X(f) keeps a constant value (white noise).

The electric functions, I(t) or V(t), as we are interested in, are different from the preceding ones that they are not deterministic, but stochastic. When the probabilities, both of the first order variables (V, I) and of the second order ones (P = VI), are constant (i.e. when the probabilities for a channel to open or close do not depend on time), one may compute a power function, S(f), from the Fourier transforms, $X_T(f)$, of n samples of the electric variable, I(t) or V(t), considered over periods of time, T :

$$S(f) = \lim \left\{ \langle |X_T(f)|^2 \rangle / T \right\} , T \to \infty \quad (3)$$

where symbol $\langle \ \rangle$ characterizes the mean. Fig. 2 represents $S_T(f)$ (Fig. 2-a) for a current I(t) (Fig. 2-b) through a membrane with a single type of channels. The channel is much easier to be characterized in Fig. 2-a than in Fig. 2-b. In a log/log system of coordinates, the values on the right part of the curve representing S(f) can be adjusted to a straight line with a slope of -2, and those on the left to a horizontal line. The abscissa, f_c, of the point where these two lines intersect each other ("cut off" frequency) is related to the mean time, τ, of opening of the channels by relation

$$f_c = 1/2 \pi \tau \quad (4)$$

The plateau value is characteristic of the mean number of channels open at a time. The curve in Fig. 2-a is termed a "Lorentzian".

MATERIAL AND METHODS

Plant Material

The preparation of the isolated vacuoles, and the general conditions of experiments are identical to those which have been explained previously (Alexandre et al., 1985 and 1986).

The Experimental Device For Obtaining $S_V(f)$

The electric variable in eq. (1) is I(t). However, performing the direct measurement of I(t) through the tonoplast would require to "voltage clamp" the isolated vacuole. This could be achieved i) either by impaling a second, voltage-clamping electrode, different from the measuring electrode, or ii) by using a switched "voltage clamp" mode with a single microelectrode. In the latter case current pulses are used to clamp the vacuole to the chosen voltage, and the membrane voltage is sampled after each pulse. The process is repeated several thousands of times per second ("switching" frequency). When neither of these two methods is available one can also derive $S_I(f)$ from $S_V(f)$, as follows.

The microelectrode inserted in the vacuole under study is connected to a high pass filter preventing fluctuations below 0.1 Hz to be recorded, then to a RC low pass filter deleting the very high frequencies. Due to the high pass filter only the flutuations, $V(t) - \bar{V}$, but not the potential itself, V(t), are recorded. Moreover the signal is filtered through an "active" low pass filter (Kemo type Butterworth 48 db) with an adjustable cut-off frequency, f. The analog signal, $V(t) - \bar{V}$, has to be stored, sampled and treated by Fourier transform. Sampling a continuous curve

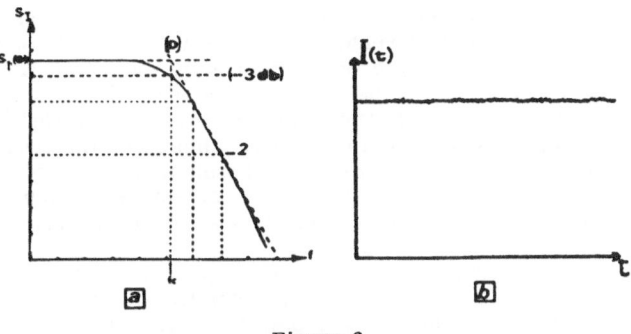

Figure 2

Typical representation (i.e. "Lorentzian")
of (a) the power intensity spectrum, $S_I(f)$, for (b) the current $I(t)$,
when the membrane contains a single type of transient channel

Both $S_T(f)$ and f are represented according to logarithmic scales; f_c is the cut-off
frequency.

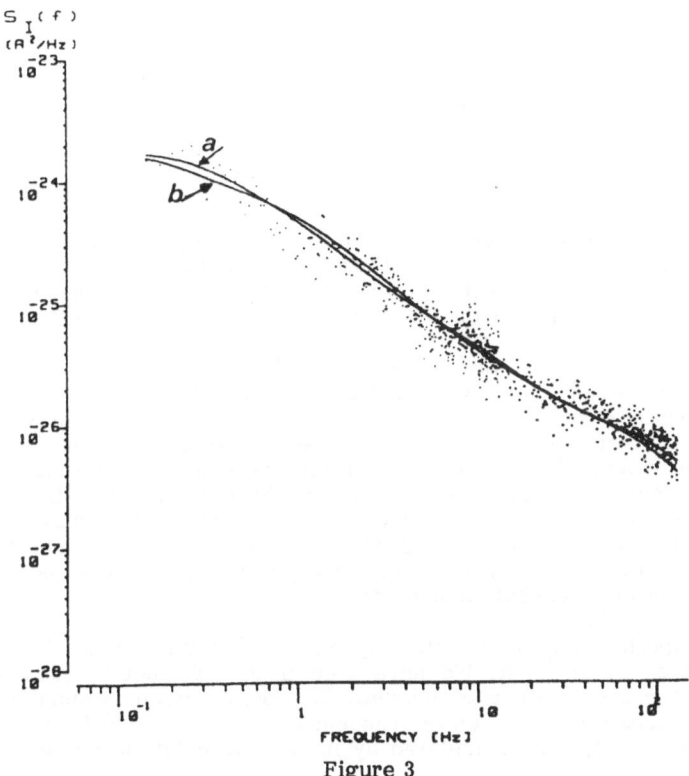

Figure 3

Spectrum of the intensity fluctuations as recorded in a red beet vacuole,
in the presence of 5 mM MgATP in the external medium

The dots correspond to the experimental measurements, and the full lines correspond
to the adjustements of two different combinations of Lorentzians to the experimental
points. The details about the two sets of Lorentzians are given in Fig. 4.

means that this curve is replaced by a discrete set of points. In order not to loose the information contained in the analog signal, it is necessary to sample it at a frequency, f_s, at least twice (in the present case, it was three times) as large as the maximum frequency existing in the signal. Moreover, it is now possible to shorten the time of computation by using the "Fast Fourier Transform" algorithm. This implies using 2^n values for each function, $x_T(t)$, sampled at the frequency f_s during time T. In the present calculations we have chosen to take $2^n = 1024$. In each of our experiments, we have used three successive cut-off frequencies, $f_1^{☆} = 10$ Hz, $F_2^{☆} = 100$ Hz, and $f_3^{☆} = 1000$ Hz, corresponding to three frequencies of signal sampling, $f_{m1} = 30$ Hz, $f_{m2} = 300$ Hz and $f_{m3} = 3000$ Hz, respectively. The corresponding times of measurement, $T = 1024/f_m$, were thus of the order of $T_1 = 34$ s, $T_2 = 3.4$ s and $T_3 = 0.34$ s. From the Fourier transform, $X_V(f)$, of the original signal, $V(t) - \overline{V}$, one finally computed the power function, $S_V(f)$, by use of equation (3).

$$S_v(f) = \langle |X(t)|^2 \rangle / T \qquad (5)$$

The mean value in eq. (5) was obtained from 17 samples.

Obtaining $S_I(f)$ from $S_V(f)$

The current power spectra, $S_I(f)$, are related to the voltage power spectra, $S_V(f)$, by the magnitude of the frequency-dependent membrane impedance, $|Z(f)|$ (De Felice, 1981) :

$$S_I(f) = S_V(f) / |Z(f)|^2 \qquad (6)$$

Hence, $S_I(f)$ can be derived from $S_V(f)$ when it is possible to estimate the value of the corresponding $|Z(f)|$. Several methods are available to measure $\|Z(f)\|$. They are all based on using the response of the vacuole to some external stimulus such as a white noise, a δ-function of Dirac, a Heaviside step, a square wave, a sine wave, etc.. For instance, applying eq. (6) to a current with a white spectrum, $S_T(f) = 1$, provides an immediate way of obtaining $|Z(f)|^2$ from $S_V(f)$ measurements. If $I(t)$ is a square wave signal with a frequency equal to f_o, then $S_I(f)$ can be computed, giving a discrete set of values whose frequencies, f, are such that f $= n f_o$ (n = 1, 2, 3, etc ..). Recording $S_V(f)$ makes it possible to calculate $|Z(f)|$ for these particular values of f. Making use of this method requires that the imposed square wave of current creates electric potentials, both, i) much larger than the electric fluctuations of the vacuole membrane (in order that these fluctuations do not disturb the measurements), and ii) much smaller than the mean membrane potential, $\overline{V}$ (in order to ensure that performing the measurements does not cause any possible potential dependent channels to open.

In our experiments, the injected square-wave of current usually had a peak to peak amplitude of 10^{-10} A. We have used three different frequencies, 0.4, 4, and 40 Hz, of the square wave of current, in order to obtain a number of values of $\|Z(f)\|$ large enough in the range of frequencies of 0.1 to 1000 Hz. For each value of the frequency, $\|Z(f)\|$ was calculated by linear interpolation between the values thus obtained.

At the lowest frequency, the capacitive effect practically does not contribute to the current transport. The impedance then becomes equivalent to the resistance. The tonoplast resistance can thus be estimated by the difference in the low frequency values of the impedances recorded inside and outside the vacuole.

REMINDER OF PREVIOUS RESULTS

A first series of results, which have already been published (Alexandre et al., 1986), can be briefly reminded. Using classical microelectrodes (filled with 3 M KCl), osmotic-like artifacts were frequently observed in the vacuoles. The vacuolar potential was none the less estimated to be in the order of 10 ± 2 mV. Using buffered microelectrodes (filled with a mixture of 0.6 M KCl, 0.6 M sorbitol, and 0.05 M Tris pH 7.7), the osmotic artifacts were no longer observed, but the vacuolar potential changes to a negative value of the order of $- 7 \pm 2$ mV. Removing the buffer from the electrode restored positive values for the potential. Whatever the type of electrode, the specific resistance of the tonoplast was always in the order of 4.1 ± 0.3 kΩ cm^2. Additions of nitrate or of MgATP to the bathing medium of the vacuoles have significantly decreased, or enhanced, respectively, the noise spectrum. This is an indication that noise analysis can be used to detect ATPase (or related enzyme) activities in vacuoles.

ORIGINAL DATA

When an ATPase is responsible for an H$^+$ influx in the vacuole, the proton-motive force thus induced causes the various ions present to tend to move through the ionic channels of the tonoplast. The noise spectrum recorded after activation of a vacuolar ATPase thus corresponds to the combination of the contributions of the various channels of the tonoplast. One might think of decomposing the total noise spectrum in a series of Lorentzians, in order to obtain the characteristics of each individual type of channel. Unfortunately, such a decomposition generally has not a unique solution. For instance, Fig. 3 and Fig. 4 show that a given overall vacuolar noise-spectrum can be described equally well by the contributions of 5 (Fig. 4-a) or of 4 (Fig. 4-b) different types of channels. Moreover, such interpretations neglect the possibility that not only the ionic channels, but also the pump might contribute to the overall noise spectrum.

In order to check the possible existence of channels, we have used vacuoles where the ATPase activities were blocked by NO$_3^-$ (Fig. 5). Moreover, we have either injected (curve a), or not (curve b), an electric current, through a microelectrode, into this vacuole, thus stimulating an active cation influx. In 5 out of 7 similar experiments, the noise spectrum could be adjusted fairly well to a single Lorentzian (Fig. 5-a), with the values of the cut-off frequency, f_c, in the range of 1.5 to 2 Hz. This corresponds to an mean time of opening, , for the corresponding channel, in the order of 80 to 100 ms. Channels with τ -values of this order of magnitude have already been observed in plant cells (Moran et al., 1985). The ion involved in this process is not known. Injecting an electric current with a microelectrode is likely to induce changes of the vacuolar ionic concentrations (H$^+$, K$^+$, Cl$^-$) in a manner different from that corresponding to the operation of a proton ATPase. By modifying the intensity of the injected current and for the ionic content of the microelectrode, and by using susbstances such as TEA (tetra-ethyl-ammonium) or DIDS (4,4'-diisothiocyanostilbene-2,2'-disulfonate) we aim to detect the possible existence of channels with shorter   values, and to identify which ions are transported by the channels.

CONCLUSION

The analysis of the "electrical noise" is a powerful method for i) measuring the membrane resistance and the force (i.e. the activity of the "pumps") driving the transmembrane ion movements, ii) determining characteristics of passive ionic channels under favourable circumstances, and iii) testing the effect of various substances on these membrane parameters. This might be especially relevant to industrial application, since it might allow one to test rapidly if a chemical considered as a possible pesticide really exerts some action on plant cell membranes.

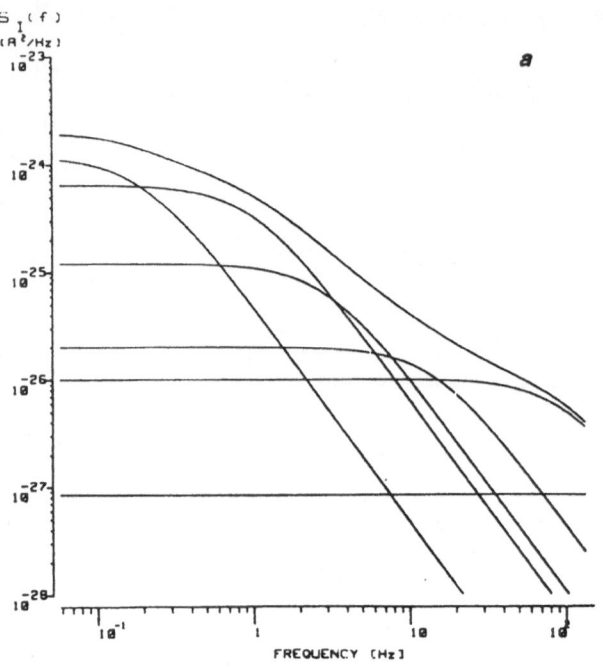

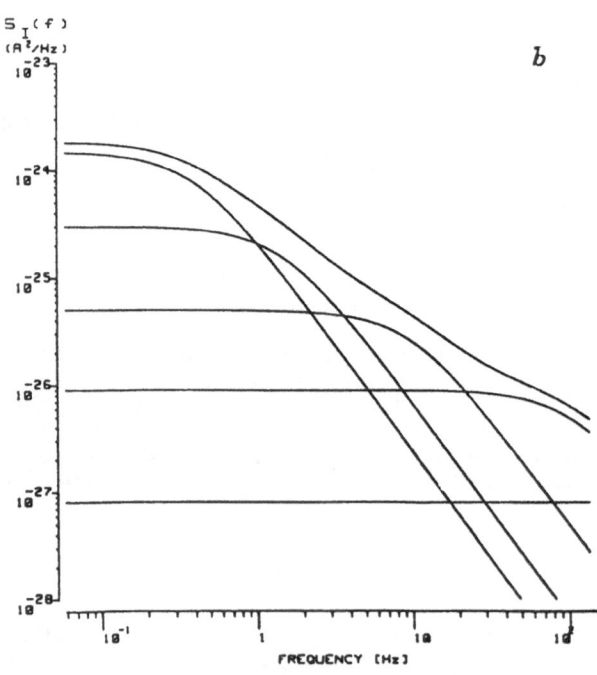

Figure 4

Adjusting the experimental data in Fig. 3
by the combination of the electronic noise and a set
of (a) five Lorentzians (of cut–off frequencies 0.2, 1, 3, 15 and 100 Hz),
or (b) four Lorentzians (of cut–off frequencies 0.4, 1.5, 10 and 80 Hz)

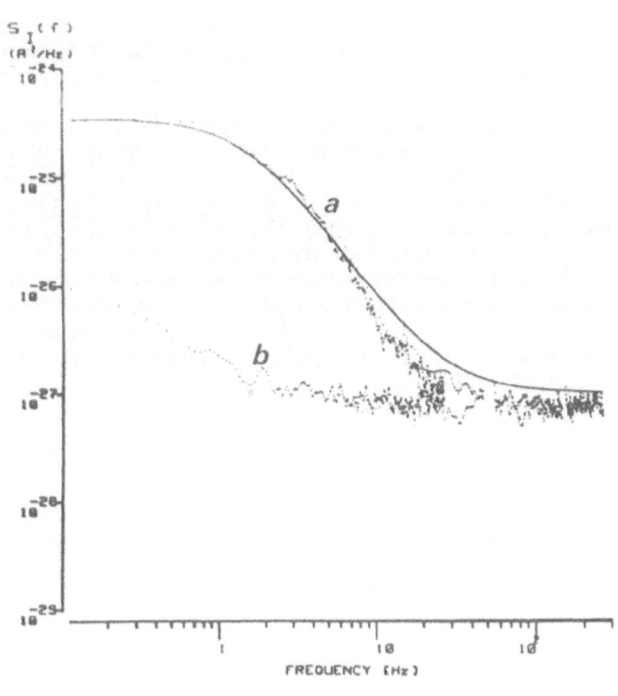

Figure 5

Noise spectrum recorded in a red beet vacuole,
with 50 mM KNO_3^- in the bathing medium,
in the presence (a), or in the absence (b) of an electric current of $2\ 10^{-10}$ A
from the outside to the inside of the vacuole

The full line corresponds to the adjustment to a Lorentzian with a cut-off frequency,
f_c, of 2 Hz.

ACKNOWLEDGEMENTS

This work was supported by grants from the CNRS (RCP No. 08726, and ARI Interface Chimie-Biologie No. 910500).

REFERENCES

Alexandre, J., Lassalles, J. P., and Thellier, M., 1985, Voltage noise in Acer pseudoplatanus cells, Plant Physiol., 79:526.

Alexandre, J., Lassalles, J. P., and Thellier, M., 1986, Electrical noise measurements on red beet vacuoles : Another way to detect the ATPase activity, Plant Physiol., in press.

De Felice, L. J., 1981, "Introduction to Membrane Noise", Plenum Press, New-York.

Gibrat, R., Barbier-Brygoo, H., Guern, J., and Grignon, C., 1985, Membrane potential difference of isolated plant vacuoles : Positive or negative ?, Biochim. Biophys. Acta, 819:206.

Leigh, R. A., 1983, Methods, progress and potential for the use of isolated vacuoles in studies of solute transport in higher plant cells, Physiol. Plant., 57:390.

Marrè, E., and Ballarin-Denti, A., 1985, The proton pumps of the plasmalemma and the tonoplast of higher plants, J. Bioenergetics Biomembranes, 17:1.

Moran, E., Ehrenstein, G., Iwasa, K., Bare, C., and Mischke, C., 1985, Ion channels in the plasmalemma of carrot root protoplasts, Biophys. J., 47:140a.

Schroeder, J. I., Hedrich, R., and Fernandez, J. M., 1984, Potassium selective single channels in guard cell protoplasts of Vicia faba, Nature, 312:361.

Thibaud, J. B., Romieu, C., Gibrat, R., Grouzis, J. P., and Grignon. C., 1984, Local ionic environment of plant membranes : Effect on membrane function, Z. Pflanzenphysiol., 114:207.

PURIFICATION AND PROPERTIES OF α-MANNOSIDASE

FROM VACUOLAR MEMBRANES OF YEAST SACCHAROMYCES CEREVISIAE

Tohru Yoshihisa, Yoshinori Ohsumi
and Yasuhiro Anraku

Department of Biology, Faculty of Science
University of Tokyo
Hongo, Bunkyo-ku 113, Tokyo, Japan

INTRODUCTION

Vacuoles of the yeast, Saccharomyces cerevisiae, play important roles in storage of amino-acids and other metabolites (Dürr et al., 1979; Hubber-Wälchi et al., 1979). They contain many hydrolases and contribute to digestive activities in the vacuolar compartments (Wiemken et al., 1979). α-mannosidase is one of such hydrolases and another marker enzyme of the vacuolar membranes (Van Der Wilden et al., 1973) than H^+-translocating ATPase (Kakinuma et al., 1981). Therefore, it is a suitable marker to study the functions and biogenesis of the vacuoles. We have succeeded in solubilizing and purifying the enzyme to near homogeneity from commercial baker's yeast and found that the purified enzyme fraction contained two isoforms. In this report, we will describe a method for solubilization and purification of yeast α-mannosidase and some properties of the two isoforms.

MATERIALS AND METHODS

Materials

Commercial baker's was purchased from Chuetsu Yeast Co. (Tokyo). 4-methyl-umbelliferyl-α-D-mannopyranoside was purchased from Koch-Light Ltd., Suffolk. The detergent octylpolyoxyethylene was kindly provided by Prof. J. P. Rosenbusch, Biozentrum, Basel University, Schwitzerland. An FPLC system and an HPLC system were purchased from Pharmacia and Toyo Soda, Tokyo, Japan, respectively.

Solubilization of α-mannosidase from vacuolar membranes

Vacuolar membranes used in solubilization experiments were prepared as described in Ohsumi and Anraku (1981) with some modifications. Solubilization mixtures (A) contained 2mg/ml protein of the vacuolar membranes, 10 mM Tris-Mes, pH 6.9, Triton X-100 or β-octylglucoside at a concentration of 0.5, 1.0, 2.0 or 4.0% (w/v), and either 5 mM $MgCl_2$ or 1 mM EDTA. Solubilization mixtures (B) contained 1.7 mg/ml protein of the membranes and 10 mM Tris-Mes, pH 6.9, Tris-Mes, pH 8.0, glycine-NaOH, pH 9.0, glycine-NaOH, pH 10.0, or Na_2CO_3, pH 11.0. Solubilized activities were recovered as supernatants by centrifugation at 180,000 x g for 40 min.

Purification of α-mannosidase

A total membrane fraction was obtained as a 3,800 - 180,000 x g fraction from a spheroplast lysate of baker's yeast in 10 mM Tris-Mes, pH 6.9, 1 mM EDTA, 0.6 M sorbitol, 0.5 mM PMSF. The membranes were solubilized by the selective extraction method (see Results). Solubilized α-mannosidase was maintained in 10 mM Tris-Mes, pH 9.0, 10% (w/v) sucrose, 0.2% (w/v) Triton X-100 or 1% (w/v) octylpolyoxyethylene and fractionated by a series of column chromatographies with 1rst DEAE-Sepharose column, a Q-Sepharose Fast Flow column, a hydroxyl-apatite column, 2nd DEAE-Sepharose column and a Sepharose CL-6B column. The final fraction was stored at - 80°C.

Separation of two isoforms of α-mannosidase

The final fraction obtained after Sepharose CL-6B column was analyzed further with a Mono Q anion exchange FPLC system. α-mannosidase activity was eluted with a linear gradient of 0 - 0.4 M NaCl in 10 mM Tris-Mes, pH 9.0, 10% (w/v) sucrose, 1% (w/v) octylpolyoxyethylene at a flow rate of 1 ml/min.

Fractionation of 73 K and 107 K polypeptides under denaturing conditions

The 73 K-rich fractions and 107 K-rich fractions in the Mono Q FPLC were concentrated by trichloracetic acid precipitation and denatured in SDS-PAGE buffer, separately. Each sample was run on a TSK-GEL SW 3000 gel filtration HPLC system at a flow rate of 0.5 ml/min.

Analytical Methods

The activity of α-mannosidase was assayed with a fluorogenic substrate 4-methylumbelliferyl-α-D-mannopyranoside (Mead et al., 1955) using 50 mM Mes-Tris, pH 6.0, as a buffer. One unit of α-mannosidase represented the amount of enzyme that catalyzed the release of 1 μmole of 4-methylumbelliferone per 1 min. SDS-PAGE and non denaturing PAGE were performed as described in Laemmli (1970) and in Davis (1964), respectively. Immunoblotting was carried out by the method of Towbin et al. (1979).

RESULTS

Solubilization of α-mannosidase

α-mannosidase was solubilized from vacuolar membranes with salt, Triton X-100, β-octylglucoside or alkaline media. Extraction with 10 mM Na_2CO_3 was most effective. 73% activity was solubilized with this method. β-octylglucoside was rather effective than Triton X-100 at a concentration above 2% (w/v) in the presence of 5 mM $MgCl_2$. Addition of 1 mM EDTA instead of 5 mM $MgCl_2$ reduced the effectiveness of these detergents. These suggested that the enzyme may be either a peripheral membrane protein or an ecto-type integral membrane protein.

From these type of results, we adopted a selective extraction method : first, the membranes were solubilized by 1% (w/v) Triton X-100 in the presence of 1 mM EDTA, and next, the α-mannosidase activity recovered in a pellet by centrifugation at 180,000 x g for 40 min was extracted with 10 mM Na_2CO_3. About 60 - 70% of the activity was routinely solubilized.

Purification of α-mannosidase

The total membrane fraction was prepared from a spheroplast lysate of commercial baker's yeast cells and α-mannosidase was solubilized from this fraction by the method described above. Alkaline extract was buffered and chromatographed by a set of columns listed in "Materials and Methods". In the chromatographies, the pH of the buffer (pH 9.0) and the presence of 0.2% (w/v) Triton X-100 or 1% (w/v) octylpolyoxyethylene were essential for keeping the enzyme in a soluble state. The final fraction (0.8 mg protein) was obtained from 400 g yeast cells and it contained α-mannosidase activity purified 4,300-fold with a yield of 5% (Table 1). SDS-PAGE patterns of samples at various stages of the purification told us that there exist three major polypeptides copurified in proportion to increase in specific activity of the enzyme, of which molecular weights were 107,000, 73,000 and 31,000, respectively. The α-mannosidase in the final fraction appeared as a single band of low mobility on non-denaturing PAGE (Davis, 1964) in the presence of 0.1% (w/v) Triton X-100, when examined both by histochemical staining and by Coomasie Brilliant Blue staining.

Table 1

Purification of α-mannosidase
from Baker's Yeast

Purification Stage	Protein (mg)	Enzyme Activity (Unit)	Specific Activity (Unit/mg)	Yield %	Purity fold
Spheroplast Lysate	68600	95.6	0.00139	100	1
Total Membrane Fraction	23900	63.6	0.00266	66.5	1.9
Triton X-100 Extract	8690	60.1	0.00692	62.9	5.0
Alkaline Extract	7730	50.5	0.00653	52.8	4.7
1st DEAE-Sepharose	1170	42.0	0.0359	43.9	25.8
Q-Sepharose Fast Flow	366	35.2	0.0962	36.8	69.2
Hydroxylapatite Bio-Gel HT	15.2	17.2	1.13	18.0	813
2nd DEAE-Sepharose	7.6	12.2	1.61	12.8	1160
Sepharose CL-6B	0.8	4.80	6.00	5.0	4320

Two Isoforms of α-mannosidase

To confirm the polypeptide composition of α-mannosidase, we analyzed further the final fraction using a Mono Q anion exchange FPLC system and then found that the activity was fractionated into two isoforms : one (Form 1) contained 73 K and 31 K polypeptides and the other (Form 2) contained 107 K polypeptide. In order to study the relation of the two isoforms, the 73 K polypeptide and 107 K polypeptide were purified with a Toyo TSK-GEL SW 3000 gel filtration HPLC system under denaturing conditions and the purified polypeptides were compared proteinchemically and immunochemically.

First, we compared the one dimensional peptide maps of the two polypeptides. The polypeptides were digested with Staphylococcus aureus V_8 protease and the peptide fragments generated were subjected to SDS–PAGE with 14% polyacrylamide gel. The peptide maps of the two polypeptides were almost identical. This suggested that the two polypeptides have homologous amino-acid sequences.

Next, we prepared anti-73 K antisera and anti-107 K antisera by immunizing rabbits by the method (Sigel et al., 1983) and the cross-reactivity with the purified polypeptides were examined by immunoblotting. As expected, the anti-73 K antiserum and the anti-107 K antiserum recognized both of the polypeptides. This result strongly supported the idea that the two polypeptides are closely related at the molecular level.

We then found that only 73 K polypeptide was detected in crude cell lysates of baker's yeast cells and laboratory strain X2180-1A cells by immunoblotting both with the anti-73 K antiserum and with the anti-107 K antiserum. Therfore, we think that Form 1 is the major isoform of α-mannosidase in Saccharomyces cerevisiae and that Form 2 is either a minor isoform or a precursor of Form 1.

Properties of the purified α-mannosidase

The α-mannosidase in the peak fraction of the Sepharose CL-6B column showed a single K_m value of 140 μM for 4-methylumbelliferyl-α-D-mannopyranoside and 48 μM for p-nitrophenyl-α-D-mannopyranoside (pNP-Man) at the optimum pH of 6.0. These properties are consistent with the previous report of Opheim (1978) studied about a crude cell lysate. Several divalent cations, such as Co^{2+}, Zn^{2+}, did not affect the activity of the enzyme but Cu^{2+} strongly inhibited the activity. 10 μM $CuCl_2$ inhibited the activity 96%.

DISCUSSION

We purified α-mannosidase of the yeast vacuolar membranes and showed that the enzyme is in two isoforms, one of which contains the 73 K and 31 K polypeptides and the other of which does the 107 K polypeptides. Recently, Jelinek-Kelly et al. (1985) demonstrated two α-mannosidases (Fraction I and Fraction II) in Saccharomyces cerevisiae : Fraction I can hydrolyze pNP-Man and is not solubilized with Triton X-100, and Fraction II can not hydrolyze pNP-Man and is solubilized with Triton X-100. We considered that the α-mannosidase purified in this report is identical with the fraction I of this α-mannosidase, judging from the substrate specificity and solubilization condition and that it is a stable marker of the vacuolar membranes. We assume that the Fraction II of this α-mannosidase which hydrolyzes a specific mannose residue of core oligosaccharide on rat liver glycoproteins will be localized on the ER or Golgi membranes from its function.

The relationship of the two isoforms in their biosynthesis is yet obscure but we obtained the anti-α-mannosidase antisera. The antisera are useful to analyze the subject in detail and, furthermore, to study the intracellular transport pathways of the vacuolar membrane proteins.

REFERENCES

Davis, B. J., 1964, Disc electrophoresis. II. Method and application to human serum proteins, Ann. N. Y. Acad. Sci., 121:404.
Dürr, M. M., Urech, K., Boller, T., Wiemken, A., Schwencke, J., and Nagy, M., 1979, Sequestration of arginine by polyphosphate in vacuoles of yeast (Saccharomyces cerevisiae), Arch. Microbiol., 121:169.
Hubber-Wälchi, V., and A., Wiemken, 1979, Differential extraction of soluble pools from the cytosol and the vacuoles of yeast (Candida utilis) using DEAE dextran, Arch. Microbiol., 120:141.

Jelinek-Kelly, S., Akiyama, T., Saunier, B., Tkacz, J.S., and Herscovics, A., 1985, Characterization of a specific α-mannosidase involved in oligosaccharide processing in Saccharomyces cerevisiae, J. Biol. Chem., 260:2253.

Kakinuma, Y., Ohsumi, Y., and Anraku, Y., 1981, Properties of H^+-translocating adenosine-triphosphate in vacuolar membranes of Saccharomyces cerevisiae, J. Biol. Chem., 256:10859.

Laemmli, U. K., 1970, Cleavage of structure proteins during the assembly of the head of bacteriophage T_4, Nature, 227:680.

Mead, J. A. R., Smith, J. N., and Williams, R. T., 1955, The biosynthesis of the glucuronides of umbelliferone and 4-methylumbelliferone and their use in in fluorimetric determination of β-glucuronidase, Biochem. J., 61:569.

Ohsumi, Y., and Anraku, Y., 1981, Active transport of basic amino-acids driven by a protonmotive force in vacuolar membrane vesicles of Saccharomyces cerevisiae, J. Biol. Chem., 256: 2079.

Opheim, D. J., 1978, -mannosidase of Saccharomyces cerevisiae. Characterization and modulation of activity, Biochim. Biophys. Acta, 524:121.

Sigel, M. B., Sinha, Y. N., and Vanderlaan, W. P., 1983, Production of antibodies in inoculation into lymph nodes, Methods In Enzymology, Vol. 93, Academic Press, New-York.

Towbin, H., Staehelin, T., and Gordon, J., 1979, Electrophoretic transfer of proteins from polyacrylamide gels to nitrocellulose sheets : Procedure and some applications, Proc. Natl. Acad. Sci. U.S.A., 76:4350.

Van Der Wilden, W., Matile, P., Schellenberg, M., Meyer, J., and Wiemken, A., 1973, Vacuolar membranes : Isolation from yeast cells, Z. Naturforsch., 28c:416.

Wiemken, A., Schellenberg, M., and Urech, K., 1979, Vacuoles : The sole compartment of sole digestive enzymes in yeast (Saccharomyces cerevisiae), Arch. Microbiol., 123:23.

SIMULTANEOUS KINETIC MEASUREMENTS OF ADENINE NUCLEOTIDES

IN ISOLATED VACUOLES BY HPLC TECHNIQUE

Max Hill, Alain Dupaix, Pierre Volfin,
Armen Kurkdjian[*] and Bernard Arrio

Institut de Biochimie, Batiment 432, Université de Paris-Sud,
Centre d'Orsay, F-91405-Orsay; and [*]Laboratoire de Physiologie
Cellulaire Végétale, C.N.R.S., F-91.190-Gif-sur-Yvette,
France

INTRODUCTION

Numerous substances are contained in plant vacuoles. Some of them are highly concentrated like sugar and organic acids which may be representative of the accumulation processes occurring in vacuoles. There are many data in the literature suggesting that transport is energy-dependent and linked to protons pumping by a tonoplast ATPase (for a review see Sze, 1985). However, in a complex system like plant vacuoles, the hydrolysis of ATP may be catalyzed not only by ATPases but also by other phosphorolytic enzymes which are not involved in an energization process of the membrane. Thus, the estimation of ATPases activities is an important and difficult point, since the results from many usual titration methods are more or less specific of the ATP hydrolysis into ADP (Fiske and Subbarow, 1925; Strehler, 1965; Adam, 1965). Whether the titration is specific or not for ATP, the data show different aspects of the adenine nucleotide hydrolysis catalyzed by tonoplast enzymes (Dupaix et al., 1986; Hill et al., 1986). It is possible to remove this uncertainty by simultaneously monitoring the evolvement of the adenine nucleotides in the presence of purified tonoplast vesicles by the suitable reverse phase HPLC technique described below.

MATERIALS AND METHODS

HPLC Technique

Chromatographic Instrumentation. The HPLC were carried out with two Gilson model 303 pumps controlled by a gradient manager 702 steered by an Apple II-e computer, a dynamic mixer and a Rheodyne injection valve 7125 with a 20 microliters filling loop. Each sample was manually injected with a 10 microliters Hamilton syringe. Nucleotides were detected at 259 nm and absorbance values were monitored with a Beckman model 165 variable wavelength detector.

Columns and mobile phases. All separations were performed at room temperature. The columns were protected by a guard column Brownlee RP18 (7 micrometers, 3.2 mm x 1.5 mm). The eluents were degassed and filtered through a 0.5 micrometer Millipore filter. Different columns and mobile phases were tested (Hill et al., 1986). The columns were equilibrated for 15 min with the eluents before the first injection and they were used all the day without regeneration. When sample

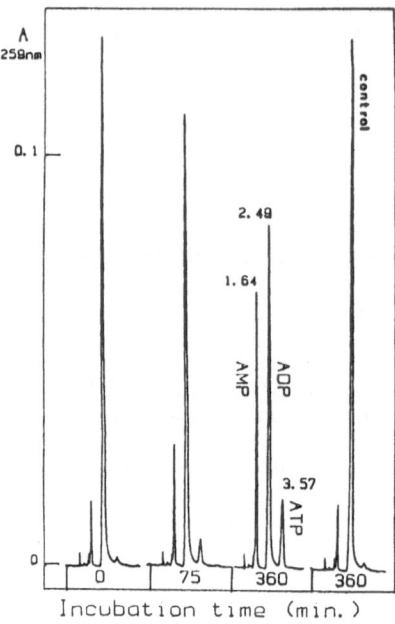

Figure 1

Sample chromatograms illustrating the hydrolysis of ADP
for various incubation times
in the presence of 15 µg tonoplast proteins
from Catharanthus roseus

Chromatographic conditions : Supelcosil LC-18-DB column (5 µm, 4.6 mm x 15 cm) with a Brownlee R.P.18 guard column (7 µm, 3.2 mm x 15 mm); 0.1 M potassium dihydrogen phosphate pH 7.0, 0.025 M tetra-n-butylammonium hydrogen sulfate containing 18% (v/v) methanol, room temperature, flow rate 2 ml/min. The retention times were 1.64, 2.49, 3.57 min. for AMP, ADP and ATP, respectively.

runs were terminated, the columns were flushed with a linear gradient elution up to 100% methanol in 60 min then, if required, returned to initial conditions in 30 min.

Chemicals and chromatographic standards. Methanol, HPLC grade, was obtained from Carlo Erba. Distilled water was deionized. Sodium dihydrogen phosphate, ammonium molybdate, magnesium chloride, potassium chloride, ammonium acetate, perchloric acid (PCA), tris(hydroxymethyl)-aminometane (Tris), tetra-n-butylammonium hydrogen sulfate, analytical grade, were purchased from Merck. 2-(N-morpholino)ethanesulfonic acid (Mes), adenosine-5'-triphosphate (ATP), adenosine-5'-diphosphate (ADP), adenosine-5'-monophosphate (AMP) were purchased from Sigma Chemicals Co.

Stock solutions. 6 mM ATP, ADP and AMP stock solutions were prepared by dissolving a weighted mass of the dried material in 50 mM Tris-Mes buffer pH 7.5, containing 5 mM $MgCl_2$, 50 mM KCl and 2 mM ammonium molybdate when specified. Nucleotide concentrations were calculated using the molar absorbance coefficient at 259 nm : 15,400 M^{-1} cm^{-1}.

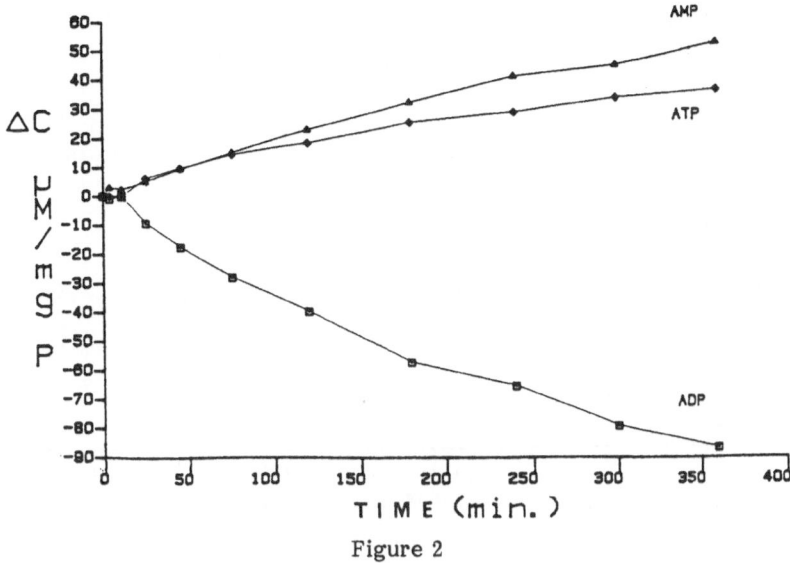

Figure 2

Kinetic analysis of ADP hydrolysis
in the presence of tonoplast proteins
from Catharanthus roseus

Incubation conditions : 3 mM ADP in 0.3 ml Tris-Mes buffer 50 mM at pH 7.5 containing 5 mM $MgCl_2$, 50 mM KCl, 1 mM ammonium molybdate and $15\mu g$ tonoplast proteins, T=20°C.

Protein concentrations. Protein concentrations were estimated by the Bradford modified method (Read and Nothcote, 1981) using bovine serum albumine as a standard.

Tonoplast Purification

Protoplasts and vacuoles from Catharanthus roseus were prepared according to the procedure of Gibrat et al. (1985) which has been modified by us (Hill et al., 1986). The contamination by protoplasts was << 1% as estimated by light microscopy. After a twice dilution in 50 mM Tris-Mes buffer pH 7.5 containing 5 mM $MgCl_2$ and 50 mM KCl, the vacuoles suspensions were homogenized with a manual Potter homogenizer. Tonoplast vesicles were recovered by two successive centrifugations at 100,000 g for 5 hours. The final pellet was suspended in $700\mu l$ of 50 mM Tris-Mes buffer at pH 7.5 containing 5 mM $MgCl_2$ and 50 mM KCl using a small size manual Potter homogenizer (Thomas Phila. U.S.A. No 50925).

Kinetic Assays

In a typical assay, the reaction was started at 20°C by mixing, under magnetic stirring in 1.5 ml conical centrifuge tubes, $150\mu l$ of suspension pellet or supernatant with $150\mu l$ of a 6 mM ATP, ADP or AMP stock solution. In due time, a $20\mu l$ aliquot was withdrawn and mixed with $20\mu l$ of 10% TCA, to stop the reaction. After centrifugation at 15,000 g for 2 min, $20\mu l$ of supernatant were adjusted to pH 5.5 with $20\mu l$ of 4 M ammonium acetate and frozen until HPLC analysis. For the nucleotides separation, $5\mu l$ of the last solution were injected.

129

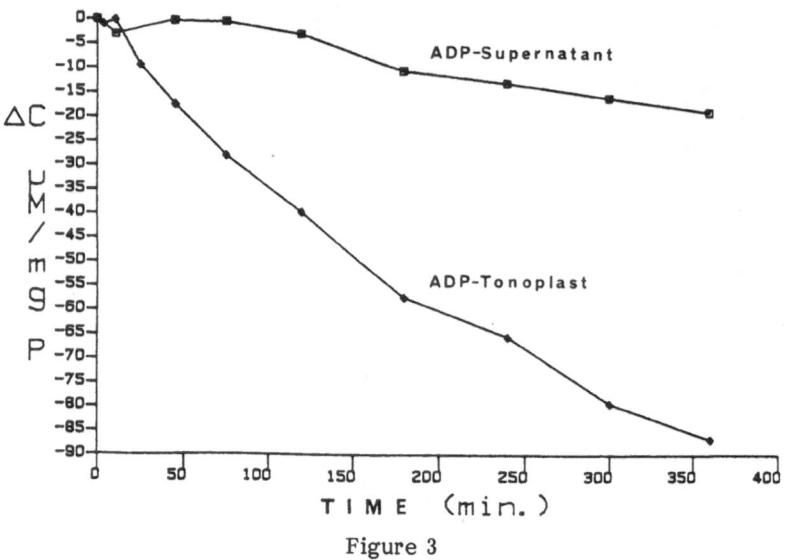

Figure 3

Comparison of tonoplast and 100,000 g supernatant effects
on ADP consumption

Incubation conditions: 3 mM ADP in 0.3 ml Tris-Mes buffer 50 mM at pH 7.5
containing 5 mM $MgCl_2$, 50 mM KCl, 1mM ammonium molybdate, 15 µg tonoplast
or 11µg supernatant proteins from <u>Catharanthus</u> <u>roseus</u>, T=20°C.

RESULTS AND CONCLUSION

An ATP-dependent proton pump has been implicated in the energization
processes of the tonoplast. Thus, data on ATP levels and other adenine nucleotides
would be useful to discriminate between different enzymes possibly located on
the tonoplast. Among the two separation methods of adenine nucleotides on RP18
(ODS) columns that we have tested (Hill et al., 1986), the method using ion-pair
reagent was chosen in this study because it was powerful to separate complex
mixtures of nucleotides and of their degradation products. The ion-pair reagent
was tetra-<u>n</u>-butylammonium hydrogen sulfate. The lowest nucleotide detectable
quantity was about two picomoles. The run to run precision of the retention times
was 1% within a day and 2% day to day. Routinely, 8 to 10 kinetic assays could
be simultaneous performed and 100 samples were daily analyzed.

As an example, the chromatograms, in Fig. 1, present a typical separation
of adenine nucleotides, ADP being the initial substrate in the presence of the
tonoplast proteins from <u>Catharanthus</u> <u>roseus</u>. The retention times were 1.64, 2.49,
3.57 min for AMP, ADP and ADP, respectively. The kinetic analysis is given in
Fig. 2. During the reaction evolvement, AMP appears then an ATP synthesis is
distinctly observed from 0 to 360 min incubation times.

The comparison of tonoplast and 100,000 g supernatant effects on ADP is
shown in Fig. 3. Even in the presence of 1 mM ammonium molybdate, which has
been described as an acid phosphatases inhibitor (D'Auzac, 1975), the ADP substrate

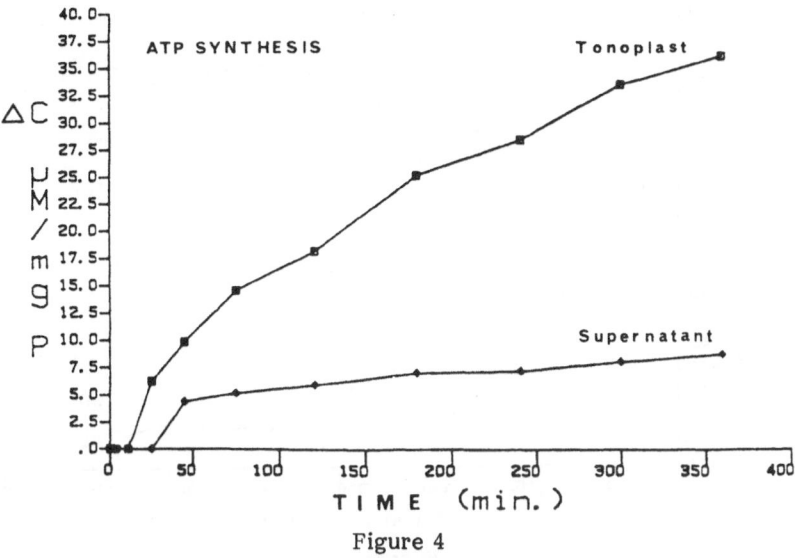

Figure 4

Comparison of Tonoplast and 100,000 g Supernatant Effects
on ATP Formation

Incubation conditions: 3 mM ADP in 0.3 ml Tris-Mes buffer 50 mM at pH 7.5 containing 5 mM $MgCl_2$, 50 mM KCl, 1 mM ammonium molybdate, 15 μg tonoplast or 11 μg supernatant proteins from Catharanthus roseus, T=20°C.

is consumed, with a concomitant ATP synthesis, as shown in Fig. 4. The relative consumption and formation of the products, ADP and ATP respectively, seem to be specific for the tonoplast rather for the 100,000 g supernatant. The ATP synthesis, from ADP and AMP in the presence of tonoplast, is reminiscent of an adenylate kinase activity.

The kinetic analysis of ATP hydrolysis in the presence of tonoplast isolated from Catharanthus roseus is given in Fig. 5. Besides the ATP hydrolysis and the ADP appearance, a slight AMP formation is observed. When AMP was used as the initial substrate, no alteration occurred in the presence of either tonoplast or 100,000 g supernatant (not shown).

All these observations emphasize the fact that complex kinetics bring different results, whether or not they are monitored by specific techniques. For example, the widely used titration of inorganic phosphate, usually at a single reaction time, cannot obviously describe the evolvement of the nucleotides. The use of RP-HPLC techniques for simultaneous kinetic measurements of adenine nucleotides reveals the presence of several enzymes which are involved in hydrolysis and synthesis of these nucleotides. For the moment, few data may be found in the literature which demonstrate such phenomena. Thus, Vigneron et al. (1982) found a nucleoside phosphatase in soluble fraction isolated from Vicia faba. Moreover, by using radiolabelled nucleotides, Thom and Komor (1984) showed that the vacuoles, isolated from sugarcane, exhibited a broad pattern of nucleotide specificity. Alibert et al. (1986) have also found a broad specificity towards nucleotides triphosphate in the case of tonoplast from Acer pseudoplatanus.

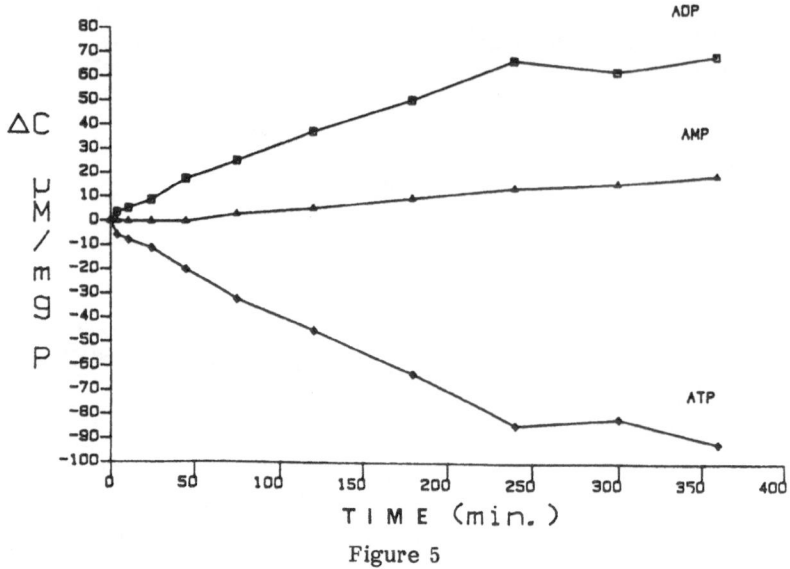

Figure 5

Kinetic analysis of ATP hydrolysis
in the presence of tonoplast proteins
from Catharanthus roseus

Incubation conditions: 3 mM ATP in 0.3 ml Tris-Mes buffer 50 mM at pH 7.5 containing 5 mM $MgCl_2$, 50 mM KCl, 1 mM ammonium molybdate and 15 μg tonoplast proteins, T=20°C.

REFERENCES

Adam, H., 1965, Adenosine-5'-diphosphate and adenosine-5'-monophosphate, in: "Methods of Enzymatic Analysis", H. U. Bergmeyer, Academic Press, New-York.

Alibert, G., Carrasco, A., and Citharel, B., 1986, Biochemical characteristics of tonoplast preparation isolated from Acer pseudoplatanus cell suspension cultures, Physiol. Vég., 24:85.

D'Auzac, J., 1975, Caractérisation d'une ATPase membranaire en présence d'une phosphatase acide dans les les lutoides d'Hevea brasiliensis, Phytochemistry, 14:671.

Dupaix, A., Hill, M., Volfin, P., and Arrio, B., 1986, Réalité de l'ATPase pompe à protons dans le tonoplaste ?, Biochimie, in press.

Fiske, C. H., and Subbarow, Y., 1925, The colorometric determination of phosphorus, J. Biol. Chem., 66:375.

Gibrat, R., Barbier-Brygoo, H., Guern, J., and Grignon, C., 1985, Membrane potential difference of isolated plant vacuoles : positive or negative ? I. Evidence for membrane binding of cationic probes, Biochim. Biophys. Acta, 819:206.

Hill, M., Dupaix, A., Volfin, P., Kurkdjian, A., and Arrio, B., 1986, HPLC technique for simultaneous kinetic measurements of adenine nucleotides in isolated vacuoles, in : "Methods in Enzymology", Academic-Press, New-York, in press.

Read, S. M., and Northcote, D. H., 1981, Minimization of variation in the response to different proteins of the Coomassie blue G dye-binding assay for protein, Anal. Biochem., 116:53.

Strehler, B. L., 1965, Adenosine-5'-triphosphate and creatine phosphate, in : "Methods of Enzymatic Analysis", H. U. Bergmeyer, ed., Academic Press, New-York.

Sze, H., 1985, H$^+$-translocating ATPases: Advances using membrane vesicles, Annu. Rev. Plant Physiol., 36:175.

Thom, M., and Komor, E., 1984, Role of the ATPase of sugarcane vacuoles in energization of the tonoplast, Eur. J. Biochim., 138:93.

Vigneron, P., Coupé, M., and D'Auzac, J., 1982, A soluble ATPase from Vicia faba: In fact a nucleoside phosphatase, Physiol. Vég., 20: 227.

COMPARISON OF TWO METHODS FOR THE ISOLATION OF VACUOLES

IN RELATION TO VACUOLAR ATPase ACTIVITY

Alain Pugin, Françoise Montrichard, Khanh Le-Quôc
and Danièle Le-Quôc

Laboratoire de Biochimie, U.A. No 04531
Faculté des Sciences, Université de Franche-Comté
16, Route de Gray, F-25.030-Besançon, France

INTRODUCTION

Several reports indicate that an anion-sensitive H^+-ATPase insensitive to vanadate occurs in the tonoplast of higher plant cells (Lin et al., 1977; Leigh and Walker, 1980; Admon et al., 1981; Marin et al., 1981; Walker and Leigh, 1981; Aoki and Nishida, 1984; Smith et al., 1984; Mandala and Taiz, 1985). Our unability to measure an ATPase activity in vacuoles of Acer pseudoplatanus cells, prepared with DEAE-dextran and dextran sulfate as described by Alibert et al. (1982) lead us to incriminate the method used for their isolation. This hypothesis was supported by the unstability of the vacuoles out of the Ficoll used to form the density gradient. Therefore a new method for the isolation of vacuoles from Acer pseudoplatanus cells has been established. These vacuoles showed an ATPase activity, the properties of which have been studied and compared to those of other well known tonoplastic ATPases.

MATERIAL AND METHODS

Acer pseudoplatanus cells were cultivated in shaken liquid medium. Protoplasts were obtained as previously described (Pugin et al., 1986) from cells taken out 8 days after the beginning of the culture.

Vacuole Isolation

The protoplasts were submitted to an ultracentrifugation in a discontinuous gradient of Ficoll 400 diluted in a buffer consisted of 12.5 mM Mes, 12.5 mM Tris, 0.7 M mannitol, 1 mM EDTA, pH 7.0. In centrifuge tubes were successively layered : 6 ml of 12% Ficoll, 6 ml of 6% Ficoll containing $12\ 10^6$ protoplasts, 8 ml of 4% Ficoll and 5 ml of buffer. After centrifugation (200,000 x g, 60 min., 4°C) the liberated vacuoles floated to the interphases between 6% and 4% Ficoll (interphase 6/4) and between 4% Ficoll and buffer (interphase 4/0). The harvested vacuoles were diluted 3 times with buffer without EDTA and collected by centrifugation (670 x g, 6 min., 4°C).

Fluorescence Microscopy

Fluorescein diacetate (FDA) and concanavalin A labelled with fluorescein

(Con A) were used to detect respectively cytoplasmic adhesions and the plasma lemma around the vacuoles.

Marker Enzymes Activities

Catalase, glucose-6-phosphate dehydrogenase and cytochrome c oxidase activities were measured according to standard techniques : α-mannosidase was measured according to Boller and Kende (1979); UDP-glucose sterol glucosyltransferase (UDPG-ST) and antimycin A insensitive NADH cytochrome c reductase activities were measured in the membranous fraction of protoplasts and in the preparation of whole vacuoles; the first according to a method described by Hartmann-Bouillon and Benveniste (1978) modified by Chanson et al. (1984) with 17.5 nC_i UDP-^{14}C-glucose per assay, the second according to Schumaker and Sze (1985).

Phosphohydrolases Activities

Phosphatase, pyrophosphatase and ATPase were asayed by measuring the liberated P_i during the hydrolysis of respectively p-NPP, $Na_4P_2O_7$ and ATP in the reactionnal medium: buffer 20 mM Mes, 20 mM Tris, 3 mM $MgSO_4$, 50 mM KCl settled to values of pH in between 5.5 and 10, 3 mM p-NPP, $Na_4P_2O_7$ and ATP with or without 100 - 200 μM ammonium molybdate, with or without 5 μg oligomycin $(10^6$ vacuoles$)^{-1}$; 4 10^5 vacuoles per assay. After 30 min incubation at 37°C, P_i was measured according to the Ames' method (1966).

RESULTS

The ultracentrifugation of protoplasts on density gradient allowed the separation of two populations of vacuoles. They floated at the interphases 6/4 (vacuoles said "heavy") and 4/0 (vacuoles said "light"). Isolated from cells taken 8 days after the beginning of the culture, the "light" vacuoles were 16.50 μm $\pm$ 1.00 ($\pm$ 2 SE) across, the "heavy" vacuoles were 10.71 μm $\pm$ 1.00 ($\pm$ 2 SE) across. The Student test indicated that these two populations were significantly different in size (P 10^{-9}). The populations changed during cells growth. For the same number of protoplasts coming from cells taken out 6, 7, 8, 9 days after the beginning of the culture, the number of large vacuoles increased as the cells grew old while the amount of the small vacuoles decreased.

After eliminating the Ficoll, the efficiency of the isolation method given by the ratio of the total α-mannosidase activity of the harvested vacuoles to the total α-mannosidase activity of the loaded protoplasts was 30% for the large vacuoles and 20% for the small ones. These preparations could be kept during 24 hours at 4°C without having a significative decrease of the number of the vacuoles which kept their ability to accumulate neutral red.

The preparations of vacuoles did not contain any protoplasts but a low ratio of vacuoplasts which was not over 4%. The vacuoles rarely showed cytoplasmic adhesions (Figures 1 and 2). The elimination of the plasmalemma was confirmed by the very low activity of UDPG-ST. The contamination by the other cell compartments did not exceed 7% (Table 1). In presence of ammonium molybdate and oligomycin the ATP hydrolysis by the large vacuoles was the highest near pH 7.5 (Figure 3). At this pH value it was inhibited neither by 100 μM and 200 μM ammonium molybdate nor by oligomycin. So it did not result from unspecific phosphatases or mitochondrial ATPase activities. Inorganic pyrophosphatase was hydrolyzed by the vacuoles but the amount of liberated P_i from ATP by pyrophosphatase was low. The hydrolysis of ATP by these vacuoles resulted from an ATPase.

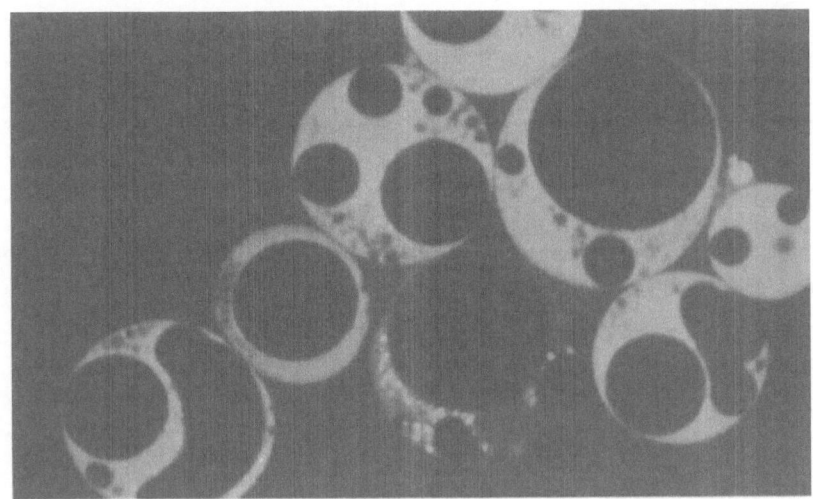

Figure 1

Protoplasts previously incubated with FDA
(x 400)

The vacuoles came out dark in the fluorescent cytoplasm.

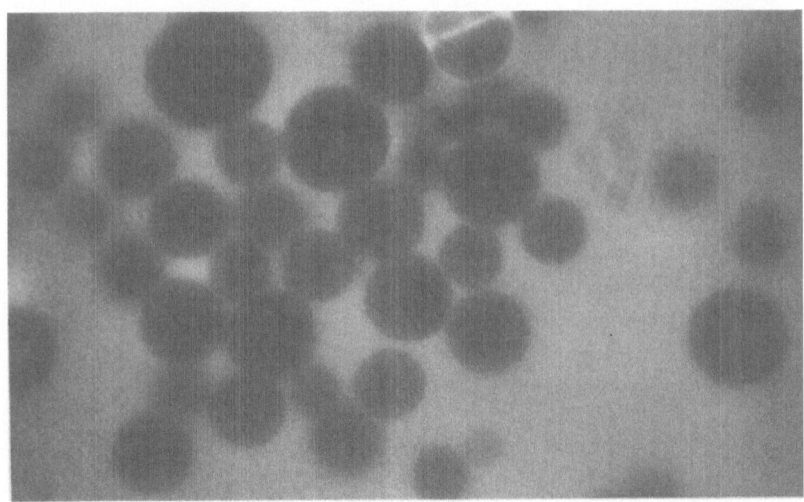

Figure 2

Population of large vacuoles
previously incubated with FDA
(x 400)

One organelle only, probably a vacuoplast showed a cap of fluorescent cytoplasm.

Table 1

Activities of marker enzymes in protoplasts
and in the preparations of large vacuoles
obtained by ultracentrifugation

	Protoplasts	Vacuoles	Vacuoles contamination (in %)[*]
Catalase	8.4×10^3	0.15×10^3	2.80
Glucose-6-P dehydrogenase	60.3	ND	0
Cytochrome c oxidase	17.74	ND	0
UDPG-ST	18.14	0.39	3.1
Antimycin A insensitive NADH–cyt c reductase	1320	57	6.7

[*]Vacuoles contamination was calculated taking into account the number of vacuoles (1.56) per protoplast isolated from 8 day old cells (Alibert et al., 1982).
ND, not detectable.

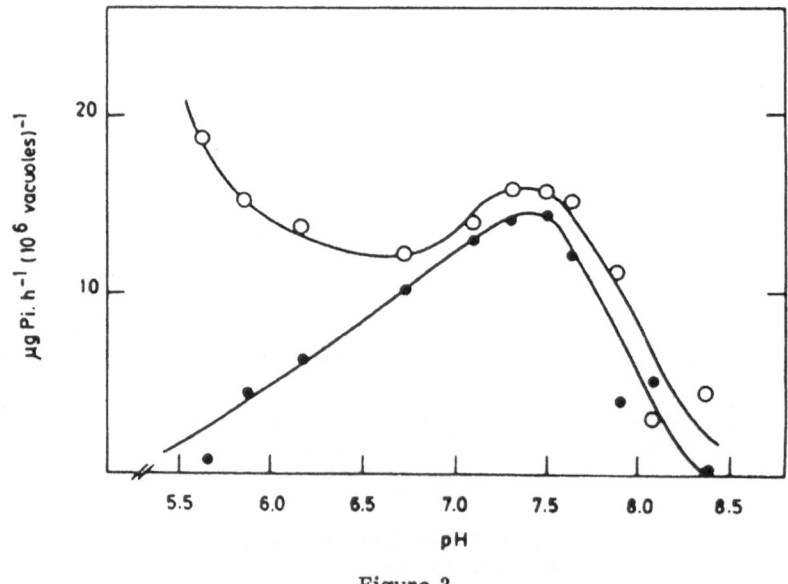

Figure 3

ATP hydrolysis without (○) or with (●) ammonium molybdate
by the vacuoles isolated by ultracentrifugation

Table 2

Some characteristics
of the <u>Acer</u> <u>pseudoplatanus</u> vacuolar ATPase

- Effect of salts (50 mM)

$MgSO_4$ (3 mM)	+	KCl	100
	+	KI	155
	+	KBr	125
	+	KNO_3	110
	+	K_2SO_4 (25 mM)	115
	+	NaCl	135
	+	NH_4Cl	140
$MgSO_4$ (3 mM) + KCl			100
+ KCl	+	KI	82
+ KCl	+	KBr	87
+ KCl	+	KNO_3	70
+ KCl	+	K_2SO_4 (25 mM)	79
+ KCl	+	NaCl	92
+ KCl	+	NH_4Cl	90

- Effect of ionophores ($1\mu M$)

none	100
CCCP	155
Gramicidin	137
Nigericin	142

- Effect of inhibitors (mM)

none	100
DES (0.1)	0
DCCD (0.1)	6
Oligomycin ($1\mu g.ml^{-1}$)	99
Molybdate (0.1)	92
Vanadate (0.05)	21
NEM (1)	42
PCMB (0.1)	17
Mersalyl (0.1)	19

- Substrate specificity[a]

ATP	100
GTP	86
UTP	59
ADP	11
AMP	35
PEP	7
pNPP	27

[a]as a percentage of the activity with ATP as substrate.

139

ATPase activity of vacuoles prepared by ultracentrifugation was inhibited by DEAE-dextran and dextran sulfate. This effect was proportional to the concentration of these two compounds. With 1 mg ml^{-1}, an efficient concentration for the isolation of vacuoles, the inhibition of ATPase activity was respectively 51% and 28% with DEAE-dextran and dextran sulfate.

Some properties of the ATPase are shown in the Table 2. These results concern the large vacuoles. The characteristics of the ATPase of the small vacuoles are now investigated. The effects of various salts were tested. The enzyme required Mg^{2+} for its activity. It was stimulated by Cl$^-$ but an eventual effect of K$^+$ could not be ruled out. In the described conditions of the assay (3 mM MgSO$_4$, 50 mM KCl), the vacuolar ATPase activity was about 30 mol h^{-1} mg^{-1} protein. NO$_3^-$ did not inhibit the enzyme activity but slightly reduced the Cl^{-1} stimulated ATPase activity.

The specificity of phosphohydrolase activity for various phosphate compounds was examined. ATP was the most effective substrate for phosphate release. The weak effect of molybdate suggests that the hydrolysis of GTP and UTP may not be due to the presence of non specific phosphatases. The presence of a separate enzyme hydrolyzing GTP and UTP is under investigation. Three ionophores able to dissipate the H$^+$ gradient were assayed. They both stimulated the ATPase activity. Inhibitors of mitochondrial ATPase (oligomycin) or non specific phosphatases (molybdate) did not inhibit the vacuolar ATPase activity but vanadate did. Both DES and DCCD were also very potent inhibitors. ATPase activity was also greatly affected by thiols reagents and particularly by mercurials (PCMB, mersalyl).

CONCLUSION

. The ultracentrifugation on density gradient of protoplasts from Acer pseudoplatanus cells allowed in one step to liberate and to separate two populations of vacuoles with a rather high yield. It was shown by fluorescence microscopy and assays of marker enzymes that there were only traces of contamination by other organelles, cytoplasmic components, and by plasmalemma in the vacuoles preparations. The vacuoles contained an ATPase with an optimum activity near pH 7.5. This enzymatic activity was inhibited by DEAE-dextran and dextran sulfate, two compounds used in an other method for the isolation of vacuoles. The absence of ATPase activity in the vacuoles prepared with DEAE-dextran and dextran sulfate is most likely due to these chemicals.

The vacuolar ATPase of Acer pseudoplatanus cells presented some similarities with other tonoplastic ATPases : alkaline optimum value, sensitivity to Cl$^-$. The increase in activity induced by protonophores suggests that the enzyme has a function as a proton pump. However, unlike most other vacuolar ATPase, this ATPase is sensitive to vanadate and only weakly inhibited by NO$_3^-$. Our and other results (Doll and Hauer, 1981; Thom et al., 1983; Wagner and Mulready, 1983; Lew and Spanswick, 1984) show that one has to be carefull with the criteria used to detect tonoplastic microsomal fractions.

REFERENCES

Admon, A., Jacoby, B., and Golschmidt, E. E., 1981, Some characteristics of the Mg-ATPase of isolated red beet vacuoles, Plant Science Letters, 22:89.

Alibert, G., Carrasco, A., and Boudet, A. M., 1982, Changes in biochemical composition of vacuoles isolated from Acer pseudoplatanus L. during cell culture, Biochim. Biophys. Acta, 721:22.

Ames, B. N., 1966, Assay of inorganic phosphate, total phosphate and phosphatases, in: "Methods in Enzymology", Vol. 8, E. F. Neufeld and V. Ginsburg, eds., Academic Press, New-York.

Aoki, K., and Nishida, K., 1984, ATPase activity associated with vacuoles and tono-plast vesicles isolated from the CAM plant <u>Kalanchoe daigremontiana</u>, Physiol. Plant., 60:21.

Boller, T. H., and Kende, H., 1979, Hydrolytic enzymes in the central vacuole of plant cells, <u>Plant Physiol.</u>, 63:1123.

Chanson, A., Mc Naughton, E., and Taiz, L., 1984, Evidence for a KCl-stimulated Mg^{2+}-ATPase on the Golgi of corn coleoptiles, <u>Plant Physiol.</u>, 76:498.

Doll, S., and Hauer, R., 1981, Determination of the membrane potential of vacuoles isolated from red-beet storage tissue, <u>Planta</u>, 152:153.

Hartmann-Bouillon, M. A., and Benveniste, P., 1978, Sterol biosynthetic capatibility of purified membrane fractions from maize coleoptiles, <u>Phytochemistry</u>, 17:1037.

Leigh, R. A., and Walker, R. R., 1980, ATPase and acid phosphatase activities asso-ciated with vacuoles isolated from storage roots of red beet (<u>Beta vulgaris</u> L.), <u>Planta</u>, 150:222.

Lew, R. R., and Spanswick, R. M., 1984, Proton pumping activities of soybean (<u>Glyci-cine max</u> L.) root microsomes : Localization and sensitivity to nitrate and vanadate, <u>Plant Science Letters</u>, 36:187.

Lin, W., Wagner, G. J., Siegelman, H. W., and Hind, G., 1977, Membrane-bound ATPase of intact vacuoles and tonoplast isolated from mature plant tissue, <u>Biochim. Biophys. Acta</u>, 465:110.

Mandala, S., and Taiz, L., 1985, Proton transport in isolated vacuoles from corn coleoptiles, <u>Plant Physiol.</u>, 78:104.

Marin, B., Smith, J. A. C., and Luttge, U., 1981, The electrochemical proton gradient and its influence on citrate uptake in tonoplast vesicles of <u>Hevea brasiliensis</u>, <u>Planta</u>, 153:486.

Pugin, A., Montrichard, F., Le-Quôc, F., and Le-Quôc, D., 1986, Incidence of the method for the preparation of vacuoles on the vacuolar ATPase activity of isolated <u>Acer pseudoplatanus</u> cells, <u>Plant Science Letters</u>, in press.

Schumaker, K. S., and Sze, H., 1985, A Ca^{2+}/H^{+} antiport system driven by the proton electrochemical gradient of a tonoplast H^{+}-ATPase from oat roots, <u>Plant Physiol.</u>, 79:1111.

Smith, J. A. C., Uribe, E. G., Ball, E. Heuer, S., and Luttge, U., 1984, Characteriza-tion of the vacuolar ATPase activity of the crassulacean-acid metabolism plant <u>Kalanchoe daigremontiana</u>, Eur. J. Biochem., 141:415.

Thom, M., Willenbrink, J., and Maretzki, A., 1983, Characteristics of ATPase from sugarcane protoplasts and vacuoles membranes, <u>Physiol. Plant.</u>, 58:497.

Wagner, G. J., and Mulready, P., 1983, Characterization and solubilization of nucleo-tide-specific Mg^{2+}-ATPase and Mg^{2+}-pyrophosphatase of tonoplast, <u>Biochim. Biophys. Acta</u>, 728:267.

Walker, R. R., and Leigh, R.A., 1981, Mg^{2+}-dependent cation stimulated inorganic pyrophosphatase associated with vacuoles isolated from storage roots of red beet, <u>Planta</u>, 153:150.

NBD-PC : A TOOL TO STUDY PHOSPHOLIPASE ACTIVITY

IN PROTOPLASTS

Lynne A. Dengler, Magaly Rincon
and Wendy F. Boss

Botany Department
North Carolina State University
Raleigh, North Carolina 27695-7612, U.S.A.

INTRODUCTION

Wild carrot cells can be grown in suspension culture so that they yield fusion permissive or fusogenic protoplasts (Boss et al., 1984). The presence of small vacuoles and high rates of inositol phospholipid metabolism were correlated with the high rates of fusion in these cells (Boss and Grimes, 1985). Increased phosphatidyl-inositol metabolism has been correlated with membrane fusion events such as secretory vesicle fusion with the plasma membrane (Laychock and Putney, 1982) and myoblast fusion during muscle maturation (Wakelam, 1983). Our hypothesis was that the rapid phosphatidylinositol headgroup metabolism found in the fusogenic protoplasts was indicative of a general increase in membrane lipid recycling as opposed to de novo synthesis and that increased lipid metabolism might contribute to the fusogenic state of the protoplasts.

In order to study lipid metabolism in the protoplasts in vivo, we used the fluorescent phospholipid, NBD-PC (1-acyl-2-(N-4-nitrobenzo-2-oxa-1,3-diazole)-aminocaproyl-phosphatidylcholine). NBD-PC has been used to study endocytosis (Sleight and Pagano, 1985) in Chinese hamster V79 lung fibroblasts. In these cells, the fluorescent phospholipid was associated with the plasma membrane at low temperature (2°C). As the cells were warmed to 37°C, NBD-PC was internalized in endosomes and became localized in the Golgi apparatus. In the animal cells, when the NBD-PC was hydrolyzed by the plasma membrane or lysosomal associated-phospholipases to lysophosphatidylcholine and NBD-caproic acid, NBD-caproic acid did not accumulate within the cells. It was found in the surrounding medium. The NBD-caproic acid had either diffused out of the cell or was transported out of the lysosomes and out of the cell. In this study we have used NBD-PC to study phospholipase A_2 activity in the fusogenic carrot protoplasts. With carrot protoplasts, the NBD-caproic acid accumulated in the vacuole.

MATERIALS AND METHODS

Cell Culture

Fusogenic and nonfusogenic carrot (Daucus carota L.) cells were grown in suspension culture as previously described (Boss et al., 1984; Boss and Grimes, 1985). Cells were subcultured every 7 days and used 4 days after transfer. Protoplasts were isolated in 2% Driselase (Plenum Sci. Co.), 0.4 molal sorbitol, 1 mM Mes (pH 5.0) on a rotary shaker (125 rpm) at 25°C for 2 h. Prior to use, the protoplasts were washed twice by centrifugation at 40 g in an osmoticum at 0.45 molal sorbitol, 1 mM Mes (pH 6.0).

Fluorescent Labeling of Protoplasts

C_6-NBD-PC (75 μl of a $CHCl_3$ solution; Avanti Polar Lipids, Inc., U.S.A.) was added to a glass test tube and dried under N_2 to form a thin film. A solution of 0.45 molal sorbitol containing 0.68 mg/ml defatted BSA (Sigma Chem. Co.) was added, and the mixture was vortexed to obtain a clear, yellow solution. To label the protoplasts, 3 ml of the NBD-PC solution was added to a flask containing 20 ml of osmoticum and approximately 0.2 ml settled volume of protoplasts. Prior to microscopic observation, an aliquot of the protoplasts was removed and washed twice in the sorbitol osmoticum. Protoplasts were labeled with NBD-caproic acid in the same manner as with NBD-PC.

Indomethacin was added at the time of addition of NBD-PC to a final concentration of 0.33 mM from a 10 mM stock solution in ETOH. Control experiments with equal volumes of ETOH showed no effect of ETOH on the uptake of distribution of NBD fluorescence. Sodium azide was prepared in the sorbitol osmoticum to a final concentration of 30 mM. The protoplasts were incubated for 20 min with azide prior to the addition of NBD-PC or NBD-caproic acid.

Lipid Extraction

Lipids were extracted according to Folch et al. (1957), dried (in vacuo) and separated on silica gel G (Fischer Sci. Co.) thin-layer plates developed in $CHCl_3$: MeOH:15 N NH_4OH (65:35:5).

Fluorescence Microscopy

A Zeiss IM 35 inverted microscope equipped with a neofluar 63x lens for differential interference contrast optics and epifluorescence optics was used. For fluorescence observations, an exciter filter BP450-495, beam splitter FT510, and barrier filter LP520 were used. Photographs were made using Kodak Kodacolor VR ASA 400 film with a Nikon UFX camera.

RESULTS

When NBD-PC was added to isolated protoplasts which were fusogenic, within 10 min there was a diffuse yellow-green fluorescence in the cytosol. After 1 to 2 h the yellow-green fluorescence was detectable in the vacuole. The fluorescence fluorescence pattern was similar for the nonfusogenic protoplasts.

Uptake of the fluorescence was inhibited by azide. If fusogenic protoplasts were incubated for 20 min with 30 mM azide and exposed to NBD-PC, at 1 min a fluorescent glow was observed on the periphery of the protoplasts. By 10 min, diffuse fluorescence was observed inside, but no fluorescence was ever detected in the vacuole even after 2 h incubation. These data suggested that uptake of NBD-PC was energy dependent. Cold temperature (10°C) also delayed the uptake of NBD-PC; however, internal fluorescence was detectable immediately upon warming.

Lipid analysis of the protoplasts indicated that whenever NBD-fluorescence was associated with the vacuoles, NBD-caproic acid was present. The question arose as to whether the NBD-caproic acid was formed in an endosome or provacuole and deposited in the vacuole by vesicle fusion, or whether it was formed by phospholipases associated with the plasma membrane and sequestered in the vacuole by diffusion or a specific transporter. To determine whether or not there was a phospholipase associated with the plasma membrane which would hydrolyze NBD-PC, the phospholipase A_2 inhibitor indomethacin (0.33 mM) was added. This treatment resulted in NBD-PC fluorescence associated only with the plasma membrane. The outer fluorescent glow persisted for up to 30 min at which time some internal fluorescence was evident. These data indicated that there was a plasma membrane-associated phospholipase A_2 which hydrolyzed NBD-PC. In the absence of the phospholipase inhibitor, this enzyme was contributing to the presence of the NBD-caproic acid in the protoplasts. The plasma membrane-associated phospholipase activity, measured by NBD-PC hydrolysis, was detected in both fusogenic and nonfusogenic protoplasts, and thus, it was not the delineating factor for the fusion permissive state of the plasma membrane.

Further studies were conducted to observe the uptake of exogenously added NBD-caproic acid. When NBD-caproic acid was added to the protoplasts, the fluorescence was seen in the endomembranes and with time, in the vacuole. The pattern of fluorescence was similar to that of the NBD-PC except the vacuolar fluorescence was always yellow-orange compared to the yellow-green seen with the phospholipid. Treatment with azide completely inhibited the uptake of the NBD-caproic acid into the vacuole. It should also be noted that single vacuoles present in the protoplast preparations did not take up either NBD-PC or NBD-caproic acid. The fluorescence associated with these vacuoles always was seen only in the tonoplast.

SUMMARY AND CONCLUSIONS

NBD-PC can be used to study phospholipase A_2 activity in protoplasts but should be used with caution when studying endocytosis in plant protoplasts. The presence of the NBD-caproic acid inside the protoplasts makes it impossible to specifically follow the uptake of the phospholipid. A phospholipase A_2 which will hydrolyze NBD-PC is associated with the plasma membrane of both fusogenic and nonfusogenic protoplasts, and thus, this enzyme does not appear to be the distinguishing factor which renders the protoplasts fusion permissive. NBD-caproic acid, which is formed either by plasma membrane or endomembrane phospholipases is sequestered into the vacuole. This is in contrast to the animal cells where the NBD-caproic acid was found in the surrounding medium, but not in the cells (Sleight and Pagano, 1985). The fact that the vacuolar fluorescence observed when the NBD-caproic acid was added to the protoplasts was a different color than when NBD-PC was added may indicate a difference in concentration or pH or a difference in the mechanism of uptake. Studies are in progress to more clearly delineate the mechanisms of uptake. In any event, whether NBD-caproic acid enters the vacuole via vesicle fusion or via an acid transport mechanism, it is clear that uptake of NBD-caproic acid into the vacuoles is energy dependent.

REFERENCES

Boss, W. F., and Grimes, H. D., 1985, Dynamics of calcium-induced fusion of fuso-genic protoplasts, in : "Beltsville Symposia on Agricultural Research. IX. Frontiers of Membrane Research in Agriculture", J. St. John, P. Jackson, and E. Berlin, eds., Rowman Allanheld Co., Totowa, N.J.

Boss, W. F., Grimes, H. D., and Brightman, A. O., 1984, Calcium-induced fusion of fusogenic wild carrot protoplasts, Protoplasma, 120:209.

Boss, W. F., and Ruesink, A. W., 1979, Isolation and characterization of concanavalin A-labeled plasma membranes from carrot suspension culture cells, Plant Physiol., 65:1005.

Folch, J., Lees, M., and Stanley, G. H. S., 1957, A simple method for isolation and purification of total lipids from animal tissues, J. Biol. Chem., 226:497.

Laychock, S. G., and Putney Jr., J. W., 1982, Roles of phospholipid metabolism in secretory cells, in : "Cellular Regulation of Secretion and Release", P. M. Conn, ed., Academic Press, New-York.

Sleight, R. G., and Pagano, R. E., 1985, Transport of a fluorescent phosphatidyl-choline analog from the plasma membrane to the Golgi apparatus, J. Cell Biol., 99:743.

Wakelam, M. J. O., 1983, Inositol phospholipid metabolism and myoblast fusion, Biochem. J., 214:77.

ON THE EXISTENCE OF A TONOPLAST-BOUND ATPASE

Alain Dupaix, Max Hill, Pierre Volfin
and Bernard Arrio

Institut de Biochimie, Bâtiment 432, Université de Paris-Sud
Centre Universitaire d'Orsay, F-91.405-Orsay, France

INTRODUCTION

In order to account for accumulation of metabolites in plant vacuoles, the existence of a proton-pumping ATPase has been widely invoked in the literature. In a complex system like plant vacuoles, the ATP hydrolysis may be catalyzed not only by ATPases but also by some other enzymes not necessarily implicated in an energization process of the tonoplast.

The methods used for ATP level estimation are more or less specific . The least specific and widely used method consists of inorganic phosphate titration by colorimetric methods derived from Fiske and Subbarow (1925). A little more specific for ATP hydrolysis are the methods using ^{32}P radioactivity, bioluminescent assays, coupled enzyme assays like pyruvate kinase/lactate dehydrogenase system. However, although the coupled enzyme method is suitable for continuous monitoring and permits rapid simultaneous assays for ATPase, it needs some care. Indeed, in all consecutive enzymatic reactions, it must always be verified that the limiting step actually corresponds to the formation or decomposition of the product which is under study, the ATP hydrolysis in the present case. Moreover, all reaction parameters, such as inhibitors, salt concentration, pH changes, and their possible effects on the coupled enzyme assay must be carefully controlled.

Depending on the titration, specificity, one can obtain different results on adenine nucleotides hydrolysis in the presence of tonoplast enzymes. Thus, we have focused our attention in developing a method which allows simultaneous kinetic measurements of these nucleotides in the presence of tonoplast proteins. This method consists of separating and titrating them by reverse-phase high performance liquid chromatography (RP-HPLC).

However, the observation of ATP hydrolysis is not a sufficient criterion to characterize an ATPase. It is also necessary to monitor the pH gradient in relation with ATP hydrolysis. This was performed by the use of quinacrine which is a pH sensitive fluorescent probe.

MATERIAL AND METHODS

All the methods concerning tonoplast purification from Catharanthus roseus, RP-HPLC technique and kinetic measurements have been already described (Hill et al., 1986; Hill et al., 1987).

pH Gradient Measurements

The quinacrine fluorescence quenching kinetics were performed at 20°C on an Eppendorf fluorimeter, equipped with interferential filters at 405 and 530 nm for excitation and emission, respectively. Vacuoles or tonoplast were added to 0.8 ml of 25 mM Hepes–NaOH buffer at pH 7.3 containing 2.5 mM DTT, 10 mM $MgCl_2$ or $MgSO_4$, 50 mM KCl and 0.55 M (for vacuoles) or 0.25 M (for tonoplast) sorbitol. After 15 minutes incubation, the reaction was initiated by addition of 5 - 10 microliters of ATP stock solutions, neutralized at pH 7.3 by solid bis-tris-propane.

RESULTS

RP-HPLC Kinetic Measurements

In Fig. 1, the RP-HPLC separation of the adenine nucleotides obtained during ATP hydrolysis in the presence of tonoplast vesicles isolated from Catharanthus roseus is shown as an illustration of the method. The kinetic analysis of these chromatograms is given in Fig. 2. In addition to ATP hydrolysis and ADP appearance, there is also AMP formation, in spite in the presence of 1 mM ammonium heptamolybdate which has been described as phosphatase inhibitor (D'Auzac, 1975). In Fig. 3, which represents the kinetic analysis of ATP hydrolysis in the presence of 100,000 g supernatant proteins, the inhibition by molybdate is also imcomplete.

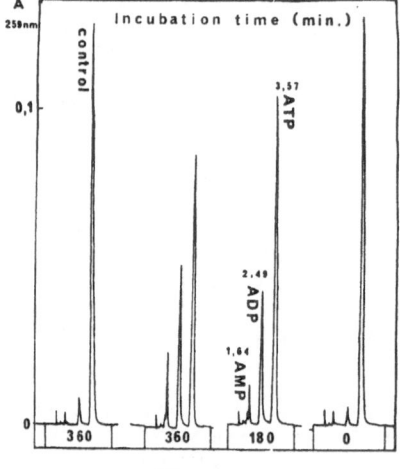

Figure 1

Separation of adenine-nucleotides during ATP hydrolysis
by 15 μg tonoplast protein from Catharanthus roseus

HPLC conditions: Five microliters of incubation medium were injected on a LC-18-DB Supelcosil column (5 μm, 4.6 mm x 15 cm, Supelco); eluents: 0.1 M KH_2PO_4 pH 7, 18% methanol (v/v), 25 mM tetra-n-butylammonium hydrogen sulfate. ATP from Sigma contained 2% ADP.

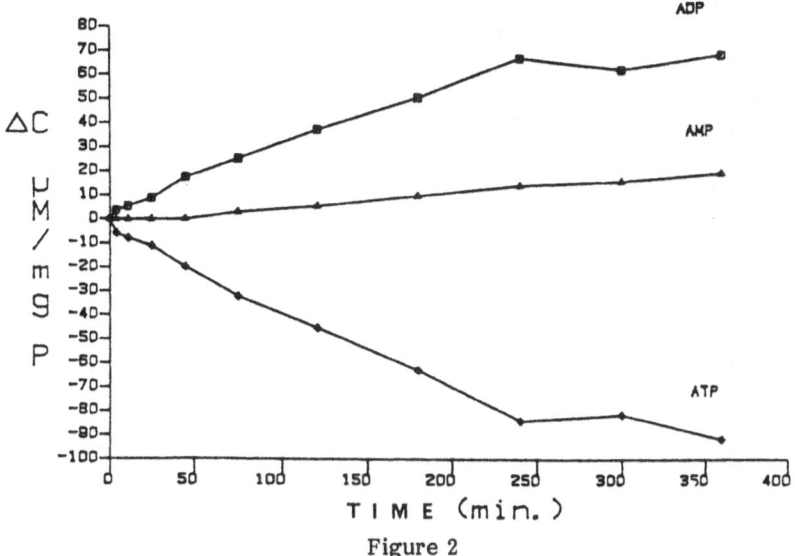

Figure 2

Variation of adenine nucleotide concentration(ΔC, in μM/mg protein)
as a function of time

Incubation conditions: 0.3 ml of 50 mM Tris-Mes buffer at pH 7.5 containing 5 mM $MgCl_2$, 50 mM KCl, 1 mM ammonium heptamolybdate and 15 μg tonoplast protein from <u>Catharanthus</u> <u>roseus</u>; the initial concentration of ATP was 3 mM.

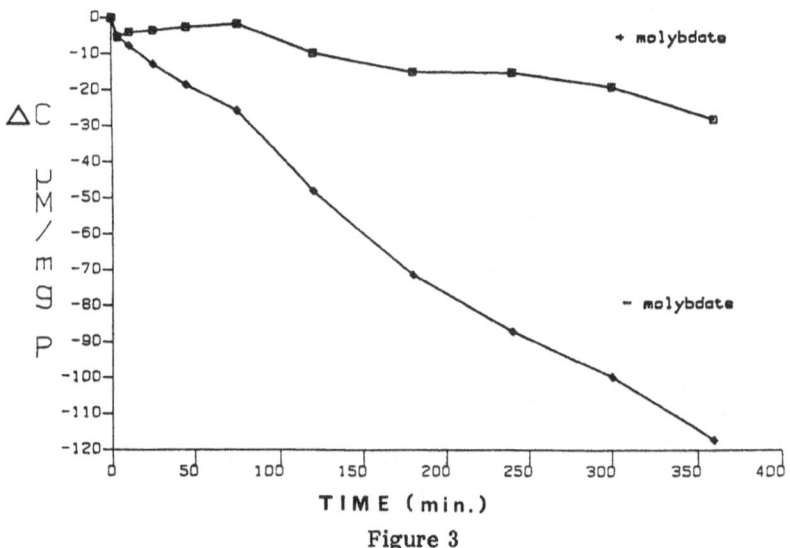

Figure 3

Variation of ATP concentration, measured by HPLC,
in the presence of 11μg protein from 100,000 g supernatant (<u>Catharanthus</u> <u>roseus</u>)
Effect of molybdate

Incubation conditions: 0.3 ml of 50 mM Tris-Mes buffer at pH 7.5 containing 5 mM $MgCl_2$, 50 mM KCl, and 11 μg supernatant protein; the initial concentration of ATP was 3 mM. (□), 1 mM ammonium heptamolybdate; (♦), without ammonium heptamolybdate.

149

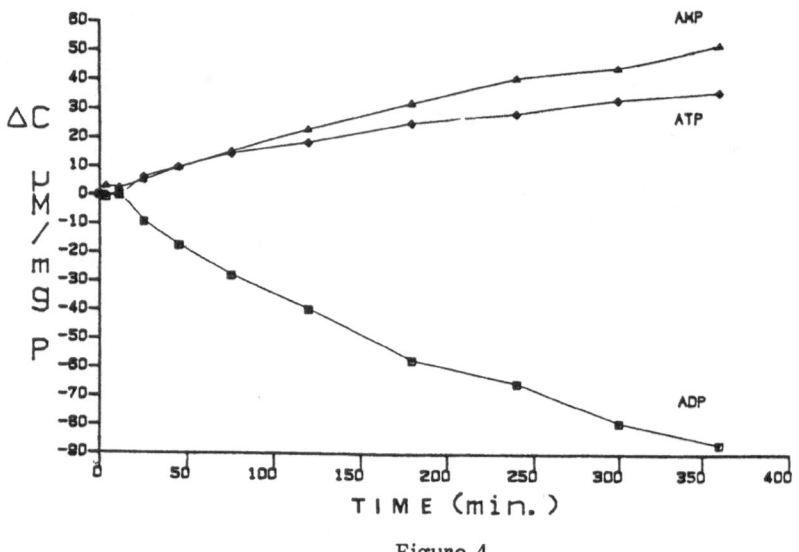

Figure 4

Variation of adenine nucleotide concentration(ΔC, in μM/mg protein)
as a function of time

Incubation conditions: 0.3 ml of 50 mM Tris-Mes buffer at pH 7.5 containing 5 mM
$MgCl_2$, 50 mM KCl, 1 mM ammonium heptamolybdate and 15 μg tonoplast protein
from Catharanthus roseus; the initial concentration of ADP was 3 mM.

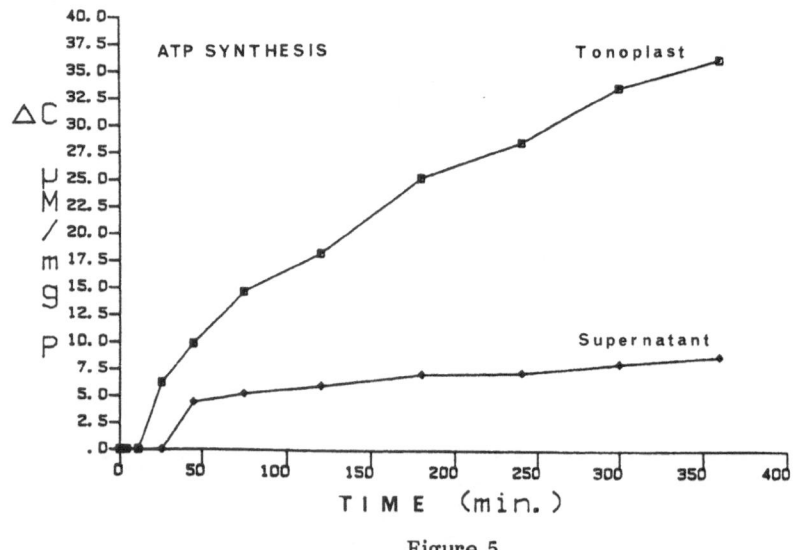

Figure 5

Comparison of the variation of ADP concentration(ΔC, in μM/mg protein)
in the presence of 100,000 g supernatant (11 μg)
or tonoplast (15 μg) protein from Catharanthus roseus

Incubation conditions: 0.3 ml of 50 mM Tris-Mes buffer at pH 7.5 containing 5 mM
$MgCl_2$, 50 mM KCl, 1 mM ammonium heptamolybdate; the initial concentration
of ADP was 3 mM.

The fate of ADP is given in Fig. 4. In the presence of tonoplast proteins, the ADP hydrolysis is accompanied by AMP and ATP formation with a rather good stoichiometry, i.e. 2 ADP→ATP + AMP. The comparison of the effect of tonoplast and 100,000 g supernatant proteins on ATP synthesis seems to indicate that this synthesis is specific for tonoplast, as shown in Fig. 5.

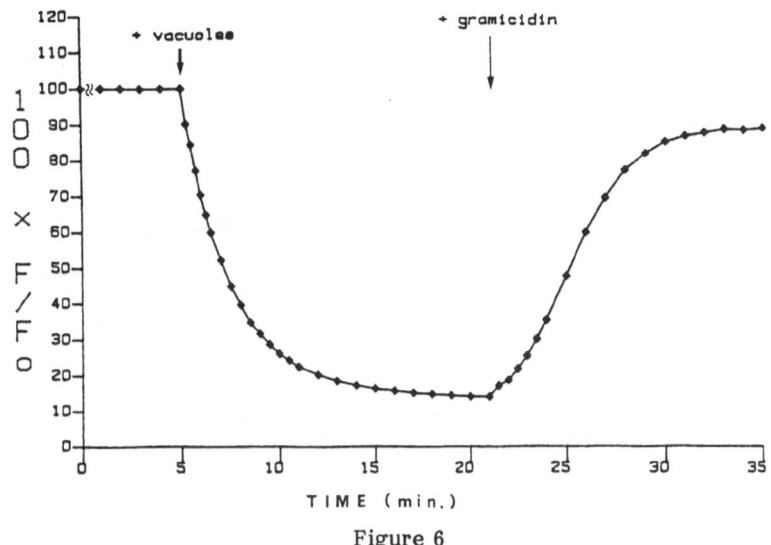

Figure 6

Measurement of pH gradient in vacuoles
isolated from <u>Catharanthus roseus</u>
by quinacrine fluorescence quenching

Experimental conditions : 0.8 ml of 25 mM Hepes-NaOH buffer at pH 7.3 containing 2.5 mM DTT, 10 mM $MgCl_2$, 50 mM KCl, 0.55 M sorbitol, 10 μM quinacrine and 25 μg protein; then 10 μM gramicidin.

Proton Transport

As shown in Fig. 6, quinacrine fluorescence quenching is observed in the presence of isolated vacuoles. This indicates that a pH gradient is formed between intravacuolar and external medium. This quenching can be reversed by adding 10 μM gramicidin. The addition of 5 mM ATP leads to a very slight quinacrine quenching (Fig. 7). The concept of proton translocation linked to a tonoplast-bound ATPase is difficult to reconcile with such a weak pH variation. "<u>A priori</u>", several hypotheses may be proposed as an explanation : i) the pre-existent pH gradient does not allow a new intravacuolar pH decrease; ii) the buffering capacity of the intravacuolar space conceals the pH variation; and iii) during vacuole preparation, a constituent, which was needed for the establishment of a pH gradient, has been lost, regardless of whether this constituent was specific for the tonoplast or consisted of an interaction with the cytosol. Fig. 8 represents the proton transport in tonoplast vesicles. Once again, the pH variation is very weak. Moreover, the activation by KCl described in the literature (Lin et al., 1977; Leigh and Walker, 1980; Walker and Leigh, 1981; Admon et al., 1981) is not clearly shown.

DISCUSSION

Several comments must be made in the light of these results. We have observed by RP-HPLC ATP and ADP hydrolysis and ATP synthesis catalyzed by

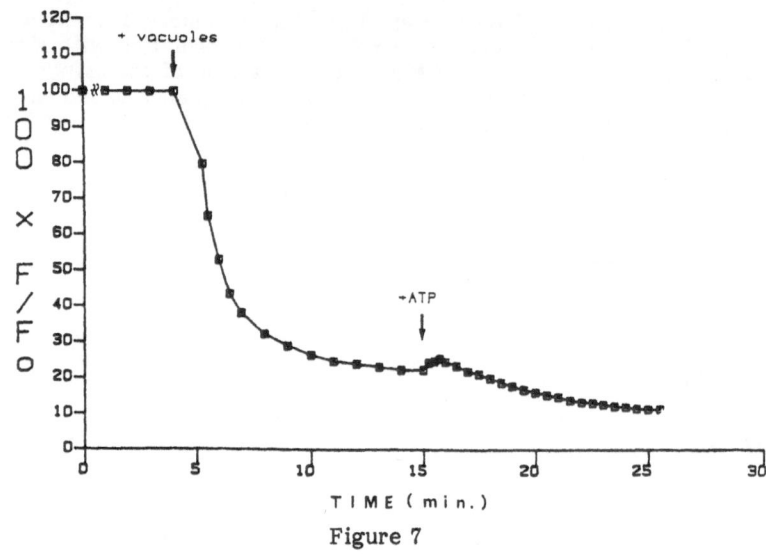

Figure 7

Effect of ATP on pH gradient in vacuoles
isolated from Catharanthus roseus

Experimental conditions: 0.8 ml of 25 mM Hepes-NaOH buffer at pH 7.3 containing 2.5 mM DTT, 10 mM $MgCl_2$, 50 mM KCl, 0.55 M sorbitol, 10 μM quinacrine and 100μg protein; then 5 mM ATP.

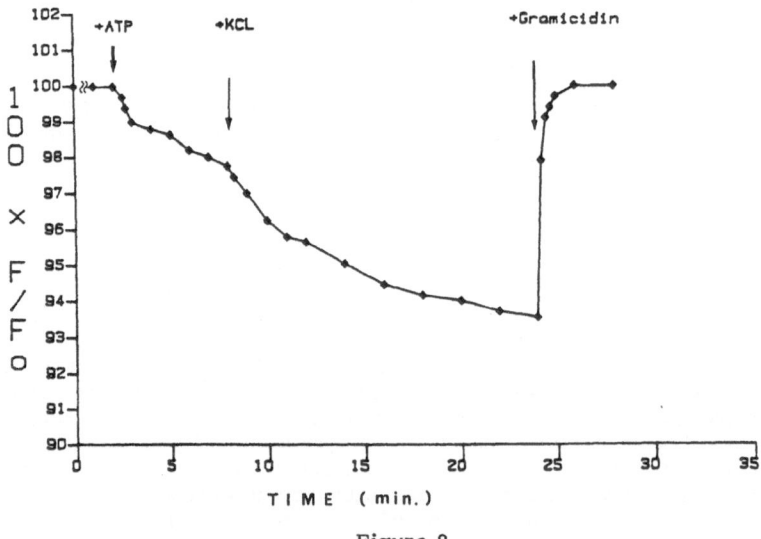

Figure 8

Measurement of pH gradient in tonoplast vesicles
isolated from Catharanthus roseus
by quinacrine fluorescence quenching
Effect of ATP

Experimental conditions: 0.8 ml of 25 mM Hepes-NaOH buffer at pH 7.3 containing 2.5 mM DTT, 10 mM $MgSO_4$, 0.25 M sorbitol, 0.15% BSA, 10 μM quinacrine and 100 μl tonoplast (about 10 μg protein); after 15 minutes incubation, the fluorescence quenching was initiated by adding 5 mM ATP, then 50 mM KCl, and 10μM gramicidin.

tonoplast proteins. The first point, i.e. the ATP hydrolysis, is not an unequivocal proof of a tonoplast-bound ATPase but rather reveals the presence of a 5'-nucleotidase. The ADP utilisation with a concomittant ATP synthesis is reminiscent of an adenylate kinase mechanism. The data found in the literature on tonoplast ATPase were mainly obtained from inorganic phosphate titration and could not be expected to reveal this type of activity. The very low levels of quinacrine fluorescence quenching kinetics, initiated in the presence of vacuoles as well as in the presence of tonoplast vesicles, cannot unequivocally distinguish the existence of a true tonoplast ATPase, in the case of <u>Catharanthus</u> <u>roseus</u>.

In the literature, two main types of experimental methods were used to reveal the existence of this ATPase. The first one consisted of separating and characterizing the different fractions obtained, from tissue homogenates, after density gradient centrifugations. The second one involved vacuole isolation after mechanical cell disruption, in some favourable examples (Leigh and Branton, 1976; Gross and Matile, 1979), or after lysis of protoplasts, which were obtained by enzymatic digestion of excised tissue or of cultured cells.

Concerning the first method, one notices some puzzling term definitions. For Sze (1984 and 1985) and others, microsomes included all the membranes except mitochondrial membranes which were eliminated by 6,000 - 13,000 g centrifugations for ten minutes. Sometimes, all the membranes were used as microsomes (Scherer and Fischer, 1985; Tognoli, 1985; Poole et al., 1984). Finally, other authors used the two definitions interchangeably (DuPont et al., 1982; Bennett and Spanswick, 1983). Moreover, the centrifugation conditions were often incompletely described : either the rotor design and relative centrifugal force were not given or centrifugation times were missing. Thus, it is difficult for the reader to be completely sure whether centrifugation was carried out under isokinetic or isopycnic conditions.

The total vacuolar proteins (cell sap + tonoplast) represent no more than five to ten percent of protoplast proteins. No more than ten percent of vacuolar proteins are from tonoplast origin. Thus, the problem consists in characterizing a membrane fraction which represents 0.5 to 1 percent of the initial proteins loaded on gradients; whose buoyant density is very close to those of Golgi and smooth endoplasmic reticulum and for which a specific biochemical marker has yet to be discovered.

According to some reports in the literature (Walker and Leigh, 1981; Admon et al., 1981), a nitrate-sensitive ATPase is bound to the tonoplast and could be a marker for this membrane (O'Neill et al., 1983); according to other authors, similar activities have been found in Golgi apparatus (Chanson and Taiz, 1985), endoplasmic reticulum or mitochondria (Hager and Helmle, 1981; Wang and Sze, 1985; Randall et al., 1985; Grubmeyer and Spencer, 1979; Churchill et al., 1983). In the case of tomato cells, the proton-pumping ATPase activity, as revealed by orange acridine fluorescence quenching, has been localized in the mitochondrial fraction and not in the microsomal fraction. Nevertheless, it is considered to be bound to the tonoplast (DuPont and De Gracia-Zabala, 1985) ? .. As evidenced by the contradictory results in the literature and taking into account, on the one hand the low percentage of tonoplast compared with other membranes and on the other hand the poor resolving power and possible artifacts of this method (Nagahashi, 1985), the use of density gradient centrifugations for characterization of the tonoplast does not seem to be the best method available at the moment. Moreover, one must be cautious concerning interpretations based on differential effects of inhibitors whose specificities are more or less well established and whose effects may vary with concentrations of inhibitors, salts and proteins.

With regard to the second method, Boller (1982) has published a review which emphasized some gaps in the literature. For instance, the results did not clearly distinguish between the effects of a possible specific ATPase and unspecific phosphatases, the presence or absence of magnesium being the unique criterion during activity measurement (Lin et al., 1977; Saunders, 1979; Doll et al., 1979; Boller and Kende, 1979). Moreover, the purity of vacuoles were not assessed (Admon et al., 1981; Wagner, 1981) or, when biochemical characterization of contamination was made, the percentage observed was sufficient to be responsible for the effects observed on ATP (Leigh and Walker, 1980).

In the authors' view, the existence of a tonoplast-bound ATPase, implicated in accumulation process, is not yet truly demonstrated. We have further presented some arguments which show why we are not convinced by the demonstration of such an activity based on subcellular fractional separation by density gradients. As for the demonstration from tonoplast, isolated from purified vacuoles, the ambiguity is mainly based on difficulty to characterize tonoplast well and to prepare it with a good yield. Our results, observed from tonoplast which was prepared according to general procedure described in the literature, did not formally preclude the existence of a tonoplast-bound ATPase but did demonstrate the presence of other enzyme species which were also responsible for the processing of adenine-nucleotides.

REFERENCES

Admon, A., Jacoby, B., and Goldschmidt, E. E., 1981, Some characteristics of the Mg-ATPase of isolated red beet vacuoles, Plant Science Letters, 22:89.
Bennett, A. B., and Spanswick, R. M., 1983, Optical measurements of ΔpH and $\Delta\Psi$ in corn root membrane vesicles. Kinetic analysis of Cl$^-$ effects on a proton-translocating ATPase, J. Membrane Biol., 71:95.
Boller, T., 1982, Enzymatic equipment of plant vacuoles, Physiol. Vég., 20:247.
Boller, T., and Kende, H., 1979, Hydrolytic enzymes in the central vacuole of plant cell, Plant Physiol., 63:1123.
Chanson, A., and Taiz, L., 1985, Evidence for an ATP-dependent proton pump on the Golgi of corn coleoptiles, Plant Physiol., 78:232.
Churchill, K. A., Holoway, B., and Sze, H., 1983, Separation of two types of electro-genic H$^+$-pumping ATPases from oat roots, Plant Physiol., 73:92.
D'Auzac, J., 1975, Caractérisation d'une ATPase membranaire en présence d'une phosphatase acide dans les lutoides d'Hevea brasiliensis, Phytochemistry, 14:671.
Doll, S., Rodier, F., and Willenbrink, J., 1979, Accumulation of sucrose in vacuoles isolated from red beet tissue, Planta, 144:407.
DuPont, F. M., Bennett, A. B., and Spanswick, R. M., 1982, Localization of a proton-translocating ATPase on sucrose gradients, Plant Physiol., 70:1115.
DuPont, F. M., and De Gracia-Zabala, M., 1985, Preparation of membrane vesicles enriched in ATP-dependent proton-transport from suspension cultures of tomato cells, Plant Physiol., 77:69.
Fiske, C. H., and Subbarow, Y., 1925, The colorimetric determination of phosphorus, J. Biol. Chem., 66:375.
Gross, K., and Matile, P., 1979, Vacuolar location of glucosinolates in horseradish root cells, Plant Science Letters, 14:327.
Grubmeyer, C., and Spencer, M., 1979, Effects of anions on a soluble ATPase from mitochondria of pea cotyledons, Plant Cell Physiol., 20:83.
Hager, A., and Helmle, M., 1981, Properties of an ATP-fueled, Cl$^-$-dependent proton pump localized in membranes of microsomal vesicles from maize coleoptiles, Z. Naturforsch. 36 c:997.
Hill, M., Dupaix, A., Volfin, P., Kurkdjian, A., and Arrio, B., 1986, H.P.L.C. technique for simultaneous kinetic measurements of adenine nucleotides in isolated vacuoles, Methods In Enzymology, in press.
Hill, M., Dupaix, A., Volfin, P., Kurkdjian, A., and Arrio, B., 1987, Simultaneous kinetic measurements of adenine nucleotides in isolated vacuoles by HPLC technique, in : "Plant Vacuoles. Their Importance in Solute Compartmentation and Their Applications in Biotechnology", B. Marin, ed., Plenum Publishing Corporation, New-York.

Leigh, R. A., and Branton, D., 1976, Isolation of vacuoles from root storage tissue of <u>Beta</u> <u>vulgaris</u> L., <u>Plant</u> <u>Physiol.</u>, 58:656.

Leigh, R. A., and Walker, R. R., 1980, ATPase and acid phosphatase activities associated with vacuoles isolated from storage roots of red beet (<u>Beta</u> <u>vulgaris</u> L.), <u>Planta</u>, 150:222.

Lin, W., Wagner, G. J., Siegelman, H.W., and Hind, G., 1977, Membrane-bound ATPase of intact vacuoles and tonoplasts isolated from mature plant tissue, <u>Biochim.</u> <u>Biophys.</u> <u>Acta</u>, 465:110.

Nagahashi, G., 1985, The marker concept in cell fractionation, <u>in</u> : "Modern Methods of Plant Analysis", H. F. Linskens and J. F. Jackson, eds., Vol. 1, Springer-Verlag, Berlin.

O'Neill, S. D., Bennett, A. B., and Spanswick, R. M., 1983, Characterization of a NO_3^--sensitive H^+-ATPase from corn roots, <u>Plant</u> <u>Physiol.</u>, 72:837.

Poole, R. J., Briskin, D. P., Kràtky, Z., and Johnstone, R. M., 1984, Density gradient localization of plasma membrane and tonoplast from storage tissue of growing and dormant red beet. Characterization of proton-transport and ATPase in tonoplast vesicles, <u>Plant</u> <u>Physiol.</u>, 74:549.

Randall, K. L., Y. Wang, and Sze, H., 1985, Purification and characterization of the soluble F_1-ATPase of oat root mitochondria, <u>Plant</u> <u>Physiol.</u>, 79:957.

Saunders, J. A., 1979, Investigations of vacuoles isolated from tobacco, <u>Plant</u> <u>Physsiol.</u>, 64:74.

Scherer, G. F. E., and Fischer, G., 1985, Separation of tonoplast and plasma membranes H^+-ATPase from Zucchini hypocotyls by consecutive sucrose and glycerol gradient centrifugation, <u>Protoplasma</u>, 129:109.

Sze, H., 1984, H^+-translocating ATPases of the plasma membrane and tonoplast of plant cells, <u>Physiol.</u> <u>Plant.</u>, 61:683.

Sze, H., 1985, H^+-translocating ATPases : Advances using membrane vesicles, <u>Annu.</u> <u>Rev.</u> <u>Plant</u> <u>Physiol.</u>, 36:175.

Tognoli, L., 1985, Partial purification and characterization of an anion activated ATPase from radish microsomes, <u>Eur.</u> <u>J.</u> <u>Biochem.</u>, 146:581.

Wagner, G. J., 1981, Enzymic and protein character of tonoplast from <u>Hippeastrum</u> vacuoles, <u>Plant</u> <u>Physiol.</u>, 68:499.

Walker, R. R., and Leigh, R. A., 1981, Characterization of a salt-stimulated ATPase activity associated with vacuoles isolated from storage roots of red beet (<u>Beta</u> <u>vulgaris</u> L.), <u>Planta</u>, 153:140.

Wang, Y., and Sze, H., 1985, Similarities and differences between the tonoplast-type and the mitochondrial H^+-ATPases of oat roots, <u>J.</u> <u>Biol.</u> <u>Chem.</u>, 260:10434.

A THIRD-CATEGORY AND A FOURTH-CATEGORY H^+-PHOSPHOHYDROLASE AT THE TONOPLAST

Philip A. Rea, Christopher J. Griffith and Dale Sanders

Department of Biology, University of York
Heslington, York Y01-5DD, England, U. K.

INTRODUCTION

Most of the membranes of plant cells, including the plasma and Golgi membranes, are known to have H^+ pumps but the tonoplast is the membrane which can be most easily isolated in the form of vesicles of low passive H^+ conductance. Tonoplast vesicles constitute an experimental system in which the pH difference (ΔpH) and electrical potential difference ($\Delta\Psi$) between the two compartments separated by the vacuolar membrane can be measured and manipulated with comparative ease (Rea et al., 1986; Blumwald et al., 1987).

The most readily demonstrated functional characteristics of tonoplast vesicles and isolated vacuoles are two primary H^+ pumps, a Cl^--stimulated, NO_3^--inhibited ("anion-sensitive") ATPase (tp-ATPase; EC 3.6.1.3) and a K^+-stimulated, Na^+-inhibited ("cation-sensitive") inorganic pyrophosphatase (tp-PPase; EC 3.6.1.1). Both pumps are capable of establishing a "reversed" (inside-acid; inside-positive) H^+-electrochemical potential difference ($\Delta\overline{\mu}_H^+$) across the tonoplast which may be utilized to drive the secondary H^+-coupled transport of a wide range of solutes (e.g. Poole and Blumwald, this volume). From a general bioenergetic and protein chemical standpoint it is these two pumps which make the tonoplast of particular interest. The tp-ATPase and tp-PPase are members of two new categories of primary pumps : the third- and fourth categories of H^+-translocating phosphohydrolase, respectively.

THIRD-CATEGORY H^+-ATPase

The Other Two Categories

Until recently, it was thought that all membrane-bound H^+-ATPases fall into two major categories. H^+-ATPases of the first category are located in the plasma membranes of eucaryotic microorganisms and plants. These so-called plasma membrane-type H^+-ATPases, collectively termed E_1E_2 to denote that they cycle between two stable states, have a H^+:ATP stoichiometry of 1, catalyse an operationally irreversible reaction and are subject to inhibition by orthovanadate. E_1E_2 enzymes consist of only one major 100 kDa subunit which forms a phosphorylated acyl-intermediate during catalysis and mediates both ATP hydrolysis and H^+-translocation. Other members of the E_1E_2 category are the Na^+,K^+-ATPase, Ca^{2+}ATPase and H^+,K^+-ATPase of animal cells which are functionally and structurally homogenous to the plasma membrane-type H^+-ATPases (Serrano et al., 1986).

The second category consists of the F_1F_0-ATPases of energy-transducing membranes. These enzymes have a H^+:ATP stoichiometry of 3, catalyse a freely reversible reaction, do not form a phosphorylated intermediate during catalysis, consist of at least nine distinct subunits, depending on the organism, and are inhibited by N_3^-.

It is now clear that the ATPase of tonoplast belongs to neither of these categories. Unlike members of the first two categories, tp-ATPases are not inhibited by either orthovanadate or N_3^-.

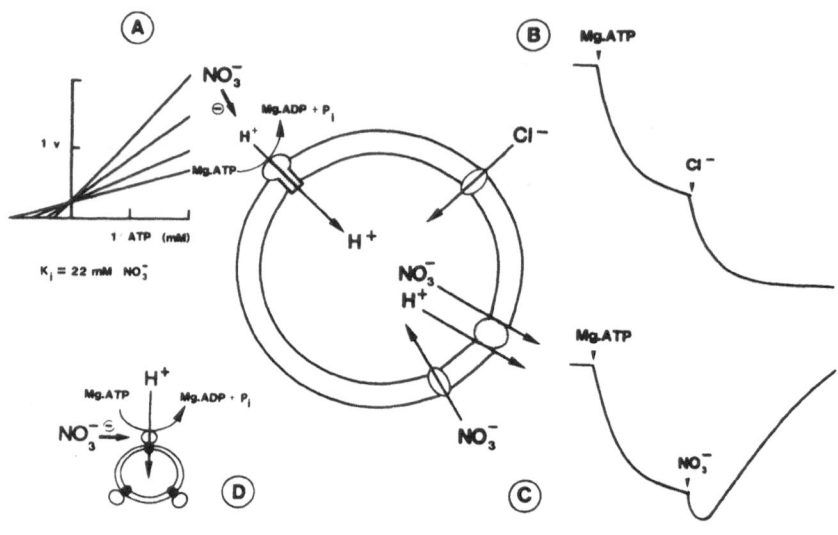

Figure 1

Schematic diagram of effects of NO_3^-
on tp-ATPase and mitochondrial F_1-ATPase

A. Pseudo-competitive inhibition of tp-ATPase-mediated ATP hydrolysis.
B. Alleviation of inside-positive $\Delta\Psi$ and stimulation of ATP-dependent H^+-transloca-
   tion by anion uniport.
C. H^+-NO_3^- symport.
D. Direct inhibition of mitochondrial F_1-ATPase.

A Note Concerning Reversible Inhibition by NO_3^-

It has been known for some time that the ATPase associated with isolated vacuoles and tonoplast vesicles is reversibly inhibited by NO_3^- (Hager and Helmle, 1981; Walker and Leigh, 1981) and this characteristic has been widely employed as a criterion of identity for tp-ATPase activity and accompanying H^+-translocation events in both homogeneous and heterogeneous membrane populations. Inhibition by NO_3^- should, however, be employed judiciously as a means of identifying tonoplast fractions and tp-ATPase-mediated H^+-translocation (Fig. 1) :

- i) The F_1F_0-ATPases of both animal and plant mitochondria (Wang and Sze, 1985) are also subject to inhibition by NO_3^- (Fig. 1-D) at concentrations similar to those necessary to inhibit the tp-ATPase. Nitrate inhibitability only has significance if insensitivity to N_3^- and/or oligomycin can also be demonstrated.

- ii) Nitrate inhibits the tp-ATPase pseudo-competitively with respect to MgATP and yields K_T values of about 39 mM and 22 mM in the absence and presence of Cl^- respectively (Fig. 1-A; Griffith et al., 1986). It is not important to state the NO_3^- concentration employed but also the MgATP and Cl^- concentrations when expressing the degree of inhibition of a particular membrane preparation.

- iii) Nitrate is subject to H^+-symport in some tonoplast vesicle preparations, for instance those from <u>Beta</u> storage root (Blumwald and Poole, 1985). Concentrations of NO_3^- insufficient to directly (sterically) inhibit the tp-ATPase increase tp-ATPase and decrease ΔpH generation (Fig. 1-C). Together with the finding that NO_3^- decreases the steady-state ΔpH generated by the tp-PPase while leaving PP_i hydrolysis unaffected (Rea and Poole, 1985 and below), these results are consistent with outward (vacuole to cytoplasm) H^+-NO_3^- symport. The existence of ΔpH-dissipating H^+-NO_3^- symporters can severely limit the applicability of NO_3-inhibited ATP-dependent H^+-translocation as a method for identifying tonoplast vesicles unless other characteristics are examined in parallel.

How NO_3^- exerts its pseudo-competitive effects is unknown. Griffith et al. (1986) initially postulated that NO_3^- is competitive because of its trigonal planar geometry and stereochemical similarity to the terminal phosphoryl group of ATP, by analogy with the mechanism proposed for the inhibition of creatine kinase by NO_3^- (Milner-White and Watts, 1971). According to the scheme proposed for creatine kinase, MgADP and NO_3^- can both occupy the active site to form a dead-end complex with stereochemically simulates the transition-state complex of MgATP. However, there is no evidence for such a mechanism for the tp-ATPase. Although both NO_3 and MgADP are competitive with respect to MgATP, they do not interact cooperatively to inhibit ATP hydrolysis and NO_3^- does not potentiate the protection conferred on the tp-ATPase by MgADP from covalent modification by N-ethylmaleimide (Griffith et al., 1986). Simple competition between Cl^- and NO_3 for a common anion-binding site is also not consistent with the published data. Nitrate is noncompetitive with Cl^- (Churchill and Sze, 1984) and both the basal and Cl-stimulated components of tp-ATPase activity are competitively inhibited by NO_3 (above). Because of the difficulty of reconciling these observations with simple stereochemical models of active site and anion-binding site interactions, the possibility of reversible chaotropic effects has been investigated (Rea et al., 1987). Thus, the fact that all of the anions capable of increasing the K_m of the tp-ATPase (SCN, ClO_4^-, and NO_3^-) are chaotropic whereas those which do not yield competitive kinetics (CH_3COO and SO_4^{2-}) are not, is compatible with this hypothesis.

Subunit Compositions

All characterized tp-ATPases comprise three distinct subunits assembled into one multimeric complex (Table 1). The components that have been identified are two polypeptides of average M_r 70 and 60 kDa and a smaller component of 16 kDa. The subunit compositions of the tp-ATPases from vacuolate cells as disparate as <u>Neurospora</u> (Bowman, 1982; Bowman and Bowman, 1985) and <u>Hevea</u> (Marin et al., 1985) are remarkably similar, and the qualitatively identical answers obtained by different laboratories using purification methods as independent as negative extraction with dichloromethane (Marin et al., 1985) or detergent solubilization followed by gel filtration (e.g. Manolson et al., 1985; Randall and Sze, 1986) is reassuring.

There is still no consensus concerning the molecular weight of the tp-ATPase. Gel filtration, which is generally an unreliable method for determining the molecular weight of membrane proteins, yields a value of between 200 and 600 kDa (Marin et al., 1985; Randall and Sze, 1985; Rea et al., unpublished). Radiation inactivation which often, but not always, yields reliable estimates of functional molecular size (see Kempner and Schlegel, 1979) gives a value of 400 kDa (Mandala and Taiz, 1985). It therefore appears that one or more of the subunits, assuming all have been identified, are present in several copies per functional molecule.

The 70 and 60 kDa polypeptides have the characteristics of catalytic and/or regulatory subunits whereas the 16 kDa component behaves in a manner consistent with it being part of H^+-channel. A tentative model of the tp-ATPase is shown in Fig. 2.

Table 1

Subunit compositions of tp-ATPases from different sources

Source	M_r, kDa	Subunits, kDa	Label(s)		Minor Components, kDa
Avena roots[1]	300–600[a]	72	^{14}C	NEM	150
			^{14}C	NBD-Cl	45
		60	–		
		14-18	^{14}C	DCCD	
Beta storage root[2]	–	67	–		150
		57	α-^{32}P	BzATP	45
		16	^{14}C	DCCD	
Hevea lutoids[3]	200[a]	66	–		50
		54	–		43
		23	–		
		13	–		
Neurospora[4]	–	70	^{14}C	NEM	
			^{14}C	NBD-Cl	
		60	–		
		16	^{14}C	DCCD	
Saccharomyces[5]	–	89	^{14}C	NEM	
			^{14}C	NBD-Cl	
		64	–		
		19.5	^{14}C	DCCD	
Zea[6]	400[b]	72	^{14}C	NEM	
			^{14}C	NBD-Cl	
		62	–		
		16	^{14}C	DCCD	
Chromaffin granule ghosts[7]	400[a]	70	–		140
		57	–		
		41	–		
		33	–		
		14	^{14}C	DCCD	

[a]Estimated by gel filtration; [b]estimated by radiation inactivation.
[1]Randall and Sze (1986 and unpublished); [2]Manolson et al. (1985); [3]Marin et al. (1985); [4]Bowman (1983), Bowman and Bowman (1985); [5]Uchida et al. (1985 and this volume); [6]Mandala and Taiz (1985), Mandala et al. (unpublished); [7]Percy et al. (1985).

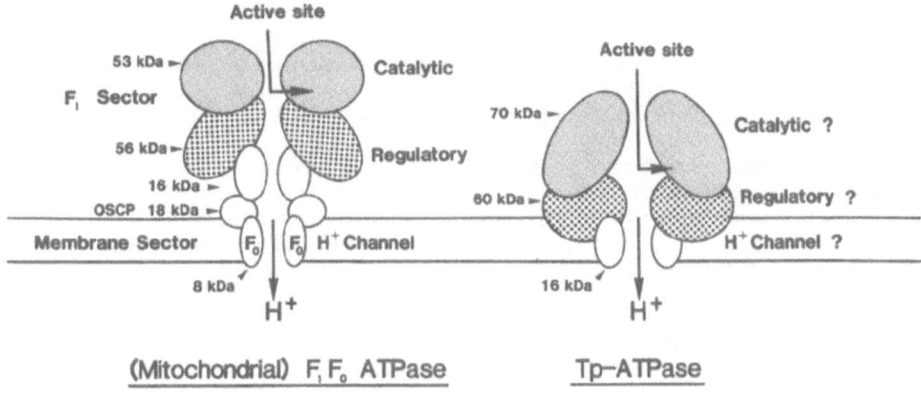

Figure 2

Tentative model of tp-ATPase

Mitochondrial F_1F_0-ATPase is shown for comparative purposes.

Catalytic/Regulatory Subunits

The two affinity probes that have been most widely applied to the identification of the 70 kDa component are N-ethylmaleimide (NEM) and 7-chloro-4-nitrobenzo-2-oxa-1,3-diazole (NBD-Cl). NEM is a sulfhydryl reagent which reacts primarily with cysteine residues whereas NBD-Cl shows specificity towards the hydroxyl-groups of tyrosyl residues. Since all of the tp-ATPases examined are susceptible to MgATP- (or MgADP-) protectable inhibition by both of these agents, it has been concluded that the enzyme contains catalytically essential cysteine and tyrosine residues proximal to MgATP (or MgADP) binding site. Because MgATP protects in the substrate (0.05 - 3 mM) concentration range, this binding site is considered to be catalytic (e.g. Wang and Sze, 1985; Griffith et al., 1986). Its ascription to the 70 kDa polypeptide is based on the finding that incubation of tonoplast vesicles or purified tp-ATPase with unlabeled NEM or NBD-Cl in the presence of MgATP, to block any non-specific binding sites, followed by incubation with ^{14}C-NEM or ^{14}C-NBD-Cl in the absence of MgATP, yields an ATP-protectable ^{14}C-labeled component of 70 kDa (e.g. Bowman and Bowman, 1985; Mandala et al., unpublished data).

The role of the 60 kDa component is still unclear. The observations that : (a) micromolar concentrations of MgATP protect this component from labeling by α-^{32}P-3-0-(4-benzoyl)benzoyladenosine-5'-triphosphate (BzATP), a photoaffinity-analog of ATP and (b) BzATP is a potent inhibitor of enzymic activity (K_I = 11 μM) suggest that the 60 kDa component carries a high-affinity nucleotide binding site which is essential for catalysis (Manolson et al., 1985). However, the finding that BzATP is not a simple competitive inhibitor of the tp-ATPase but causes the enzyme to display positive cooperative kinetics with respect to MgATP indicates that it might contain an ATP-binding site distinct from the catalytic site, possibly a regulatory site (Manolson et al., 1985).

An alternative explanation of the current affinity-labeling data is that the 70 kDa and 60 kDa polypeptides participate in catalysis in a manner similar to that proposed for mitochondrial F_1-ATPase (Williams and Coleman, 1982). This is depicted in Fig. 2 by indicating that the catalytic site comprises the cleft between the α and β subunits of F_1 and the 70 and 60 kDa subunits of the tp-ATPase respectively.

DCCD-Binding Protein

The 16 kDa component has the characteristics of a proteolipid. It is not readily visualized with either Coomassie Blue or silver stain after SDS-PAGE but is labeled by ^{14}C-N,N'-dicyclohexylcarbodiimide (DCCD) (Table 1).

The lipophilic carboxyl reagent DCCD is a potent inhibitor of all known H$^+$-ATPases. Incubation of tonoplast vesicles or partially purified tp-ATPase with this reagent inhibits activity to the 50 % level at concentrations of between 1 and 20 μM depending on the preparation, incubation conditions and membrane protein:inhibitor ratio (see references in Table 1) and ^{14}C-DCCD labels only one polypeptide of 16 kDa which copurifies with the enzyme (e.g. Manolson et al., 1985; Randall and Sze, 1986). Ethyl-3-(3-dimethyl-aminopropyl)carbodiimide (EDAC), a hydrophilic carbodiimide, on the other hand, is a poor inhibitor and does not label the 16 kDa component.

The greater efficacy of lipophilic over hydrophobic carbodiimides in inhibiting and covalently modifying the tp-ATPases implies that, at least, a segment of the 16 kDa polypeptide, presumably that portion proximal to the labeled glutamyl and/or aspartyl residue(s), is hydrophobic and carries the site responsible for irreversible inhibition of the tp-ATPase by DCCD.

Relationship of Tp-ATPase to Other H$^+$-ATPases

The identification of three polypeptides of 70, 60 and 16 kDa in the tp-ATPase complex clearly distinguishes it from the E_1E_2 H$^+$-ATPases on a structural basis. Likewise, the major components of the tp-ATPase are sufficiently different in M_r from the catalytic (β), regulatory (α) and DCCD-binding (c) subunits of the F_1F_0ATPase to be distinguished (see Fig. 2). Investigations of antigenic cross-reactivity corroborate these conclusions (Mandala et al., unpublished data).

The F_1F_0- and tp-ATPases though clearly different in subunit composition do nonetheless have a number of functional similarities. Both enzymes :

- 1) Are insensitive to orthovanadate but similarly sensitive to inhibition by DCCD, 4,4'-diisothiocyanostilbene-2,2'-disulfonate (DIDS), NEM, NBD-Cl, quercetin and NO$_3^-$.

- 2) Are subject to anion activation although the orders of activation for the tp-ATPase and F_1F_0-ATPase are Cl$^-$, Br$^-$, I$^-$ > HCO$_3^-$ > IDA > SO$_4^{2-}$ >> none and HCO$_3^-$ > Cl$^-$, Br$^-$ > I$^-$ > none > SO$_4^{2-}$, IDA respectively (Randall and Sze, 1986).

- 3) Appear to comprise distinct functional sectors composed of discrete subunits of different hydrophobicity. The sectors of the F_1F_0-ATPase are easily resolved by washing membranes at very low ionic strength in the presence of EDTA. This releases the F_1 component of the enzyme, which may be purified as a soluble ATPase, but leaves the F_0 sector in the membrane (Senior, 1985). The F_1 sector is extrinsic to the membrane, is composed of water-soluble subunits and contains the catalytic and regulatory subunits. The F_0 sector is intrinsic and largely composed of hydrophobic subunits, including the DCCD-binding proteolipid. While EDTA washes in low-salt media do not detach the 70 and 60 kDa subunits of the tp-ATPase (Uchida et al., 1985; Rea et al., 1987), the 67 and 57 kDa components of the enzyme from Beta are dissociated from the membrane by chaotropic anions (Rea et al., 1987). This finding, in conjunction with their accessibility to substrate and other hydrophiles like NEM, NBD-Cl and BzATP, agrees with the notion of a more extrinsic association of the 70 and 60 kDa components with the enzyme complex. Conversely, the methanol : chloroform-solubility of a peptide of M_r similar to that of the DCCD-binding protein (Rea et al., unpublished) might imply that this component is highly lipophilic and deeply seated in the phospholipid bilayer. The tp-ATPase may therefore have structure/function partitioning similar to that of the F_1F_0 complex, albeit less pronounced.

Other Third-Category ATPases ?

Lysosomes (Ohkuma et al., 1982), chromaffin granules and synaptosomes (Cidon et al., 1983), clathrin-coated vesicles (Xie et al., 1984) and endosomes (Galloway et al., 1983) also contain H^+-ATPases which are insensitive to both azide and orthovanadate and consequently appear to be distinct from the E_1E_2- and F_1F_0- categories of H^+-ATPase. It has therefore been postulated (e.g. Manolson et al., 1985; Percy et al., 1985) that the tp-ATPase is only one member of a wider class of third-category ATPases. There are insufficient data for a firm conclusion, but the limited structural and functional information do not conflict with this suggestion.

The current status of the argument is as follows :

- 1) All of the H^+-ATPases concerned are associated with endomembranes with an acid interior generated by ATP-dependent electrogenic H^+-translocation. The reversed $\Delta\bar{\mu}_H+$ is implicated in the sequestration of organic solutes and maintenance of lysosomal-type functions, primarily endocytosis and intracellular digestion.

- 2) All of the H^+-ATPases are stimulated by halides, Cl^- in particular, except for the lysosomal enzyme which is only slightly stimulated (Cidon et al., 1983). It should be noted that the tp-ATPase of Neurospora is not markedly stimulated by Cl^- and is only inhibited by high levels of NO_3^- (Bowman and Bowman, 1982; Bowman, 1983). Consequently, anion-sensitivity may be an unreliable indicator of identity even between members of the tp-subclass.

- 3) Where the inhibitor-sensitivities of the enzymes have been examined systematically (pig platelet granule (Dean et al., 1984) and chromaffin granule ghosts (Percy et al., 1985)), inhibition by DCCD, NEM, NBD-Cl, tributyltin and quercetin have been demonstrated.

- 4) The chromaffin granule H^+-ATPase, perharps the best characterized non-plant, non-E_1E_2, non-F_1F_0 enzyme, has an apparent M_r of 400 kDa, comprises six major polypeptides of 70, 57, 41, and 33 kDa and a DCCD-binding lipohilic polypeptide of 16 kDa (Percy et al., 1985; also see Table 1).

Suitability of DCCD-Binding Proteins for Studies of Sequence Homology and Reconstitution

The extent of the third-category and the relatedness of its members with each other and members of the F_1F_0-category can only be quantitated reliably from sequence data. Since all of these enzymes appear to have a DCCD-binding protein in common, this component seems to be the best suited for this purpose.

The 16 kDa component of the tp-ATPase has five major advantages for phylogenic comparisons :

- a) If it is indeed a proteolipid and intrinsic membrane protein, it should be relatively easy to isolate by extraction with apolar solvents because of its extreme hydrophobicity.

- b) It is readily identified on the basis of labeling with ^{14}C-DCCD.

- c) Sequence analyses should be facilitated by its low M_r.

- d) There are already extensive sequence data for the 8 kDa DCCD-binding (c) peptide of F_0. Thus, it may be possible to elucidate whether the 16 kDa polypeptide of the tp-ATPase complex and the 8 kDa proteolipid of F_0 have a common ancestry. For instance, did the 16 kDa component arise by duplication of a gene coding for a "proto-8 kDa protein" which was also an antecedent of the 8 kDa subunit of F_0 ?

- e) The amino-acid sequence of the (c) peptide of F_0 has been highly conserved during evolution. This is attributed to its small size and the fact that it must interact with the other components of the F_1F_0 complex so that only a limited number of amino-acid substitutions are functionally permissible. If the same applies to the 16 kDa component of tp-ATPases, sequence homologies should be pronounced.

In the shorter term, isolation of the 16 kDa polypeptide might enable its reconstitution into liposomes and a direct test of whether it constitutes a H^+-channel by determining if it elicits a DCCD-inhibitable increase in passive H^+-conductance across the phospholipid bilayer. The association of the 16 kDa polypeptide with a H^+-channel will otherwise remain hypothetical.

FOURTH CATEGORY H^+-PHOSPHOHYDROLASE

PPi-Dependent H^+-Translocation

For a number of years it has been known that microsomes from higher plants contain a Mg^{2+}-dependent inorganic pyrophosphatase in addition to a NO_3^--inhibited, azide-insensitive ATPase. Karlsson (1975) initially demonstrated a K^+-stimulated PPase in the 25,000-30,000 g membrane fraction from sugar beet (Beta vulgaris L.) roots and cotyledons and Walker and Leigh (1981) subsequently characterized a Mg^{2+}-dependent PPase with qualitatively similar kinetics associated with vacuoles isolated from storage root of red beet (Beta vulgaris L.). However, it was several years before the transport function of the enzyme was delineated. While Churchill and Sze (1983) and Bennett and Spanswick (1984) had shown that PP_i will support H^+-translocation by tonoplast vesicles isolated from Avena and Beta roots respectively, it was not known if PP_i-dependent H^+-translocation results from the activity of the PPase described by Karlsson (1975) and Walker and Leigh (1981) or whether the tp-ATPase is capable of utilizing PP_i as substrate under some circumstances.

The initial indication that PP_i-dependent H^+-translocation is mediated by the tp-PPase and not the tp-ATPase came from a direct comparison of the substrate, mineral ion and effector-sensitivites of PP_i hydrolysis and PP_i-dependent H^+-translocation (Rea and Poole, 1985). In summary :

- i) ATP and PP_i hydrolysis by uncoupled tonoplast vesicles do not interact. The inclusion of PP_i in the tp-ATPase assay medium does not change the relationship between reaction velocity and MgATP providing that excess Mg^{2+} is present. Likewise, MgATP has no effect on the kinetics of PP_i hydrolysis.

- ii) PP_i hydrolysis and PP_i-dependent H^+-translocation are maximally stimulated by K^+ and inhibited by Na^+ whereas the tp-ATPase has a specific requirement for halides (Fig. 3-A).

- iii) PP_i hydrolysis is not subject to inhibition by NO_3^-.

The applicability of criterion (iii) to PP_i-dependent H^+-translocation depends on the tonoplast preparation concerned. Tonoplast vesicles from Beta show NO_3^--inhibitable PP_i-dependent H^+-translocation (Rea and Poole, 1985) and this is now known to result from H^+-NO_3^- symport by this membrane preparation rather than inhibition of the tp-PPase per se (Blumwald and Poole, 1985 and above). By contrast, in membranes from Zea (Chanson et al., 1985) and Avena (Wang and Sze, 1985) both PP_i hydrolysis and PP_i-dependent H^+-translocation are maximal in rate and extent with NO_3^-.

Physical Resolution of Tp-PPase from Tp-ATPase

Wagner and Mulready (1983), on the basis of the differential detergent sensitivities of the PPase and ATPase activies of Tulipa vacuolar membranes, initially

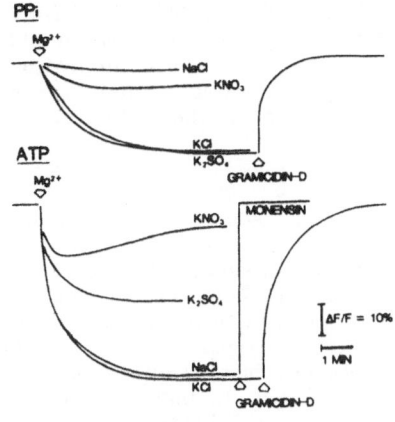

Figure 3-A

Inorganic ionic requirements of PP_i- and ATP-dependent H^+-translocation
by tonoplast vesicles from Beta

PP_i and ATP were added at concentrations of 0.6 and 1.0 mM respectively.

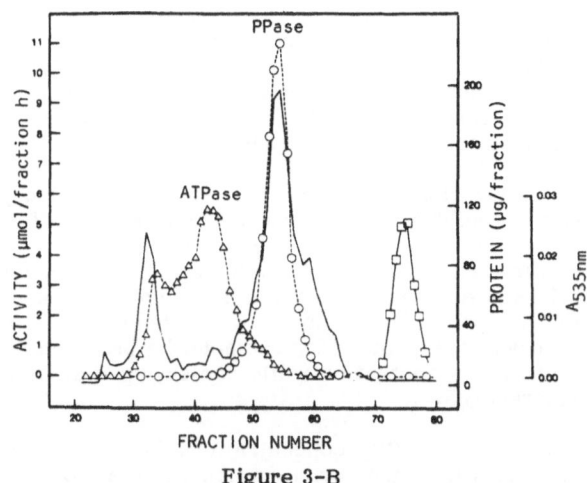

Figure 3-B

Chromatography of Triton X-100-solubilized tonoplast
on Sephacryl-400

Protein (—), betanin (□), ATPase (△) and PPase (○) activity.

suggested that the two enzymes are physically distinct and this has recently been
confirmed by their chromatographic resolution (Rea and Poole, 1986; Fig. 3-A).

The separated tp-PPase is specific for PP_i as substrate and the separated
tp-ATPase shows essentially no activity towards this compound although both
enzymes show identical ionorganic ion requirements to their corresponding activities

in native tonoplast (Rea and Poole, 1986). This finding and the differential susceptibilities of the tp-PPase and tp-ATPase to irreversible inhibition by chaotropic anions (whether activity is measured as substrate hydrolysis or H^+ translocation; Rea et al., this volume) clearly demonstrates that PP_i-dependent H^+ translocation by tonoplast vesicles is solely mediated by the tp-PPase.

Ubiquity of Tp-PPases

H^+-translocating tp-PPases seem to be ubiquitous in higher plants. PP_i-dependent H^+-translocation, K^+-stimulated and/or MoO_4^{2-}-insensitive PP_i hydrolysis have been demonstrated in isolated tonoplast vesicles or isolated vacuoles from Beta storage root (Walker and Leigh, 1981; Rea and Poole, 1985 and 1986), immature Beta roots and cotyledons (Karlsson, 1975; Rea et al., unpublished), Zea coleoptiles (Chanson et al., 1985), Avena roots (Wang and Sze, 1985), leaves of Kalanchoe (Smith et al., 1984), petals of Tulipa (Wagner and Mulready, 1983) and EGTA-permeabilized cells of Nitella (Shimmen, personal communication). Thus, both monocots and dicots, C_3, C_4 and CAM plants, leaves and roots, and coenocytic algae contain the enzyme. It is also noteworthy that Ohsumi et al. (1985) show a clear separation of the solubilized ATPase from vacuoles of Saccharomyces from an inorganic PPase associated with the same membrane fraction. Lower as well as higher plants may therefore contain the enzyme.

PPi as a Physiologically Significant Substrate

Cytoplasmic MgATP concentrations are sufficient to drive ATPase-mediated H^+-translocation but are $MgPP_i$ concentrations sufficient for the tp-PPase ?

First, it should be appreciated that all of the major biosynthetic pathways generate PP_i. Examples are : the acylation of CoA in fatty acid synthesis; the aminoacylation of tRNA in polypeptide synthesis; phosphodiester bond formation in polynucleotide synthesis; and sugar activation in polysaccharide synthesis. Other important PP_i-forming reactions are the synthesis of nucleosides and lipids, and adenosine phosphosulfate formation from ATP and SO_4^{2-}.

Second, the assumption that the PP_i formed during biosyntheses undergoes hydrolysis to shift the equilibria of the reactions concerned in favor of net synthesis (Kornberg, 1963) may be too restrictive an interpretation of the role of PP_i in biological energy transduction. While formally correct, in so far as many of the biosynthetic reactions are freely reversible, it is not imperative that the energy liberated upon PP_i hydrolysis be simply lost as heat. The same thermodynamic driving force for biosyntheses would, for example, be provided by a tp-PPase which lowers the cytosolic PP_i concentration while conserving the energy of the anhydride bond as a transmembrane $\Delta \overline{\mu_H}^+$.

Third, where cytosolic PP_i concentrations have been measured, they are several orders of magnitude greater than simple thermodynamic considerations would indicate. Thus, cytoplasmic PP_i has been calculated to be approximately 7 nM in hepatocytes, given an intracellular P_i concentration of 4 mM and assuming that the soluble PPase reaction is near equilibrium i.e.

$$\Delta G^\circ (PP_i \quad P_i) = -19.2 \text{ KJ.mol}^{-1} = -RT \ln K_{eq}$$

$$K_{eq} = [P_i]^2 / [PP_i] = 2320$$

(Guynn et al., 1974). Direct measurements of the PP_i content, however, give values of between 1 and 10 M (Guynn et al., 1974). Similarly, the measured PP_i contents of Zea and Pisum shoots and roots fall in the range 5 - 39 nmol/g fresh weight (Smyth and Black, 1984; Edwards et al., 1984) which is 3-4 orders of magnitude greater than predicted. If all of the measured PP_i were restricted to the cytosol, the tp-PPase would experience a PP_i concentration of between 50 and 390 μM, assuming a cytoplasm:vacuole ratio of 0.1 and a tissue density of unity. This is more sufficient to maintain maximal tp-PPase activity as the enzyme has a K_m of 15-20μM (Rea and Poole, 1985).

The cytoplasmic P_i concentration of plant cells has recently been estimated to be 5-6 mM by ^{31}P-NMR (Rebeille et al., 1984) which, if soluble PPases were to predominate, indicates nM PP_i concentrations. It is therefore evident that the soluble PPases are not at equilibrium with the bulk of the PP_i pool so that this compound probably has greater metabolic versatility than was once thought (see Reeves, 1976 for an overview of energy-conserving PP_i-dependent metabolism), one of which might be energy-conservation at the tonoplast.

Do the Tp-PPase and Tp-ATPase Reside in the Same Membrane ?

Crucial to an understanding of the physiological roles of the tp-PPase and tp-ATPase and their interactions is to know if the two enzymes reside in the same membrane. It is now clear that both activities are predominantly associated with isolated intact vacuoles and the tonoplast fraction from whole tissue homogenates but it does not automatically follow that any one vacuole or vesicle contains both pumps. Indeed, Chanson et al. (1985) have suggested that the tonoplast fraction from Zea coleoptiles contains two subpopulations of vesicles : one containing the tp-PPase and another the tp-ATPase. They find that the initial rate of H^+-translocation in the presence of PP_i is about 0.5 times the rate with ATP whereas the rate with both PP_i and ATP is approximately 1.5 times the rate with ATP alone. Since each enzyme does not appear to respond to the $\Delta\overline{\mu_H}^+$ generated by the other – i.e. the two interact in a simple additive manner – they deduce that the tp-PPase and tp-ATPase do not reside on the same subpopulation of vesicles. However, assuming that $\Delta\overline{\mu_H}^+$ = 0 at zero-time, initial rates say little about two independent contributions to a common $\Delta\overline{\mu_H}^+$ since the H^+-gradient will not have formed at zero-time. What is critical is the steady-state when pumping is rate-limiting by $\Delta\overline{\mu_H}^+$? When this is examined, the electrical potential and ΔpH generated by the tp-PPase and tp-ATPase are non-additive (Kaestner and Sze, 1986; Rea et al., unpublished), indicating coresidence in the same population of vesicles and kinetic control by a common $\Delta\overline{\mu_H}^+$. This is consistent with the strictly linear relationship between PP_i- and ATP-dependent H^+-translocation on any one linear dextran or sucrose gradient (Rea et al., unpublished) and the finding that single vacuoles of EGTA-permeabilized cells of Nitella are capable of both PP_i- and ATP-elicited intravacuolar acidification (Shimmen, personal communication).

Pump Stoichiometry and Physiological Poise

Why two pumps in one membrane ? Do the tp-ATPase and tp-PPase operate in parallel ? Both questions remain unanswered but are probably interconnected. If it is accepted that the tp-PPase and tp-ATPase reside on the same membrane, the key points are : (i) the steady-state $\Delta\overline{\mu_H}^+$ values which can be established by each pump individually; (ii) their H^+:substrate stoichiometries; and (iii) their respective physiological poises.

The information available suggest that the two pumps have similar capacities. From a comparison of the tp-PPase and tp-ATPase of Beta at optimal substrate concentrations, steady-state ΔpH values of 1.4 and 2.0 respectively have been obtained in vitro (Rea and Poole, unpublished). In Zea and Avena (Chanson et al., 1985; Leigh and Pope, 1987) the tp-PPase can generate similar ΔpH values to those achieved by the tp-ATPase. It should be emphasized, however, that all of these measurements were made fluorimetrically. Consequently, their strict quantitative significance is contingent on an assessment of the proportion of actively pumping (right-side-out) sealed vesicles and of whether the same proportion and/or same vesicles contain both H^+-phosphohydrolases.

If the tp-PPase and tp-ATPase have similar pump capacities in vitro, do they have the same physiological poise ? That is, would both enzymes be expected to operate in the forward direction (uphill H^+-translocation driven to substrate hydrolysis) or might one or both operate in the reverse mode (substrate synthesis driven by downhill H$^+$-translocation) under some circumstances ?

Equilibrium thermodynamics enable an assessment of the feasibility of pump reversal in the intact vacuole. If n represents the H$^+$:ATP or H$^+$:PP$_i$ coupling ratio, then the conditions under which the tp-ATPase and tp-PPase will be at equilibrium – i.e. no net synthesis or hydrolysis – are given by expressions (1) and (2).

$$\Delta G = n\,F\Delta\Psi - RT\left[\ln k_{ATP} + \ln\frac{[ATP]}{[ADP][P_i]} + n\ln\frac{[H^+]_c}{[H^+]_v}\right] = 0 \qquad (1)$$

$$\Delta G = n\,F\Delta\Psi - RT\left[\ln k_{PP_i} + \ln\frac{[PP_i]}{[P_i]^2} + \ln\frac{[H^+]_c}{[H^+]_v}\right] = 0 \qquad (2)$$

where k_{ATP} and k_{PPi} are the equilibrium constants for ATP and PP$_i$ hydrolysis, F is the Faraday and subscripts c and v refer to the cytosolic and vacuolar compartments, respectively. Note that as n increases in the term for transmembrane pH difference, the overall tendency for reversal increases i.e. ΔG becomes more positive. Thus, if $\Delta\Psi$ is assumed to be + 20 mV (a typical value) the zero-point for a plot of ΔG against increasing ΔpH yields an estimate of the minimum ΔpH required for switching the pump from a hydrolytic to a synthetic mode. Such plots for the tp-ATPase and tp-PPase, assuming coupling ratios of 1, 2 and 3, as shown in Fig. 4. Representative values for $\ln k_{ATP}$, $\ln k_{PPi}$, [ATP], [ADP], [PP$_i$] and [P$_i$] were taken from the literature (see legend to Fig. 4).

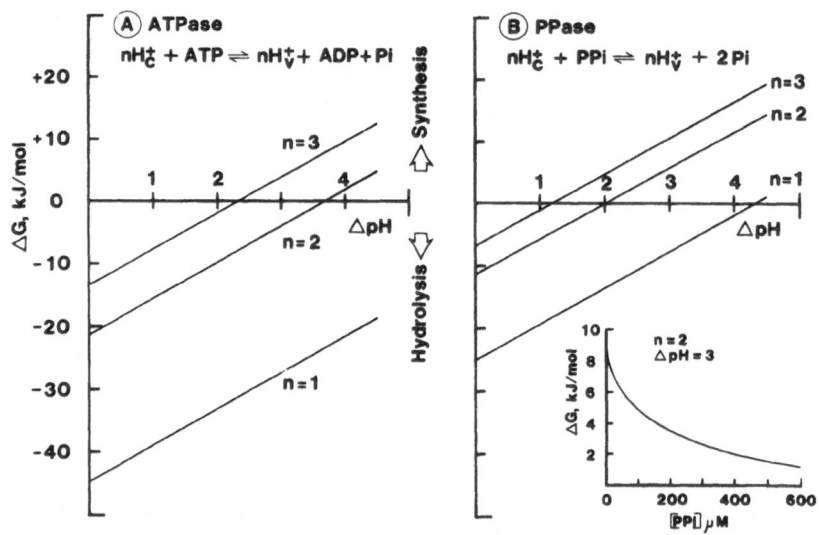

Figure 4

Influence of transtonoplast ΔpH
on reversibilities of tp-ATPase (A) and tp-PPase (B)
assuming coupling ratios (n values) of 1, 2 and 3

The relationships between ΔG and ΔpH were calculated from expressions (1) and (2). pH$_c$ = 7.5, $\Delta\Psi$ = + 20 mV, [ATP] = 3 mM, [ADP] = 0.5 mM, [PP$_i$] = 50 M, [P$_i$] = 5 mM. $\ln k_{ATP}$ = 11.5 (Rosing and Slater, 1972); $\ln k_{PPi}$ = 10.12 (Reeves, 1976). Inset: Influence of PP$_i$ concentration on reversibility of tp-PPase when n = 2 and ΔpH = 3.

The estimated H^+:ATP stoichiometry of the tp-ATPase is 2 (Bennett and Spanswick, 1984). Thus, from expression (1), the enzyme would be expected to operate in a hydrolytic mode except at the most acidic intravacuolar pH values (Fig. 4-A) such as those shown by the CAM plant Kalanchoe at the end of a dark fixation period when vacuolar pH's as low as 3 units have been estimated (Smith, this volume). In marked contrast, only a stoichiometry of 1 will poise the tp-PPase in favor of hydrolysis (Fig. 4-B). Proton:PP_i stoichiometries of 2 or 3, on the other hand, result in net PP_i synthesis from P_i, at ΔpH values as low as 2 units. Since the tp-ATPase is capable of establishing a   pH of more than 2 units, this enzyme clearly has the capacity to provide the requisite free energy for PP_i synthesis by establishing a transmembrane $\Delta\mu_H^+$ of sufficient magnitude to reverse the tp-PPase if H^+:PP_i $\geq$ 2. Moreover, the fact that the tendency for pump reversal increases with decreasing PP_i concentration (Fig. 4-B, inset) would enable tight control of cytosolic PP_i levels by the concerted actions of the tp-ATPase and tp-PPase in ATP hydrolysis and PP_i synthesis respectively.

CONCLUDING REMARKS

It is now clear that the vacuolar membrane contains two functionally and physically distinct phosphohydrolases which catalyse electrogenic H^+-translocation. Both enzymes are novel and belong to neither the E_1E_2- nor F_1F_0-categories of primary cation pump. Research priorities for the tp-ATPase are studies directed at understanding the roles of the 70 and 60 kDa subunits in catalysis and regulation; the involvement of the 16 kDa subunit in transmembrane H^+-conduction; and investigations of F_1- and F_0-like structure/function partitioning. In the longer term, comparisons of sequence homology between the DCCD-binding proteins from different sources may enable elucidation of the evolutionary relatedness of the tp-ATPase with other putative third-category H^+-translocases. The most important questions concerning the tp-PPase are the problems of pump stoichiometry and physiological poise in that reversal of the tp-PPase is only thermodynamically feasible if H^+:PP_i $\geq$ 2. Only when we know the answer to this question can we hazard a guess for why two H^+-translocating phosphohydrolases coexist in one membrane.

REFERENCES

Bennett, A. B., and Spanswick, R.M., 1984, H^+-ATPase activity from storage tissue of Beta vulgaris. I. Identification and characterization of anion-sensitive H^+-ATPase, Plant Physiol., 74:538.

Bennett, A. B., and Spanswick, R. M., 1984, H^+-ATPase activity from storage tissue of Beta vulgaris. II. H^+/ATP stoichiometry of an anion-sensitive H^+-ATPase, Plant Physiol., 74:545.

Blumwald, E., and Poole, R. J., 1985, Nitrate storage and retrieval in Beta vulgaris : Effects of nitrate and chloride on proton gradients in tonoplast vesicles, Proc. Natl. Acad. Sci. U.S.A, 82:3683.

Blumwald, E., Rea, P. A., and Poole, R. J., 1987, Preparation of tonoplast vesicles : Applications to H^+-coupled secondary transport in plant vacuoles, Methods In Enzymology, in press.

Bowman, E. J., 1983, Comparison of the vacuolar membrane ATPase of Neurospora crassa with the mitochondrial and plasma membrane ATPases, J. Biol. Chem. 258:15238.

Bowman, E. J., and Bowman, B. J., 1982, Identification and properties of an ATPase in vacuolar membrane of Neurospora crassa, J. Bacteriol., 151:1326.

Bowman, E. J., and Bowman, B. J., 1985, The H^+-translocating ATPase in vacuolar membranes of Neurospora crassa, in : "Biochemistry and Function of Vacuolar Adenosine-Triphosphatase in Fungi and Plants", B. Marin, ed., Springer-Verlag, Berlin, Heidelberg, New-York, Tokyo.

Chanson, A., Fichmann, J., Spear, D., and Taiz, L., 1985, Pyrophosphate-driven proton transport by microsomal membranes of corn coleoptiles, Plant Physiol., 79:159.

Churchill, K. A., and Sze, H., 1983, Anion-sensitive, H^+-pumping ATPase in membrane vesicles from oat roots, Plant Physiol., 71:610.

Churchill, K. A., and Sze, H., 1984, Anion-sensitive H^+-pumping ATPase from oat roots : Direct effects of Cl^-, NO_3^- and a disulfonic stilbene, Plant Physiol., 76:490.

Cidon, S., Ben-David, H., and Nelson, N., 1983, ATP-driven proton fluxes across membranes of secretory organelles, J. Biol. Chem., 258, 11684.

Dean, G. E., Fischkes, H., Nelson, P. J., and Rudnick, G., 1984, The hydrogen ion pumping adenosine-triphosphatase of platelet dense membrane. Differences from F_1F_0- and phosphoenzyme-type ATPases, J. Biol. Chem., 259:9569.

Edwards, J., Ap Rees, T., Wilson, P. M., and Morrell, S., 1984, Measurement of the pyrophosphate in tissues of Pisum sativum L., Planta, 162:188.

Galloway, C. J., Dean, G. E., Marsh, M., Rudnick, G., and Mellman, I., 1983, Acidification of macrophage and fibroblast endocytic vesicles in vitro, Proc. Natl. Acad. Sci. U.S.A., 80:3334.

Griffith, C. J., Rea, P. A., Blumwald, E., and Poole, R. J., 1985, Mechanism of stimulation and inhibition of tonoplast H^+-ATPase of Beta vulgaris by chloride and nitrate, Plant Physiol., 81:120.

Guynn, R. W., Veloso, D., Lawson, J. W. R., and Veech, R. L., 1974, The concentration and control of cytoplasmic free inorganic pyrophosphate in rat liver in vivo, Biochem. J., 140:369.

Hager, A., and Helmle, M., 1981, Properties of an ATP-fueled, Cl^--dependent proton pump localized in membranes of microsomal vesicles from maize coleoptiles, Z. Naturforsch., 36 c:997.

Kaestner, K. H., and Sze, H., 1986, Potential-dependent anion transport across tonoplast vesicles from oat roots, Plant Physiol., 80 : S-428.

Karlsson, J., 1975, Membrane-bound potassium and magnesium ion stimulated-inorganic pyrophosphatase from roots and cotyledons of sugar beet (Beta vulgaris L.), Biochim. Biophys. Acta, 399:356.

Kempner, E. S., and Schlegel, W., 1979, Size determination of enzymes by radiation inactivation, Anal. Biochem., 92:2.

Kornberg, A., 1963, On the metabolic significance of phosphorylytic and pyrophosphorylytic reactions, in : "Horizons in Biochemistry", M. Kasha and B. Pullman, eds., Academic Press, New-York.

Leigh, R. A., and Pope, A. J., 1987, Understanding tonoplast function : Some emerging problems, in : "Plant Vacuoles. Their Importance in Solute Compartmentation and Their Applications in Biotechnology", B. Marin, ed., Plenum Publishing Corporation, New-York.

Mandala, S., and Taiz, L., 1984, Solubilization and partial purification of a tonoplast ATPase from corn coleoptiles, Plant Physiol., 78:327.

Manolson, M. F., Rea, P. A., and Poole, R. J., 1985, Identification of 3-O-(4-benzoyl)-benzoyladenosine-5'-triphosphate- and N,N'-dicyclohexylcarbodiimide-binding subunits of a higher plant H^+-translocating tonoplast ATPase, J. Biol. Chem., 260:12273.

Marin, B., Preisser, J., and Komor, E., 1985, Solubilization and purification of the ATPase from the tonoplast of Hevea, Eur. J. Biochem., 151:131.

Milner-White, E. J., and Watts, D. C., 1971, Inhibition of adenosine-5'-triphosphate-creatine phosphotransferase by substrate-anion complexes. Evidence for the transition state organization of the active site, Biochem. J., 122:727.

Ohkuma, S., Moriyama, Y., and Takano, T., 1982, Identification and characterization of a proton pump on lysosomes by fluorescein isothiocyanate-dextran fluorescence, Proc. Natl. Acad. Sci. U.S.A., 79:2758.

Ohsumi, Y., Uchida, E., and Anraku, Y., 1985, The H^+-translocating ATPase in vacuolar membranes of Saccharomyces cerevisiae, in : "Biochemistry and Function of Vacuolar Adenosine-Triphosphatase in Fungi and Plants", B. P. Marin, ed., Springer-Verlag, Berlin, Heidelberg, New-York, Tokyo.

Percy, J. M., Pryde, J. G., and Apps, D. K., 1985, Isolation of ATPase I, the proton pump of chromaffin-granule membranes, Biochem. J., 231:557.

Randall, S. K., and Sze, H., 1986, Properties of the partially purified tonoplast H^+-pumping ATPase from oat roots, J. Biol. Chem., 261:1364.

Rea, P. A., Blumwald, E., and Poole, R. J., 1986, Tonoplast vesicles – a model system for the study of H^+-coupled transport, in : "Models in Plant Physiology/Biochemistry/Technology", D. W. Newman and K. G. Wilson, eds., CRC Press, West Palm Beach, Florida, in press.

Rea, P. A., and Poole, R. J., 1985, Proton-translocating inorganic pyrophosphatase in red beet (Beta vulgaris L.) tonoplast vesicles, Plant Physiol., 77:46.

Rea, P. A., and Poole, R. J., 1986, Chromatographic resolution of H^+-translocating pyrophosphatase from H^+-translocating ATPase of higher plant tonoplast, Plant Physiol., 81:126.

Rebeille, F. R., Bligny, R., and Douce, R., 1984, Is the cytoplasmic P_i concentration a limiting factor for plant cell respiration ? Plant Physiol., 74:355.

Reeves, R. E., 1976, How useful is the energy in inorganic pyrophosphate ? Trends Biochem. Sci., 1:53

Rosing, J., and Slater, E. C., 1972, The value of ΔG^o for the hydrolysis of ATP, Biochim. Biophys. Acta, 267:275.

Schneider, D. L., 1981, ATP-dependent acidification of intact and disrupted lysosomes. Evidence from an ATP-driven proton pump, J. Biol. Chem., 256:3858.

Senior, A. E., 1985, The proton ATPase of Escherichia coli, Current Top. Membr. Transport, 23:135.

Serrano, R., Kielland-Brandt, M. C., and Fink, G. R., 1986, Yeast plasma membrane ATPase is essential for growth and has homology with $(Na^+ + K^+)$, K^+ and Ca^{2+}-ATPases, Nature, 319:689.

Smith, J. A., Uribe, E. G., Ball, E., Heuer, S., and Lüttge, U., 1984, Characteristics of the vacuolar ATPase activity of the crassulacean-acid-metabolism plant Kalanchoe daigremontiana, Eur. J. Biochem., 141:415.

Smyth, D. A., and Black, C. C., 1984, Measurement of the pyrophosphate content of plant tissues, Plant Physiol., 75:862.

Uchida, E., Ohsumi, Y., and Anraku, Y., 1985, Purification and properties of H^+-translocating Mg^{2+}-adenosine-triphosphatase from vacuolar membranes of Saccharomyces cerevisiae, J. Biol. Chem., 260:1090.

Wagner, G. J., and Mulready, P., 1983, Characterization and solubilization of nucleotide-specific Mg,ATPase and Mg,pyrophosphatase of tonoplast, Biochim. Biophys. Acta, 728:267.

Walker, R. R., and Leigh, R. A., 1981, Mg^{2+}-dependent cation-stimulated inorganic pyrophophatase associated with vacuoles isolated from storage roots of red beet (Beta vulgaris L.), Planta, 153:150.

Wang, Y. H., and Sze, H., 1985, H^+-pumping pyrophosphatase in tonoplast vesicles of oat roots, Plant Physiol., 77:S-833.

Williams, N., and Coleman, P. S., 1982, Exploring the adenine nucleotide binding sites on mitochondrial F_1-ATPase with a new photoaffinity probe, 3'-0-(benzoyl)benzoyladenosine-5'-triphosphate, J. Biol. Chem., 257:2834.

Xie, X.-S., Stone, D. K., and Racker, E., 1984, Activation and partial purification of the ATPase of clathrin-coated vesicles and reconstitution of the proton pump, J. Biol. Chem., 259:11676.

STRUCTURE AND FUNCTION OF THE SUBUNITS OF THE VACUOLAR

MEMBRANE H$^+$-ATPase OF <u>SACCHAROMYCES CEREVISIAE</u>

Yasuhiro Anraku, Etsuko Uchida and Yoshinori Ohsumi

Department of Biology, Faculty of Science, University of Tokyo
Hongo, Tokyo 113, Japan

INTRODUCTION

We have found a novel H$^+$-translocating ATPase in vacuolar membranes of the yeast <u>Saccharomyces</u> <u>cerevisiae</u> (Kakinuma et al., 1981; Uchida et al., 1985). Using a preparation of right-side-out vacuolar membrane vesicles of high purity (Ohsumi and Anraku, 1981), we showed that the H$^+$-ATPase generates an electrochemical potential difference of protons across the membrane of 180 mV, interior acid (Kakinuma et al., 1981), and that the enzyme serves as a common energy-donating system for seven n H$^+$/amino-acid antiport systems and one H$^+$/Ca^{2+} antiport, which all are present in the vacuolar membrane (Sato et al., 1984; Ohsumi and Anraku, 1983; Anraku, 1987).

In this paper we describe the enzymatic properties, molecular organization and catalytic function of the vacuolar membrane H$^+$-ATPase. We also propose the molecular mechanism of ATP hydrolysis by the enzyme, which was studied under non steady state conditions. Several lines of basic evidence that the vacuolar membrane H$^+$-ATPase should be regarded as a third type of H$^+$-translocating ATPase are presented.

RESULTS AND DISCUSSION

Enzymatic Properties of Vacuolar Membrane H$^+$-ATPase

The vacuolar membrane H$^+$-ATPase was solubilized efficiently by a zwitterionic detergent ZW-3.14 (<u>N</u>-tetradecyl-<u>N</u>,<u>N</u>-dimethyl-3-ammonio-1-propane-sulfonate) and purified 38-fold by glycerol density gradient centrifugation (Uchida et al., 1985). The purified enzyme required phospholipids for maximal activity and was free of acid and alkaline phosphatase activities (Uchida et al., 1985). The enzyme in the presence of phospholipids had the same pH optimum (pH 6.9) and K$_m$ value for ATP hydrolysis (0.21 mM) as the native membrane-bound enzyme (Kakinuma et al., 1981; Uchida et al., 1985). It hydrolyzed ATP, GTP, UTP, and CTP with this order of preference. ADP was not hydrolyzed, and inhibited the enzyme purity noncompetitively, with a K$_i$ value of 0.31 mM (Uchida et al., 1985).

Table 1 summarizes the effects of various inhibitors of H$^+$-translocating ATPases on the vacuolar membrane H$^+$-ATPase activity. The activity of the purified enzyme was strongly inhibited by DCCD (<u>N</u>,<u>N</u>'-dicyclohexylcarbodiimide), NBD-Cl (7-chloro-4-nitrobenzo-2-oxa-1,3-diazole), tributyltin, and SITS (4-acetamide-

4'-isothiocyanatostilbene-2,2'-disulfonic acid), which are inhibitors of F_0F_1-type ATPases from mitochondria and/or chloroplasts, but not by sodium azide or oligomycin. The enzyme was not also affected by sodium vanadate or miconazole, which are inhibitors of the dephosphorylation and phosphorylation of plasma H^+-ATPase from the yeast Schizosaccharomyces pombe, but, interestingly, it was inhibited by DES and quercetin, which are inhibitors of Na^+/K^+-ATPases.

Table 1

Effects of Inhibitors on Vacuolar Membrane H^+-ATPase

Inhibitor	mM	Relative Activity (%)	
		Native Enzyme	Purified Enzyme
None	–	100	100
DCCD	0.001	63	36
NBD-Cl	0.1	27	23
Tributyltin	0.1	45	14
SITS	0.004	44	36
Sodium azide	2.0	110	95
Oligomycin	0.047	74	96
Sodium vanadate	0.1	96	95
Miconazole	0.2	109	106
DES	0.1	48	30
Quercetin	0.1	67	37
KNO_3	50	55	57
KSCN	50	69	28
$CaCl_2$	0.1	98	101
$CuCl_2$	0.5	12	10

Activities were determined in the standard reaction mixture containing 25 mM Mes/Tris (pH 6.9), 5 mM ATP, 5 mM $MgCl_2$, and additions as indicated. For assay of purified activity, sonicated soybean phospholipids (0.1 mg/ml) were added to the reaction mixture, which resulted in at amost 3-fold stimulation of the activity (Uchida et al., 1985).

The enzyme activity depended solely on Mg^{2+}, and did not require Ca^{2+}. It was sensitive to nitrate and thiocyanate anions. The activity was highly sensitive to Cu^{2+}, like that of yeast plasma membrane H^+-ATPase, but unlike that of yeast mitochondrial F_0F_1-ATPase. Obviously, both the purified and native vacuolar membrane H^+-ATPases shared common enzymological properties of substrates specificities (Kakinuma et al., 1981; Uchida et al., 1985), and inhibitor sensitivities (Table 1), and were clearly distinct from the two established types of H^+-ATPases in the yeast Saccharomyces cerevisiae by their unique characteristics in nitrate-sensitive but azide- and vanadate-insensitive mode of ATP hydrolysis (Uchida et al., 1985). As known well, the F_0F_1-type ATP synthase is sensitive to azide and the plasma membrane E_1E_2-type H^+-ATPase is sensitive to vanadate, but neither is affected by nitrate.

Subunits of Vacuolar Membrane H^+-ATPase

The polypeptide composition of the purified vacuolar membrane H^+-ATPase was determined electrophoretically on polyacrylamide gel in the presence of 15%

sodium dodecylsulfate (Uchida et al., 1985). The enzyme is composed of three sub-units a (89 kDa), b (64 kDa), and c (20 kDa), with a molar ratio of $[a, b]_m[c]_n$ (Table 2). Subunits a and b are apparently present in equimolar amount (Uchida et al., 1985). The smallest subunit, subunit c, was scarcely stained by Coomassie brilliant blue and it was difficult to estimate its molar ratio to subunits a and b.

Table 2

Subunits of Vacuolar Membrane H^+-ATPase and Their Functions

Subunit	Molecular Weight	Function	Comment
a	89 kDa	ATP hydrolysis	It has a catalytic site which contains single NBD-Cl sensitive tyrosine residue and a binding site for 8-azido-ATP. Under nonsteady state conditions, ATP initially binds in the NBD-Cl-sensitive catalytic site to form an enzyme-ATP complex and then bound ATP is hydrolyzed to ADP and P_i.
b	64 kDa	not known	It is apparently present in equimolar amount with subunit a.
c	20 kDa	proton channel	It is a DCCD-binding subunit and is poorly stained with Coomassie blue.

Function of Subunits of Vacuolar Membrane H^+-ATPase

The subunit c has been identified as a DCCD-binding polypeptide of the purified enzyme (Uchida et al., 1985). Similar results showing that the smallest subunit of vacuolar membrane H^+-ATPases isolated from fungal and plant cells can bind DCCD have been reported (Bowman et al., 1986; Manolson et al., 1985). These consistent findings indicate that subunit c constructs a channel for proton translocation in the vacuolar membrane H^+-ATPase complex. The function of the subunit b has not yet been identified.

Recently, we found that subunit a serves a catalytic site for ATP hydrolysis (Uchida et al., 1987). As NBD-Cl was found to be a potent inhibitor of the enzyme activity (Uchida et al., 1985) and it is a preferential chemical probe capable of modifying certain amino-acid residues covalently (Ferguson et al., 1975), we used it for identifying the catalytic site and/or an ATP binding site. Major results obtained from kinetic and protein-chemical studies were as follows (Uchida et al., 1987) and are summarized in Table 2 : i) The enzyme activity was inactivated by NBD-Cl with a concentration of half-maximal inactivation of about 1 μM in the neutral pH range; ii) This inactivation was protected by the presence of ATP, ADP, or AMP-PNP; iii) The inactivated enzyme could be restored to the original level of activity by 2-mercaptoethanol; iv) Kinetic and chemical studies on the inactivation showed that chemical modification of a single tyrosine residue per molecule of the enzyme responded to inactivation of the activity; and v) When the enzyme was inactivated with ^{14}C-NBD-Cl, subunit a was specifically labeled and this labeling was completely prevented by the presence of ATP or other substrate nucleotides.

Furthermore, we found that 8-azido-α-^{32}P-ATP bound to subunit $\underline{a}$ specifically (Uchida et al., 1987). This photoaffinity labeling was saturable with an apparent dissociation constant of 10^{-5} M and was decreased by the presence of ATP and ADP. From these results we concluded that subunit $\underline{a}$ of the yeast vacuolar membrane H$^+$-ATPase has a catalytic site and/or an ATP binding site that contains a single tyrosine residue essential for catalytic function.

Molecular Mechanism of ATP Hydrolysis

We analyzed the initial step of ATP hydrolysis by the purified enzyme, applying the rapid equilibrium method (Grubmeyer et al., 1982; Cross et al., 1982; Penefsky, 1985) under conditions where micromolar excess enzyme was incubated with γ-^{32}P-ATP. We found that ATP initially bound in the NBD-Cl-sensitive catalytic site to form an enzyme-ATP complex and then hydrolysis took place but the products, ADP and P$_i$, remained bound in the catalytic site (Uchida et al., 1987). These observations indicated that the enzyme catalyzes ATP hydrolysis by following sequential mechanisms :

$$E + ATP \rightleftharpoons E \bullet ATP \rightleftharpoons E {\bullet ADP \atop \bullet P_i} \rightleftharpoons E + ADP + P_i$$

The molecular mechanism shown above is similar to that of mitochondrial F_0F_1-type ATP synthase (Grubmeyer et al., 1982; Cross et al., 1982; Penefsky, 1985) but differs from that of plasma membrane E_1E_2-type H$^+$ATPase. Characteristic features of these three types of H$^+$-ATPases are summarized in Table 3.

Table 3

Characteristic Features of F_0F_1-, E_1F_2- and V-H$^+$-ATPases

	F_0F_1	E_1E_2	V
Ion transported	H$^+$ only	H$^+$, M$^+$, M^{2+}	H$^+$ only
No. of Subunits :			
- peripheral	5 (E. coli)	0	0
- integral	3 (E. coli)	1, 2, or 3	3 (Yeast)
Covalent Intermediate	None	Asp P	None
Anion Sensitivity	Azide	Vanadate	Vanadate
Physiological Poise	Hydrolysis Synthesis	Hydrolysis –	Hydrolysis –
Cellular Distribution :			
- Prokaryotes	Plasma Membrane	Plasma Membrane	None
- Eukaryotes	Mitochondria Chloroplasts	Plasma Membrane	Vacuoles

CONCLUSION

We presented consistent evidence that yeast vacuoles possess a third type of H$^+$-translocating ATPase. Several recent papers have described the partial purification of a new group of nitrate-sensitive H^+-ATPases from plant and fungal vacuolar membranes (Marin et al., 1985; Tognoli, 1985; Randall and Sze, 1986; Bowman and Bowman, 1982; Lichko and Okorokov, 1985). All these H^+-ATPases bear structural resemblance one another and appear to be composed of three major polypeptides, like the enzyme from Saccharomyces cerevisiae (Uchida et al., 1985).

Studies of molecular mechanism of ATP hydrolysis by the yeast enzyme indicated that the formation of an enzyme-ATP is prerequisite to the hydrolysis of bound ATP to ADP and P_i in the NBD-Cl-sensitive catalytic site. Comparative biochemical studies on the mechanisms of mitochondrial and vacuolar membrane types of proton pumping reactions coupled with ATP hydrolysis should provide further information on the elementary but versatile ways in biological energy-transforming processes.

ACKNOWLEDGEMENTS

This study and the other report (Anraku, 1987) presented at this Workshop were supported in part by a Grant-in-Aid from the Ministry of Education, Science, and Culture of Japan, a Special Coordination Fund for Promotion of Science and Technology from the Science and Technology Agency of Japan, and grants from the Sankyo Foundation of Life Science and the Inamori Foundation, Japan.

REFERENCES

Anraku, Y., 1986, Active transport of amino-acids and calcium ions in fungal vacuoles, in : "Plant Vacuoles. Their Importance in Plant Cell Compartmentation and their Applications in Biotechnology", B. Marin, ed., Plenum Publishing Corporation, New-York.
Bowman, E. J., and Bowman, B. J., 1982, Identification and properties of an ATPase in vacuolar membranes of Neurospora crassa, J. Bacteriol., 151: 1326.
Bowman, E. J., Mandala, S., Taiz, L., and Bowman, B. J., 1986, Structural studies of the vacuolar membrane ATPase from Neurospora crassa and comparison with the tonoplast membrane ATPase from Zea mays, Proc. Natl. Acad. Sci. U.S.A., 83:48.
Cross, R. L., Grubmeyer, C., and Penefsky, H. S., 1982, Mechanism of ATP hydrolysis by beef heart mitochondrial ATPase : Rate enhancements resulting from cooperative interactions between multiple catalytic sites, J. Biol. Chem., 257:12101.
Ferguson, S. J., Lloyd, W. J., Lyons, M. H., and Radda, G. K., 1975, The mitochondrial ATPase : Evidence for a single essential tyrosine residue, Eur. J. Biochem., 54:117.
Grubmeyer, C., Cross, R. L., and Penefsky, P. S., 1982, Mechanism of ATP hydrolysis by beef heart mitochondrial ATPase : Rate constants for elementary steps in catalysis at a single site, J. Biol. Chem., 257:12092.
Kakinuma, Y., Ohsumi, Y., and Anraku, Y., 1981, Properties of H^+-translocating adenosine-triphosphatase in vacuolar membranes of Saccharomyces cerevisiae J. Biol. Chem., 256:10859.
Lichko, L. P., and Okorokov, L. A., 1985, What family of ATPase does the vacuolar H^+-ATPase belong to ?, FEBS Letters, 187:349.
Manolson, M. F., Rea, P. A., and Poole, R. J., 1985, Identification of 3-0-(4-benzoyl)-benzoyl-adenosine-5'-triphosphate- and N,N'-dicyclohexyl-carbodiimide-binding subunits of higher plant H^+-translocating tonoplast ATPase, J. Biol. Chem., 260:12273.
Marin, B. P., Preisser, J., and Komor, E., 1985, Solubilization and purification of the ATPase from the tonoplast of Hevea, Eur. J. Biochem., 151:131.
Ohsumi, Y., and Anraku, Y., 1981, Active transport of basic amino-acids driven by a proton-motive force in vacuolar membrane vesicles of Saccharomyces cerevisiae, J. Biol. Chem., 256:2079.

Ohsumi, Y., and Anraku, Y., 1983, Calcium transport driven by a proton-motive force in vacuolar membrane vesicles of Saccharomyces cerevisiae, J. Biol. Chem., 258:5614.

Penefsky, H. S., 1985, Reaction mechanism of the membrane-bound ATPase of submitochondrial particles from beet heart, J. Biol. Chem., 260:13728.

Randall, S. K., and Sze, H., 1986, Properties of partially purified tonoplast H^+-pumping ATPase from oat roots, J. Biol. Chem., 261:1364.

Sato, T., Ohsumi, Y., and Anraku, Y., 1984, Substrate specificities of active transport systems for amino-acids in vacuolar membrane vesicles of Saccharomyces cerevisiae : Evidence of seven independent proton/amino-acid antiport systems, J. Biol. Chem., 259:11505.

Tognoli, L., 1985, Partial purification and characterization of an anion activated-ATPase from radish microsomes, Eur. J. Biochem., 146:581.

Uchida, E., Ohsumi, Y., and Anraku, Y., 1985, Purification and properties of H^+-translocating, Mg^{2+}-adenosine-triphosphatase from vacuolar membranes of Saccharomyces cerevisiae, J. Biol. Chem., 260:1090.

Uchida, E., Ohsumi, Y., and Anraku, Y., Characterization and function of a catalytic subunit a of H^+-translocating adenosine-triphosphatase from vacuolar membranes of Saccharomyces cerevisiae: A study with 7-chloro-4-nitrobenzo-2-oxa-1,3-diazole, J. Biol. Chem., submitted.

DIFFERENTIAL SUSCEPTIBILITIES OF TONOPLAST ATPASE AND PPASE

TO IRREVERSIBLE INHIBITION BY CHAOTROPIC ANIONS

Philip A. Rea, Christoffer J. Griffith and Dale Sanders

Department of Biology, University of York
Heslington, York Y01-5DD, England, U. K.

INTRODUCTION

Isolated intact vacuoles and tonoplast vesicles are distinguished from other endomembranes by their possession of a NO_3^--inhibited, azide-insensitive ATPase (tp-ATPase; Rea et al., 1987). However, information concerning the mode of action of NO_3^- is scant.

Nitrate exerts pseudo-competitive kinetics of inhibition with respect to MgATP and potently chaotropic anions, such as SCN^- and ClO_4^-, can simulate the effect (Griffith et al., 1986). It has therefore been proposed that NO_3^- exerts its reversible inhibitory effects chaotropically (Griffith et al., 1986).

Circumstantial evidence for a chaotropic mode of action are :

- 1) The tp-ATPase is a multimeric complex with an apparent functional M_r of 400 kDa (Mandala and Taiz, 1985) comprising at least three distinct subunits (Manolson et al., 1985). Chaotropes destabilize membranes and enzyme complexes by the disruption of hydrophobic interactions between subunits (Hatefi and Hanstein, 1974).

- 2) Relatively high (millimolar) concentrations of NO_3^- are required for marked inhibition of the tp-ATPase.

- 3) Chaotropic anions, mimic NO_3^- whereas the antichaotropic anions, CH_3COO^- and SO_4^-, and the putative NO_3^--analog, ClO_4^-, do not (Griffith et al., 1986).

- 4) The partially purified, solubilized tp-ATPase is approximately twice as sensitive to NO_3^- as the enzyme in native vesicles (Manolson et al., 1985). Susceptibility to chaotropic effects becomes more pronounced when membrane proteins are solubilized (Hatefi and Hanstein, 1974).

Here, we examine the relationship between reversible and irreversible inhibition of the tp-ATPase by chaotropic anions to determine if the inhibitions seem follow the same rank orders of potency. Non-specific chaotropic effects (e.g. oxidative damage; Hatefi and Hanstein, 1974) are distinguished from specific effects by comparing the tp-ATPase with the tonoplast inorganic pyrophosphatase (tp-PPase) which appears to reside on the same membrane but is insensitive to NO_3^- (Rea et al., 1987).

MATERIALS AND METHODS

Tonoplast vesicles were isolated from fresh red beet (Beta vulgaris L.) storage root as detailed elsewhere (Rea and Poole, 1985).

The membranes were treated with chaotrope by the addition of 4 ml chaotrope solution to 1 ml of tonoplast suspension (1 - 3 mg/ml membrane protein). The mixture was left on ice for 30 min, diluted 20-fold with suspension medium (1.1 M glycerol, 0.5 mM butylated hydroxytoluene and 5 mM DTT buffered to pH 8.0 with 5 mM BTP-Mes) and the membranes were pelleted by centrifugation at 100,000 g for 30 min or 200,000 g for 1 h.

One hundred ml volumes of the 200,000 g supernatants from the chaotrope treatments were prepared for SDS-PAGE by dialysis for 48 h at 4°C against one change of 2 l of 2 mM Tris-Cl (pH 8.0). The dialysates were lyophylized to dryness at - 5°C, resuspended in deionized water and denatured at 60°C for electrophoresis. One dimensional SDS-PAGE on concave exponential gradient (9 - 14%) gels was performed as described by Manolson et al. (1985).

ATP- and PP_i-dependent intravesicular acidification were assayed fluorimetrically with the monoamine dye acridine orange. Fluorescence (F) or fluorescence quench (Q) were monitored with a Perkin-Elmer LS-5 Luminescence spectrometer interfaced to an IBM-PC/XT microprocessor. Instantaneous rates of quench (% Q/min) were computed by a non-linear least squares curve fitting routine (Jennings et al., 1987).

RESULTS

Influence of Pretreatment of Microsomes with 0.24 M KI

In some membrane preparations from Beta, pretreatment of the microsomal pellet with 0.24 M KI is necessary to minimize contamination of the tonoplast fraction with F_1-ATPase and mitochondrial membrane fragments (cytochrome c oxidase; Rea et al., unpublished). Potassium iodide has also been employed to diminish contamination of tonoplast fractions by acid phosphatase (Poole et al., 1984). The standard KI treatment was therefore retained to guard against contamination by F_1-ATPase and/or acid phosphatase. It was however important to assess the effect of this standard treatment on the tp-ATPase and tp-PPase prior to investigating the response to chaotropes in general since I^- is itself chaotropic.

Pretreatment of microsomes from fresh red beet with 0.24 M KI had no effect on NO_3^--sensitive, azide-insensitive ATPase activity but caused a 20% elevation of K^+-stimulated, Na^+-inhibited PPase activity (Table 1). Contamination by MoO_4^{2-}-sensitive- (non-specific phosphatase-), orthovanadate-sensitive- (plasma membrane-) and azide-sensitive- (F_1-) ATPase activities was negligible in both the KI-treated and non-treated tonoplast fractions. The standard KI treatment has no apparent inhibitory effect on the specific activities of the tp-ATPase and tp-PPase.

Differential Sensitivities of Tp-ATPase and Tp-PPase
to Irreversible Inhibition by Chaotropic Anions

The tp-ATPase and tp-PPase had markedly different susceptibilities to irreversible inhibition by chaotropic anions whether activity was measured as ATP or PP_i-dependent H^+-translocation (Fig. 1 and 2-A) or as ATP or PP_i hydrolysis (Fig. 2-B). Thiocyanate, ClO_4^-, I^- and NO_3^- yielded k_{50} values for the irreversible inhibition of ATP hydrolysis of 0.29, 0.35, 0.43 and > 1.25 M respectively (Fig. 2-B). The corresponding values for the inhibition of ATP-dependent H^+-translocation

Table 1

Influence of Treatment of Microsomes with 0.24 M KI
before sucrose density gradient centrifugation
on specific activities and inhibitor–sensitivities of ATP and PP_i hydrolysis
by tonoplast vesicles collected from 10/23% (v/v) sucrose interface
after centrifugation at 80,000 g for 2 h

Fifty mM KCl was included in all of the assay media except when substituted by 50 mM NaCl. Potassium nitrate (100 mM), Na_2MoO_4 (200 μM), Na_3VO_4 (100 μM) or NaN_3 (2 mM) were added as indicated.

	Effectors	Activity			
		μmol/mg.h		(% control)	
		ATPase		PPase	
– KI	None	71.4	(100.0)	17.5	(100.0)
	KNO_3	17.4	(24.3)	–	–
	Na_2MoO_4	70.9	(99.3)	–	–
	Na_3VO_4	68.5	(95.9)	–	–
	NaN_3	75.9	(106.2)	–	–
	KNO_3 + NaN_3	15.4	(21.5)	–	–
	NaCl	–	–	4.7	(27.0)
+ KI	None	69.6	(100.0)	21.6	(100.0)
	KNO_3	17.4	(25.0)	–	–
	Na_2MoO_4	66.8	(96.0)	–	–
	Na_3VO_4	72.6	(104.3)	–	–
	NaN_3	71.4	(102.6)	–	–
	KNO_3 + NaN_3	17.1	(24.6)	–	–
	NaCl	–	–	5.1	(23.6)

were 0.3 (ClO_4^-) and 0.4 M (I^-) (Fig. 2-A). The tp-PPase, on the other hand, was not inhibited by ClO_4^-, I^- or NO_3^- and only appreciably inhibited by SCN^- concentrations in excess of 0.5 M (Fig. 2-B). Chaotrope concentrations in excess of the k_{50} for inhibition of the tp-ATPase did however have non–specific effects on H^+-translocation. High concentrations of chaotropes inhibited both ATP- and PP_i-dependent H^+-translocation (Fig. 2-A) but had no effect, or only a small effect, on PPase-mediated substrate hydrolysis (Fig. 2-B).

The non–chaotropic anions SO_4^- (data not shown) and CH_3COO^- (Fig. 2) affected neither the tp-ATPase nor tp-PPase (Fig. 2).

Because it yielded the widest separation between the onsets of inhibition of the tp-ATPase and tp-PPase with increase in chaotrope concentration, KI was employed as the chaotrope for all subsequent manipulations. The potential for the control and imposition of specific alterations of membrane function and composition without oxidative damage (Hatefi and Hanstein, 1974) was considered to be greater for a chaotrope of intermediate rather than high potency.

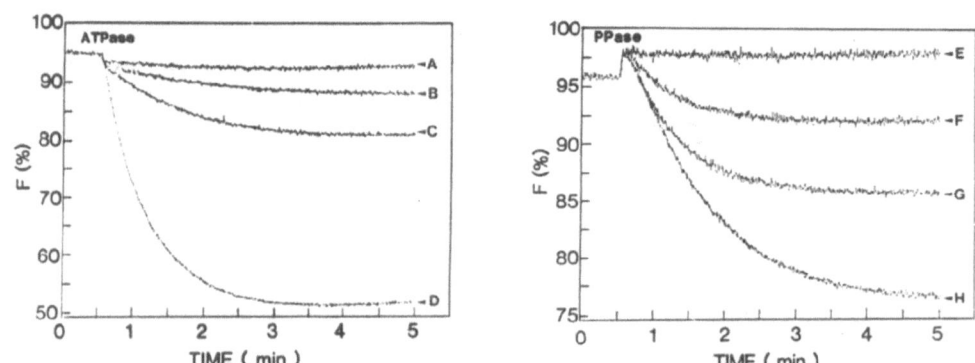

Figure 1
Differential sensitivities of ATP- and PP_i-dependent H^+-translocation
to irreversible inhibition by KI

The tonoplast vesicles were treated with 1.0 M (A, F), 0.6 M (B, H), 0.5 M (C, H), 0.25 and 0.00 M (D, H), 0.8 M (G) and 1.5 M KI (E). The vesicles were washed free of chaotrope before measuring ATP- or PP_i-dependent intravesicular acidification.

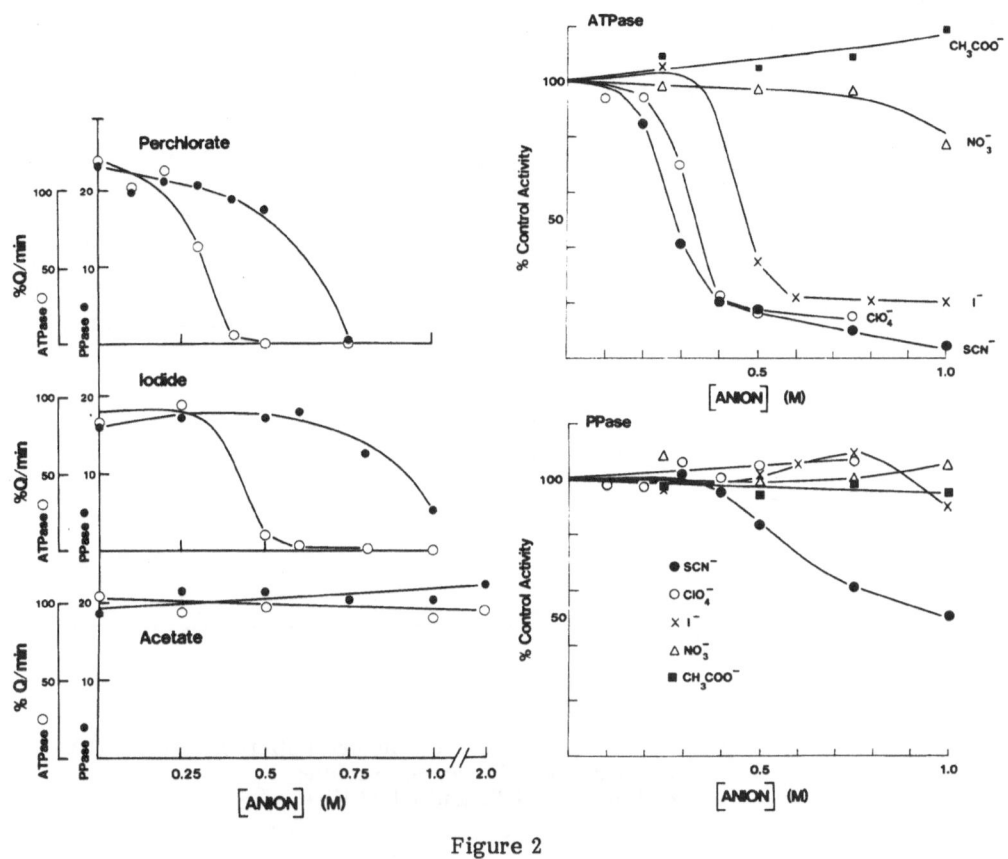

Figure 2

Differential sensitivities of tp-ATPase and tp-PPase
to irreversible inhibition
by SCN$^-$, ClO$_4^-$, I$^-$, NO$_3^-$ or CH$_3$COO$^-$

A. Initial rate of H$^+$-translocation (% Q/min) by the tp-ATPase (O) and tp-PPase
(●) respectively. B. Rate of substrate hydrolysis.

Differential Dissociation of Membrane Proteins

SDS-PAGE of the 100,000 g membrane pellets recovered after treatment with KI revealed a diminution of the intensity of two bands with M_r values of 67 and 57 kDa (Fig. 3-A). Tonoplast vesicles treated with 0.5, 0.75 and 1.0 M KI showed a progressive decrease in the intensity of the 67 and 57 kDa bands whereas these two bands showed little or no alteration in vesicles with or 0.25 M KI (data not shown). The KI concentrations necessary for the onset of irreversible inhibition of the tp-ATPase and attenuation of the 67 and 57 kDa bands (ca. 0.5 M) therefore correspond.

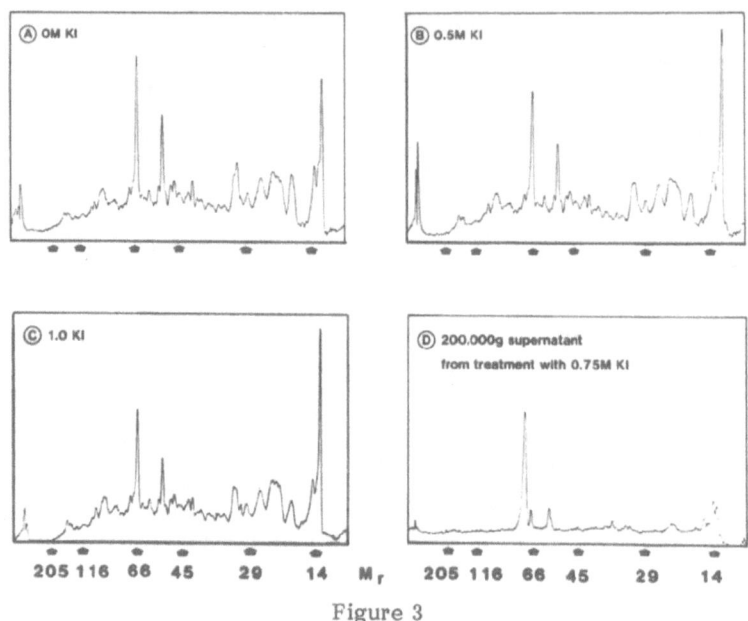

Figure 3

SDS-PAGE analysis of pelletable (A, B and C) and non-pelletable protein (D)
after treatment of tonoplast vesicles
with 0.0, 0.5, 0.75 and 1.0 M KI

Analysis of the 200,000 g supernatant recovered from the treatment with 0.75 M KI demonstrated one major band of 67 kDa and two minor bands at 66 and 57 kDa respectively (Fig. 3-D).

DISCUSSION

There are a number of important conclusions stemming from the results presented above :

- 1) Caution is required when using chaotropic anions for the removal of loosely associated proteins from the tonoplast (Poole et al., 1984; Rea and Poole, 1985) because the tp-ATPase is itself sensitive to irreversible chaotropic effects. While a KI concentration of 0.24 M KI, as employed by Poole et al. (1984) and Rea and Poole (1985), has no discernible effect on tp-ATPase activity, higher concentrations are inhibitory.

- 2) The differential sensitivities of the tp-ATPase and tp-PPase to irreversible inhibition by chaotropic anions directly demonstrates that the two activities are physically and functionally distinct both during H^+-translocation and substrate hydrolysis, without resort to kinetic (Rea and Poole, 1985) or chromatographic analyses (Rea and Poole, 1986).

- 3) Chaotrope concentrations sufficient to maximally inhibit the tp-ATPase but leave the tp-PPase unaffected preferentially detach two polypeptides of M_r 67 and 57 kDa from the tonoplast.

Solubilization and partial purification of the tp-ATPase from higher plant tissues has shown it to consist of 67-70 kDa polypeptide which is labeled by ^{14}C-NEM in the absence of MgATP but not in the presence of substrate (Mandala and Taiz, 1985), a 57-60 kDa polypeptide which labels with α-^{32}P-BzATP (Manolson et al., 1985) and a 16 kDa component which is labeled by ^{14}C-DCCD (Manolson et al., 1985). Preferential detachment of two major bands at 67 and 57 kDa from the tonoplast as indicated by attenuation of their instensity in the membrane pellet and their appearance in the 200,000 g supernatant at chaotrope concentrations sufficient to maximally inhibit the tp-ATPase but not appreciably inhibit the tp-PPase therefore indicates relatively specific dissociation of the 67 and 57 kDa subunits of the tp-ATPase from the membrane. Although it is not possible to establish an unambiguous correlation between loss of enzymatic activity and dissociation of subunits, because of the possibility of non-dissociative denaturation, the treatment of tonoplast vesicles with chaotropic anions might constitute a facile, one-step procedure for the partial purification of the 67 and 57 kDa subunits of the ATPase complex.

Table 2

Comparison of Sensitivity of Tp-ATPase
to Reversible and Irreversible Inhibition by Anions

The values shown are the K_i values for competitive inhibition with respect to MgATP (data from Griffith et al., 1986) and the k_{50} values for the irreversible inhibition of hydrolytic activity (data from Fig. 2-B) respectively. ND = not determined.

Anion	K_I, M	k_{50}, M
SCN^-	0.038	0.29
ClO_4^-	0.054	0.35
ClO_3^-	0.120	ND
I^-	Stimulatory	0.43
NO_3^-	0.022	> 1.25
CH_3COO^-	Non-Inhibitory	Non-Inhibitory
SO_4^{2-}	Non-competitive	Not inhibitory

The reversible, competitive inhibitions exerted by SCN^-, ClO_4^- and NO_3^- and their irreversible, chaotropic effects do not appear to be related. The reversible, competitive inhibitions imposed by the anions tested follow the sequence NO_3^- > SCN^- > ClO_4^- >> CH_3COO^- whereas their irreversible inhibitory potencies fall in the sequence SCN^- > ClO_4^- > NO_3^- >> CH_3COO^-, SO_4^- (see Table 2). Hence, NO_3^- is an anomalously potent competitive inhibitor if reversible chaotropism is the principal cause of the inhibitions seen in the low concentration range. Although there are numerous examples of weak chaotropes, such as NO_3, being more potent destabilizing agents than potent ones (Hatefi and Hanstein, 1974), in that electrostatic as well as hydrophobic interactions may participate in the maintenance of enzyme functional integrity (see Griffith et al., 1986 for further discussion), this does not explain the behavior of the tp-ATPase. Any relationship between competitive inhibition in the low and irreversible inhibition in the high concentration range is refuted. Since we have already shown that NO_3^- does not exert its competive effect by stereochemically simulating the trigonal planar geometry of the terminal phosphoryl-group of ATP to form a dead-end complex with MgADP (Griffith et al., 1986), the mode of action of this inhibitory anion remains unresolved.

REFERENCES

Griffith, C. J., Rea, P. A., Blumwald, E., and Poole, R. J., 1986, Mechanism of stimulation and inhibition of tonoplast H^+-ATPase of Beta vulgaris by chloride and nitrate, Plant Physiol., 81:120.

Hatefi, Y. H., and Hanstein, W. G., 1974, Destabilization of membranes by chaotropic anions, Methods In Enzymology, 31:770.

Mandala, S., and Taiz, L., 1985, Partial purification of a tonoplast ATPase from corn coleoptiles, Plant Physiol., 78:327.

Manolson, M. F., Rea, P. A., and Poole, R. J., 1985, Identification of 3-O-(4-benzoyl)-benzoyl-adenosine-5'-triphosphate- and N,N'-dicyclohexylcarbodiimide binding subunits of a higher plant H^+-translocating tonoplast ATPase, J. Biol. Chem., 260:12273.

Poole, R. J., Briskin, D. P., Kràtky, Z., and Johnstone, R. M., 1984, Density gradient localization of plasma membrane and tonoplast from storage tissue of growing and dormant red beet. Characterization of transport ATPase in tonoplast vesicles, Plant Physiol., 74:549.

Rea, P. A., and Poole, R. J., 1985, Proton-translocating inorganic pyrophosphatase in red beet (Beta vulgaris L.) tonoplast vesicles, Plant Physiol., 77:46.

Rea, P. A., and Poole, R. J., 1986, Chromatographic resolution of H^+-translocating-pyrophoshatase from H^+-translocating ATPase of higher plant tonoplast, Plant Physiol., 81:126.

GENERATION OF A pH-GRADIENT ACROSS THE TONOPLAST OF VACUOLES ISOLATED FROM GREEN SUSPENSION CELLS OF CHENOPODIUM RUBRUM

Bernd Hoffmann, Margrit Gogarten-Boekels
and Friedrich-Wilhelm Bentrup

Botanisches Institut I der Justus-Liebig-Universität
Senckenbergstrasse 17, D-6300-Giessen, F.R.G.

INTRODUCTION

Elucidation of solute transport across the tonoplast requires reliable information on how this membrane is energized. For this purpose we have studied ATP-dependent generation of the electrical membrane potential and conductance by means of the patch-clamp technique (Bentrup et al., 1985; Bentrup et al., 1986), and in parallel experiments, formation of the transtonoplast pH-gradient.

MATERIALS AND METHODS

Protoplasts and vacuoles were isolated from a photoautotrophic and phytohormone-independent cell suspension culture of Chenopodium rubrum L. as described previously (Bentrup et al., 1985; Bentrup et al., 1986). For the spectrophotometrical assay vacuoles were suspended in a medium composed of 300 mM mannitol, 25 mM KCl, 2 mM MgCl$_2$, 1 mM dithiotreitol, and 25 mM Hepes/KOH (pH 7).

RESULTS AND CONCLUSIONS

Vacuoles isolated from Chenopodium rubrum suspension cells show a strong ATP-dependent proton translocation across the tonoplast. Typically, within 25 minutes ATP-generated vacuolar acidification reaches two pH units, i.e. pH 5 (Fig. 1). Acidification is calculated from the vacuolar accumulation of acridine orange; i.e. based upon known vacuolar and extra-vacuolar volumina, a general shift of the vacuolar pH can be calculated from the absorbance loss in the sample at 490 nm (Gogarten-Boekels et al., 1985). Proton translocation is ATP-specific; addition of equimolar (2 mM) amounts of AMP, ADP, GTP or ITP fails to produce a detectable vacuolar acidification (Fig. 2).

The ATP-dependent acidification is quickly reversed upon addition of 16 μM of the uncoupler FCCP (Fig. 1). The final absorbance level reaches the initial level prior to the addition of ATP, indicating that the observed accumulation of acridine orange is not caused by adsorption of the dye to vacuolar constituents. Furthermore, 25 mM KNO$_3$ also reverse the ATP-generated acidification, but at a five-fold slower rate than FCCP. Thus KNO$_3$ might act either with the membrane.

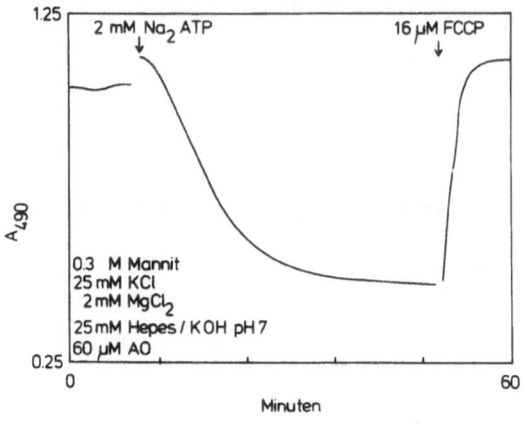

Figure 1

Kinetics of accumulation and subsequent release of acridine orange (AO)
by isolated vacuoles of Chenopodium rubrum
upon addition of ATP and FCCP, respectively

The depicted maximum absorbance shift at 490 nm reflects about two pH units
(see Gogarten-Boekels et al., 1985)
ATPase or might cause proton efflux from the vacuole via a proton-nitrate symport.

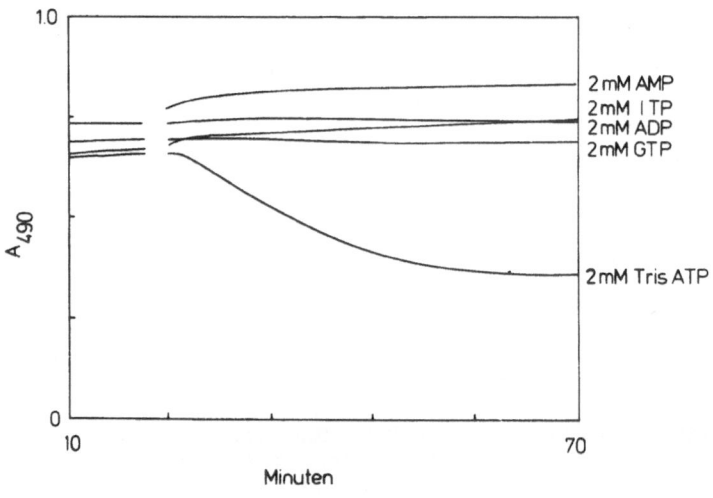

Figure 2

Kinetics of accumulation of acridine orange
by isolated vacuoles of Chenopodium rubrum
upon addition of the different nucleotides indicated

For other details see Fig. 1

Our electrophysiological experiments (Bentrup et al., 1985; Bentrup et al., 1986), using the whole-cell-recording mode of the patch-clamp technique, show that 2 mM ATP hyperpolarize the tonoplast to 15 - 20 mV (vacuole positive). This voltage is also dissipated by 16 μM FCCP.

We conclude that a proton-pumping ATPase in the tonoplast of Chenopodium rubrum creates a small voltage and a big pH gradient according to a pump-and-leak model we have developed recently for a tonoplast-containing vesicle preparation from the Chenopodium cells (Gogarten-Boekels et al., 1985).

ACKNOWLEDGEMENTS

This work was supported by the Deutsche Forschungsgemeinschaft (Grant Be 466/21).

REFERENCES

Bentrup, F.-W., Hoffmann, B., Gogarten-Boekels, M., Gogarten, J. P., and Baumann, C., 1985, A patch clamp study of tonoplast electrical properties in vacuoles isolated from Chenopodium rubrum suspension cells, Z. Naturf., 40 c:886.
Bentrup, F.-W., Hoffmann, B., Gogarten-Boekels, M., Gogarten, J. P., and Baumann, C., 1986, ATP-dependent acidification and tonoplast hyperpolarization in isolated vacuoles from green suspension cells of Chenopodium rubrum, Proc. Natl. Acad. Sci. U.S.A., 83:2431.
Gogarten-Boekels, M., Gogarten, J. P., and Bentrup, F.-W., 1985, Kinetics and specificity of ATP-dependent proton translocation measured with acridine orange in microsomal fractions from green suspension cells of Chenopodium rubrum L., J. Plant Physiol., 118:309.

Mg, ATP-DEPENDENT H$^+$ TRANSPORT IN VACUOLES AND TONOPLAST VESICLES

FROM ACER PSEUDOPLATANUS CELLS

Roberta Colombo, Raffaela Cerana and Piera Lado

Centro C. N. R. per la Biologia Cellulare e Molecolare delle Piante, Dipartimento di Biologia "Luigi Gorini", Università di Milano, Via Celoria, 26, I-20133-Milano (Italy)

INTRODUCTION

Large evidence has accumulated about the existence of an ATPase associated with the vacuolar membrane, with well-defined characteristics different from those of the plasmalemma ATPase. The tonoplast ATPase is stimulated by Cl$^-$, insensitive to monovalent cations, inhibited by nitrate and unaffected by vanadate and has a broad pH optimum between pH 6 and 8 (Sze, 1984; Marrè and Ballarin-Denti, 1985).

The hydrolyzing activity of this ATPase is coupled to an electrogenic transport of protons. This capability of the tonoplast ATPase to catalyze H$^+$ transport has been shown, in most cases, in vesicles either derived from microsomal preparations ("light fraction") or prepared from isolated vacuoles (Churchill and Sze, 1983; De Michelis et al., 1983; Bennett et al., 1984; Hager and Biber, 1984; Poole et al., 1984; Mandala and Taiz, 1985; Thom and Komor, 1985) and only in few cases H$^+$ transport has been shown in intact vacuoles (Mandala and Taiz, 1985; Thom and Komor, 1985; Wagner and Lin, 1982; Jochem et al., 1984; Chrestin et al., 1985; John and Miller, 1986).

In this paper we studied whether it is possible to measure an ATP-dependent H$^+$ transport using the optical pH acridine orange (AO) in intact vacuoles from Acer pseudoplatanus cells, or whether the buffering capacity and the low pH of the vacuolar sap present this measurement. Thus, we had compared the ATP-dependent H$^+$ transport in intact vacuoles isolated by two different methods and in vesicles prepared from the same vacuoles.

MATERIALS AND METHODS

Plant Material

Cell suspension cultures of Acer pseudoplatanus derived from a strain kindly supplied by Prof. J. Guern were grown as described by Leguay and Guern (1975). The cells were harvested at the 8th day after subculture (exponential phase) when the density was about 4 10^5 cells/ml. The protoplasts were prepared by enzymatic digestion (Kurkdjian and Barbier-Brygoo, 1984).

Vacuole Isolation

The vacuoles (A) were isolated from the protoplasts by polybase-induced lysis of plasmalemma on a Ficoll gradient according to the method described by Alibert et al. (1982), slightly modified as follows : the pH of Mes-Tris buffer was raised to pH 7.5 and 1 mM DTT was added to all solutions. The collected vacuoles (about 2 - 3 10^6/ml) were in a medium containing 25 mM Mes-Tris buffer (pH 7.5), 0.6 M mannitol, 5% (w/v) Ficoll 400, 1 mg/ml dextran sulphate, 1 mM DTT. This vacuole suspension was added with 0.5% BSA, immediately after collection.

The vacuoles (B) were obtained by osmotic shock. The protoplast suspension (2 10^6/ml) was diluted 1:1 with 25 mM Mes-Tris buffer (pH 7.5), containing 2 mM EDTA, 2 mM DTT, 0.1% BSA and incubated under low agitation (25 rpm) at room temperature. After 25 min 7 volumes of a solution containing 0.55 M mannitol, 8% Ficoll 400, 1 mM EDTA, 1 mM DTT, 0.05% BSA (pH 7.3), were gently added. Centrifuge tubes were filled with 12 ml of the vacuole suspension, previously filtered on 53 m pore nylon cloth. Then 0.8 ml of 2 mM Mes-Tris buffer (pH 7.3) containing 0.6 M mannitol, 1 mM DTT and 0.25% BSA were stratified on the top. After 15 min centrifugation at 160 g at 4°C the vacuoles were collected at the interface.

Preparation of Tonoplast Vesicles

The vesicles were obtained by passing the vacuole suspension through a 26 gauge needle with a syringe, until no vacuoles were observed under light microscopy. This procedure was performed either directly on the vacuole suspension or in the case of Table 4 after 1:1 dilution with an appropriate medium, in order to obtain the desired pH or buffer concentration.

Proton Transport Measurements

H^+ transport was measured by the rate of decrease in absorbance of AO at 492 nm with 550 nm as the reference wavelength by a Sigma ZWS II dual wavelength spectrophotometer equipped with a recorder.

The reaction mixture (1.5 ml) contained 0.5 - 1 10^6 vacuoles or corresponding tonoplast vesicles, 25 mM Mes-tris buffer at pH 7.9 (unless otherwise specified), 1 mM EGTA, 100 mM KCl, 500 μM sodium molybdate (to inhibit acid phosphatases), 4 mM $MgSO_4$, 2.5 μM AO and mannitol to a final 0.6 M osmolarity.

The ATP-dependent transport of H^+ measured in these conditions is proportional to the vacuole concentration in the range 0.3/1.5 10^6 vacuoles/sample.

Each experiment was performed several times in triplicate.

Protein Determination

The vacuole suspension in the absence of BSA was added with a 20-fold volume of 1 mM $MgSO_4$ and centrifuged at 106,000 g for 1 h. Protein content of the pellet was determined by the Lowry method as modified by Markwell et al. (1978) with BSA as a standard. The protein content of 10^6 vacuoles ranged from 6 to 12 μg.

Chemicals

AO, purchased from BDH, was recrystallized according to Pal and Schubert (1963). Fatty acid free BSA (fraction V) and MgATP were obtained from Sigma. Purified sodium salt of ATP was obtained passing Na_2-ATP (Sigma) solution through Dowex 50 cation exchanger resine and then neutralizing the eluted ATP with NaOH. FCCP and oligomycin, purchased from Sigma, was dissolved in ethanol.

RESULTS

ATP-dependent H$^+$ Transport in Intact Vacuoles

The vacuoles were isolated from protoplast of <u>Acer</u> cells either by polybase-induced plasmalemma lysis on a Ficoll gradient (A) or by osmotic shock (B). In both preparations the isolated vacuoles retained at least in part their original acidity, as indicated by the accumulation of neutral red.

Vacuoles (A)

The vacuoles suspended in the assay medium accumulated AO and the addition of FCCP (Fig. 1-A) only slightly affected the AO distribution, indicating that in isolated vacuoles the protons are close to their electrochemical equilibrium (Jochem et al., 1984; Weigel and Weis, 1984). Addition of 10 mM ammonium sulphate induced a complete release of accumulated AO (data not shown).

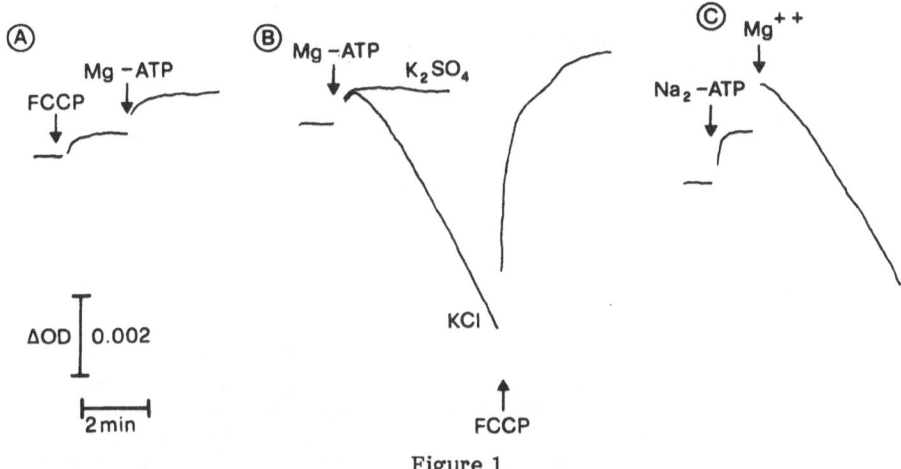

Figure 1

ATP-dependent H$^+$ transport in vacuoles (A)

KCl: 100 mM; K$_2$SO$_4$: 50 mM; FCCP: 10 μM; MgATP, Na$_2$ATP: 6 mM. Mg^{2+} addition in C: 10 mM. pH of the assay medium: 7.9. 10^6 vacuoles/sample. Traces are from one representative experiment.

Fig. 1-B shows that the addition of MgATP to the vacuole suspension in the presence of a permeating anion led to a decrease of AO absorbance indicating an acidification of the intravacuolar space and that the amount of AO accumulated in the presence of MgATP was completely released by 10 μM FCCP. The addition of MgATP in the presence of FCCP was ineffective (Fig. 1-A). ATP disodium salt did not induce H$^+$ transport and the addition of Mg^{2+} was required to induce H$^+$ (Fig. 1-C). MgADP was ineffective.

Table 1 shows the effects of inhibitors of various ATPases (oligomycin, vanadate, NO$_3$$^-$) on H$^+$ transported measured at pH 7.3 and 7.9. At both pHs oligomycin (2 μg/ml) had a very small inhibiting effect, thus ruling out a significant contribution of submitochondrial particles to the measured H$^+$ transport. A significant inhibition by 50 M vanadate was observed at pH 7.3, while at pH 7.9 vanadate was ineffective. This indicates that contaminating plasmalemma vesicles, able to transport H$^+$, are present in the vacuole preparation and that the contribution of this contamination to H^+ transport dissappears at high pH in agreement with the reported pH dependence of the plasmalemma ATPase (Sze, 1984; Marrè and Ballarin-Denti, 1985; Cocucci and Marrè, 1984).

Table 1

Effects of Inhibitors and EGTA
on ATP-dependent H^+-transport (vacuoles A)

	H^+-transport[a]	
	pH 7.3	pH 7.9
Control	2.50	1.80
+ 50 mM KNO_3	0.98	0.08
+ 50 μM Vanadate	1.51	1.60
+ 2 μg/ml oligomycin	2.32	1.64
- EGTA	1.92	0.80

[a]The figures represent the initial rate of decrease in AO absorbance after the addition of MgATP and are expressed as $(\Delta OD \times min^{-1} \times 10^6 \text{ vacuoles}^{-1}) \times 10^3$.

In the presence of 50 mM NO_3^-, an inhibitor of the tonoplast ATPase used to distinguish this ATPase from that of plasmalemma, H^+ transport was strongly inhibited (60% at pH 7.3 and 100% at pH 7.9). The rates of nitrate-sensitive H^+ transport were similar at the two pHs in agreement with the finding that in tonoplast vesicles from red beet H^+ transport is little sensitive to pH changes between pH 7 and 8 (Poole et al., 1984).

These data indicate that H^+ transport measured at pH 7.9 is driven by the ATPase localized on the vacuolar membrane.

The data of Table 1 also show that in the absence of 1 mM EGTA the rate of H^+ transport is lower at both pHs. It has been demonstrated that while plasmalemma ATPase is strongly inhibited by Ca^{2+} (Cocucci and Marrè, 1984), tonoplast ATPase is not affected by this cation (Tognoli, 1985). As the stimulation of H^+ transport by EGTA was even higher at pH 7.9, when H^+ transport driven by the plasmalemma ATPase was negligeable, the most reasonable interpretation of the effect of EGTA might be that EGTA prevents a partial dissipation of the H^+ gradient due to a Ca^{2+}/H^+ antiport at tonoplast level (Chrestin et al., 1985; Schumaker and Sze, 1985; Blumwald and Poole, 1986).

Vacuoles (B)

The rates of H^+ transport in vacuoles (A) were small, although reproducible in all experiments. To study whether the low rates of H^+ transport could depend on the isolation procedure, we measured H^+ transport in vacuoles isolated by osmotic shock (B) (Table 2) H^+ transport in these vacuoles was, in most cases, hardly detectable (Thom and Komor, 1985; John and Miller, 1986). A serious damage to the H^+ transport system during the isolation is unlikely as the vesicles derived from the same vacuoles were able to transport H^+.

Table 2

ATP-dependent H^+ Transport (pH 7.9)
in Intact Vacuoles (A and B)
and in Vesicles derived from them

	Sodium molybdate (500 μM)		Ammonium molybdate (100 μM)	
	(A)	(B)	(A)	(B)
Vacuoles	1.71	0.05	0.85	0.30
Vesicles	5.10	11.50	2.85	6.05

The values are expressed as (ΔOD x min$^-$ x 10^6 vacuoles^{-1}) x 10^3. (A) : isolation by polybase-induced lysis; (B) : isolation by osmotic shock.

Table 2 shows the results obtained by substituting sodium molybdate with ammonium molybdate in the assay medium. H^+ transport of vacuoles (A) was inhibited as expected if the NH_3 present in the medium in equilibrium with NH_4^+ permeates into the vacuoles partially neutralizing the ATP-dependent ΔpH. On the contrary in vacuoles (B) the presence of NH_4^+ made evident a small H^+ transport (see Wagner and Lin, 1982). This effect in vacuoles (B) might depend on an increase in pH of the vacuolar sap, which could make measurable the activity of the H^+ pump, otherwise undetectable because of the strong buffering capacity of the vacuolar sap. The increase in intravacuolar pH could also reduce a possible inhibition of the H^+ pump by the low internal pH.

The more reasonable interpretation of the different behavior of vacuoles (A) and (B) is that the vacuoles (B) better retained the original acidity and buffering capacity of vacuolar sap, while the membrane of vacuoles (A) became more leaky. This is supported also by the fact that the amount of AO which accumulated into vacuoles (B) incubated with the probe before the addition of ATP was larger than that into vacuoles (A) (data not shown). We cannot rule out the possibility that the vacuole (A) suspension contained a certain amount of vesicles arising from the membrane of vacuoles broken during the isolation procedure and that the measured H^+ transport occurred mainly in these vesicles.

H^+ Transport in Vesicles from Vacuoles (A) and (B)

Tonoplast vesicles derived from both vacuoles (A) and (B) showed an ATP-dependent H^+ transport (Table 2). The rate values were higher than those in intact vacuoles (2-3 fold for vesicles (A) and much more for vesicles (B) ins spite of the fact that some of the vesicles might be inside-out. In the following experiments vesicles derived from vacuoles (B), which could be easily manipulated, were used to characterize H^+ transport.

Table 3 shows that H^+ transport (measured at pH 7.9) in vesicles was inhibited by nitrate, unaffected by vanadate and oligomycin and stimulated by the presence of EGTA, this picture being similar to that observed in vacuoles (A).

To check the possibility that the intravacuolar pH controls the activity of the tonoplast H^+-ATPase, vesicles at different internal pHs (5.3; 5.9; 6.9; 7.9) were prepared by diluting (1:1) the vacuole suspension in 2 mM Mes-Tris buffer with

Table 3

Effects of Inhibitors and EGTA
on ATP-dependent H^+ Transport in Vesicles (B)

Control	12.5
+ 50 mM KNO_3	5.2
+ 50 μM Vanadate	2.0
+ 2 μg/ml oligomycin	12.5
- EGTA	8.5

The values are expressed are expressed as (ΔOD x min^{-1} x 10^6 $vacuoles^{-1}$) x 10^3. pH of the assay medium : 7.9.

HCl or NaOH (see also Materials and Methods). Due to the presence of buffering substances released during vacuole isolation besides Mes-Tris buffer, the buffering capacities of the obtained vesicle preparations, measured by titration, resulted to be the same at each considered pH in the range $\pm$ 0.3 pH units and were lower than that of the cell sap measured by titration ("low buffer").

We do not know whether the intravesicular pH changed because of a possible H^+ leakage, when the vesicles were added to the assay medium (pH 7.9). However the equilibrium distributions of AO were different in the vesicles sealed on different pHs and remained constant during a period of time equal to that of the assay of the ATP-dependent H^+ transport.

Table 4 shows that H^+ transport was slightly affected by changes of intravesicular pH between 7.9 and 5.9, whereas it was markedly inhibited lowering the pH to 5.3. These data seems to indicate an inhibition of the H^+ pump activity by a low vacuolar pH. Further work is required to establish if the control of the H^+-pumping tonoplast ATPase by internal pH can be explained in thermodynamic and/or kinetic terms.

Table 4

Effect of Intravesicular pH and Buffering Capacity
on ATP-dependent H^+ Transport in Vesicles (B)

Intravacuolar pH	H^+ transport[a]	
	Low buffer"	"High buffer"
5.3	5.25	0.53
5.0	10.25	–
6.9	12.25	–
7.8	10.70	–

[a]The values are expressed as (ΔOD x min^{-1} x 10^6 $vacuoles^{-1}$) x 10^3. For "low buffer" and "high buffer" see the text. pH of the assay medium : 7.9.

In another experiment (Table 4) the vesicles were sealed on solutions containing either 50 mM Mes adjusted at pH 5.3 with Tris or 55 mM Mes-Tris buffer at pH 6.9 ("high buffer"), whose buffering capacities were the same and similar to that of the cell sap. At pH 6.9, the rate of H^+ transport is relevant, whereas at pH 5.3 H^+ transport is hardly detectable.

These data indicate the difficulty to measure H^+ transport in intact vacuoles depends not only on the strong buffering capacity but also on the control of the H^+ transport activity by the intravacuolar pH.

CONCLUSIONS

The main conclusions suggested by the reported data are the following :

a) The failure to measure a relevant H^+ transport in intact vacuoles depends not only on the strong buffering capacity of the vacuolar sap which makes undetectable the operation of the H^+ pump but also on an inhibition of the H^+ pump by the low intravacuolar pH.

b) The isolation procedure of the vacuoles influences the possibility to measure H^+ transport, since the different methods can alter the original state of the vacuolar sap in different ways. This fact has to be considered not only in the case of H^+ transport measurements, but also in the case of the analysis of the vacuolar sap and of the direct electrophysiological measurements in isolated vacuoles.

ACKNOWLEDGEMENTS

We would like to thank E. Marrè, F. Rasi-Caldogno, M. I. De Michelis for the critical reading of the manuscript and A. Kurkdjian for the informations and suggestions for protoplast and vacuole isolation.

REFERENCES

Alibert, G., Carrasco, A., and Boudet, A. M., 1982, Changes in biochemical composition of vacuoles isolated from Acer pseudoplatanus L. during cell culture, Biochim. Biophys. Acta, 721:22.
Bennett, A. B., O'Neill, S. D., and Spanswick, R. M., 1984, H^+-ATPase activity from storage tissue of Beta vulgaris. I. Identification and characterization of an anion-sensitive H^+-ATPase, Plant Physiol., 74:538.
Blumwald, E., and Poole, R. J., 1986, Kinetics of a Ca^{2+}/H^+ antiport in isolated tonoplast vesicles from storage tissue of Beta vulgaris L., Plant Physiol., 80:727.
Chrestin, H., Gidrol, X., D'Auzac, J., Jacob, J. L., and Marin, B., 1985, Cooperation of a "Davies type" biochemical pH-stat and the tonoplastic bioosmotic pH-stat in the regulation of the cytosolic pH of Hevea latex, in : "Biochemistry and Function of Vacuolar Adenosine-Triphosphatase in Fungi and Plants", B. P. Marin, ed., Springer-Verlag, Berlin, Heidelberg, New-York and Tokyo.
Churchill, K. A., and Sze, H., 1983, Anion-sensitive, H^+-pumping ATPase in membrane vesicles from oat roots, Plant Physiol., 71:610.
Cocucci, M. C., and Marrè, E., 1984, Lysophosphatidylcholine-activated, vanadate-inhibited, Mg^{2+}-ATPase from radish microsomes, Biochim. Biophys. Acta, 771:42.
De Michelis, M. I., Pugliarello, M. C., and Rasi-Caldogno, F., 1983, Two distinct proton-translocating ATPases are present in membrane vesicles from radish seedlings, FEBS Letters, 162:85.
Hager, A., and Biber, W., 1984, Functional and regulatory properties of H^+ pumps at the tonoplast and plasma membranes of Zea mays coleoptiles, Z. Naturforsch., 39:927.

Jochem, P., Rona, J.-P., Smith, J. A. C., and Lüttge, U., 1984, Anion-sensitive ATPase activity and proton transport in isolated vacuoles of species of the CAM genus Kalanchöe, Physiol. Plant., 62:410.

John, P., and Miller, A. J., 1986, Electrogenic proton translocation by the adenosine-triphosphatase of intact vacuoles isolated from beet (Beta vulgaris L.), J. Plant Physiol., 122:1.

Kurkdjian, A. C., and Barbier-Brygoo, H., 1984, A hydrogen ion-selective liquid membrane microelectrode for measurement of the vacuolar pH of plant cells in suspension culture, Anal. Biochem., 132:96.

Leguay, J. J., and Guern, J., 1975, Quantitative effects of 2,4-dichlorophenoxyacetic acid on growth of suspension-cultured Acer pseudoplatanus cells, Plant Physiol.,56:356.

Mandala, S., and Taiz, L., 1985, Proton transport in isolated vacuoles from corn coleoptiles, Plant Physiol., 78:104.

Markwell, M. A. K., Haas, S. M., Bieber, L. L., and Tolbert, N. E., 1978, A modification of the Lowry procedure to simplify protein determination in membrane and lipoprotein samples, Anal. Biochem., 87:206.

Marrè, E., and Ballarin-Denti, A., 1985, The proton pumps of the plasmalemma and the tonoplast of higher plants, J. Bioenergetics Biomembranes, 17:1.

Pal, M. K., and Schubert, M., 1963, Measurement of the stability of metachromatic compounds, J. Am. Chem. Soc., 84:4384.

Poole, R. J., Briskin, D. P., Krátky, Z., and Johnstone, R. M., 1984, Density gradient localization of plasma membrane and tonoplast from storage tissue of growing and dormant red beet, Plant Physiol., 74:549.

Schumaker, K. S., and Sze, H., 1985, A Ca^{2+}/H^+ antiport system driven by the proton electrochemical gradient of a tonoplast H^+-ATPase from oat roots, Plant Physiol., 79:1111.

Sze, H., 1984, H^+-translocating ATPases of the plasma membrane and tonoplast of plant cells, Physiol. Plant., 61:683.

Thom, M., and Komor, E., 1985, Electrogenic proton translocation by the ATPase of sugarcane vacuoles, Plant Physiol., 77:329.

Tognoli, L., 1985, Partial purification and characterization of an anion-activated ATPase from radish microsomes, Eur. J. Biochem., 146:581.

Wagner, G. J., and Lin, W., 1982, An active proton pump of intact vacuoles isolated from Tulipa petals, Biochim. Biophys. Acta, 689:261.

Weigel, H. J., and Weis, E., 1984, Determination of the proton concentration difference across the tonoplast membrane of isolated vacuoles by means of fluorescence, Plant Science Letters, 33:163.

THE EFFECTS OF ANIONS ON PP$_i$-DEPENDENT PROTON TRANSPORT

IN TONOPLAST VESICLES FROM OAT ROOTS

Andrew J. Pope and Roger A. Leigh

Rothamsted Experimental Station
Harpenden, Hertfordshire, U. K.

INTRODUCTION

The majority of investigations of anion transport at the tonoplast have involved examination of the effects of anions on ATP-dependent H$^+$-transport in sealed tonoplast vesicles (e.g. Bennett and Spanswick, 1983; Churchill and Sze, 1984; Blumwald and Poole, 1985; Lew and Spanswick, 1985). These studies have shown that anions stimulate ΔpH formation and dissipate membrane potential ($\Delta\Psi$) and that these effects are sensitive to anion channel blockers such as DIDS and SITS. This is thought to indicate that anions move trough selective channels in the tonoplast in response to the $\Delta\Psi$ generated by ATPase, causing dissipation of the $\Delta\Psi$ and thus an increase in ΔpH, as expected from the chemiosmotic hypothesis (Mitchell, 1966).

Unfortunately, further study of this process is difficult because the tonoplast ATPase is directly stimulated by Cl$^-$ and inhibited by both NO$_3^-$ and DIDS (Walker and Leigh, 1981 a; Bennett and Spanswick, 1983; Churchill and Sze, 1984). This makes it difficult to separate effects on ΔpH formation that are due to secondary anion transport from those due to direct alteration of the ATPase activity (Bennett and Spanswick, 1983; Churchill and Sze, 1984). However, this problem can be overcome by studying the effects of anions on the H$^+$-translocating inorganic pyrophosphatase (PPase) that is associated with the tonoplast. This enzyme is dependent on Mg^{2+} and K$^+$ for activity but is not directly affected by anions (Walker and Leigh, 1981 b; Rea and Poole, 1985; Wang et al., 1986). Therfore any effects of anions on pyrophosphate (PP$_i$)-dependent H$^+$-transport should be due only to secondary transport phenomena.

MATERIALS AND METHODS

Growth of seedlings of oat (_Avena sativa_ L., cv. Bulwark or Trafalgar) and preparation of sealed tonoplast vesicles were as described by Wang et al. (1986). Formation of an acidic pH or a positive membrane potential within the vesicles was followed by measuring the quenching of the fluorescence of quinacrine or Oxonol V, respectively. Conditions for these assays were similar to those described by Wang et al. (1986) except that the PP$_i$ concentration was 150 mmol m^{-3}. This is close to the optimum PP$_i$ concentration for both H$^+$-transport and PP$_i$ hydrolysis (Leigh and Pope, 1987).

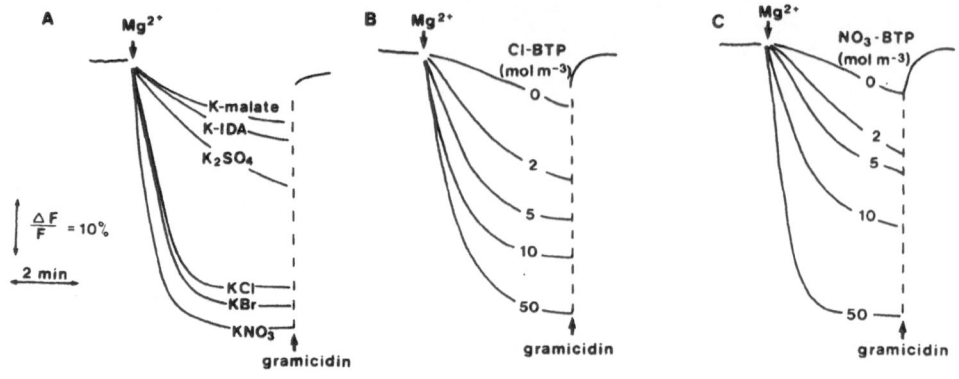

Figure 1

The effect of anions on PP$_i$–dependent ΔpH formation
measured by quinacrine fluorescence quenching

The anion concentration in (A) was 50 mol m^{-3}.

RESULTS

Pyrophosphate–driven ΔpH formation was stimulated by anions (Fig. 1) and the order of effectiveness of the anions was :

$$NO_3^- > Br^- > Cl^- \gg SO_4^{2-} > malate = iminodiacetate (IDA)$$

Both the initial rate and the total extent of quenching were stimulated indicating that anions altered both the rate of H$^+$–transport and the final ΔpH achieved. Stimulation by Cl$^-$ showed saturation kinetics but that by NO$_3^-$ was better described by a two-phase response consisting of saturable and linear components (Fig. 2). The saturable phases for both Cl$^-$ and NO$_3^-$ had K$_m$ values of approximately 1 mol m^{-3}.

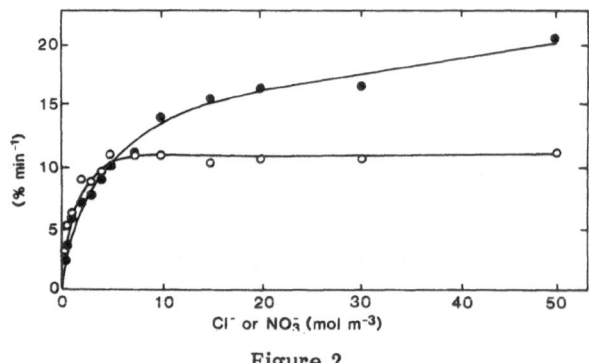

Figure 2

The effect of various concentrations of Cl$^-$ (○) and NO$_3^-$ (●)
on the initial rate of PP$_i$–dependent H$^+$–pumping

The simplest explanation for the effects of anions on PP$_i$–dependent H$^+$–transport is that NO$_3^-$, Cl$^-$ and Br$^-$ are permeable anions that dissipate the generated $\Delta\Psi$ by the PPase and so stimulate both the rate and extent of ΔpH formation (see Introduction). It follows from this that the order of effectiveness of anions at dissipating $\Delta\Psi$ should be similar to their ability to stimulate ΔpH. To test this we examined the effects of the anions on $\Delta\Psi$ using the optical probe, Oxonol-V.

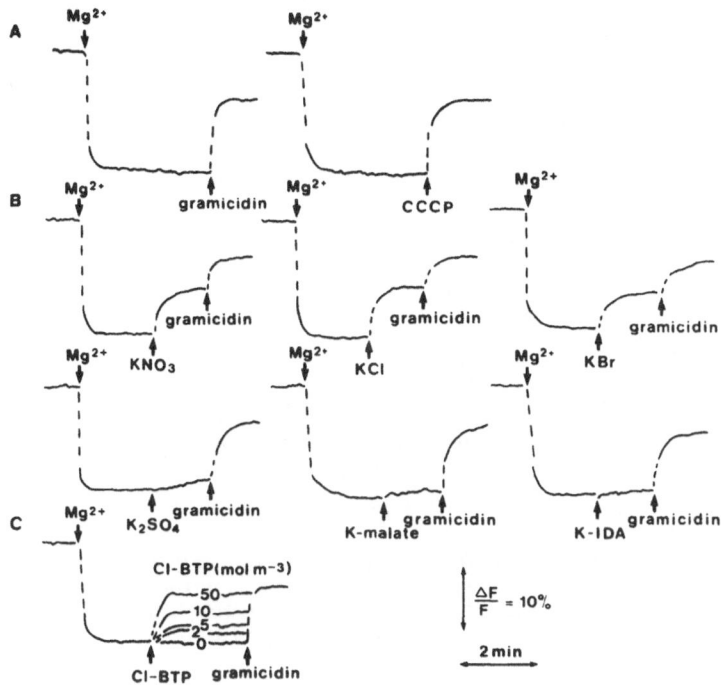

Figure 3

The effects of anions on PP_i-dependent
measured by quenching of Oxonol-V fluorescence

A: Addition of gramicidin-D or CCCP at a concentration of 2.5 μg cm^{-3} dissipates $\Delta\Psi$; B: 10 mol m^{-3} KNO$_3$, KCl and KBr partially dissipate $\Delta\Psi$ but K$_2$SO$_4$, K-malate and K-IDA do not; C: Extent of dissipation of $\Delta\Psi$ increases with Cl$^-$ concentration.

When the PP_i-dependent H$^+$-pump was activated by the addition of Mg^{2+}, the fluorescence of Oxonol-V was quenched, indicating the formation of a positive $\Delta\Psi$ within the vesicles (Fig. 3). This was reversed by gramicidin-D and the protonophore, carbonyl cyanide m-chlorophenylhydrazone (CCCP), indicating that H$^+$-transport was responsible for the $\Delta\Psi$ (Fig. 3-A).

Of the anions tested, only NO$_3^-$, Cl$^-$ and Br$^-$ reversed the quenching, indicating that they were able to relieve $\Delta\Psi$ (Fig. 3-B), consistent with their ability to stimulate ΔpH.

The observation that only Cl$^-$, Br$^-$ and NO$_3^-$ were able to significantly stimulate ΔpH formation and dissipate $\Delta\Psi$ suggests that a selective anion channel might be involved in $\Delta\Psi$-dependent transport of anions across the tonoplast (see also Lew and Spanswick, 1985). To test this we examined the effects of the anion channel blocker, DIDS. Initial experiments indicated that 50 mmol m^{-3} DIDS inhibited PPase activity by only 20% (not shown), yet 20 mmol m^{-3} DIDS completely prevented the stimulation of ΔpH formation by Cl$^-$ and by low concentrations of NO$_3^-$ (Fig. 4). With 50 mol m^{-3} NO$_3^-$ a proportion of the stimulation was insensitive to DIDS. This DIDS-insensitive component showed a linear NO$_3^-$ concentration dependence (cf. Fig. 2). Similar results were obtained with SITS except that higher concentrations were needed. Experiments are in progress to determine the effect of DIDS on the dissipation of membrane potential by anions.

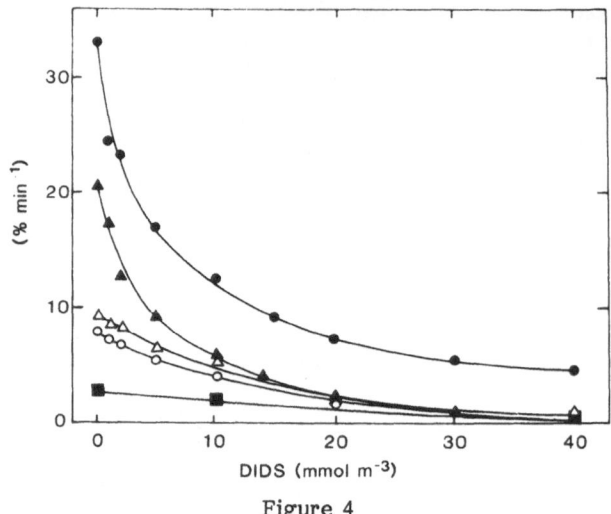

Figure 4

The effect of DIDS on PP_i-dependent ΔpH formation
measured in the absence of anions (■), or in the presence
of 2 mol m^{-3} Cl$^-$ (△), 2 mol m^{-3} NO$_3^-$ (○),
50 mol m^{-3} Cl$^-$ (▲) or 50 mol m^{-3} NO$_3^-$ (●)

DISCUSSION

The stimulation of ΔpH and dissipation of $\Delta\Psi$ by Cl$^-$, Br$^-$ and NO$_3^-$ suggests that these anions move into the vesicles in response to the $\Delta\Psi$ generated by the PP_i-dependent H$^+$-pump, causing a decrease in $\Delta\Psi$ and an increase in ΔpH, as expected from the chemiosmotic hypothesis (Mitchell, 1966). The lack of effect of other physiologically important anions such as SO$_4^{2-}$ and malate indicates that the anion transport system is selective. The sensitivity to DIDS and SITS indicates that an anion channel is probably involved. Essentially similar conclusions were reached by Kaestner and Sze (1986) in a preliminary report of a study of the effect of anions on PP_i-dependent $\Delta\Psi$, also with oat root tonoplast vesicles.

For NO$_3^-$, there was evidence for a second transport pathway which showed non-saturation kinetics (Fig. 2). A non-saturable phase has been demonstrated for Cl$^-$ dependent stimulation of H$^+$-transport into maize root tonoplast vesicles (Bennett and Spanswick, 1983) and for ATP-dependent Cl$^-$ uptake into barley mesophyll vacuoles (Martinoia et al., 1986).

The K_m for both Cl$^-$ and NO$_3^-$ was about 1 mol m^{-3}. Previously, there has been some difficulty in determining the K_m for secondary anion transport because ATP was used as the energy source and it was difficult to separate effects due to secondary anion transport from direct alteration of the ATPase activity (Bennett and Spanswick, 1983; Churchill and Sze, 1984).

In the vesicles used in this study, NO$_3^-$ stimulated PP_i-dependent H$^+$-transport as effectively as Cl$^-$. This contrasts with previous results for both oat and beet tonoplast vesicles which indicated that NO$_3^-$ was less stimulatory than Cl$^-$ (Rea and Poole, 1985; Wang et al., 1986). It was suggested that this might be due to the operation of a H$^+$/NO$_3^-$ symport that could be involved in the export of NO$_3^-$ from the vacuole (Blumwald and Poole, 1985). This symport is either absent from the vesicles used in this study or the rate of H$^+$ efflux through it is low compared to the rate of PP_i-dependent H$^+$-transport into the vesicles. The reason for the difference between the present results and those obtained in the other studies is unclear. Chanson et al. (1985) found that PP_i-dependent ΔpH formation in tonoplast vesicles from corn coleoptiles was equally stimulated by Cl^- and NO_3^- indicating that this membrane also lacks significant symport activity.

In conclusion, the effects of anions on PP_i-dependent H^+-transport confirm that the tonoplast possesses a DIDS-sensitive anion channel that allows the transport of anions across this membrane in response to the $\Delta\Psi$ generated by electrogenic H^+-transport. Whether this can fully account for anion accumulation within vacuoles in vivo depends on the size of the anion concentration gradient and $\Delta\Psi$ across the tonoplast in intact cells. Such a mechanism could explain Cl^- accumulation in Avena coleoptile vacuoles. Bates et al. (1982) found that the tonoplast $\Delta\Psi$ is at least + 40 mV in this tissue and, at this value, Cl^- would probably be in electrochemical equilibrium across the tonoplast (Pierce and Higinbotham, 1970). However, in giant algae, such as Nitella and Chara, Cl^- is not at electrochemical equilibrium across the tonoplast and some other form of energy-dependent transport must be invoked (MacRobbie, 1970). The nature of these other anion transport systems is unclear but further studies of anion transport at the tonoplast in vitro should help to identify the possible mechanisms.

REFERENCES

Bates, G. W., Goldsmith, M. H. M., and Goldsmith, T. H., 1982, Separation of tonoplast and plasma membrane potential and resistance in cells of oat coleoptiles, J. Membrane Biol., 66:15.
Bennett, A. B., and Spanswick, R. M., 1983, Optical measurement of ΔpH and $\Delta\Psi$ in corn root membrane vesicles : Kinetic analysis of Cl^- effects on a proton-translocating ATPase, J. Membrane Biol., 71:95.
Blumwald, E., and Poole, R. J., 1985, Nitrate storage and retrieval in Beta vulgaris L. : Effects of nitrate and chloride on proton gradients in tonoplast vesicles, Proc. Natl. Acad. Sci. U.S.A., 82:3683.
Chanson, A., Fichmann, J., Spear, D., and Taiz, L., 1985, Pyrophosphatase-driven proton transport by microsomal membranes of corn coleoptiles, Plant Physiol., 79:159.
Churchill, K. A., and Sze, H., 1984, Anion-sensitive, H^+-pumping ATPase of oat roots. Direct effects of Cl^-, NO_3^-, and a disulfonic stilbene, Plant Physiol., 76:490.
Kaestner, K. H., and Sze, H., 1986, Potential-dependent anion transport across tonoplast vesicles from oat roots, Plant Physiol., 80 S:81.
Leigh, R. A., and Pope, A. J., 1987, Understanding tonoplast function : Some emerging problems, in : "Plant Vacuoles. Their Importance in Solute Compartmentation and Their Applications in Biotechnology", B. Marin, ed., Plenum Publishing Corporation, New-York.
Lew, R. R., and Spanswick, R. M;, 1985, Characterization of anion effects on nitrate-sensitive ATP-dependent proton pumping activity of soybean (Glycine max L.) seedling root microsomes, Plant Physiol., 77:352.
MacRobbie, E. A. C., 1970, The active transport of ions in plant cells, Quart. Rev. Biophys., 3:251.
Martinoia, E., Schramm, M. J., Kaiser, G., Kaiser, W. M., and Heber, U., 1986, Transport of anions in isolated barley vacuoles. I. Permeability to anions and evidence for a Cl^- uptake system, Plant Physiol., 80:895.
Mitchell, P., 1966, "Chemiosmotic Coupling In Oxidative And Photosynthetic Phosphorylation", Glynn Research, Bodmin, Cornwall, U. K.
Pierce, W. S., and Higinbotham, N., 1970, Compartments and fluxes of K^+, Na^+ and Cl^- in Avena coleoptiles, Plant Physiol., 46:666.
Rea, P. A., and Poole, R. J., 1985, Proton-translocating inorganic pyrophosphatase in red beet (Beta vulgaris L.) tonoplast vesicles, Plant Physiol., 77:46.

Walker, R. R., and Leigh, R. A., 1981 a, Characterization of a salt-stimulated ATPase activity associated with vacuoles from storage roots of red beet (Beta vulgaris L.), Planta, 153:140.

Walker, R. R., and Leigh, R. A., 1981 b, Mg^{2+}-dependent, cation-stimulated inorganic pyrophosphatase associated with vacuoles from storage roots of red beet (Beta vulgaris L.), Planta, 153:150.

Wang, Y., Leigh, R. A., Kaestner, K. H., and Sze, H., 1986, Electrogenic H^+-pumping pyrophosphatase in tonoplast vesicles from oat roots, Plant Physiol., 81:497.

CHANGES OF VACUOLAR AND CYTOPLASMIC pH ASSOCIATED

WITH THE ACTIVATION OF THE H^+ PUMP IN ELODEA LEAVES

F. Albergoni, M. T. Marrè, V. Trokner, N. Beffagna,
G. Romani and Erasmo Marrè

Centro del C.N.R. per la Biologia Cellulare e Molecolare delle Piante, Dipartimento di Biologia, Università di Milano, Via Celoria 26, I-20133-Milano (Italy)

INTRODUCTION

The widespread presence of an ATP-driven H^+ pump operating at the plasmalemma of higher plant cells is well established by in vivo and by in vitro work (Marrè and Ballarin-Denti, 1985). In recent investigations attention has been centered on the relationships between the activity at this pump, the activity of the biochemical PEP-malate pH-stat and the regulation of intracellular pH (Marrè et al., 1986 b).

Some of the natural plant hormones (auxin, brassinolide, abscisic acid) and natural compounds, such as the phytotoxin fusicoccin (FC), are known to induce well defined changes in the activity of the pump (seen as electrogenic H^+ extrusion) on one hand, and on the rate of dark fixation and malate accumulation, on the other (Marrè, 1985).

The cause-effect relationships between these effects is still open to investigation. Two main hypotheses have been proposed :

- 1) The activation of the pump by FC (and possibly also by the hormones) would depend on a primary effect on the cytoplasm, leading to an acidification of the cytosol, which would stimulate the proton pump (Bertl and Felle, 1985; Brummer et al., 1984; Hager and Moser, 1985) (in fact, cytosol acidifying treatments induce a stimulation of electrogenic H^+ extrusion (Romani et al., 1985)).

- 2) FC would activate the pump directly at plasmalemma level, thus inducing an alkalinization of the cytosol (Marrè et al., 1986 b; Marrè et al., 1985). This conclusion is based on : a) the demonstration of an activating effect of FC on electrogenic H^+ transport in native plasmalemma vesicles (Rasi-Caldogno et al., 1986), and b) some recent results, obtained in maize or barley root segments by either the ^{31}P-NMR method (Reid et al., 1985) or the DMO distribution method (Marrè et al., 1986 b; Marrè et al., 1985), which indicate a significant, although modest, alkalinization of the cytosol of the FC-treated cells.

In this investigation the problem of the relationships between the activity of the ATP-driven, vanadate-sensitive plasmalemma H^+ pump and the regulation of cytosolic and vacuolar pH has been approached by utilizing Elodea densa leaves, a material which appeared very suitable for this type of study, in as much as all cells of the thin, cytologically uniform leaf blade are in direct contact with the medium. The results presented in this paper are in full agreement with previous results indicating that FC and K^+ primarily stimulate the ATP-dependent plasmalemma H^+ pump, thus inducing a cytosolic pH rise and the consequent activation of the PEP-malate pH-stat. They also indicate a close relationship between photosynthetic CO_2 fixation and the regulation of the activity of the proton pump.

MATERIAL AND METHODS

Young, almost fully grown Elodea leaves were excised from healthy plants cultured as described elsewhere (Albergoni et al., 1986). Batches of 25 leaves (about 135 mg fresh weight) were transported in 50 ml Erlenmeyer flasks and preincubated for 2 h in 0.5 mM $CaSO_4$, in the light (10 Klux).

Methods for photosynthesis (Albergoni et al., 1986), transmembrane potential (Albergoni et al., 1986), cell sap preparation (Marrè et al., 1986 b), H^+ extrusion (Marrè et al., 1986 a), malate (Romani et al., 1983), citrate (Dagley, 1974) and ATP (Novacky et al., 1978) determinations were as described in previous papers.

Intracellular pH values were measured by the weak acid or weak base distribution method, and calculated as by Kurkdjian et al. (1981) and by Guern et al. (1982).

The accumulation in the tissue of both the weak acid (5,5-dimethyl-oxazolidin-2,4-dione, "DMO") and the weak base (benzylamine) used reached in the controls a constant level after 30-45 minutes of incubation, indicating equilibration and thus allowing the calculation of vacuolar pH. Cytoplasmic pH was then calculated from the DMO distribution data, by the following equation (Marrè et al., 1986 b):

$$pH_{cyt} = pK + \log\left\{\frac{1}{V_c}\left[C_t - V_v \cdot \left(AH + 10^{pH_v - pK} + \log AH\right)\right] - AH\right\} - \log AH$$

where the volume of intracellular water was taken as equal to the fresh weight x 0.72 (correction for 20% free space and 8% dry weight). A cytosol/vacuole volume ratio of 1/9 was assumed in these calculations on the basis of measurements on electron microscope sections.

Preliminary experiments had shown that neither DMO nor benzylamine are metabolized by the tissue during the incubation period considered.

RESULTS

Electrogenic H^+ Extrusion by Elodea Leaves and its Responses to FC, Cations, Light and Plasmalemma ATPase Inhibitors

The data of Table 1 show the operation in Elodea leaves, in the dark, of a proton extruding mechanism, characterized by : a) a strong requirement for a permeating cation (K^+, Na^+, but not Cs^+, Li^+ and Ca^{2+}) in the medium; b) a marked activation by fusicoccin in the presence of K^+ or Na^+; c) a strong inhibition by orthovanadate and by erythrosine B (EB) (Cocucci and Marrè, 1986), two inhibitors of the H^+ transporting plasmalemma ATPase. Illumination of the leaves with white or red (but not green) light induced a marked K^+-dependent, vanadate-inhibited increase in H^+ extrusion, which was completely suppressed by the presence of 5 μM DCMU (Marrè et al., 1986 a). This specific photosynthesis inhibitor did not influence H^+ extrusion in the dark. As shown in Fig. 1-A, the activation of H^+ extrusion by either FC or light was associated with a hyperpolarization of the transmembrane potential of about - 30 mV. Once again, DCMU completely suppressed

Table 1

Influence of Some Effectors on H^+ Extrusion

	H^+ extrusion (μmol x g^{-1} fr. wt. x h^{-1})	
	- FC	+ FC
Control (dark)	0.09	0.05
K^+	2.19	8.16
Na^+	0.05	1.52
Li^+	0.00	0.05
Ca^{2+}	- 0.03	- 0.02
Cs^+	0.01	0.02
K^+ + DCMU	2.21	8.33
K^+ + Na_3VO_4	- 0.03	2.79
K^+ + EB	0.03	1.20
Control (light)	- 0.08	- 0.07
Light + K^+	5.94	9.43
Light + K^+ + Na_3VO_4	0.70	3.30
Light + K^+ + DCMU	2.23	7.98

The effects of 0.1 mM FC, 2 mM cations, 0.1 mM vanadate, 0.3 mM EB, 5 μM DCMU and white light (10 Klux) were measured by titration of the medium (135 mg leaves per 10 ml medium) under bubbling N_2; 0.5 mM $CaSO_4$ and 0.5 mM Mes/Bis-Tris-propane buffer, pH 6 present in basal medium.

the light-induced hyperpolarization without influencing the effect of FC in the dark.

The effects of varying the external concentration of K^+ on PD, in the light and in the dark, are shown in Fig. 1-B and 1-C : the rapid K^+-induced depolarization is quite marked at the lower K^+ concentrations (0.1 to 0.5 mM) and progressively decreases for further increases in the concentration of K^+ in the medium.

This behavior is roughly parallel to that of the effect of external K^+ on both FC-induced and light-induced H^+ extrusion (Fig. 2), and suggests that the activation of the pump by K^+ depends on its depolarizing activity (Marrè and Ballarin-Denti, 1985). These data indicate the operation in Elodea leaves of an ATP-driven, FC-sensitive H^+ pump, and the capacity of photosynthesis to activate it.

Changes in Intracellular pH accompanying FC- and K^+-induced Activation of the H^+ Pump

The data of Fig. 3 and 4 show that the (FC + K^+)-induced increase in H^+ extrusion was accompanied by a steady increase in pH of the cell sap (Fig. 3), a decrease in benzylamine accumulation (Fig. 4-A) and an increase in DMO accumulation (Fig. 4-B). Intracellular DMO and benzylamine levels in the control leaves reaches a maximum, steady value (indicating equilibration) within about 60 minutes. The (FC + K^+)-induced decrease in weak base, or increase in weak

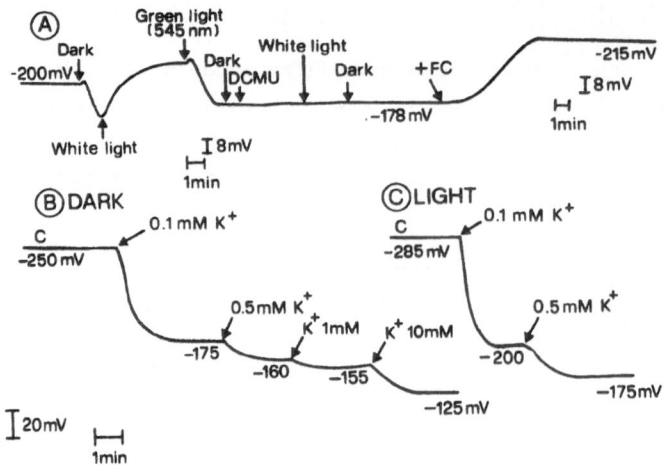

Figure 1

Transmembrane electrical potential changes
in Elodea densa leaf cells

As induced: (A) by white or green light, by 5 μM DCMU and by 0.1 mM FC; (B) by the addition of K_2SO_4 at increasing concentrations in the dark and (C) in the light (10 Klux, white light).

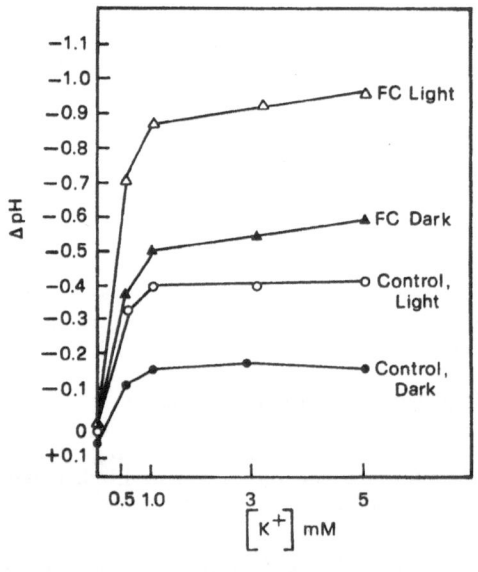

Figure 2

Effects of the external K^+ (as sulphate) concentration
on the pH of the medium of incubation of Elodea leaves

Experimental conditions as in Table 1. pH values were measured after removal of CO_2 by bubbling N_2.

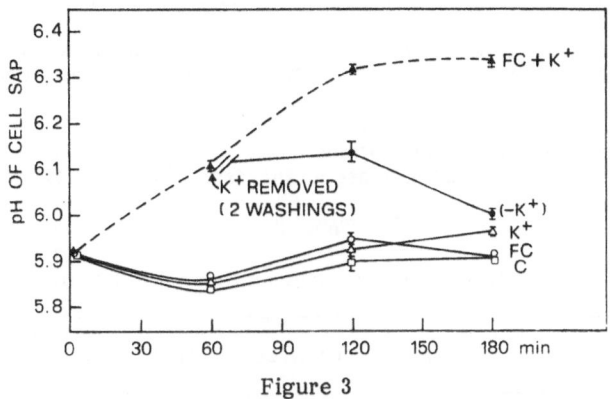

Figure 3

FC- and K^+-induced changes in cell sap pH
and effect of the substitution of the (FC + K^+) medium
with K^+-free medium

The samples (300 mg leaves) were incubated under agitation at 20°C in 20 ml of
0.5 mM $CaSO_4$, 20 mM Mes/Na, pH 5.5 and 5 μM DCMU. FC and K_2SO_4, when
added, were 0.1 mM and 5 mM, respectively.

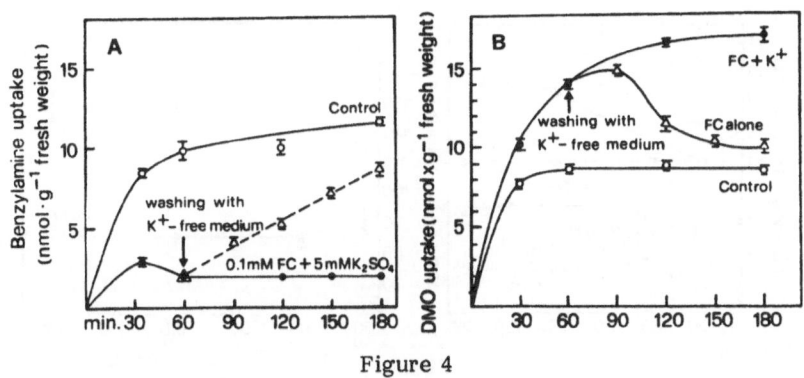

Figure 4

Effects of (FC + K^+) and of the (FC + K^+) containing concentration medium
with K^+-free medium (arrow)
on the accumulation of $(7\text{-}^{14}C)$-benzylamine (A) and of $(1\text{-}^{14}C)$-DMO (B)

Benzylamine or DMO initial concentrations were 10 μM and 6 μM, respectively.
Samples (50 mg leaves per 10 ml medium) were incubated at 20°C in 0.5 mM $CaSO_4$,
10 mM Mes/Na; pH 5.5 and 5 M DCMU in the presence or not of 0.1 mM FC and
5 mM K_2SO_4.

acid accumulation was largely reversed by the removal of K^+ from the medium (Fig. 4), a treatment which almost completely blocks H^+ extrusion. When fed separately neither FC nor K^+ did significantly influence benzylamine or DMO accumulation. Substantially similar results were obtained with nicotine (instead of benzylamine) and with isobutyric acid (instead of DMO).

The results of the calculation of vacuolar and then of cytoplasmic pH values from the data of Fig. 4 (see Methods) are shown in Table 2. According to these calculations, the simultaneous presence of K^+ and FC appears to have induced an increase in pH of about 0.39 units in "bulk cytoplasm" and of 0.71 units in the vacuole, whereas the direct measurement of cell sap pH (comprehending both cytoplasmic and vacuolar fluids) indicate an increase of about 0.42 units.

Table 2

(FC + K^+)-induced Changes in Intracellular pH

	Control	(FC + K^+)	Δ
Cell sap	5.90	6.32	+ 0.42
Vacuole (benzylamine)	5.31	6.02	+ 0.71
Cytoplasm (DMO)	7.50	7.89	+ 0.39

The pH values of cell sap, vacuole and "bulk cytoplasm" were calculated from (7-^{14}C)-benzylamine (pK = 9.33) or (2-^{14})-DMO (pK = 6.4) distribution after 2 h of incubation. DMO or methylamine initial concentrations were 6 μM or 10 μM, respectively. Experimental conditions as in Fig. 4.

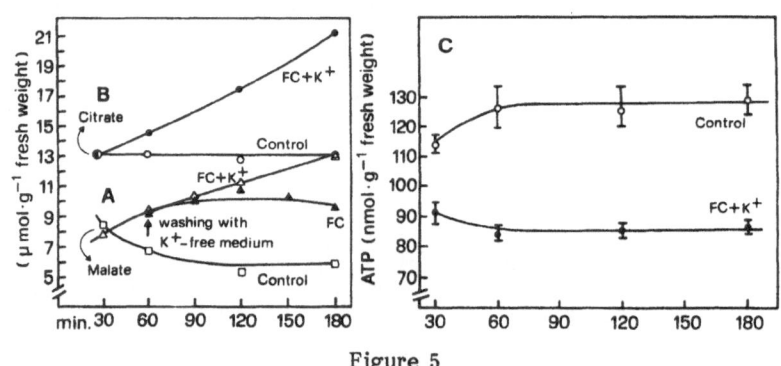

Figure 5

Changes in malate (A), citrate (B) and ATP (C) levels
induced by (FC + K^+)

Experimental conditions as in Fig. 4.

Metabolic Changes Associated With FC-induced Stimulation Of H^+ Extrusion

Previous results have shown that FC- and also hormone-induced stimulation of H^+ extrusion in stem and root segments is associated with important metabolic changes among which : a) an early increase in malate level, interpreted as due to increased activity of PEP carboxylation (biochemical pH-stat (Davies, 1973; Raven and Smith, 1974)); b) a slight but significant decrease in ATP level, interpreted as depending on the enhancement of ATP utilization by the plasmalemma H^+ pump (Marrè and Ballarin-Denti, 1985). Thus we measured the changes induced by the addition of FC and K^+ (maximum activation of the pump) on the levels of malate and other organic acids and of ATP in Elodea leaves. The results reported in Fig. 5-A show that (FC + K^+) induced an early progressive increase in malate. This was completely inhibited by the substitution of the medium with fresh K^+-free medium, a treatment which blocks H^+ extrusion and reverses (FC + K^+)-induced changes in weak acid or weak base accumulation (compare Fig. 4). Unexpectedly enough, (FC + K^+) also induced a marked, steady accumulation of citrate (Fig. 5-B). Minor changes were observed for other organic acids, among which a slight increase in pyruvate, detectable after 30-60 minutes of treatment.

The (FC + K^+)-induced increase in H^+ extrusion was also associated with an early decrease in the level of ATP, which reached after 60 min of treatment a steady value by about 25% lower than in the controls (Fig. 5-C). This early decrease of ATP to a lower stationary level seems in agreement with the view that FC-induced H^+ extrusion is mediated by a mechanism utilizing the energy released by ATP hydrolysis.

CONCLUSIONS

The results of this paper can be summarized, as follows :

1 - Elodea densa leaves are endowed with a proton extruding mechanism activated by the presence of K^+ in the medium, inhibited by the plasmalemma ATPase inhibitors vanadate and erythrosine B, and markedly stimulated by FC. The stimulation of H^+ extrusion by FC is associated with the hyperpolarization of the transmembrane electrical potential and with an early decrease of ATP to a lower steady state level. These data are interpreted as indicating the operation in this material of an electrogenic ATP-driven, FC-sensitive H^+ pump similar to that demonstrated in stem and root segments of several other plant species.

2 - White light induces in Elodea leaves, together with photosynthesis, an hyperpolarization of the transmembrane electrical potential and an increase in H^+ secretion. Both of these effects of light are suppressed by DCMU, which specifically blocks photosynthetic O_2 evolution and CO_2 fixation, and which does not influence basal or FC-induced H^+ secretion in the dark. These data are interpreted as suggesting that some product(s) of photosynthetic activity is (are) able to increase the rate of operation of the plasmalemma H^+ pump (see, for the possible physiological meaning of H^+ extrusion for photosynthesis of aquatic plants; Lucas and Berry, 1985).

3 - The activation of H^+ extrusion by FC and K^+ is associated with a conspicuous increase in the rate of accumulation of the weak acid DMO, and with a similarly marked decrease in that of the weak base benzylamine. Both DMO and benzylamine reach in the controls steady levels in the tissue after 1 h of incubation, and no appreciable metabolism of any of the two compounds is observed during the treatment. Calculations from the weak acid or base distribution values indicate that (FC + K^+) increases vacuolar pH by 0.71 units, and cytoplasmic pH by 0.39 units. Direct measurements of expressed cell sap pH show a (FC + K^+)-induced pH increase of 0.42 units. These data are interpreted as indicating that FC (in the presence of the permeating cation K^+) stimulates H^+ extrusion by activating the pump at plasmalemma level, thus inducing an alkalinization of both the cytoplasm and the vacuole.

4 - (FC + K$^+$)-induced H$^+$ extrusion is associated with an early, consistent increase in the levels of both malate and citrate. The increase in malate is interpreted as depending on an enhancement of PEP carboxylation and as a consequence of the FC-induced rise of cytoplasmic pH. The increase in citrate obviously represents a metabolic response counteracting cytoplasm alkalinization. Its mechanism and its role in the regulation of intracellular pH remains open to further investigation.

REFERENCES

Albergoni, F., Marrè, M. T., and Marrè, E., 1986, Preliminary observations on the integration of photosynthesis with solute transport. I. Elodea densa. Planch as a suitable experimental material, Rend. Acc. Naz. Lincei, in press.

Bertl, A., and Felle, H., 1985, Cytoplasmic pH of root hair cells of Sinapis alba recorded by a pH-sensitive microelectrode. Does FC stimulate the proton pump by cytoplasmic acidification ?, J. Exp. Bot., 36:1142.

Brummer, B., Felle, H., and Parish, R. W., 1984, Evidence that acid solutions induce plant cell elongation by acidifying the cytosol and stimulating the proton pump, FEBS Letters, 174:223.

Cocucci, M. C., and Marrè, E., 1986, Erythrosin B as an effective inhibitor of electrogenic H$^+$ extrusion, Plant Cell Environment, in press.

Dagley, S., 1974, Citrat UV-spectrophotometr. Best., in : "Methoden der Enzymastischen Analyse", Vol. 2, A. U. Bermeyer, ed., Verlag-Chemie Weinheim, Berlin.

Davies, D. D., 1973, Control of and by pH, in : "Symp. Soc. Exp. Biol.", Vol. 27, pp. 513-524.

Guern, J., Kurkdjian, A., and Mathieu, Y., 1982, Hormonal regulation of intracellular pH : Hypotheses versus facts, in : "Plant Growth Substances", P. F. Wareing, ed., Academic Press, London.

Hager, A., and Moser, I., 1985, Acetic acid esters and permeable weak acids induce active proton extrusion and extension growth of coleoptile segments by lowering the cytoplasmic pH, Planta, 163:391.

Kurkdjian, A., Morot-Gaudry, J. F., Wuilleme, S., Lamant, A., Jolivet, E., and Guern, J., 1981, Evidence for an action of fusicoccin on the vacuolar pH of Acer pseudoplatanus cells in suspension culture, Plant Science Letters, 23:233.

Lucas, W. J., and Berry, J. A., 1985, Inorganic carbon transport in aquatic photosynthetic organism, Physiol. Plant., 65:539.

Marrè, E., 1985, Fusicoccin- and hormone-induced changes of H$^+$ extrusion : Physiological implications, in : "Frontiers of Membrane Research in Agriculture", J. B. Saint-John, E. Berlin and P. C. Jackson, eds., Rowman and Allanheld, Totowa.

Marrè, E., and Ballarin-Denti, A., 1985, The proton pumps of the plasmalemma and the tonoplast of higher plants, J. Bioenergetics Biomembranes, 17:1.

Marrè, E., Marrè, M. T., and Romani, G., 1985, Effects of plant hormones on transport and metabolism : Involvement of changes in cytoplasmic pH, in : "Proceedings of the 16th Meeting of FEBS Moscow 1984", VNU Science Press, Utrecht.

Marrè, M. T., Albergoni, F., and Marrè, E., 1986, Preliminary observations on the integration of photosynthesis with solute transport. II. Photosynthesis-induced activation of the vanadate-sensitive electrogenic proton pump in Elodea densa Planch leaves, Rend. Acc. Naz. Lincei, in press.

Marrè, M. T., Romani, G., Bellando, M., and Marrè, E., 1986, Stimulation of weak acid uptake and increase in cell sap pH as an evidence for FC and K$^+$ induced cytosol alkalinization, Plant Physiol., in press.

Novacky, A., Ullrich-Eberius, C. I., and Lüttge, U., 1978, Membrane potential changes during transport of hexoses in Lemna gibba G1, Planta, 138:263.

Rasi-Caldogno, F., De Michelis, M. I., Pugliarello, M. C., and Marrè, E., 1986, H$^+$-pumping driven by the plasma membrane ATPase in membrane vesicles from radish : Stimulation by fusicoccin, Plant Physiol., in press.

Raven, J. A., and Smith, F. A., 1974, Significance of hydrogen ion transport in plant cells, Can. J. Bot., 52:1035.

Reid, R. J., Field, L. D., and Pitman, M. G., 1985, Effects of external pH, fusicoccin and byturate on the cytoplasmic pH in barley root tips measured by ^{31}P-nuclear magnetic resonance spectroscopy, Planta, 166:341.

Romani, G., Marrè, M. T., Bellando, M., Alloatti, G., and Marrè, E., 1985, H$^+$ extrusion and potassium uptake associated with potential hyperpolarization in maize and wheat root segments treated with permeant weak acids, Plant Physiol., 79:734.

Romani, G., Marrè, M. T., and Marrè, E., 1983, Effects of permeant weak acids on dark CO_2 fixation and malate level in maize root segments, Physiol. Vég., 21:867.

INHIBITION OF ATP-DEPENDENT PROTON PUMPING IN PLANT TONOPLAST-
AND PLASMALEMMA-ENRICHED FRACTIONS CAUSED BY MYCOCHROMONE

Franscesco Macri[*], Angelo Vianello[*]
and Maria-Cecilia Cocucci[**]

[*]Istituto di Biosintesi Vegetali, Sezione di Padova, C. N. R.,
Corso Stati Uniti, 4, I-35020-Padova; and [*]Istituto di Defesa
delle Piante, Università di Udine, Plaza Le Kolbe, 4, I-33100-
Udine, and [**]Centro di Studio per la Biologia Cellulare e Molecolare
delle Piante, C. N. R., Via Celoria, 26, I-20133-Milano, Italy

INTRODUCTION

The presence of H^+-pumping ATPases in nonmitochondrial plant membranes, thought to be localized on plasmalemma and tonoplast, has been described (Marin, 1985; Marrè and Ballarin-Denti, 1985; Serrano, 1983; Sze, 1985). The H^+-ATPases generate electrochemical gradients which supply the driving force for the fluxes of the major ions into cytoplasm and vacuoles.

Mycochromone is a dimethyl 2-(2-ethenyl-5-hydroxy-4H-1-benzopyran-4-onyl-3)-malonate produced by Mycosphaerella rosigena (Assante et al., 1979).

The present report deals with the study of the effect of mycochromone on ATP-dependent proton translocation on pea stem tonoplast- and radish seedling plasma membrane mediated-enriched preparations.

MATERIALS AND METHODS

Pea stem (Pisum sativum L., cv. Alaska) tonoplast-enriched fractions were isolated as previously described (Macri et al., 1986). Radish seedling (Raphanus sativus L., cv. Tondo Rosso Quarantino) plasma-enriched fractions were prepared according to Rasi-Caldogno et al. (Rasi-Caldogno et al., 1985). Solubilized and partially purified radish plasma membrane H^+-ATPase was obtained as described by Cocucci and Marrè (Cocucci and Marrè, 1984), except that the last step was omitted (solubilization with lysolecitihin).

Acridine orange uptake was monitored as decrease of absorbance at 495 nm by a double beam Perkin-Elmer, model 554, spectrophotometer. The incubation mixtures for tonoplast and plasma membrane vesicles were as described by Macri et al. (1986) and Rasi-Caldogno et al. (1985), respectively. For the experiments with nigericin, the medium was an in Macri et al. (1986), except that the pH was 7.2 and KCl 1mM.

ATPase activity, evaluated as release of P_i (Cross et al., 1978), was measured in incubation mixtures as described previously for pea tonoplast (Macri et al., 1986) and for radish plasma membrane (Cocucci and Marrè, 1984), respectively. Solubilized H⁺-ATPase activity was assayed in 25 mM Pipes-Tris (pH 6.4), 3 mM $MgCl_2$, 25 mM KCl, 2% dimethyl-sulfoxide and the reaction started by 3 mM ATP.

Protein concentration was measured by the biuret method (Gornall et al., 1949), using bovine serum albumin as a standard, after washing the samples with 5 mM $MgCl_2$.

Mycochromone was a generous gift of Prof. L. Nasini, Milan. The compound was dissolved in dimethyl-sulfoxide to give a 24 10^{-3} M stock solution.

RESULTS AND DISCUSSION

Fig. 1-A shows that mycochromone (12 10^{-5} M) inhibits the ATP-dependent acridine orange uptake of ca. 75% in pea tonoplast-enriched vesicles (traces a and b). Conversely, the compound does not alter the nigericin-induced acridine uptake (Fig. 1-B, traces a and b). Fig. 1-C (traces a and b) shows that mycochromone, at the same concentration, is also capable of inhibiting the ATP-dependent acridine orange uptake in radish plasma membrane-enriched preparations.

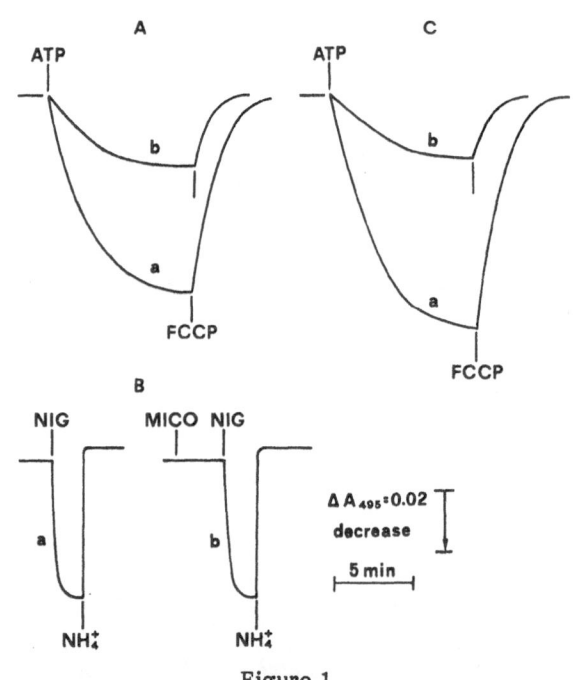

Figure 1

Effect of mycochromone on :
A) ATP-dependent and B) nigericin-induced acridine orange uptake
in pea stem tonoplast-enriched vesicles
and C) ATP-dependent acridine orange uptake
in radish plasma membrane-enriched vesicles

a: control; b: mycochromone-treated. Protein in the incubation medium was 0.15 mg/ml for tonoplast- and 0.6 mg/ml for plasma membrane-enriched vesicles, respectively. Additions were: 10 mM Tris-ATP, 6 μM FCCP, 12 10^{-5} M mycochromone, 2 μM nigericin and 10 μl of a saturated solution of $(NH_4)_2SO_4$.

The decrease of the initial rate of ATP-dependent acridine orange uptake in both types of preparations is linked to the time of preincubation with mycochromone. At least 5 to 6 min are necessary to obtain the maximum inhibitory effect. The inhibition increases rapidly by increasing mycochromone concentration up to $20 \cdot 10^{-5}$ M and reaches a maximum at $40 \cdot 10^{-5}$ M. In addition, mycochromone does not collapses the ATP-generated proton gradient of the vesicles.

The double reciprocal plot of the mycochromone-induced decrease of the initial rate of acridine orange uptake (6 and $12 \cdot 10^{-5}$ M) in tonoplast-enriched preparations, shows that the maximun rate is unaffected, while the K_m value is increased, indicating that the inhibitory effect of mycochromone is competitive. The calculated K_i is approximately $5 \cdot 10^{-5}$ M (Fig. 2). Another linear transformation of the velocity curves according to the Hanes-Wolf plot, provides the same information.

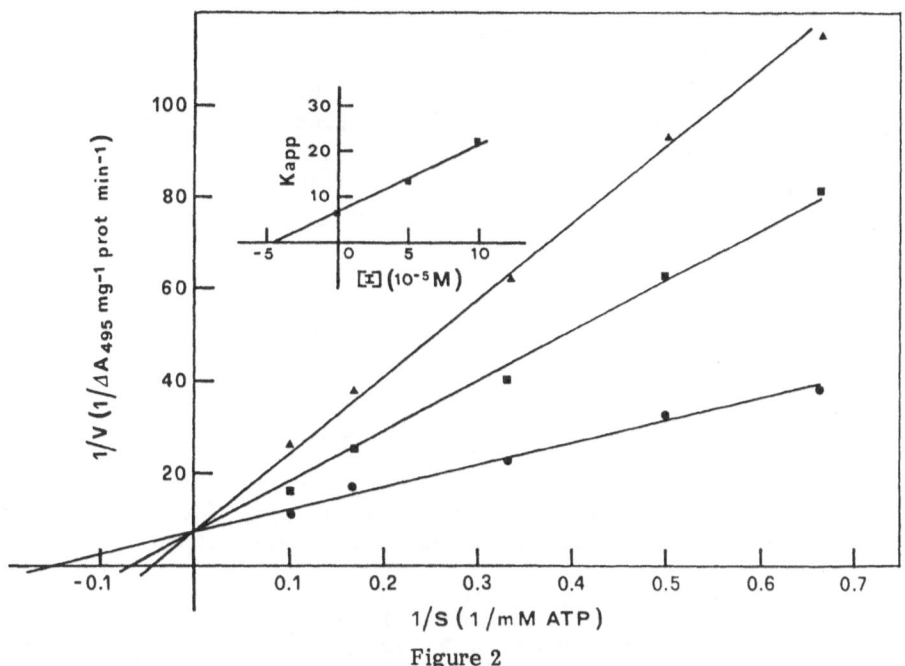

Figure 2

Double reciprocal plot of the initial rate
of ATP-dependent acridine orange uptake in pea stem tonoplast-enriched fractions
as a function of Tris-ATP concentration
at two levels of mycochromone
($\bullet$, control; $\blacksquare$, $\blacktriangle$, 6 and $12 \cdot 10^{-5}$ M mycochromone)

Mycochromone does not inhibit the ATPase activity of pea tonoplast- and radish plasma membrane-enriched preparations, and the activity of the H^+-ATPase solubilized from radish evaluated as release of P_i (Table 1).

The results show that the major effect of mycochromone on pea tonoplast and radish plasma membrane-enriched vesicles is the inhibition of the ATP-dependent proton translocation, although it is unable to induce a parallel inhibition of ATPase activity on the same preparations. In addition, mycochromone neither modifies the H^+/K^+ exchange induced by nigericin, nor collapses the ATP-generated proton gradient of the vesicles, indicating that mycochromone does not alter the permeability of the vesicles to protons. Mycochromone, hence, appears a useful tool to separate the scalar from the vectorial function of H^+-ATPases.

Table 1

Effect of Mycochromone on ATPase Activity of Pea Tonoplast, Radish Plasma Membrane and of a solubilized and partially purified H^+-ATPase from radish preparations.

Additions	pea tonoplast	radish plasma membrane	solubilized radish H^+ATPase
	μmol P_i x mg^{-1} prot. x h^{-1}		μmol P_i x mg^{-1} prot. x min^{-1}
None	23.6	5.97	3.06
$7 . 10^{-4}$ M Myco.	25.0	5.65	2.88
$14 . 10^{-4}$ M Myco.	22.8	4.89	2.95
100 μM Na_3VO_4	-	1.32	1.01

REFERENCES

Assante, G., Camarda, L., Merlini, L., and Nasini, G., 1979, Mycochromone and mycoxanthone : Two new metabolites from Mycosphaerella rosigena, Phytochemistry, 18:311.

Cocucci, M. C., and Marrè, E., 1984, Lysophosphatidylcholine-activated, vanadate-inhibited, Mg^{2+}-ATPase from radish microsomes, Biochim. Biophys. Acta, 771:42.

Cross, J. W., Briggs, W. R., Dohrmann, U. C., and Rayle, P. M., 1978, Auxin-receptors of maize coleoptile membranes do not have ATPase activity, Plant Physiol., 61:581.

Gornall, A. G., Bardawill, C. J., and David, M. M., 1949, Determination of serum proteins by means of the biuret reaction, J. Biol. Chem., 177:751.

Macri, F., Vianello, A., and Pennazio, S., 1986, Salicylate-collapsed membrane potential in pea stem mitochondria, Physiol. Plant., 67:136.

Marin, B. P., 1985, "Biochemistry and Function of Vacuolar Adenosine-Triphosphatase in Fungi and Plants", Springer-Verlag, Berlin.

Marrè, E., and Ballarin-Denti, A., 1985, The proton pumps of the plasmalemma and the tonoplast of higher plants, J. Bioenerg. Biomembranes, 17:1.

Rasi-Caldogno, F., Pugliarello, M. C., and, De Michelis, M. I., 1985, Electrogenic transport of protons driven by the plasma-membrane ATPase in membrane vesicles from radish. Biochemical characterization, Plant Physiol., 77:200.

Serrano, R., 1983, Purification and reconstitution of the proton-pumping ATPase of fungal and plant plasma membranes, Arch. Biochem. Biophys., 227:1.

Sze, H., 1985, H^+-translocating ATPases : Advances using membrane vesicles, Annu. Rev. Plant Physiol., 36:175.

TRANSPORT OF INORGANIC IONS IN TONOPLAST VESICLES

Ronald J. Poole and Eduardo Blumwald

Centre for Plant Molecular Biology,
Biology Department, McGill University
1205, Doctor Penfield Avenue
Montreal, P. Q., Canada H3A 1B1

INTRODUCTION

Since the tonoplast proton pumps create a positive membrane potential in the vacuole, the maintainance of high cytoplasmic K^+ and the accumulation of Cl^- and NO_3^- in the vacuole requires only passive transport mechanisms (channels or uniports) for these ions, energized by $\Delta\Psi$. On the other hand, the vacuolar accumulation of Na^+ and Ca^{2+}, and the retrieval of stored nutrient anions such as NO_3^- for metabolic use requires active transport mechanisms, which we now believe to be energized by proton efflux from the vacuole. This paper will discuss our work on these proton-coupled secondary transport mechanisms, with emphasis on their physiological significance, the technical problems involved, and the many questions remaining to be investigated.

MATERIALS AND METHODS

The material which we have used, storage tissue of red beet and sugar beet, has two advantages for this work. First, the relative densities of the various membranes allow a good purification of tonoplast membranes by sucrose density gradient centrifugation. Secondly, the main physiological function of this material is the vacuolar storage and retrieval of nutrients. In addition, as a halophyte, beet is particularly appropriate for studying sodium transport. Work with tonoplast vesicles takes advantage of their high surface:volume ratio compared with intact vacuoles. This allows more sensitive measurement of fluxes, and also makes it easy to change solute concentrations on either side of the membrane.

The internal pH of tonoplast vesicles can be adjusted by centrifuging and/or incubating them in buffer solutions. Known pH gradients can then be created simply by diluting the vesicles into a buffer of different pH. The reliability of the assumed pH values can be confirmed by the distribution of a weak base or weak acid if this behaves in a theoretically ideal manner : This can be shown to be true for some spin labelled probes (Poole et al., 1985). The fluorescent weak base acridine orange is convenient for following rapid changes in ΔpH with time. Although this probe does not behave in an ideal way, fluxes can be followed on an arbitrary scale, or the response can be calibrated.

The results with spin-labelled probes showed that after a pH jump at low temperature all of the vesicles in a tonoplast preparation could hold a known ΔpH long enough for experimental measurement. At room temperature in the presence

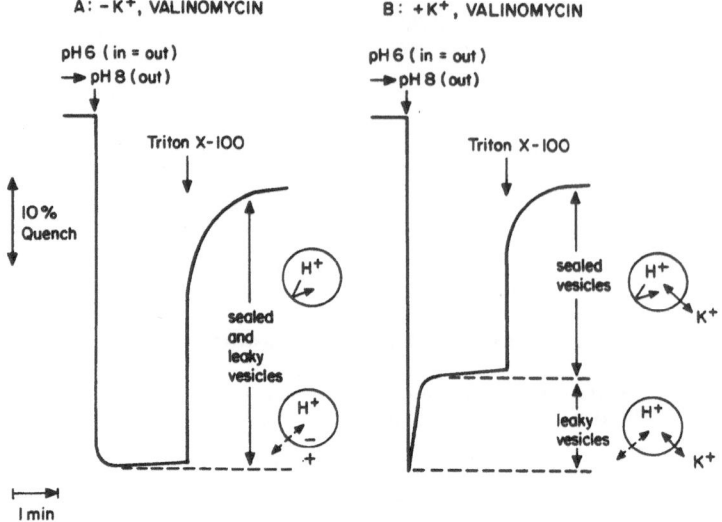

A: – K⁺, VALINOMYCIN

B: + K⁺, VALINOMYCIN

Figure 1

Time course of acridine orange fluorescence after pH jump
Resolution of H⁺-sealed and H⁺-leaky (buffer ion-sealed) vesicle populations
in presence of K⁺ and valinomycin

of diffusible ions, at least some vesicles were leaky to protons. It is important to evaluate this ion permeability in order to assess the usefulness of tonoplast vesicles as a model system for studying transport into the vacuole. The following analysis suggests that proton leaks may be largely confined to a sub-population of damaged vesicles.

Resolution of H⁺-sealed and H⁺-leaky Vesicle Populations

The diagram in Fig. 1 shows a typical time course of acridine orange fluorescence after a pH jump in presence or absence of K⁺ (inside and outside the vesicles) and valinomycin. On formation of a ΔpH (inside acid) the weak base accumulates in the vesicles and there is an immediate quenching of fluorescence. The rate of fluorescence recovery shows the rate of decay of the ΔpH due to H⁺ efflux from the vesicles. In the absence of readily-permeable ions (Tris-Mes buffer ions only) the pH gradient shows only a very slow decay until the addition of gramicidin. However, in presence of K⁺ + valinomycin, proton efflux after the pH jump shows two sharply distinct phases. One component of the ΔpH is completely discharged in less than 1 minute. The other decays only very slowly until discharged by gramicidin. This suggests the existence of two distinct populations of vesicles as illustrated by the vesicle diagrams in Fig. 1. In the absence of other permeable ions, diffusion of H⁺ will create a membrane potential which prevents rapid decay of ΔpH. When this potential is discharged by permeation of K⁺ assisted by the K⁺-ionophore valinomycin, the H⁺-leaky population of vesicles loses its proton gradient in less than one minute, while the H^+-sealed vesicles are little affected. We have noticed that as the vesicle preparation ages, the leaky component increases progressively at the expense of the sealed component. This supports the above interpretation, and also shows that the proton leak is a result of damage to the vesicles, and not a normal property of the tonoplast.

SODIUM TRANSPORT

The above discussion is relevant to our study of Na^+ transport in tonoplast vesicles (Blumwald and Poole, 1985 a). In the presence of K^+ and valinomycin, i.e. in the sealed fraction of vesicles, H^+ efflux is induced by addition of Na^+ (e.g. Fig. 2-B), even though the membrane potential is clamped at zero. The Na^+-dependent H^+ efflux shows Michaelis-Menten kinetics, with an apparent K_m for Na^+ of 7.5 mM. In addition, the diuretic compound amiloride acts as a competitive inhibitor of the Na^+/H^+ antiport, as it does for Na^+/N^+ antiport in mammalian kidney epithelium.

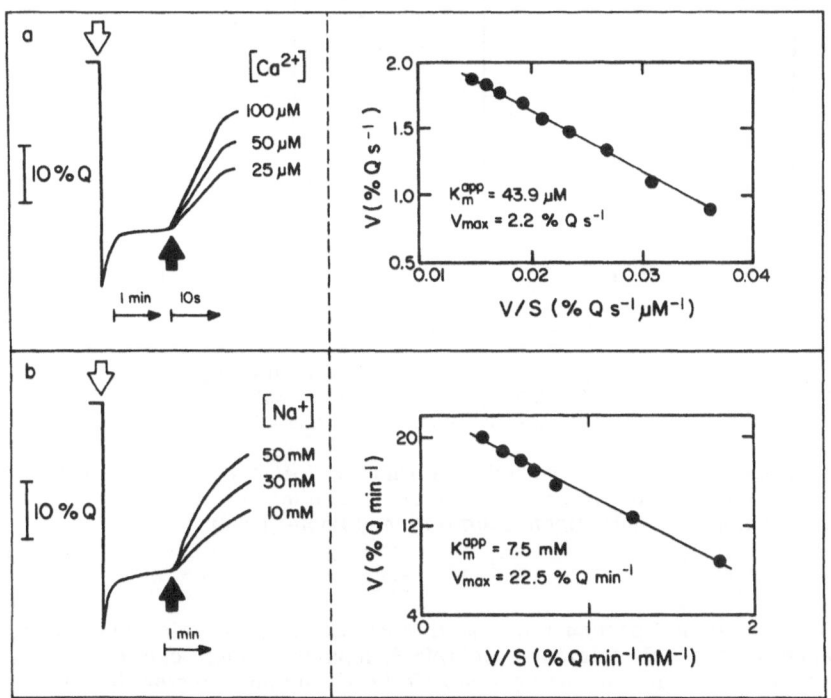

Figure 2

Kinetics of (a) Ca^{2+}/H^+ and (b) Na^+/H^+ antiports

Left : Effect of Ca^{2+} and Na^+ on rate of recovery of acridine orange fluorescence after a pH jump in presence of K^+ and valinomycin. Note difference in time scale between A and B. These also differ in buffer concentration. A : 20 mM Tris-Mes. B : 5 mM Tris-Mes. Right : Eadie-Hofstee plots and kinetic coefficients based on initial rate of change of acridine orange quenching induced by various concentrations of Ca^{2+} and Na^+ respectively. Data from Blumwald and Poole (1985 a and 1986).

In the absence of K^+ and valinomycin, we observed an additional component of Na^+-dependent H^+ effluent, linear with Na^+ concentration, which we attributed to passive permeability (Blumwald and Poole, 1985 a). Now, we can see that this passive permeability belongs to the leaky population of vesicles, and doesn't represent a true property of the tonoplast. This makes sense physiologically, since a passive permeability to Na^+ would greatly increase the energy cost of maintaining a sodium gradient at the tonoplast.

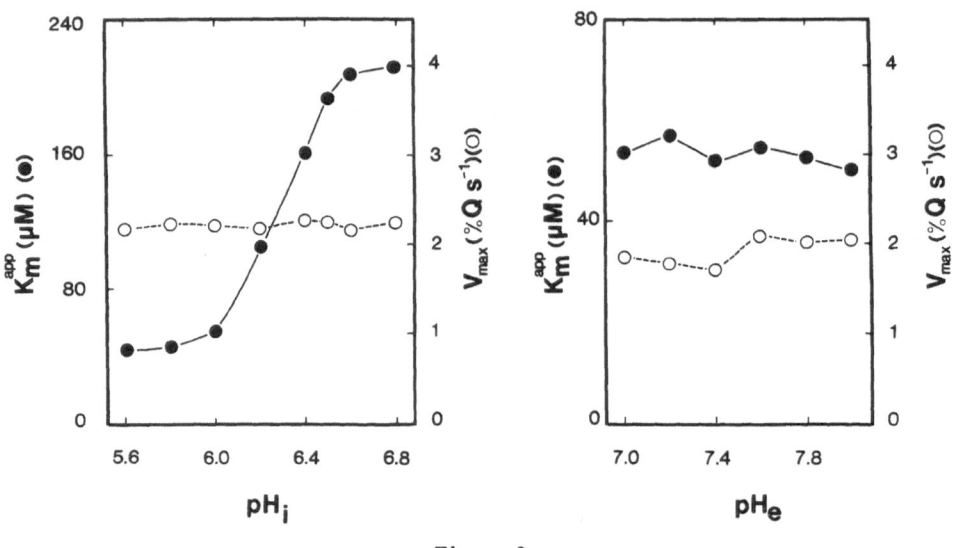

Figure 3

pH dependence of apparent K_m and V_{max}
for Ca^{2+}-dependent H^+ efflux
(change in fluorescence quench)

When internal pH (pH_i) was varied, external pH (pH_e) was held at pH 8. Similar results were obtained if pH was held at pH 6. Similar results were obtained if ΔpH was held constant at 2 units (from Blumwald and Poole, 1986).

The Na^+/H^+ antiport seen in sealed tonoplast vesicles is clearly of fundamental importance in the salt-tolerance of this halophyte, which can accumulate high concentrations of sodium in the vacuole. We have only made a start in characterizing this transport system, and much more work is needed to understand its function. For example, we have shown that the antiport is reversible, but we have not determined the apparent K_m for Na^+ efflux (Blumwald and Poole, 1985 a). This information is relevant to the question of the magnitude of the sodium gradient that can be maintained across the tonoplast. More information is also needed about the physiological regulation of this transport system. The communication by P. A. Rea and D. Sanders in this volume shows that the Na^+/H^+ antiport changes in kinetic properties during development.

The sensitivity of the Na^+/H^+ antiport to amiloride suggests the use of amiloride derivatives for the future isolation of the protein and eventually the gene for this transport system. This may provide one approach toward the genetic engineering of salt tolerance in other species.

CALCIUM TRANSPORT

Calcium is becoming well–established as a second messenger mediating plant cell responses to external stimuli. In many respects this role of calcium parallels that in animal cells. The low level of calcium in the cytosol (about 10^{-7} M) enables the influx of relatively few Ca^{2+} ions to rapidly change the internal concentration and brings about rapid physiological responses. In both plants and animals, the intracellular membranes play a large role in the release and reabsorption of Ca^{2+} ions. In animal cells, the endoplasmic reticulum and mitochondria may be the major organelles which store and release calcium. However, in plants, our results suggest that the tonoplast plays the dominant role in intracellular Ca^{2+} transport (Blumwald and Poole, 1986)..

In the presence of K^+ and valinomycin, i.e. in sealed tonoplast vesicles with zero membrane potential, we demonstrated a Ca^{2+}–dependent H^+ flux, as well as a ΔpH–dependent Ca^{2+} flux with identical kinetics (Blumwald and Poole, 1986).

Fig. 2 compares the kinetics of the Ca^{2+}/H^+ antiport with the Na^+/H^+ antiport in tonoplast vesicles of red beet. Not only is the K_m for Ca^{2+} 200–fold lower than that for Na^+, but the V_{max} is much greater. In terms of change in % Quench/min, V_{max} for Ca^{2+} is 6–fold faster. However, taking account of the 4–fold higher concentration of buffer, the V_{max} for H^+–transport is about 24–fold higher for the Ca^{2+}/H^+ antiport than for Na^+/H^+ antiport, and this is in a halophyte where Na^+ transport is particularly important ! The V_{max} for Ca^{2+}/H^+ antiport is comparable to that for the tonoplast H^+-ATPase or H^+-PPase. Thus in some circumstances Ca^{2+} transport may consume a very substantial fraction of the total proton-motive force production at the tonoplast. This also suggests that a large activation of Ca^{2+} fluxes by stimuli such as light could affect cytoplasmic and vacuolar pH values. Although the K_m for calcium transport at the tonoplast is higher than that reported for the endoplasmic reticulum (Blumwald and Poole, 1986), the V_{max} is so much greater that tonoplast transport would dominate even at low cytoplasmic levels of calcium.

The same study of Ca^{2+}/H^+ exchange showed an interesting effect of pH on the kinetics of the antiport (Fig. 3). Surprisingly enough, the V_{max} was little affected by either internal or external pH over the ranges examined, nor was it affected by ΔpH over the range of 1 to 2 pH units. However, the apparent K_m for external Ca^{2+} was changed 5–fold over a narrow range of internal pH. This effect is much more sharply dependent on pH than a normal dissociation curve, and suggests a cooperative effect with a Hill coefficient of 4. This could mean that 4 proton-dissociating groups on the internal surface of the membrane have a transmembrane effect on Ca^{2+} binding.

As in the case of Na^+ transport, much more work is needed to really under-stand this transport system. At present, we do not know whether our vesicles all have the same orientation. We know that the Na^+/H^+ antiport is freely reversible, but we have not shown that the antiports are kinetically symmetrical in the membrane. The technique of creating an artificial proton-motive force by a sudden change in external pH leaves open the question of vesicle orientation. One way of dealing with this problem is to work with gradients generated by the proton pumps. In this case, we know that only normally–oriented vesicles are energized. This approach was used in the following study.

NITRATE TRANSPORT

As a physiologically important ion, nitrate, like sugars, amino–acids and malate, must not only be stored in the vacuole but also be retrieved from the vacuole for metabolic use. As noted in the Introduction, influx of NO_3^- into the vacuole probably requires only passive transport in response to the interior–positive membrane

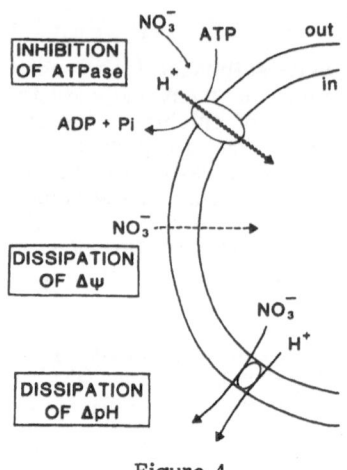

Figure 4

Model for effects of nitrate at the tonoplast
(from Blumwald and Poole, 1985 b)

potential. On the other hand efflux of NO_3^- from the vacuole against the membrane potential is likely to require co-transport with protons. Studies on this two-way transport are further complicated by the fact that nitrate has an inhibitory effect on the tonoplast H^+-ATPase. In Fig. 4, these various postulated activities of nitrate are summarized in a model of the tonoplast membrane. Note that influx of NO_3^- should be recognized by a dissipation of $\Delta\Psi$, whereas proton–coupled efflux should cause dissipation of ΔpH, and these effects should be separated from the inhibition of the ATPase.

First, to characterize the direct effect of NO_3^- on the ATPase, it is necessary to abolish its effect on $\Delta\Psi$ and ΔpH. This can be done by studying the effect of NO_3 on ATPase activity in presence of gramicidin to collapse the proton gradient. Under these conditions, NO_3^- acts as a competitive inhibitor of ATP hydrolysis (Griffith et al., 1986) with a fairly high K_I (22 mM and 39 mM in the presence and absence of Cl^- respectively). Since other chaotropic ions, SCN^- and ClO_4^-, also act as competitive inhibitors, we tentatively ascribe the inhibition of the ATPase to a reversible destabilization of the enzyme (Griffith et al., 1986). Because of the high K_I for nitrate, low concentrations (1 or 2 mM NO_3^-) have essentially no inhibitory effect on the ATPase. In this concentration range, it is found (Blumwald and Poole, 1985 b) that NO_3^- dissipates both $\Delta\Psi$ and ΔpH as predicted by the model, whereas Cl^-, which presumably does not need to be retrieved from the vacuole in this material, dissipates only $\Delta\Psi$.

These fragmentary studies give support to the idea that efflux from the vacuole may well be by a different transport system than influx, just as metabolic pathways are frequently reversed by a different route. They also emphasize the need for regulatory systems to prevent futile cycling, and call for more studies on the regulation of transport across the tonoplast.

REFERENCES

Blumwald, E., and Poole, R. J., 1985 a, Na^+/H^+ antiport in isolated tonolast vesicles from storage tissue of Beta vulgaris, Plant Physiol., 78:163.

Blumwald, E., and Poole, R. J., 1985 b, Nitrate storage and retrieval in Beta vulgaris : Effects of nitrate and chloride on proton gradients in tonoplast vesicles, Proc. Natl. Acad. Sci. U.S.A., 82:3683.

Blumwald, E., and Poole, R. J., 1986, Kinetics of Ca^{2+}/H^+ antiport in isolated tonoplast vesicles from storage tissue of Beta vulgaris L., Plant Physiol., 80: 727.

Griffith, C. J., Rea, P. A., Blumwald, E., and Poole, R. J., 1986, Mechanism of stimulation and inhibition of tonoplast H^+-ATPase of Beta vulgaris by chloride and nitrate, Plant Physiol., 81:120.

Poole, R. J., Mehlhorn, R. J., and Packer, L., 1985, A study of transport in tonoplast vesicles using spin-labelled probes, in : "Biochemistry and Function of Vacuolar ATPase in Fungi and Plants", B. P. Marin, ed., Springer-Verlag, Berlin, Heidelberg, New-York, Tokyo.

WASH-ACTIVATED Na^+/H^+ EXCHANGE IN TONOPLAST VESICLES

FROM BETA VULGARIS STORAGE ROOT DISKS

Philip A. Rea, Roger A. Leigh and Dale Sanders

Department of Biology, University of York
Heslington, York Y01-5DD, England

INTRODUCTION

The presence of a large central vacuole in higher plant cells enables the accumulation and storage of osmotica which would otherwise interfere with cytoplasmic processes. Such compartmentation is especially necessary in the case of Na^+, which at the concentrations often found in the vacuole is a general inhibitor of protein synthesis (Wyn Jones et al., 1979). By implication, then, it might be expected that those plants utilizing Na^+ as an osmoticum (particularly halophytes) are capable of energy-dependent Na^+ transport across the tonoplast. This expectation has been confirmed by Blumwald and Poole (1985), who have recently demonstrated Na^+/H^+ antiport in tonoplast vesicles from red beet (Beta vulgaris L.). The prevailing pH gradient across the tonoplast (ΔpH = 2 units) would therefore be capable of energizing a vacuolar/cytoplasmic Na^+ concentration ratio of 100-fold for the minimum expected H^+/Na^+ stoichiometry of 1.

In many storage tissues, including red beet, the capacity for uptake of a number of solutes is enhanced when the tissue is sliced and washed (or "aged") for a period of days in distilled water (Poole, 1976). However, the mechanisms by which transport capacity is controlled during aging are poorly understood, and constitute a major unsolved problem in the field of plant membrane transport. In the work reported here, we show that Na^+/H^+ exchange by isolated tonoplast vesicles is specifically stimulated during aging. Moreover, the enhancement of antiport appears to result from activation of a preexisting transport system or systems, rather than de novo synthesis of transport protein.

MATERIALS AND METHODS

Storage tissue of freshly-harvested (greenhouse-grown) red beet was sliced to 2 - 3 mm thickness in a food processor. The slices were aged in vigorously-aerated distilled water (approx. 33 g tissue/l) which was renewed daily. Tonoplast vesicles were isolated on a 6% dextran cushion as described by Briskin et al. (1985) and the compositions of the homogenisation and suspension media were those described by Rea and Poole (1985). The appropriate intravesicular ionic composition was achieved by dilution of the vesicles into medium of the desired composition, incubation on ice for 1 h, centrifugation at 80,000 g for 30 min and resuspension to a final membrane protein concentration of 2 - 5 mg/ml. Normally, fresh preparations were used for experiments, though preliminary work involved membranes which had undergone 1 freeze/thaw cycle.

Vesicular ΔpH was monitored in a Perkin-Elmer LS-5 fluorescence spectrometer using the monoamine dye acridine orange as indicator. The reactions were performed in a 2 ml volume which was rapidly stirred and thermoregulated at 25°C. The fluorescence signal was calibrated by subjecting vesicles to a pH jump (i.e. rapid dilution of the concentrated vesicle suspension into medium of higher pH which contained acridine orange) as described by Bennett and Spanswick (1983). Experiments which involved either Na^+-dependent changes in intravesicular pH or $^{22}Na^+$ influx were also initiated by a pH jump. A microcomputer-based method was used for the collection and storage of fluorescence data and for computation of the initial rates of Na^+dependent H^+ efflux (Jennings et al., 1987).

Influx of $^{22}Na^+$ was assayed in a reaction volume of 1 ml. The vesicles were collected on Sartorius nitrocellulose filters (0.22 μm pore size) and washed with 1 ml ice-cold non-radioactive medium. Radioactivity on the filters was measured with a Packard Auto–Gamma 500γ-counter.

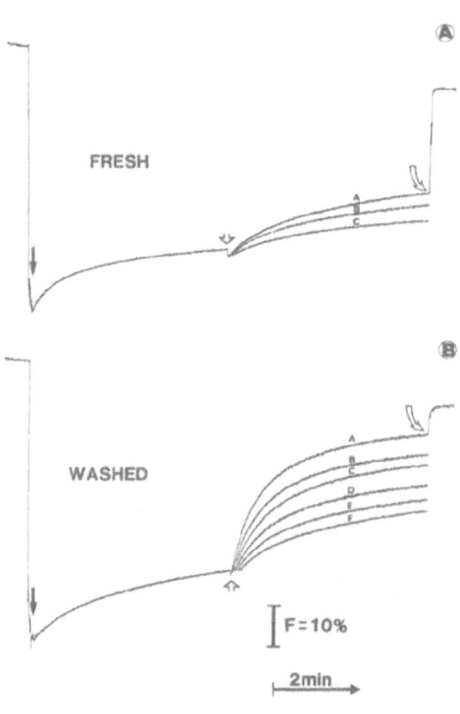

Figure 1

Recovery of acridine orange fluorescence
by addition of Na^+ to tonoplast vesicles
from (A) fresh tissue and (B) tissue washed for 6 days

Fluorescence quenching was initiated by a pH jump from 6 to 8 (black arrows). White arrows indicate addition of Na_2SO_4 to bring the final (Na^+) to A, 30 mM; B, 20 mM; C, 10 mM; D, 5 mM; E, 2.5 mM; F, 1.25 mM. Complete reversal of ΔpH-dependent fluorescence quench was achieved by the addition of 5 mM NH_4^+ (curved white arrows).

RESULTS

Wash–activated Na$^+$/H$^+$ Exchange

After a pH jump from 6 to 8, acridine orange fluorescence in tonoplast vesicles prepared from freshly-harvested storage root is quenched by 65% (Fig. 1-A). The fluorescence recovers by 16% in a non–linear manner over the ensuing 5 min (t½ = 1 min). Addition of Na$_2$SO$_4$ after this period results in a stimulation of the rate of fluorescence recovery, as described by Blumwald and Poole (1985), and is indicative of Na$^+$/H$^+$ exchange. Fluorescence could subsequently be restored to within 10% of the initial value by dissipation of ΔpH with NH$_4^+$.

Tissue which had been washed for 6 days showed an almost identical fluorescence response during and after the pH jump (Fig. 1-B), indicating a similar background proton permeability. However, on addition of Na$^+$ there is a marked difference between the two pretreatments. Both the initial rate and extent of Na$^+$dependent fluorescence recovery in vesicles from washed tissue are greater than those found with vesicles from fresh tissue.

A summary of the results from a more detailed experiment is shown in Fig. 2. The initial rate of fluorescence recovery is a saturable function of (Na$^+$), and is approximated by a single Michaelis-Menten function. The most noteworthy feature of these results is that the V$_{max}$ of the initial rate of fluorescence recovery is increased by only 35% as a result of washing the tissue, whereas the K$_m$ is decreased by a factor of 3.

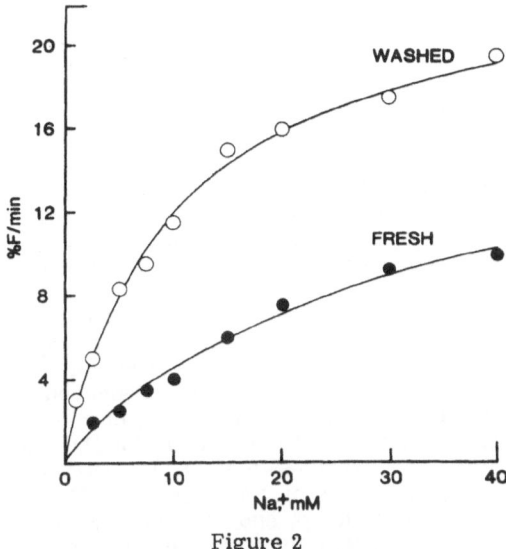

Figure 2

Kinetics of the initial rate of acridine orange
fluorescence (% F/min) in tonoplast vesicles
from fresh tissue and tissue washed for 6 d

Data are derived from an experiment similar to that shown in Fig. 1. Solid lines show Michaelis-Menten fits estimated by a non-linear least squares (Marquardt, 1963). Parameter values : Fresh, V$_{max}$ = 17.8 ± 1.6 %/min, K$_m$ = 29.6 ± 4.7 mM; Washed, V$_{max}$ = 24.0 ± 0.9 %/min, K$_m$ = 10.0 ± 0.9 mM.

The major decline in K$_m$ for Na$^+$/H$^+$ antiport occurs during the first day of washing (Fig. 3), while the change in V$_{max}$ over the whole period is only just significant. The absence of a wash–dependent change in the activity of the tonoplast H$^+$-ATPase indicates that the effects on Na$^+$ transport are relatively specific.

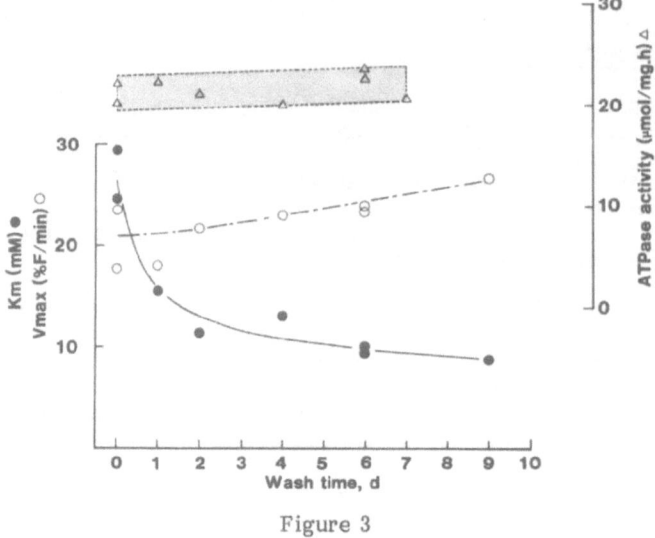

Figure 3

Time-course of kinetic changes
in tonoplast Na^+/H^+ exchange
during washing of beet tissue

The K_m and V_{max} for Na^+-dependent acridine fluorescence recovery were estimated as in Fig. 1 and 2. The V_{max} data approximate a linear function of slope 0.72 ± 0.24 % F min^{-1} day^{-1}. Tonoplast H^+-ATPase activity was measured in the same membrane preparations for comparative purposes.

Under open-circuit conditions (no voltage clamp), Na^+/H^+ exchange might be mediated in one of the ways : by a stoichiometrically-coupled antiporter, or simply by two distinct Na^+ and H^+ uniporters coupled electrophoretically. In order to eliminate an electrically-mediated contribution, Blumwald and Poole (1985) clamp the membrane potential to zero mV with valinomycin and a transmembrane (K^+) ratio of 1. However, in our hands at least, this treatment is not satisfactory: the associated driving force for H^+ efflux becomes so great on depolarization of the open-circuit (inside-negative) membrane potential that ΔpH cannot be sustained long enough for accurate measurement of the Na^+-dependent fluorescence recovery. We have therefore adopted the slightly modified procedure of clamping the membrane potential at the equilibrium potential for H^+ (E_H). For the 2 unit pH jumps normally used, this involved intra- and extravesicular (K^+) = 40 and 0.4 mM, respectively. In these conditions, the initial rates of Na^+-dependent fluorescence recovery were very similar to those without a clamp, and we conclude from the results - as well as from the fact that Na^+-dependent fluorescence recovery shows simple Michaelian kinetics with respect to (Na^+) (Fig. 2) - that even under open-circuit conditions, the only significant pathway for Na^+ is via an antiporter. In contrast to Blumwald and Poole (1985), we have no evidence to suggest a significant contribution by electrophoretic coupling.

Reversal of Na^+/H^+ Exchange

Dilution of Na^+-loaded vesicles into an Na^+-free medium provides an alternative method of measuring exchange rate. The results are shown in Fig. 4. Panel A shows that dilution of vesicles preloaded with Na_2SO_4 into an equimolar (Na^+) medium results in a small amount of non-specific fluorescence quenching. Imposition of an outwardly-directed Na^+ gradient, on the other hand, causes rapid fluorescence quenching (complete within 0.3 min) amounting to 39% in this experiment. Incorporation of the antibiotic Na^+/H^+ exchanger monensin into the medium results in a even larger initial quench (67%) which relaxes close to the quench obtained without ionophore (43%), presumably as the proton gradient itself dissipates as a result of passive H^+ movement through other transport pathways.

If SCN^- rather than $SO_4{}^{2-}$ is used as the counter-ion during Na^+-preloading, the experiment in panel A can be repeated, with broadly similar results (steady-state quench : 34% - Fig. 4, Panel B). Although the second, more minor phase of fluorescence quenching is obviously a slower process, this result implies that the dominant pathway for Na^+/H^+ exchange is via an antiporter, since the presence of the permeant SCN^- anion will off-set or reverse any negative membrane potential caused by electrophoretic leakage of Na^+. Electrophoretic coupling of the Na^+ and H^+ fluxes therefore appears to constitute an insignificant component of the observed exchange.

It should be noted that two points relating to the operation of the antiporter are not properly addressed by this experiment (cf. Blumwald and Poole, 1985). The first concerns the stoichiometry of the exchange reaction. It is tempting to speculate that Na^+/H^+ antiport must be electroneutral since the exchange reaction is not sensitive to the membrane potential. Detailed analysis of the reaction kinetics of carrier-type transport systems shows, however, that even electrophoretic exchanges mediated by a single system can be voltage-insensitive over large spans of membrane potential (Sanders et al., 1984). The second point relates to the kinetic behavior of the transport system. It is not justifiable to conclude from data of the type in Fig. 4 that the antiporter is freely reversible, since uniform orientation of the transport protein is not assured. Indeed, such information as is available from latency studies on the H^+-ATPase indicates a **random** (approx 1:1 inside:rightside-out) orientation of the population of vesicles (C. J. Griffith, personal communication).

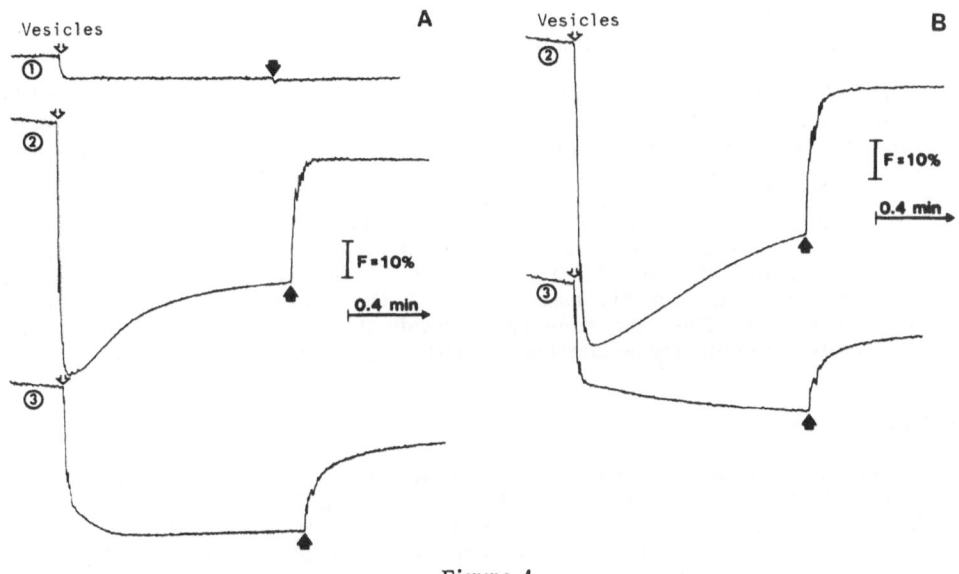

Figure 4

Acridine orange fluorescence quenching
driven by reversal of Na^+/H^+ exchange

Vesicles prepared from tissue washed for 6 days were loaded with 40 mM Na_2SO_4 (Panel A) or 80 mM NaSCN (Panel B) at pH 8 and diluted into an identical medium (A-1), Na^+-free buffer (A-3; B-3) or Na^+-free buffer containing 2 μM monensin (A-2; B-2), all at pH 8. Dark arrows indicate the addition of 5 mM $NH_4{}^+$ to dissipate ΔpH.

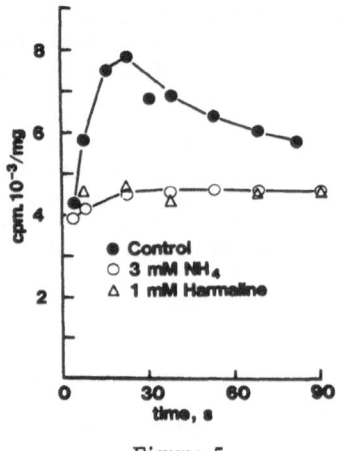

Figure 5

$^{22}Na^+$ accumulation into tonoplast vesicles
driven by 2 unit pH jump

Vesicles were prepared from storage tissue disks washed for 6 days, and the assays
were performed as described in Materials and Methods.

Na$^+$ Accumulation Driven by pH Jump

It follows from the experiments described so far that imposition of an inside-
acid pH gradient should drive accumulation of Na$^+$. This conclusion was tested
by following $^{22}Na^+$ influx into vesicles for various times after a standard pH jump
from 6 to 8. The results are shown in Fig. 5. A clear overshoot in Na$^+$ uptake is
observed after the pH jump, peaking between 20-30 s. Relaxation is presumably
due to leakage from the vesicles as the pH gradient decays. If the pH jump is
performed in the presence of NH$_4^+$ (which abolishes the pH gradient) or harmaline
(a diuretic and inhibitor of Na$^+$/H$^+$ exchange in animal cells), no overshoot is
observed, and the radioactivity associated with the vesicles is presumably due to
non-specific binding. Thus, the standing pH gradient of 2 units across the tonoplast
of intact cells would clearly be capable of energizing vacuolar Na$^+$ accumulation.

DISCUSSION

Na$^+$/H$^+$ exchange is demonstrable in tonoplast vesicles of red beet by several
methods : H$^+$ efflux driven by external Na$^+$ and H$^+$ influx driven by internal Na$^+$
(both measured as changes in fluorescence of the ΔpH dye acridine orange), as
well as Na$^+$ influx driven by an inside-acid pH gradient. The coupled fluxes are
catalysed by an antiporter, rather than by an electrophoretically-mediated exchange
through uniporters. Washing the slices of storage tissue for one day or more results
primarily in a decrease in the K$_m$ of the antiporter, which militates against an
interpretation for enhancement of ion flux during washing by de novo synthesis
of carriers. Instead, it is possible that a preexisting carrier undergoes some form
of covalent modification. A burgeoning literature on Na$^+$/H$^+$ antiport in animal
cells also indicates that the antiporter can be covalently modify, probably by protein
kinase C (Moolenaar, 1986). We are currently investigating the possibility that
phosphorylation of the Na$^+$/H$^+$ antiporter of tonoplast might be responsible for
wash-activated Na$^+$/H$^+$ exchange.

ACKNOWLEDGEMENTS

This work was supported by funds from the Agricultural and Food Research Council (U. K.) and the University of York.

REFERENCES

Bennett, A. B., and Spanswick, R. M., 1983, Optical measurements of ΔpH and $\Delta\Psi$ in corn root membrane vesicles : Kinetic analysis of Cl^- effects on a proton-translocating ATPase, J. Membrane Biol., 71:95.

Briskin, D. P., Thornley, W. R., and Wyse, R. E., 1985, Membrane transport in isolated vesicles from sugar beet taproot. I. Isolation and characterization of energy-dependent, H^+-transporting vesicles, Plant Physiol., 78:865.

Blumwald, E., and Poole, R. J., 1985, Na^+/H^+ antiport in isolated tonoplast vesicles from storage tissue of Beta vulgaris, Plant Physiol., 78:163.

Marquardt, D. W., 1963, An algorithm for least squares estimation of non-linear parameters, J. Soc. Indust. Appl. Math., 11:431.

Moolenaar, W. H., 1986, Effects of growth regulators on intracellular pH regulation, Annu. Rev. Physiol., 48:363.

Poole, R. J., 1976, Transport in cells of storage tissues, in : "Encyclopedia of Plant Physiology", Vol. 2-A, M. G. Pitman and U. Lüttge, eds., Springer-Verlag, Berlin.

Rea, R. A., and Poole, R. J., 1985, Proton-translocating inorganic pyrophosphatase in red beet (Beta vulgaris L.) tonoplast vesicles, Plant Physiol., 77:46.

Sanders, D., Hansen, U.- P., Gradmann, D., and Slayman, C. L., 1984, Generalized kinetic analysis of ion-driven cotransport systems : A unified interpretation of selective ionic effects on Michaelis parameters, J. Membrane Biol., 77: 123.

Wyn-Jones, G., Brady, C. J., and Spiers, J., 1979, Ionic osmotic relations in plant cells, in : "Recent Advances in the Biochemistry of Cereals", D. L. Laidman, and R. G. Wyn Jones, eds., Academic Press, London.

ISOLATION OF TONOPLAST VESICLES FROM KIWI FRUIT CELLS :

ELECTROCHEMICAL INVESTIGATION OF K$^+$ AND H$^+$ MEMBRANE TRANSPORT

LINKED TO A MAGNESIUM-ATP HYDROLYTIC ACTIVITY

Jean-Pierre Rona, F. Chedhomme, M. Convert
and Michèle Monestiez

Laboratoire d'Electrophysiologie des Membranes
Unité Associée C.N.R.S. No. 1180
Université de Paris VII, 2, Place Jussieu
75251-Paris-Cedex 05, France

INTRODUCTION

Mechanical breaking of giant cells of the peripheral layer of Actinidia chinensis fruits produces in the extraction juice, many large vesicles that appear like free vacuoles or protoplasts (Chedhomme and Rona, 1984). The latter are of a much smaller size than the initial cells; these pseudoprotoplasts are always devoid of nuclei as in the case with "vacuoplasts" of Poterioochromonas malhamensis (Jochem et al., 1983).

Several findings (Chedhomme and Rona, 1984 and 1986) suggest that these membrane structures are due to fragmentation followed by revesiculation of the tonoplast surrounding the cytoplasmic strands of giant cells of this zone.

The study of these structures with Nomarski's interference phase contrast microscopy, shows that revesiculation occurs either directly in the endovacuolar medium through fragmentation of the central vacuole (type A vesicles), or on the layer of cytoplasmic strands which surrounds the largest vacuoles (type B and C vesicles). During the resealing of the tonoplast arround the cytoplasm of the strands, small secondary vacuoles may remain trapped, giving the vesicle its protoplast aspect (type B vesicle, Chedhomme and Rona, 1986).

The diversity of formation patterns of vesicles implies a change in the tonoplast interface orientation with respect to their usual contact medium : depending on whether the membrane reforms around the vacuolar juice or around the cytoplasmic strands, the tonoplast surface initially in contact with the endovacuolar medium will either remain in contact with this medium, as for intact isolated vacuoles (Rona et al., 1980), or, by contrast, be in contact with the outer environment, when the cytoplasm is trapped within the vesicles.

The purpose of the present investigation is to confirm our previous results pointing to the tonoplast as the membrane limiting the different types of vesicles present in Kiwi juice. This study uses electrophysiological data pertaining to the structures involved and the specific ability (anion sensitive vacuolar ATPase) of the vacuolar membrane to hydrolyse MgATP (Spanswick et al., 1984).

MATERIAL AND METHODS

Vesicles were collected from the cellular juice that appeared after excision of the peripheral layer of the fruit of <u>Actinidia chinensis</u> var. Hayward. Fruit conservation and juice collection methods have been described previously (Chedhomme and Rona, 1986).

Preparation of the Biological Material

After collection of Kiwi juice, the suspension was divided into two batches. In the first batch, the vesicles were studied in the juice directly, about 15 minutes after extraction from the fruit (measurement of PD and K^+, H^+ internal ionic activity), the vesicles of the second batch were purified by centrifugation in discontinuous gradient (Ficoll 5/10/20%) and resuspended in an artificial medium with pre-controlled ionic status and osmotic potential (Fig. 1); the purified vesicle suspension was used for electrophysiological measurements in the artificial medium at pH 6.5 or at pH 8 (Hepes-Tris, 25 mM); 0.7 M mannitol was added to maintain the osmotic potential around - 2.2 MPa, close to the cellular juice potential (Chedhomme and Rona, 1986).

Approximately 60% of type A vesicles settled at the 5% - 10% Ficoll interface, and 50% of type B and C vesicles settled at the 10% - 20% Ficoll interface.

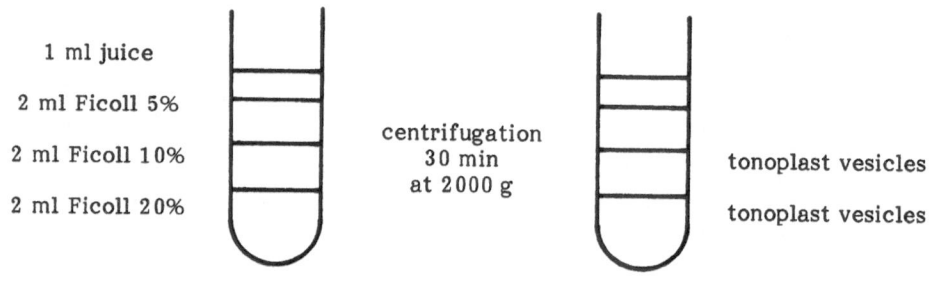

1 ml juice

2 ml Ficoll 5%

2 ml Ficoll 10%

2 ml Ficoll 20%

centrifugation
30 min
at 2000 g

tonoplast vesicles

tonoplast vesicles

Figure 1

Procedure for purification of tonoplastic vesicles
in buffered medium
(Hepes-Tris 25 mM, mannitol 0.7 M, pH 6.5)

ATP-Hydrolysis

For measurements of enzymatic activity, fruits were peeled and the outer pericarp cut into small pieces, was homogenized in a laminar grinder with saccharose (0.25 M), Hepes-Tris 25 mM (pH 7.4) and 0.5% β-mercaptoethanol. The homogenate was strained through cheesecloth and centrifuged at 8000 g/10 min, 13000 g/15 min and 30,000 g/90 min, consequently.

1 ml of membrane preparation was layered onto a 10 ml discontinuous saccharose gradient and the fractions were collected at the interfaces d = 1.10 - 1.15 and 1.15 - 1.20 after centrifugation at 110,000 g for 60 min.

The release of P_i (Peterson, 1978) in the membrane fraction as a result of ATP hydrolysis was measured in 100 μM sodium orthovanadate, 100 μM ammonium molybdate, and 0.2 mM azide with a Perkin Elmer, Lambda 5 spectrophotometer.

The ATP was in the form of Tris–ATP to which $MgSO_4$ was added equimolarly (3 mM/3 mM). Measurements were carried out at 20°C. After a 2 h-hydrolysis, activities were expressed on the basis of membrane protein content. The effects of Cl^- (KCl 50 mM) and NO_3^- (KNO_3 50 mM) on tonoplast ATPase activity (O'Neill et al., 1983) were simultaneously tested in the presence of inhibitors of ATPase activity in the plasmalemma and mitochondrial membranes, as well as an inhibitor of acid phosphatases.

Electrophysiological Measurements

Transmembrane electrical potential (E) and electrical resistance (R) were measured as previously described (Rona et al., 1980) using high impedance glass microelectrodes to minimize the risk of electric membrane shunts (Cornel et al., 1983). Measurements of PD gave the endovesicular potential related to the external reference medium (electrical zero) with no indication as to the nature of the orientation of limiting membranes (E_{vo} and E_{co} : vacuole and cytoplasm related to external medium; E_{vc}: vacuole in relation to cytoplasm).

The K^+ endovesicular activity was measured with doubled barreled micro-electrodes (Rona and Cornel, 1985). Glass micropipettes (Clark, 2GC 150F type) were drawn out vertically (Narishige PE2). One of the capillaries was filled with NaCl (0.5 M) to be used as an electric reference. The other was filled with KCl (0.5 M), and contained in its tip the K^+ liquid ion exchanger (100 to 400 μl; reference : Corning 477317). The two capillaries were linked to the differential electrometer (F.223.A.WPI) by a silver chloride thread (0.25 mM in diameter). The second reference electrode (calomel electrode : Radiometer K 401) was immersed in the suspension medium and functionned as an electric reference for simultaneous measurements of transmembrane potentials.

Internal pH measurement were made with a single barrel microelectrode with a H^+ liquid ion exchanger (Amman et al., 1981; Kurkdjian and Barbier-Brygoo, 1983). Practically speaking, after the capillaries (Clark, GC 150 F type) had been elongated, their tip was made hydrophobic by treatment with silicone (Tri-N-butyl-chlorosilane, 25 μl and α-mono-chloronaphtalene, 1 ml solution). The solution was evaporated at 120°C for 2 h. The tip (100 to 300 μm) was then filled with the H^+ ion exchange liquid using a capillary. The micropipette was then completely filled with a KCl solution (3 M). the resin was made up of a mixture of 30.3 mg of tri-dodecylamine, 2.1 mg of sodium tetraphenylborate and 267.6 mg of nitrophenyl-octylether, stirred in darkness for 3 h. The distilled ion exchange liquid could be preserved for about one month at 4°C with no significant loss of selectivity for H^+, under the condition that it will be regenerated, 12 h before use, by saturation in darkness with CO_2 in a closed box with carbonic ice. The measured slope was 64 mV per pH unit between pH 4 and pH 7, at 25°C.

Endovesicular H^+ activity was calculated from values obtained with the selective microelectrode, connected of membrane potential, which was measured separately on a control batch, just before ionic activity measurements.

RESULTS AND DISCUSSION

Mg-ATP Hydrolytic Activity

A significant evolution of tonoplast specific MgATP-hydrolytic activity of <u>Actinidia chinensis</u> cells has been observed during the two months of maturation in the "Deliciosa variety". This activity, which is weak at the onset of ripening, increases after one month maturation (Fig. 2). It disappears after two months (mature fruits), as in the "Hayward variety" which is stored three to six months after harvesting. This tonoplastic specific activity (measured in the presence of molybdate (100 μM), vanadate (100 μM) and azide (0.2 mM)) was Cl^--stimulated. It was only inhibited by NO_3^- when the tonoplast MgATP-hydrolytic activity was high (more than $0.2\,\mu$mol P_i mg^{-1} protein h^{-1}).

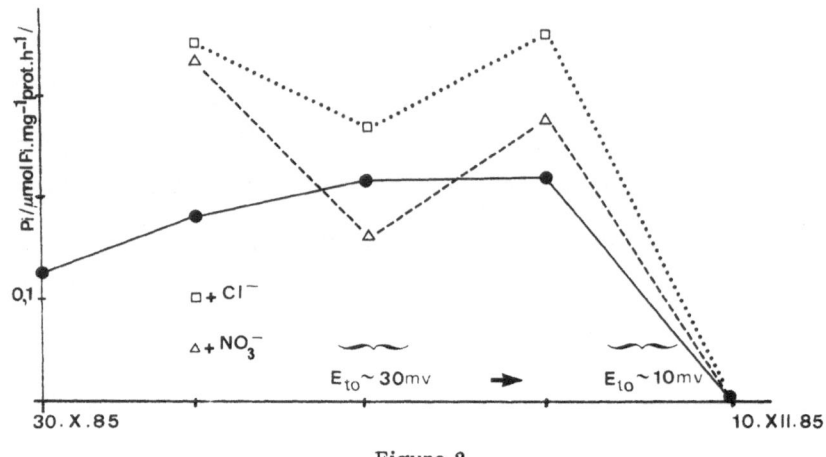

Figure 2

Evolution of tonoplast MgATP-hydrolytic activity
during maturation of Kiwi cells

KNO_3 ( ) and KCl ( ) (50 mM) effects; control ( )

For the "Hayward variety", results are more difficult to interpret. This difficulty stems from two ill defined factors : Fruit age after harvesting which cannot be determined accurately and the period of storage in the industrial cold room, which seems to influence the tonoplast specific Mg-ATP activity differently depending on the fruit batch studied. NO_3^- inhibition is not always significant; no effect has even been observed in certain experiments.

In the "Deliciosa and Hayward varieties", the values of tonoplast specific ATP-hydrolyzing activity relative to the same content in membrane protein, are low in comparison to previously reported values (Sze, 1984). However, this slight tonoplast specific ATP-hydrolyzing activity (O'Neill et al., 1983; Sze, 1984) can be invoked as an argument for the hypothesis of a vacuolar origin for membranes limiting the three types of vesicles suspended in the Kiwi juice.

Electrophysiological Data

Vesicle Polarization. Table 1 contains the results obtained on vacuolar vesicles (A type), cytoplasmic vesicles (C type) and pseudoprotoplasts (B type), all in thermodynamic equilibrium with their respective medium.

As for free intact vacuoles, the electric profile recorded for type A vesicles, shows only one positive plateau for each measurement in a homogeneous vesicle population (50 ± 10 m in diameter); the respective amplitudes of electrical gradient and specific membrane resistance are $E_{vo} = 10 \pm 2$ mV and r = 0.4 kΩ cm^2, these values are close to those generally observed with the microelectrode technique across the tonoplasts of free intact vacuoles (Rona and Cornel, 1979; Barbier et al., 1985).

Table 1

Extremes Values of Endovesicular PD Measured in Cellular Juice and Artificial Medium (0.01 mM K^+ and Na^+, 0.05 mM KCl)

Type of medium	Type of vesicles	PD (mV)
Cellular juice	"A"	$+ 6 < E_{vo} < + 15$
	"B"	$- 13 < E_{co} < 0$
	"B"	$+ 5 < E_{vo} < + 12$
	"C" (b)	$- 50 < E_{co} < - 20$
	"C" (a)	$- 23 < E_{co} < - 20$
Artificial medium	"A"	$+ 8 < E_{vo} < + 17$
	"B"	$- 25 < E_{int} < - 5$
	"C"	$- 26 < E_{co} < - 13$

Varieties are mentioned when measured PD amplitudes are significantly different : "Hayward variety" (a); "Deliciosa variety" (b). E_{co} and E_{vo} are respectively the cytoplasmic and vacuolar potentials with respect to the external medium; E_{int} is internal potential with respect to the external medium.

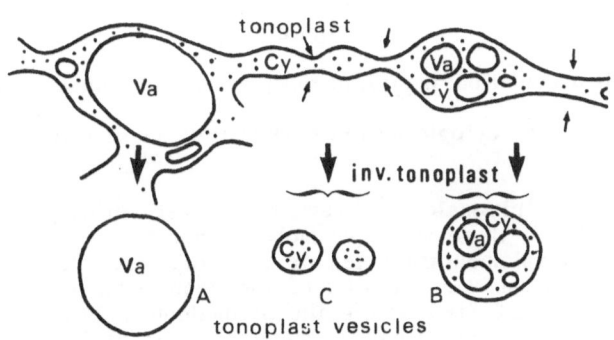

tonoplast

inv. tonoplast

tonoplast vesicles

Figure 3

Revesiculation of tonoplast cytoplasmic strands and example of three types of vesicles (A, B and C)

Arrows show the suggested orientations of the revesiculation within a cell; Cy: cytoplasm; Va: vacuole; inv. tonoplast: inverted tonoplast.

Measurements on type C (cytoplasmic) vesicles are not as straight forward owing to the vesicles' small size (10 to 20 μm in diameter). However our trials show their polarization, in spite of its heterogeneity, to be negative ($E_{co} = - 20$

+ 3 mV, with respect to external medium); vesicles from freshly harvested fruits ("Delicosia variety"), are markedly more electronegative ($\simeq$ - 48 mV) than those from cold stored Kiwi fruits ("Hayward variety") (see Table 1).

For type B vesicles (pseudoprotoplasts), there are two series of values depending on how deeply the microelectrode was inserted (Table 1); the frequency of occurrence of profiles with two plateaus is higher for vesicles where the cytoplasmic layer and the vacuole are well differentiated (Chedhomme and Rona, 1986). As soon as the electrode is inserted, the first group of potentials, when measured, always displays negative values.

The second group (deeper insertion and higher electrical resistance) can have electronegative as well as electropositive values, depending on the kind of processing and the type of pseudoprotoplast studied.

Thus, with the exception of clearly negative values, the electrical characteristics of revesiculated pseudoprotoplasts are similar to those of intact protoplasts (Rona et al., 1980), whereas the limiting membranes do not appear to be of plasmic origin (plasmalemma), as shown in Fig. 3 and the photograph in Fig. 4.

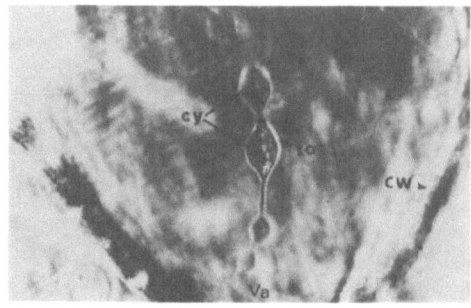

Figure 4

Revesiculation, in situ, of tonoplast cytoplasmic strands

Cy : cytoplasm; to : tonoplast; cw : cell well

Actually, the plasmalemma remains almost entirely attached to the pectocellulose walls, if there is no plasmolysis prior to the rupture as is done for other materials. Complementary attempts to obtain protoplasts from Kiwi cells, have shown that the osmotic potential of the juice (- 2.2 to - 2.0 MPa) may be decreased to about - 4.5 MPa before the plasmalemma detaches itself from the wall (Fig. 5). Intact protoplasts have a diameter 5- to 20-fold greater than those of vesicles B or C; moreover, the latter are easily observed inside the protoplasts, and for that reason, contamination of vesicle membranes by plasmalemma seems highly unlikely. In addition, the mechanical effect of breaking the cells in this way is less rough than classical disruptions used to obtain total membrane fractions (Bennett and Spanswick, 1983).

MgATP-Hydrolytic Activity and Tonoplast Polarization. With type A vesicles, which are naturally positively polarized in the artificial medium at pH 8, the addition of MgATP (9 mM, concentration in the final medium), causes a positive hyperpolarization of the limiting membrane from 8 to 12 mV (Fig. 6-A). This effect has already been observed under similar conditions in intact vacuoles isolated from other

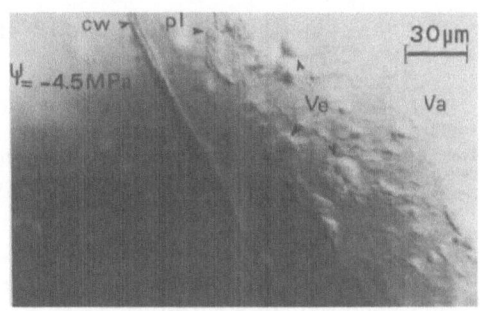

Figure 5

Plasmolysed cell in which the future intact Kiwi protoplast
has detached itself from the wall

The osmotic potential of the medium is about - 4.5 MPa. cw: cell wall; pl: plasmalemma; Va: vacuole; Ve: vesicle.

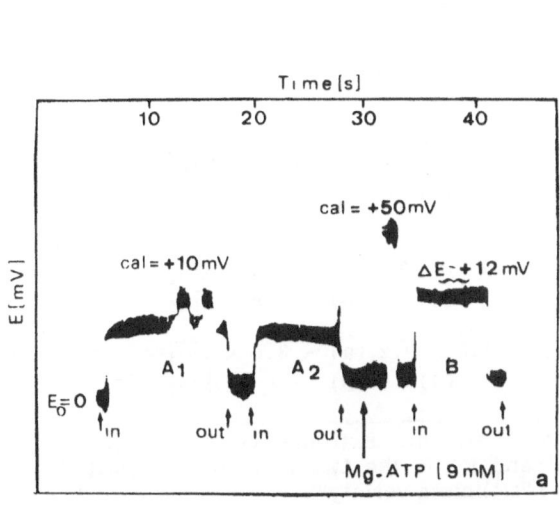
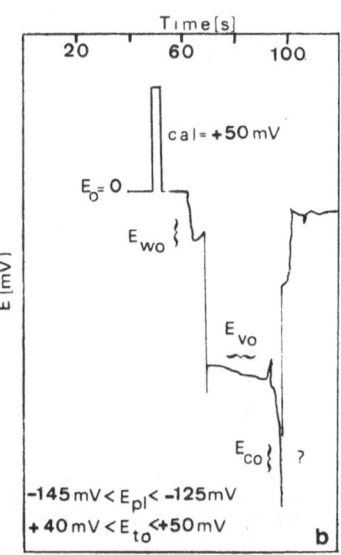

Figure 6

Oscilloscope traces of measurements
of transtonoplast electrical potentials
of isolated vacuoles (a) and of a cell (b)

(a) Electrical recording on type A vesicle; A_1 and A_2: controls; B: with MgATP
(final concentration 9 mM); E: tonoplast hyperpolarisation with MgATP (pH 8);
in, out: tip of electrode in or out of the vesicle.
(b) Transvesicular electrical profile for Kiwi cells ("Deliciosa variety"); E: potential
difference; wo: wall with respect to outside; vo: vacuole with respect to outside;
pl: plasmalemma; to: tonoplast; cal: calibration.

biological material (Wagner and Lin, 1982; Rona et al., 1984; Jochem et al., 1984; Monestiez et al., 1986).

Moreover, the measurement of transtonoplastic potential in intact Kiwi cells ("Deliciosa variety"), suspended in the previous medium (Fig. 6-B), indicates that the electropositive gradient, E_{VC}, across the vacuolar membrane has an amplitude of about + 30 mV, a value closer to those observed in energized type A vesicles (+ MgATP).

K+- and H+-electrochemical Investigation. Measurements of K^+ and H^+ activities are made using specific microelectrodes on different types of vesicles extracted from the "Hayward and Delicosia varieties".

These measurements are made aver 15 to 20 min for each sample. The values of K^+ and internal pH activities for type A and C vesicles are shown in Table 2. There are no significant differences between the two varieties when measurements for the same ion are compared, even in the most polarized vesicles ("Delicosia variety").

Table 2

Ionic Activities (H+ and K+) Measured
in Vacuolar and Cytoplasmic Vesicles in the Kiwi Juice (steady state)

Vesicle type		Ionic activities (mM)		$\Delta\overline{\mu}_i$ (kJ.mol^{-1})
		external	internal	
Vacuolar vesicles (A)	K^+	64.5 $\pm$ 6.2 (6)	35.5 $\pm$ 1.6 (6)	$- 0.87 < \Delta\overline{\mu}_{K^+} < 0$
	pH	4.1 $\pm$ 0.1 (3)	4.38 $\pm$ 0.2 (3)	$- 0.98 < \Delta\overline{\mu}_{H^+} < 0.12$
Cytoplasmic vesicles (C)	K^+	67 $\pm$ 6.5 (9)	104 $\pm$ 18.4 (a) (9)	(a) $- 1.4 < \Delta\overline{\mu}_{K^+} < - 0.6$ (b) $- 3.8 < \Delta\overline{\mu}_{K^+} < - 3.3$
	pH	4.06 $\pm$ 0.2 (4)	4.3 $\pm$ 0.2 (a) (4)	(a) $- 3.6 < \Delta\overline{\mu}_{H^+} < - 1.5$ (b) $- 6 < \Delta\overline{\mu}_{H^+} < - 5.4$

Values are expressed as mean $\pm$ sd (n). Extreme values of the electrochemical potential for H^+ and K^+ in the two types of tonoplast vesicles (a : "Hayward variety"; b : "Deliciosa variety").

In type A vesicles, the internal K_+ activity is lower by about 50% than the external medium; this means that the endovesicular potassium is very close to thermodynamic equilibrium ($\Delta\overline{\mu_K}^+ = - 0.5$ kJ mol^{-1}). This observation is similar to previously reported results on vacuoles of "low salt" barley root cells in situ, where the potassium concentration of the vacuolar sap is less than that of the cytoplasm (Pitman and Saddler, 1967; Pitman et al., 1981). Similarly, in cambial cells of the sycomore maple the endovesicular potassium is less concentrated than the cytoplasmic potassium; there, as for type A vesicles, steady state K^+ remains very close to thermodynamic equilibrium($\Delta\overline{\mu_K}^+ = - 0.4$ kJ mol^{-1}) (Rona et al., 1982).

ΔpH measurements show that the acidity level of these vesicles is close to that of cellular sap. This indicates on the one hand that H^+ concentration is very near the thermodynamic equilibrium, and on the other hand that vacuolar pH imposes its own acidity to cellular juice. This last observation is an additional confirmation to the hypothesis of a vacuolar origin for type A vesicles.

In Table 2, one can see that, unlike the previous case, in type C vesicles, K^+ internal activity is remarkably higher than that of the cellular sap; its order of magnitude (about 100 mM) can be compared to concentrations of this ion in the cytoplasm of several plant cells, this being true regardless of the measurement technique used (Leigh and Wyn-Jones, 1984). Unexpectedly the pH is near that of type A vesicles, whereas in a cytoplasmic phase, we would expect it to close to neutral.

However, vesicles of the Delicosia variety, which are distinctively electronegative with respect to the external medium (E_{co} = - 48 mV) have a lower H^+ concentration than that of thermodynamic equilibrium. On that basis, in those vesicles, H^+ may be supposed to be expelled actively from the cytoplasm out across the tonoplast.

CONCLUSION

To sum up, membrane vesicles of cellular juice extracted from Actinidia chinensis fruits, appear to originate from fragmentation of the central vacuoles of these cells, followed by revesiculation of the tonoplast. These vesicles can either include the vacuolar sap or cytoplasmic strands or both (pseudoprotoplasts). It appears that when cytoplasm is trapped in these vesicles, the orientation of the endovesicular interface of the membrane is reversed in relation to the medium which is the electrical reference.

Preliminary assays of the membrane enzymatic activity show a tonoplast specific MgATP-hydrolytic activity (Spanswick et al., 1984). Stimulation by Cl^- and inhibition by NO_3^- when tonoplast ATP hydrolytic activity is high enough are also apparent.

These results indicate that the electrical characteristics of the tonoplast of Kiwi vesicles are very similar to those observed in isolated and in situ vacuoles of protoplasts, meaning positive hyperpolarization, mainly of diffusive origin when the energy source provided by the cytoplasm (MgATP) is weak or non-existent. H^+ and K^+ ions are both close to the thermodynamic equilibrium in vacuolar and cytoplasmic vesicles of the Hayward variety; this implies that it is not a proton pump which polarizes these vesicles positively or negatively, but rather the direction and amplitude of the K^+ gradient across the membrane.

The diffusive character of measured PD can result from two combined factors : One is the important mobilization of total organic acids in the Kiwi cell vacuole throughout maturation (Heatherbell, 1975; Reid et al., 1982) and the role that they appear to play in the passive positive PD across the tonoplast of isolated vacuoles ($P_{COO^-} \gg P_{K^+}$) (Gibrat et al., 1985), the other is the loss or modification of certain enzymatic activities at maturity and senescence, especially after a long storage period (Hartmann, 1983); the activity of tonoplastic ATPase may have been among these affected.

In the "Deliciosa variety", where the harvesting period is very short, the presence of a weak electronegative component (E_{co} = - 20 to - 30 mV) due to proton efflux across the tonoplast ($\overline{\Delta \mu_{co}}$ = - 6 kJ mol^{-1}) cannot be ruled out. The sign of the PD is reversed by comparison with intact vacuoles, along with tonoplast orientation towards the external medium and cytoplasmic vesicles may provide ATP as an energy source to a possible Mg-ATPase pump, located in the membrane.

ACKNOWLEDGEMENTS

This work was supported by grants from the C.N.R.S. (U.A. 1180) and from the "Ministère de la Recherche et de la Technologie" (83380). Kiwi fruits were kindly provided by Mr. C. Desmoulins (Ste Roland Lacour). We thank Mme U. Bousquet for help in the preparation of the manuscript.

REFERENCES

Ammann, D., Lanter, F., Steiner, R. A., Schulthess, P., Shijo, Y., and Simon, W., 1981, Neutral carrier based hydrogen ion selective microelectrode for extra and intracellular studies, Anal. Chem., 53:2267.

Barbier-Brygoo, H., Gibrat, R., Renaudin, J. P., Brown, S. C., Pradier, J. M., Grignon, C., and Guern, J., 1985, Membrane potential difference of isolated plant vacuoles : positive or negative ?, Biochim. Biophys. Acta, 819:215.

Bennett, A. B., and Spanswick, R. M., 1983, Optical measurements of ΔpH and $\Delta\Psi$ in corn root membrane vesicles. Kinetic analysis of Cl$^-$ effects on a proton translocating ATPase, J. Membrane Biol., 71:95.

Chedhomme, F., and Rona, J. P., 1984, Suitability of the spontaneously formed vesicles in Kiwi fruit juice for investigating the electrical properties of the tonoplast, in : "Abstracts of the IVème Congrès de la F.S.E.P.V. (Strasbourg, France, July 29/August 3 1984", pp. 478-479, Université de Strasbourg.

Chedhomme, F., and Rona, J. P., 1986, Isolation and electrical characterization of tonoplast vesicles from the Kiwi fruit (Actinidia chinensis), Physio. Plant., 67:29.

Cornel, D., Grignon, C., Rona, J. P., and Heller, R., 1983, Measurements of intracellular potassium activity in protoplasts of Acer pseudoplatanus : Origin of their electropositivity, Physiol. Plant., 57:208.

Gibrat, R., Barbier-Brygoo, H., Guern, J., and Grignon, C., 1985, Transtonoplast potential difference and surface potential of isolated vacuoles, in : "Biochemistry and Function of Vacuolar ATPase in Fungi and Plants", B. P. Marin, ed., Springer-Verlag, Berlin, Heidelberg, New-York, and Tokyo.

Hartmann, C., 1983, Quelques aspects de la senescence du fruit, Physiol. Vég., 21:137.

Heatherbell, D. A., 1975, Identification and quantitative analysis of sugars and non-volatile organic acids in Chinese gooseberry fruits (Actinidia chinensis Planch.), J. Sci. Fd. Agric., 26:815.

Jeschke, W. D., and Stelter, W., 1976, Measurements of longitudinal ion profiles in single roots of Hordeum and Atriplex by use of flameless atomic spectroscopy, Planta, 128:107.

Jochem, P., Thomson, K. S., and Schab, D., 1983, Isolation of 'vacuoplasts' from Poterioochromonas malhamensis, Plant Physiol., 73:418.

Jochem, P., Rona, J. P., Smith, J. A. C., and Lüttge, U., 1984, Anion-sensitive ATPase activity and proton transport in isolated vacuoles of species of the CAM genus Kalanchoe, Physiol. Plant., 62:410.

Kurkdjian, A., and Barbier-Brygoo, H., 1983, A hydrogen ion-selective liquid membrane microelectrode for measurement of the vacuolar pH of plant cells in suspension culture, Anal. Biochem., 132:96.

Leigh, R. A., and Wyn-Jones, R. G., 1984, A hypothesis relating critical potassium concentrations for growth to the distribution and functions of this ion in the plant cell, New Phytol., 97:1.

Monestiez, M., Belabed, A. M., Pennarum, A. M., Convert, M., Cornel, D., and Rona, J. P., 1986, Some characteristics of tonoplast NO$_3^-$ transport processes on Acer pseudoplatanus L. cells, in : "Plant Vacuoles. Their Importance in Solute Compartmentation and Their Appications in Biotechnology", B. Marin, ed., Plenum Publishing Corporation, New-York.

O'Neill, S. D., Bennett, A. B., and Spanswick, R. M., 1983, Characterization of a NO$_3^-$-sensitive H$^+$-ATPase from corn roots, Plant Physiol., 72:837.

244

Pitman, M. G., and Saddler, H. D. W., 1967, Active sodium and potassium transport in cells of barley roots, Proc. Natl. Acad. Sci. U.S.A., 57:44.

Peterson, G. L., 1978, A simplification of the protein assay method of Lowry et al. which is more generally applicable, Anal. Biochem., 83:346.

Pitman, M. G., Lauchli, A., and Stelzer, R., 1981, Ion distribution in roots of barley seedlings measured by electron probe X-ray microanalysis, Plant Physiol., 68:673.

Reid, M. S., Heatherbell, D. A., and Pratt, H. K., 1982, Seasonal patterns in chemical composition of the fruit of Actinidia chinensis, J. Amer. Soc. Hort. Sci., 107:316.

Rona, J. P., and Cornel, D., 1979, Résistances electriques chez les cellules libres, les protoplastes et les vacuoles isolées d'Acer pseudoplatanus L., Physiol. Vég., 17:1.

Rona, J. P., and Cornel, D., 1985, An electrogenic proton pump on the tonoplast of Acer pseudoplatanus L. free cells and isolated vacuoles, in : "Biochemistry and Function of Vacuolar ATPase in Fungi and Plants", B. P. Marin, ed., Springer-Verlag, Berlin, Heidelberg, New-York, and Tokyo..

Rona, J. P., Cornel, D., Chedhomme, F., and Heller, R., 1984, The contribution of the plasmalemma and the tonoplast to the electrical properties of Acer pseudoplatanus L. cells, in : "Membrane Transport in Plants", K. Janacek, R. Rybova, and K. Sigler, eds., Czech. Acad. Sci. Press, Prague.

Rona, J. P., Cornel, D., Grignon, C., and Heller, R., 1982, The electrical potential difference across the tonoplast of Acer pseudoplatanus cells, Physiol. Vég., 20:459.

Rona, J. P., Van De Sype, G., Cornel, D., Grignon, C., and Heller, R., 1980, Plasmolysis effect on electrical characteristics of free cells and protoplasts of Acer pseudoplatanus L., J. Electroanal. Chem., 116:377.

Spanswick, R. M., O'Neill, S. D., and Bennett, A. B., 1984, Plasma membrane and tonoplast ATPases : Characteristics, H^+ transport, and reconstitution, in : "Membrane Transport in Plants", K. Janacek, R. Rybova, and K. Sigler, eds., Czech. Acad. Sci. Press, Prague.

Sze, H., 1984, H^+-translocating ATPases of the plasma membrane and tonoplast of plant cells, Physiol. Plant., 61:683.

Wagner, G. J., and Lin, W., 1982, An active proton pump of intact vacuoles isolated from Tulipa petals, Biochim. Biophys. Acta, 689:261.

SOME CHARACTERISTICS OF TONOPLAST NO_3^- TRANSPORT PROCESSES

ON ACER PSEUDOPLATANUS L. CELLS

Michèle Monestiez, Abdel Magid Belabed, Anne-Marie Pennarum,
Monique Convert, Daniel Cornel and Jean-Pierre Rona

Laboratoire d'Electrophysiologie des Membranes
Unité Associée C.N.R.S. No 1180, Université de Paris VII
2, Place Jussieu, 75251-Paris-Cedex 05, France

INTRODUCTION

Data on the NO_3^- accumulation mechanisms at the tonoplast has remained widely unknown. NO_3^- uptake induced a depolarization or an hyperpolarization of cells (Ullrich and Novacky, 1981; Thibaud and Grignon, 1981). This can be classically attributed to an electrogenic OH^-/NO_3^- antiport or H^+/NO_3^- symport at the plasmalemma (Rona and Cornel, 1984; Thibaud and Grignon, 1981). But the existence of an NO_3inhibited and Cl^--stimulated Mg-ATPase on tonoplast is now well established (Sze, 1985). Furthermore, NO_3^- vacuolar net accumulation could give rise to depolarization of the tonoplastic membrane. Nitrate induced hyperpolarization in cells could result partly from the depolarization of the tonoplast. The aim of these experiments was to investigate some of the NO_3^- vacuolar uptake characteristics and those of related electrical events.

MATERIALS AND METHODS

Plant material

Free cells of Acer pseudoplatanus L. were grown in Heller's medium.

Isolation of Vacuoles and Tonoplast Vesicles

Protoplasts were isolated from suspension cultures as previously described (Rona and Cornel, 1981). Vacuoles were obtained by osmotic shock of washed protoplasts pellet. The lysate was layered over a discontinuous Ficoll gradient in 0.6 M mannitol, Hepes-Tris 25 mM pH 8.0, DTT 0.5 mM with (or without when indicated) BSA 0.1%, KCl 10 mM, NaCl 10 mM, $CaCl_2$ 0.5 mM, $MgCl_2$ 0.5 mM, saccharose 44 mM. Intact vacuoles were collected at the 5%-7.5% and 7.5%-10% Ficoll interfaces (after 100,000 g, 30 min centrifugation). For ATP-hydrolytic assays, the vacuoles were vesiculated with about five vigourous strokes in a glass Teflon tissue homogenizer in Hepes-Tris pH 8.0. After 100,000 g, 30 min centrifugation, the pellet was resuspended in a minimal volume of Hepes-Tris 25 mM, pH 8.0. For electrical measurements, intact vacuoles were resuspended (5%, v/v) in the Ficoll deprivated medium.

Electrical Measurements

Electrical measurements were performed as described previously (Rona and Cornel, 1981). Results were expressed as mean $\pm$ SD and values in the brackets (n), were the number of independant experiments. On intact cells, when there is a double oscilloscope trace, two electrodes are simultneously implanted, one in the vacuole (inferior trace), the other in the cytoplasm (superior trace). The reference electrode was localized in the external medium : the transtonoplastic potential was deduced by the difference between the two values measured on the graph ($E_{VC} = E_{VO} - E_{CO}$). E_{CO} = PD (cytoplasm with respect to outside); E_{VO} = PD (vacuole with respect to outside); E_{VC} = PD (vacuole with respect to cytoplasm).

Microscopic observation of Neutral Red Accumulation

Cells, protoplasts or isolated vacuoles were prepared in the basic medium which contained Hepes-Tris 25 mM at the indicated pH, mannitol 0.6 M, DTT 0.5 mM, BSA 0.1%, KCl 10 mM, NaCl 10 mM, $CaCl_2$ 0.5 mM, $MgCl_2$ 0.5 mM, saccharose 44 mM. Neutral red was added at 0.05%. Vacuolar coloration was observed by Nomarski interference microscopy technique.

Nitrate Endocellular Contents and Nitrate Net Uptake

Nitrate endocellular contents and nitrate net uptake ware measured by NO_3^- medium depletion and/or NO_3^- endocellular content increase, by colorimetric method.

ATP-Hydrolytic Activity

ATP hydrolytic activity was assayed according to the method of Peterson (1978).

RESULTS

Electric Potentials

(a) In Cells. On nitrate deprivated cells (20 days old), transtonoplastic potentials E_{VC} decreased rapidly after $NaNO_3$ addition (9 mM) ΔE_{VC} = 8 mV (short time effect) (Photo 1). In these conditions, assuming that in 8 min the absorbed NO_3^- was localized only in the cytoplasm, the cytoplasmic NO_3^- concentration was 1 mM (results given from net influx measurements, as far as nitrate reductase was not induced during this time).

Depolarisation of tonoplast, ΔE_{VC} = 10 mM, was also observed after 6 days in 10 mM NO_3^- enriched Heller' growth medium (long time effect) (Table 1).

(b) In Isolated Vacuoles (pH = 8). Short time depolarization of tonoplast occurs after $NaNO_3$ or NaCl 20 mM addition (Table 2). MgATP (10 mM) increased the PD with or without NaCl (20 mM) (ΔE_{VO} = 12 mV). No such effect was observed when $NaNO_3$ is the added salt (Table 3).

Neutral Red Penetration

On 11 day old cells, neutral red induced after 3 min red vacuolar accumulation with a pH 8 external medium. This processus was not observed at pH 4 or 5.5. Identical phenomena could be shown on protoplasts and isolated vacuoles (Photo 2-A and -B). These facts suggested that pH (acid in) was necessary for the neutral red accumulation as it can be shown in others materials. On protoplasts, MgATP

Control
+ 28 mV < E_{VC} < + 34 mV

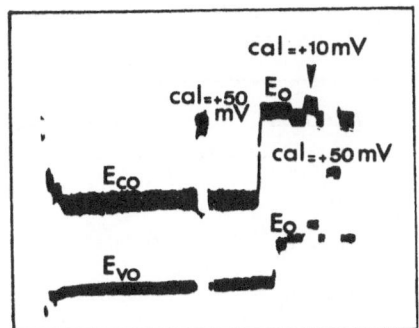

8 min after addition of NaNO$_3$
+ 22 mV < E_{VC} < + 24 mV

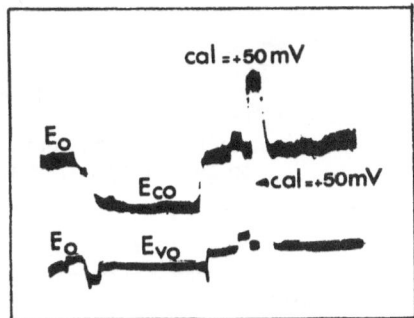

Photo 1

Evolution of cytoplasmic and vacuolar potentials
after addition of NaNO$_3$ (9 mM)
on nitrate deprivated cells (20 day old cells)

Simultaneous measurements of vacuolar and cytoplasmic potentials were made
on one single cell.

Table 1

Transplasmalemma and Transtonoplastic Potentials
on six day old cells grown on Heller's medium (control)
and on 10 mM nitrate Heller's medium (enriched)

Medium	Electrical Cells Potentials (mV)	
	E_{CO}	E_{VC}
control	− 50.0 ± 7.2 (6)	+ 25.2 ± 5.5 (5)
enriched	− 45.5 ± 12 (4)	+ 15.2 ± 1.9 (4)

Table 2

Effect of 20 mM NaCl or 20 mM NaNO$_3$
on Transtonoplastic Potentials from Isolated Vacuoles

Addition	Transtonoplastic potentials E_{vo} (mV)
none	23.15 ± 2.38
NaCl	14.04 ± 6.52
NaNO$_3$	13.64 ± 4.09

The basic medium had a minimum composition : Mannitol 0.6 M, Hepes-Tris 25 mM, pH 8.0.

Table 3

Effect of 10 mM MgATP and 20 mM NaCl and 20 mM NaNO$_3$
on electrical potentials of isolated vacuoles

addition	Transtonoplastic potentials (mV)	
	$-$	+ MgATP
none	11.3 ± 1.8 (11)	23.5 ± 3.9 (16)
none NaCl	10.7 ± 1.5 (3) 7.1 ± 1.7 (8)	19.5 ± 2.8 (11)
none NaNO$_3$	10.9 ± 5.0 (16) 9.7 ± 2.2 (9)	9.4 ± 0.7 (8)

The basic medium contained Hepes-Tris 25 mM pH 8.0, mannitol 0.6 M, DTT 0.5 mM, BSA 0.1%, KCl 10 mM, NaCl 10 mM, CaCl$_2$ 0.5 mM, MgCl$_2$ 0.5 mM and saccharose 44 mM.

(5 mM) addition at pH 5.5 induced after 5 to 10 min red coloration of vacuoles (Photo 2-C). This ATP effect could be interpreted as a consequence of a vacuolar acidification (Hager et al., 1980).

ATP-Hydrolytic Activity

Tonoplast membrane vesicles obtained from intact purified vacuoles exhibited a Mg-dependent ATP-hydrolytic activity. It was stimulated by NO$_3^-$ and Cl$^-$ and inhibited by SO$_4^{2-}$. To determine whether these ions directly affected the activity they were assayed on Triton-stimulated ATP-hydrolytic activity. Thus, NO$_3^-$ appeared to inhibit, and Cl- and SO$_4^{2-}$ appeared without effect (Tables 4 and 5).

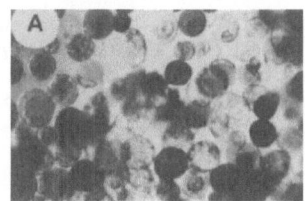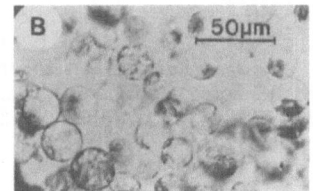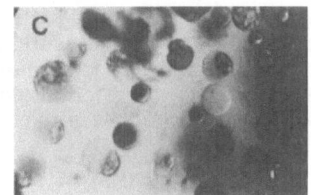

Photo 2

Effect of neutral red addition (0.05%) on protoplasts

The basic medium contained Hepes-Tris 25 mM at the indicated pH, mannitol 0.6 M, DTT 0.5 mM, BSA 0.1%, KCl 10 mM, NaCl 10 mM, CaCl$_2$ 0.5 mM, MgCl$_2$ 0.5 mM, saccharose 44 mM. Photo 2-A: pH 8; photo 2-B: pH 5.5; photo 2-C: pH 5.5 + MgATP.

Table 4

Effect of anion salt on MgATP hydrolytic activities
of tonoplast vesicles

added salt	ATP-hydrolytic activities (μmol P$_i$ mg^{-1} h^{-1})	
	Mg–dependent	Anion–dependent
none	0.29	0
KNO$_3$	0.53	+ 0.24
KCl	0.64	+ 0.35
K$_2$SO$_4$	0.24	– 0.05

MgATP hydrolytic activity was determined as the difference between with or without 1 mM MgSO$_4$. Basic medium contained ATP 1 mM, 25 mM Hepes-Tris pH 8, 0.2 mM NaN$_3$, 100 μM sodium othovanadate, 100 μM ammonium molybdate. Anion-hydrolytic activity determined as the difference in activity in the presence and absence of anion salt.

NO$_3^-$ Electrochemical Potentials and Permeability Coefficients

6/8 day old cells were used : They were accumulating NO$_3$ from Heller's medium, their NO$_3^-$ contents were important (640 μeq g^{-1} DW). External medium was 0.2 mM. In that conditions, cell walls content was negligeable and total content could be assimilated to endocellular content.

In such conditions, nitrate electrochemical potentials $\Delta\overline{\mu}_{vc}$ have been calculated. Two cases must have been envisaged (Table 6). At first, the cytoplasmic concentration is 0.1 mM (Belton et al., 1985), thus $\Delta\overline{\mu}_{vc}$ was + 12 kJ mol^{-1}. Consequently, it will exist NO$_3^-$ vacuolar active influx. Secondly, assuming that nitrate cytoplasmic levels were from 1% (Aslam and Oaks, 1975) to 10% (Marigo et al., 1985) of endocellular contents μ_{vc} could be nul : Nitrate appeared near thermodynamic equilibrium in the vacuole with respect to the cytoplasm. Permeability coefficient could be estimated (from Goldman equation and net influx measurements : 3.57 μeq g^{-1} DW). P was between 2 and 6 10^{-1} cm s^{-1}.

Table 5

Effect of Triton X-100 (0.01%)
on Anion–ATP–hydrolytic Activity

Addition	Anion–ATP–hydrolytic activities (μmol P$_i$ mg$^-$ h^{-1})		
	NO$_3^-$	Cl$^-$	SO$_4^{2-}$
none	+ 0.09	+ 0.16	– 0.13
Triton	– 0.11	+ 0.02	0

Triton–sensitive ATPase activity was determined as the difference between total activity (measured in the presence of MgSO$_4$ 1mM, ATP 1 mM, 25 mM Hepes-Tris pH 8, 0.2 mM N$_3$Na, 100μM sodium orthovana-date, 100 μM ammonium molybdate) and total activity in the presence of Triton. In that experiment, Triton–stimulated activity was + 0.13 μmol mg^{-1} h^1.

Table 6

Nitrate Electrochemical Potentials($\Delta \overline{\mu_{vc}}$)
and Tonoplast Permeability Coefficients (P)
After Endocellular Content Estimations

Q$_c$/Q$_t$ (%)	C$_c$	C$_v$	$\Delta \overline{\mu_{vc}}$	P	Consequences
<0.1	0.1	45	+14	–	active influx
1	3.7	44	+4	–	active influx
2–5	7.5–18	42	0	negative	passive fluxes
5–10	18–35	41	0–(–2)	2–4 10^{-9}	passive fluxes

C$_c$ and C$_v$, in mM; $\Delta \overline{\mu_{vc}}$ in kJ mol^{-1}; P in cm s^{-1}. Cell content : 640 μeq g^{-1} dry weight.

CONCLUSION

The existence of Mg-ATPase proton pump was demonstrated on the tonoplast of _Acer_ cells. It was inhibited by NO$_3^-$. The short time effect could be interpreted as a direct action of NO_3^- on this enzyme. NO_3^- uptake induced a depolarization of the tonoplast which may be the result of two processes occurring simultaneously or not : a diffusion into the vacuole (long time effect) and a reduction of electrogenic component of the transtonoplastic potential by the inhibition of the Mg-ATPase proton pump (short time effect).

According to Nernst's criteria, NO_3^- could enter passively in the vacuole. But it may be also an active influx if cytoplasmic concentration was very low (which may be the case when nitrate reductase was functioning).

ACKNOWLEDGEMENTS

This work was supported by grants from CNRS (UA 1180). Tha authors thanks Claude Grignon for helpfull discussions and Ulrike Bousquet for preparation of manuscript.

REFERENCES

Aslam, R., and Oaks, A. N., 1975, Effect of glucose on the induction of nitrate-reductase in corn roots, Plant Physiol., 56:634.

Belton, P. S., Lee, R. B., and Ratcliffe, R. G., 1985, A ^{14}N nuclear magnetic resonance study of inorganic metabolism in barley, maize and pea roots, J. Exptl. Bot., 36:190.

Hager, A., Frenzel, R., and Laible, D., 1980, ATP-dependent proton transport into vesicles of microsomal membranes of Zea mays coleoptiles, Z. Naturforsch., 35c:783.

Marigo, G., Bouyssou, H., and Belkoura, H., 1985, Vacuolar efflux of malate and its influence on nitrate accumulation in Catharanthus roseus cells, Plant Sci. Letters, 39:97.

Peterson, G. L., 1978, A simplified method for analysis of inorganic phosphate in the presence of interferring substances, Anal. Biochem., 84:164.

Rona, J. P., and Cornel, D., 1985, An electrogenic proton pump on the tonoplast of Acer pseudoplatanus L. free cells and isolated vacuoles, in : "Biochemistry and Physiology of Vacuolar Adenosine-Triphosphatase in Fungi and Plants", Springer-Verlag, Berlin, Heidelberg, New-York and Tokyo.

Sze, H., 1985, H^+-translocating ATPases. Advances using membrane vesicles, Annu. Rev. Plant Physiol., 36:175.

Thibaud, J. P., and Grignon, C., 1981, Mechanism of nitrate uptake in corn roots, Plant Sci. Letters, 22:279.

Ulrich, W. R., and Novacky, A., 1981, Nitrate-dependent membrane potential changes and their induction in Lemna gibba G_1, Plant Sci. Letters, 22:211.

ACTIVE TRANSPORT OF AMINO-ACIDS AND CALCIUM IONS

IN FUNGAL VACUOLES

Yasuhiro Anraku

Department of Biology, Faculty of Science
University of Tokyo, Hongo, Tokyo 113, Japan

INTRODUCTION

Yeasts are favored model systems of eukaryote cells in investigations on the intracellular compartments of ions and inorganic metabolites. Vacuoles are the largest organelles in yeast cells and they are postulated to function as lysosomes and storage compartments (Matile, 1978; Wiemken and Nurse, 1973; Wiemken and Dürr, 1974; Huber-Wälchli and Wiemken, 1979). We have demonstrated the presence in the vacuolar membrane of the yeast Saccharomyces cerevisiae of an H^+-translocating ATPase (Kakinuma et al., 1981; Ohsumi et al., 1985 b; Uchida et al., 1985) and several active transport systems, which are specific for Ca^{2+} (Ohsumi and Anraku, 1983), arginine (Ohsumi and Anraku, 1981), and other nine amino-acids (Sato et al., 1984 a; Sato et al., 1984 b). Using a preparation of right-side-out vacuolar membrane vesicles of high purity, we showed that the H^+-ATPase generates an electrochemical potential difference of protons across the membrane of 180 mV, interior acid (Kakinuma et al., 1981), and that the electrochemical proton gradient formed drives all the above transport by a mechanism of n H^+/substrate antiport (Ohsumi and Anraku, 1983; Ohsumi and Anraku, 1981; Sato et al., 1984 a).

To understand the mechanism and physiological roles of amino-acid transport and pool formation in the vacuolar compartment in situ, it is essential to develop a diagnostics for analyzing the contents of in vivo vacuolar constituents quantitatively. Wa have found a brief treatment of yeast cells with 200 μM $CuCl_2$ makes the plasma membrane permeable, remaining the vacuolar membrane intact (Ohsumi et al., 1985 a; Ohsumi and Anraku, 1987). This "Cu^{2+}-method" enabled us to determine the vacuolar and cytosolic amino-acid pools differentially (Kitamoto et al., 1984).

This paper describes the kinetic properties of secondary active transport systems for amino-acids and Ca^{2+} in the vacuolar membrane of the yeast Saccharomyces cerevisiae. Evidence that the vacuolar amino-acid pool changes dynamically with change of nitrogen source for growth and the vacuoles are in fact regulating the nitrogen metabolism and cell division by ensuring homeostasis of amino-acid pools in the cytoplasm are presented.

MATERIAL AND METHODS

Strain and Culture Condition

The haploid strain of Saccharomyces cerevisiae, X2180-1A, from the Yeast Genetic Stock Center, Berkeley, was used throughout. Cells were grown in YEPD medium containing 1% yeast extract (Difco), 2% polypeptone, and 2% glucose at 30°C in a 10 liter Magnaferm fermentor (New Brunswick Scientific Co., Inc.) that was aerated at an air flow of 6 liters/min and agitation speed of 200 rpm.

Preparation of Vacuolar Membrane Vesicles

Right-side-out vacuolar membrane vesicles which were essentially free of mitochondrial contamination were prepared as described previously (Kakinuma et al., 1981; Ohsumi et al., 1985 b; Uchida et al., 1985).

Standard Transport Assays

The assay mixture for amino-acid transport (100 μl) consisted of 20 mM Mes/Tris (pH 6.9), 4 mM $MgCl_2$, 20 mM KCl, 0.5 mM ATP, ^{14}C-amino-acid (1-10 $mC_i/mMol$), and 20-60 μg of protein of vacuolar membrane vesicles (Sato et al., 1984 a). The assay mixture for Ca^{2+} transport (100 μl) consisted of 20 mM Mes/Tris (pH 6.7), 4 mM $MgCl_2$, 20 mM KCl, 0.3 mM ATP, 0.5 mM $CaCl_2$ (2 $mC_i/mmol$), and 10-30 μg of protein of vacuolar membrane vesicles (Ohsumi and Anraku, 1983).

The above mixtures without substrate were preincubated for 1 min at 25°C, and the reaction was strated by adding labeled substrate and stopped by diluting the reaction mixtures with 5 ml of cold buffer consisting of 25 mM Mes/Tris (pH 6.9), 5 mM $MgCl_2$, and 25 mM KCl (for amino-acid transport assays) or 10 mM Mes/Tris (pH 6.7), 5 mM $MgCl_2$, 25 mM KCl (for Ca^{2+} transport assay). Vesicles were recovered quickly on a membrane filter (Sartorius Type S11, 0.45 μ) and washed with 10 ml of the above buffers, respectively. The radioactivities taken up by the vesicles were determined in a Beckman LS-9000 liquid scintillation counter. The initial rates of amino-acid and Ca^{2+} uptakes were determined from the difference between the substrate uptakes in the first 5 and 20 s, and were expressed as nanomoles of substrate taken up per min.

Differential Extraction of the Amino-Acid Pools from Vacuolar and Cytosolic Compartments ("Cu^{2+}-Method")

Cells grown in YEPD medium were harvested, washed twice with distilled water, and suspended in 1.5 ml of 2.5 mM phosphate buffer (pH 6.0), 0.6 M sorbitol, 10 mM glucose, and 200 μM $CuCl_2$ at a cell density of 2 10^8 cells/ml. The mixture was incubated for 10 min at 30°C and 1 ml of the cell suspension was filtered and the cells were subsequently washed with 2 ml of 2.5 mM phosphate buffer (pH 6.0)- 0.6 M sorbitol. The filtrate and washing (total 3 ml) were combined and used as a sample of the cytosolic amino-acid pool. The washed cell pellet was suspended in 3 ml of distilled water and boiled for 15 min. The supernatant obtained after centrifugation was used as an extract of the vacuolar amino-acid pool (Kitamoto et al., 1984). These amino-acid pools were determined with a Hitachi automatic amino-acid analyser 835.

Other Methods and Chemicals

ATP-dependent change in fluorescence quenching of quinacrine was recorded as Kakinuma et al. (1981). Assays of marker enzyme activities and chemicals used were described previously (Kakinuma et al., 1981; Ohsumi et al., 1985 b; Uchida et al., 1985).

RESULTS AND DISCUSSION

Cellular Amino-Acid Transport and Formation of Differential Amino-Acid Pools in Yeast Cells : An Overview

In 1970, Crabeel and Grenson demonstrated that transport of amino-acids across the plasma membrane of Saccharomyces cerevisiae was unidirectional and did not show efflux of incorporated amino-acids even under conditions causing exchange-diffusion in prokaryotic cells (Crabeel and Grenson, 1970). This finding was the first indication that suggested the presence of differential amino-acid pools in the yeast cells. Chemical analyses of the distribution of amino-acids in several yeast cells showed that the bulk of basic amino-acids, lysine, arginine, and histidine, are localized in the vacuole against a large concentration gradient. While glutamate and aspartate are almost completely excluded from this compartment (Matile, 1978; Wiemken and Nurse, 1973; Wiemken and Dürr, 1974; Huber-Wälchli and Wiemken, 1979). The amounts of constituent amino-acids in the vacuolar and cytosolic pools changed dynamically depending on nitrogen sources for growth (Messenguy et al., 1980). These findings indicate that the vacuole is a metabolically active organelle and must have specific transport systems for basic amino-acids (Boller et al., 1975).

In Saccharomyces cerevisiae, cellular uptakes of amino-acids across the plasma membrane are catalyzed by several active transport systems. The general amino-acid transport system with a broad substrate specificity specifically mediates the active transport of various amino-acids except proline (Grenson et al., 1970). Specific transport systems for lysine, arginine, histidine, methionine, proline, and glutamate are also present in the plasma membrane (see ref. in Eddy, 1980). These transport systems mediate the active influxes by a nH^+/amino-acid symport mechanism with high affinities for substrates, e.g. the K_t values in the order of 10^{-6} M (Eddy, 1980; Cooper, 1982). Thus, yeast cells can scavenge nitrogenous compounds very efficiently from the surrounding medium and maintain them in vacuolar compartment against large concentration gradients.

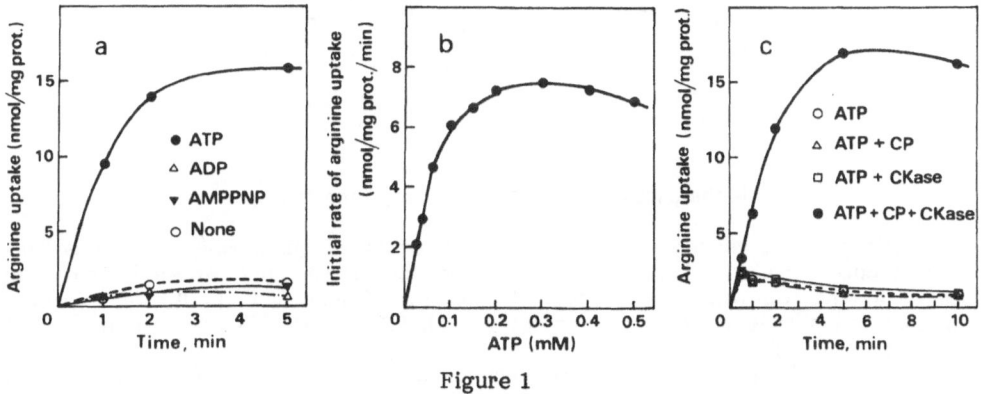

Figure 1

Arginine transport

A, ATP-dependent arginine transport by vacuolar membrane vesicles. Assays were carried out under standard conditions (see Methods) but with 0.3 mM each of the nucleotide indicated.

B, Dependence of the initial rate of arginine uptake.

C, Requirement of ATP hydrolysis for arginine transport. Assays were carried out as above, but with a suboptimal concentration of 0.03 mM ATP in the presence of either 1 mM creatine phosphate (CP) or 1 μg of creatine kinase (CKase or both).

ATP-dependent Arginine Transport The vacuolar membranes vesicles from Saccharomyces cerevisiae could take up arginine actively only in the presence of ATP and established a concentration gradient of 40-fold (Fig. 1-A). No transport of arginine was observed without ATP, or with 0.3 mM each of ADP or AMP-PNP instead of ATP. The initial rate of arginine uptake reached a plateau level at 0.3 mM ATP (Fig. 1-B). At a suboptimal concentration of 0.03 mM ATP, the steady level of arginine accumulation decreased greatly, however, the addition of an ATP-regenerating system (creatine phosphate and creatine kinase) enhanced both the initial rate of uptake and steady level of accumulation (Fig. 1-C). These results indicated that the active arginine transport is driven by energy provided by hydrolysis of ATP (Ohsumi and Anraku, 1981).

Mechanism of energy coupling. As we have already discovered a novel H$^+$-translocating ATPase in the vacuolar membrane (Kakinuma et al., 1981), this active arginine transport appeared to be coupled with this ATP hydrolysis-dependent formation of a proton-motive force across the vacuolar membrane. To prove this idea and elucidate further the reaction mechanism, following experiments were carried out. Fig. 2-A shows that the fluorescence signal of quinacrine was quenched by incubating the vacuolar membrane vesicles with ATP, reflecting uptake of protons and formation of ΔpH. DCCD inhibited the proton uptake.

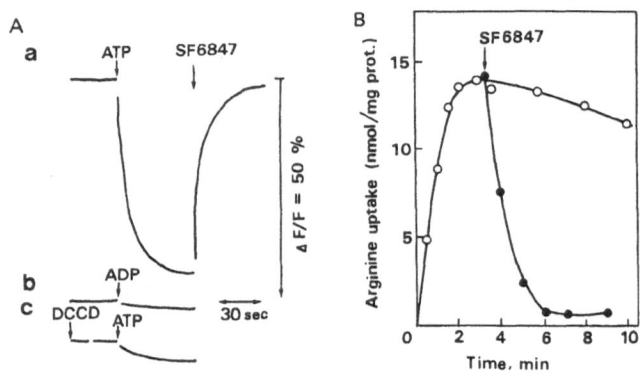

Figure 2

Mechanism of ATP-dependent arginine transport

A, ATP-dependent change in fluorescence quenching of quinacrine. Reactions were carried out as described under Methods. Additions were made where indicated by arrows at the following final concentrations: a, 0.5 mM ATP; b, 0.5 mM ADP; c, Vacuolar membrane vesicles were treated with 0.1 mM DCCD for 20 min at 25°C before addition of 0.5 mM ATP.
B, Efflux of ^{14}C-arginine from the vesicles by addition of SF 6847. The uptake reaction was started in standard assay mixture (2 ml) and, at the times indicated, samples of 100 μl were filtered as described. After incubation for 3 min, half of the reaction mixture was mixed with ethanol at a final concentration of 1% (open circle), and the other half with 1 μM SF 6847 in ethanol (closed circle).

Quantitative determination by a flow-dialysis method showed that the electrochemical potential difference of protons generated under these conditions was 180 mV, with contribution of 1.7 pH units, interior acid, and of a membrane potential of 75 mV, interior positive (Kakinuma et al., 1981).

Fig. 2-A also shows that the protonophore SF 6847 dissipated the electrochemical proton gradient completely. In accordance with this observation, SF 6847 could initiate rapid efflux of arginine from the vesicles and uncoupled the ATP-dependent active arginine uptake (Fig. 2-B). From these results we concluded that the active arginine transport in the vacuolar membrane vesicles is mediated by a mechanism of n H^+/arginine antiport (Ohsumi and Anraku, 1981).

Substrate Specificities of Seven n H^+/amino-acid Antiport Systems   Twenty amino-acids were each incubated with vacuolar membrane vesicles in the presence of 0.5 mM ATP to examine whether they were taken up actively. We found that ten amino-acids were taken up actively against, at least, 5-fold concentration gradients (Sato et al., 1984 a). Kinetic analyses of these transport activities indicated the presence of seven independent n H^+/amino-acid antiport systems in the vacuolar membrane vesicles. Those are arginine, arginine-lysine, histidine, phenylalanine-tryptophan, tyrosine, glutamine-asparagine, and isoleucine-leucine transport systems and their properties are summarized in Table 1 (Sato et al., 1984 a).

Table 1

Seven n H^+/Amino-Acid Antiport Systems
in Vacuolar Membrane Vesicles of <u>Saccharomyces</u> <u>cerevisiae</u>

System	Substrate	K_t (mM)	Optimal pH
Arg	Arginine	0.40	7.5
Arg-Lys	Arginine	1.5	7.5-7.9
	Lysine	0.56	7.5-7.9
His	Histidine	1.2	6.5
Phe-Trp	Phenylalanine	1.0	6.4-6.7
	Tryptophan	3.0	6.4-6.7
Tyr	Tyrosine	5.0	6.5
Gln-Asn	Glutamine	8.8	6.4
	Asparagine	3.0	6.4
Ile-Leu	Isoleucine	5.0	6.3-7.0
	Leucine	30	6.3-7.0

These n H^+/amino-acid antiport systems in the vacuole differ from those n H^+/amino-acid symport systems of the plasma membrane in the following characteristics : i) The transport systems of the vacuolar membrane have narrow substrate specificities; ii) Their K_t values are mostly in the orders of 10^{-3} M, which are 1 – 2 orders of magnitude higher than those of the transport systems of the plasma membrane (Eddy, 1980; Cooper, 1982) and iii) The uptake activities of the vacuolar membrane show an optimal pH in the neutral range (Table 1), whereas those of the plasma membrane are in the acidic range (Cooper, 1982).

An Arginine/Histidine Exchange Transport System

Stimulation of Arginine Uptake by Histidine   A high-affinity arginine uptake activity in the vacuolar membrane vesicles could be measured in the presence of a low concentration of arginine and excess histidine (Sato et al., 1984 b). Fig. 3-

A shows a typical example of stimulation of the initial rate of arginine uptake and increase in the steady level of its accumulation by histidine. Interestingly, addition of histidine to the reaction mixture after the steady state had been attained also induced net influx of arginine into the vacuolar membrane vesicles (Fig. 3-B). Kinetic examinations indicated that the stimulating effect of histidine was saturable and that it was notable only when the concentration of arginine used (53 μM) was much lower than the K_T value for the arginine-specific transport system (400 μM, see Table 1).

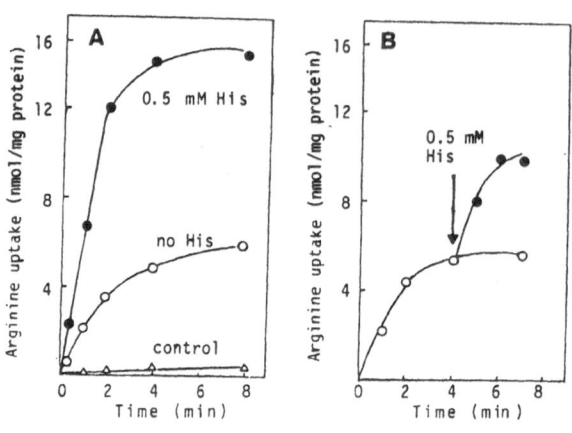

Figure 3

Stimulation of arginine uptake by histidine

A, Effects of histidine on the initial rate of arginine uptake and steady level of its accumulation. Assays were carried out under standard conditions but with 53 μM arginine and 0.5 mM ATP in the presence or absence of 0.5 mM histidine. Control represents an experiment without ATP and histidine.
B, Enhancement of the steady level of arginine accumulation by histidine. Assays were carried out as above and at the time indicated by the arrow, 0.5 mM histidine was added.

Trans-effect of Histidine on Arginine Uptake   These findings mentioned above suggested the existence of a high-affinity arginine transport activity with an exchange mechanism for histidine. To elucidate the mechanism that is operating in the vesicles, we examined the effect of preloading with histidine on arginine uptake. Vacuolar membrane vesicles were preincubated with 0.1 mM histidine in the presence of ATP and an ATP-regenerating system, and then the initial rate of arginine uptake and steady level of its accumulation were determined by addition of 53 μM ^{14}C-arginine to the reaction mixture. It was found that the initial rate was stimulated 6-fold and the steady level was enhanced 4-fold by preloading, indicating the operation of a trans-effect of histidine on arginine efflux (Sato et al., 1984 b).

Mechanism of Arginine/Histidine Exchange Transport   Since the vacuolar membrane vesicles have two n H^+/arginine antiport systems and one n H^+/histidine antiport system as shown in Table 1, the kinetic properties of this novel arginine/histidine exchange transport system can be studied only under limited conditions and in the absence of ATP. We found that when the vesicles were preloaded with 10 mM histidine in the absence of ATP for 6 min at 25°C and then diluted 50-fold at 15°C with the same solution containing 53 μM ^{14}C-arginine instead of 10 mM histidine, transient uptake of arginine took place with concomitant downhill efflux of histidine from the vesicles. Using these non-energized conditions, we analyzed the transport mechanism kinetically. All the data so far obtained indicated that the chemical potential of histidine was the sole source of energy that drove the arginine efflux into the vacuolar membrane vesicles. No active uptake of histidine was detected when the same vesicles were preloaded with 10 mM arginine. We concluded that this unique transport catalyzes exchange of 1 mol of arginine outside with 1 mol of histidine inside the vesicles. The K_T(entry) for arginine and the K_T (exit) for histidine were determined to be 0.1 mM and 1.1 mM, respectively (Sato et al., 1984 b).

The arginine/histidine exchange transport is a third arginine sequestering system in Saccharomyces cerevisiae (Table 2). If this exchange transport system functions in vivo, arginine may be accumulated efficiently in the vacuole by dissipation of a chemical gradient of histidine across the vacuolar membrane. Wiemken and his coworkers reported that arginine is accumulated in the vacuoles against large concentration gradients in yeast cells (Wiemken and Dürr, 1974; Boller et al., 1975).

Table 2

Basic Amino-Acid Transport Systems
in Vacuolar Membrane Vesicles of Saccharomyces cerevisiae

System	Substrate	K_t (mM)	Mechanism
Arg	Arginine	0.40	nH^+/Arg antiport
Arg–Lys	Arginine	1.5	nH^+/Arg antiport
	Lysine	0.56	nH^+/Lys antiport
His	Histidine	1.2	nH^+/His antiport
Arg/His	Arginine$_{(out)}$	0.10	Arg$_{(out)}$/His$_{(in)}$–
	Histidine$_{(in)}$	1.1	exchange transport

Differential Determination in situ of Vacuolar and Cytosolic Amino-Acid Pools

"Cu^{2+}-method" as in situ diagnostics for measuring vacuolar and cytoplasmic amino-acid pools   We have established a brief method for determining the amounts of pool amino-acids in the vacuolar and cytosolic compartments differentially (Kitamoto et al., 1984). Treatment of intact yeast cells with 200 μM $CuCl_2$ (see Methods) induced changes in the permeability barrier of the plasma membrane but not of the vacuolar membrane (Ohsumi and Anraku, 1987).

Vacuolar Amino-Acid Pool and its Role in Nitrogen Metabolism   Using this "Cu^{2+}-method" we could examine the nature of in vivo vacuolar amino-acid pools formed under several physiological conditions. Table 3 shows constituent amino-acids in the vacuolar and cytosolic pools in cells of Saccharomyces cerevisiae, strain X2180-1A, which were grown in YEPD medium (Kitamoto et al., 1984). Histidine, arginine, lysine, tyrosine, and phenylalanine were rich in the vacuolar compartment and for all these amino-acids the vacuolar membrane has specific transport systems (see Table 1). Aspartate and glutamate are the cytosolic amino-acids and they were ecluded almost completly from the vacuolar compartment (Table 3). These results suggested strongly that the vacuolar amino-acid transport systems are functioning physiologically as sequestration devices and relate to the formation of the vacuolar amino-acid pool in vivo.

Several lines of evidence have indicated that the vacuolar amino-acid pool is metabolically active and feeds the cells with nitrogen sources for growth upon the starvation (Kitamoto et al., 1984). For instance, yeast cells were grown in a nitrogen-rich medium, harvested, washed, and transferred in a nitrogen-starved conditions, these cells could grow about two generations and ceased growing when the vacuolar basic amino-acids had been used up. More important was the observation that during starvation the decrease in the vacuolar basic amino-acid pool was tightly coupled with the counter influx of K^+ into the vacuolar compartment, possibly maintaining the vacuolar turgor homeostatically.

Table 3

Distribution of Amino-Acids
in Cells of Saccharomyces cerevisiae[a]

Type	Amino-Acid	Ratio of Vacuole/Cytosol	Total Pool[b]
Vacuole rich	His	11.5	39.8
	Arg	8.1	70.5
	Lys	5.9	123
	Tyr	1.6	1.4
	Phe	1.3	29.0
Even Distribution	Thr	0.98	47.4
	Val	0.94	11.4
	Leu	0.84	13.8
	Met	0.83	6.4
	Ile	0.70	8.4
Cytosol rich	Asp	0.14	28.0
	Glu	0.12	161

[a]Cells of Saccharomyces cerevisiae, strain X2180-1A were grown in YEPD medium and harvested at the exponential phase of growth. Vacuolar and cytosolic amino-acid pools were extracted differentially by "Cu^{2+}-method" and the constituent amino-acids were determined as described in Methods.
[b]Total pool represents nanomoles amino-acid/10^8 cells.

These observations suggest that the vacuolar amino-acid transport systems function as chemical sensing systems for recognized the difference of chemical gradients of substrate amino-acids and operate emergent outflows of the sequestered amino-acids by some unknown regulatory mechanisms.

An n H^+/Ca^{2+} Antiport System in Vacuole

Homeostasis of Ca^{2+} in the Cytoplasm In many organisms, Ca^{2+} is essential for proliferation and plays important physiological roles as a second messenger in the regulation of cellular enzyme systems and motility. The cytosolic concentration of Ca^{2+} is maintained below that of the environment (Silver, 1977) by the transport systems in the plasma membranes and other organelles (Stroobant and Scarborough, 1979; Nicholls and Akerman, 1982).

Properties of n H^+/Ca^+ Antiport System Fig. 4-A shows ATP-dependent uptake of Ca^{2+} into the vacuolar membrane vesicles. At the steady level of accumulation, a 150-fold concentration gradient was established, which was four times that of arginine uptake. The active Ca^{2+} transport showed saturation kinetics with the K_t value of 0.1 mM and optimal pH of 6.4 (Fig. 4-B). The transport system takes up Sr^{2+} as commun substrate but shows no affinity for Mg^{2+}. We found again that the active Ca^{2+} transport is mediated by a mechanism of n H^+/Ca^+ antiport (Ohsumi and Anraku, 1983).

The vacuolar Ca^{2+} transport activity was found to be active in "Cu^{2+}-treated" cells (Ohsumi and Anraku, 1987). When 0.5 mM $^{45}CaCl_2$ was incubated with cells of the yeast Saccharomyces cerevisiae, strain X21801A, in the presence of 100 μM $CuCl_2$, most of $^{45}Ca^{2+}$ was taken up rapidly in the vacuole and formed a large

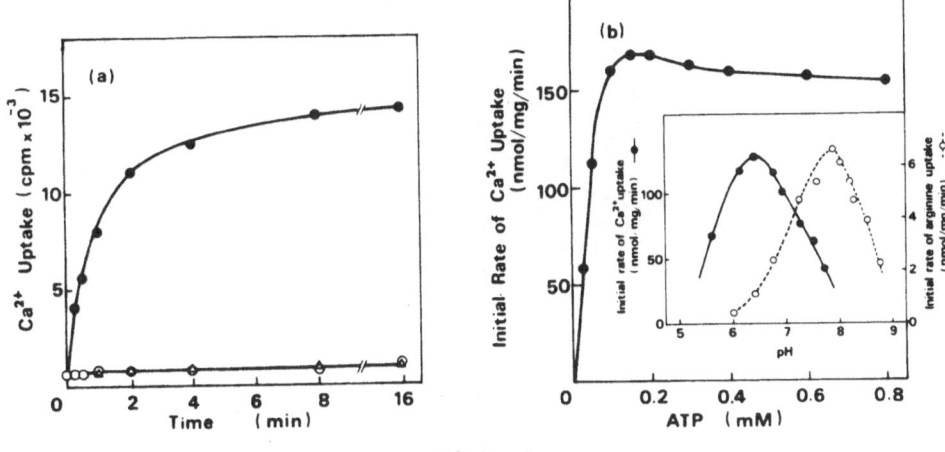

Figure 4

Calcium transport

A, ATP-dependent calcium transport. Assays were carried out under standard conditions with 0.3 mM ATP (closed circle) or AMP-PNP (triangle) or without ATP (open circle).
B, Dependence of the initial rate of calcium uptake. Insert: pH dependence of the initial rate of calcium (closed circle) or arginine uptake (open circle).

inclusion made of Ca^{2+}-polyphosphate complexes, which was easily detected by light microscopy (Ohsumi and Anraku, 1987). Eilam examined the properties of Ca^{2+} efflux systems in Saccharomyces cerevisiae and suggested the presence of two exchangeable intracellular Ca^{2+} pools which turned over at different rates (Eilam, 1982). It is quite likely that the vacuole may represent the main compartment responsible for the slow turnover pool and ensure homeostasis of Ca^{2+} in the cytoplasm of the yeast cells.

CONCLUSION

We established a simple method for purifying intact vacuoles and vacuolar membrane vesicles with right-side-out orientation from Saccharomyces cerevisiae. Both preparations were essentially free of mitochondrial and other organelle contaminations, judging from assays of marker enzyme activities.

We found that a novel H^+-translocating ATPase, seven n H^+/amino acid antiport systems, an arginine/histidine exchange transport system, and an n H^+/Ca^{2+} antiport system are located in the vacuolar membrane. Their biochemical functions in the cell are digrammatically shown in Fig. 5.

We have developed a brief "Cu^{2+}-method" as an <u>in situ</u> diagnostics for measuring <u>in vivo</u> functions of the vacuole. Our observations suggested that the vacuole is an active organelle and is able to regulate the intracellular nitrogen metabolism and cell division. Most vacuolar transport systems mentioned above seemed to be responsible for the differential formation of vacuolar and cytosolic solute compartments and to ensure homeostasis of amin-acid and Ca^{2+} pools in the cytoplasm.

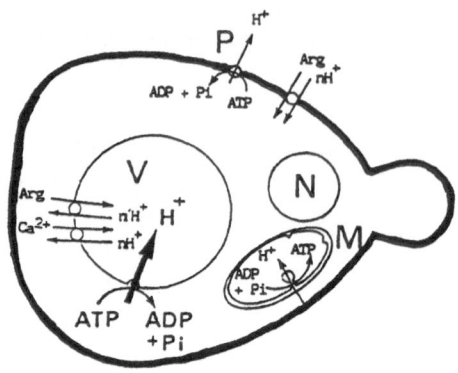

Figure 5

Diagrammatic representation
of vacuolar functions
in <u>Saccharomyces cerevisiae</u>

V, P, M, and N represent the vacuole, plasma membrane, mitochondrion, and nucleus, respectively. For details of the vacuolar functions depicted, see text.

REFERENCES

Boller, T., Dürr, M., and Wiemken, A., 1975, Characterization of a specific transport system for arginine in isolated yeast vacuoles, <u>Eur. J. Biochem.</u>, 54:81.

Cooper, T. G., 1982, Transport in <u>Saccharomyces cerevisiae</u>, <u>in</u> : "The Molecular Biology in the Yeast <u>Saccharomyces</u> : Metabolism and Gene Expression", J. Strathern, E. W. Jones, and J. R. Broach, eds., Cold Spring Harbor Laboratory, New-York.

Crabeel, M., and Grenson, M., 1970, Regulation of histidine uptake by specific feedback inhibition of two histidine permeases in <u>Saccharomyces cerevisiae</u>, <u>Eur. J. Biochem.</u>, 14:197.

Eddy, A. A., 1980, Some aspects of amino-acid transport in yeast, <u>in</u> : "Microorganisms and Nitrogen Sources", J. W. Payne, ed., Wiley, Chichester, New-York, Brisbane, Toronto.

Eilam, Y., 1982, The effect of monovalent cations on calcium efflux in yeasts, <u>Biochim. Biophys. Acta</u>, 687:8.

Grenson, M., Hou, C., and Crabeel, M., 1970, Multiplicity of the amino-acid permeases in <u>Saccharomyces cerevisiae</u> : IV. Evidence for a general amino-acid permease, <u>J. Bacteriol.</u>, 103:770.

Huber-Wälchli, V., and Wiemken, A., 1979, Differential extraction of soluble pools from the cytosol and the vacuoles of yeast <u>Candida utilis</u> using DEAE–dextran, <u>Arch. Microbiol.</u>, 120:141.

Kakinuma, Y., Ohsumi, Y., and Anraku, Y., 1981, Properties of H^+-translocating adenosine-triphosphatase in vacuolar membranes of <u>Saccharomyces cerevisiae</u> <u>J. Biol. Chem.</u>, 256:10859.

Kitamoto, K., Ohsumi, Y., and Anraku, Y., 1984, Quantitative determination of vacuolar and cytosolic amino-acid pools in <u>Saccharomyces cerevisiae</u>, <u>in</u> : "Abstract of the Annual Meetings of Agricul. Chem. Soc. Japan".

Matile, P., 1978, Biochemistry and function of vacuoles, Annu. Rev. Plant Physiol., 29:193.

Messenguy, F., Colin, O., and Ten Have, J.-P., 1980, Regulation of compartmentation of amino-acid pools in Saccharomyces cerevisiae and its effects on metabolic control, Eur. J. Biochem., 108:439.

Nicholls, D., and Akerman, K., 1982, Mitochondrial calcium transport, Biochim. Biophys. Acta, 683:57.

Ohsumi, Y., and Anraku, Y., 1981, Active transport of basic amino-acids driven by a proton-motive force in vacuolar membrane vesicles of Saccharomyces cerevisiae, J. Biol. Chem., 256:2079.

Ohsumi, Y., and Anraku, Y., 1983, Calcium transport driven by a proton-motive force in vacuolar membrane vesicles of Saccharomyces cerevisiae, J. Biol. Chem., 258:5614.

Ohsumi, Y., and Anraku, Y., 1987, On the nature of changes induced by cupric ion in the permeability barrier of yeast plasma membrane, Biochim. Biophys. Acta, submitted.

Ohsumi, Y., Anraku, Y., and Kitamoto, K., 1985 a, Yeast vacuoles : Their functions and applications in brewing (in Japanese), J. Brewing Soc. Japan, 80:11.

Ohsumi, Y., Uchida, E., and Anraku, Y., 1985 b, The H^+-translocating ATPase in vacuolar membranes of Saccharomyces cerevisiae, in : "Biochemistry and Function of Vacuolar Adenosine-triphosphatase in Fungi and Plants", B. P. Marin, ed., Springer-Verlag, Berlin, Heidelberg, New-York, and Tokyo.

Sato, T., Ohsumi, Y., and Anraku, 1984 a, Substrate specificities of active transport systems for amino-acids in vacuolar membrane vesicles of Saccharomyces cerevisiae : Evidence of seven independent proton/amino-acid antiport systems, J. Biol. Chem., 259:11509

Sato, T., Ohsumi, Y., and Anraku, Y., 1984 b, An arginine/histidine exchange transport system in vacuolar membrane vesicles of Saccharomyces cerevisiae, J. Biol. Chem., 259:11509.

Silver, S., 1977, Calcium transport in microrganisms, in : "Microorganisms and Minerals", E. D. Weinberg, ed., Dekker, New-York.

Stroobant, P., and Scarborough, G. A., 1979, Active transport of calcium in Neurospora plasma membrane vesicles, Proc. Natl. Acad. Sci. U.S.A., 76:3102.

Uchida, E., Ohsumi, Y., and Anraku, Y., 1985, Purification and properties of H^+-translocating, Mg^{2+}-adenosine-triphosphatase from vacuolar membranes of Saccharomyces cerevisiae, J. Biol. Chem., 260:1090.

Wiemken, A., and Dürr, M., 1974, Characterization of amino-acid pools in the vacuolar compartment of Saccharomyces cerevisiae, Arch. Microbiol., 101:45.

Wiemken, A., and Nurse, 1973, Isolation and characterization of the amino-acid pools located within the cytoplasm and vacuoles of Candida utilis, Planta, 109:293.

ATP-DEPENDENT UPTAKE OF MALATE IN VACUOLES

FROM CAM PLANT, KALANCHOE DAIGREMONTIANA

Kojiro Nishida and O. Tominaga

Botanical Institute, Faculty of Sciences
Kanazawa University, Kanazawa 920, Japan

INTRODUCTION

Intracellular translocation of malate during crassulacean acid metabolism (CAM) is a complex phenomenon. The key enzyme phosphoenolpyruvate carboxylase catalyzes the synthesis of malate in cytoplasm has been established (Kluge and Tin, 1978). In the dark period the malate synthesized in the cytoplasm is transported into vacuoles and accumulated in its. In the light period the malate is effluxed from the vacuoles into cytoplasm, and then deacidification proceed.

Using the leaf slices of Kalanchoe daigremontiana submerged in solution, Lüttge et al. (1975) and Lüttge and Ball (1979) showed that the malate efflux from the vacuoles into suspension medium is a passive process. The other most important flux involved in a malate influx into the vacuoles. Lüttge and Ball (1979), Lüttge et al. (1982) have proposed that the operation of an ATPase mediating H^+-transport across the tonoplast coupled with the transport of malate. Direct evidence that malate accumulation into the vacuoles is an energy-dependent process was shown by Nishida and Hayashi (1979) using the leaves of Kalanchoe pinnata. According to their results, vacuoles released malate extremely into cytoplasm till the leaf cells became death due to the effluent acid when acidified leaf was placed in N_2 in the dark.

Isolation of vacuoles from the CAM cells was established by Buser and Matile (1977) and Kenyon et al. (1978), and malate uptake was examined using the isolated vacuoles (Buser-Suter et al., 1982). Recently, the presence of tonoplast ATPase has been reported from two laboratories (Aoki and Nishida, 1984; Smith et al., 1984).

In this experiments, we prepared small vacuoles from the isolated ones to study the uptake of malate by the vacuoles. The reason is that the isolated vacuoles are extremely big (diameter, about 80 μm to 90 μm in our reaction medium) and vacuoles are quite fragile in the reaction medium, while the small vacuoles prepared in this experiments are stable and survive about 100 % without rupturing for the short term experiments. Evidence is presented for an MgATP-dependent malate uptake against a transmembrane concentration gradient.

MATERIAL AND METHODS

Plant Material

Kalanchoe daigremontiana Hamet et Perr. which was cultivated in a glass house in natural photoperiod was used for the material. About 20 g of the 5th to 7th leaves counted from the top leaf was harvested at 10:00 a.m., at which time malate deacidification was still in progress.

Protoplast Isolation

The leaves were prepared by washing with distilled water, removing the midribs and peeling off the lower epidermis. The leaf was cut into about 2 mm^3 with a razor blade and washed with the medium containing 0.33 M sorbitol, 1 mM $CaCl_2$ and 25 mM Mes/NaOH (pH 5.6), then immersed in 75 ml of digestion medium containing 0.33 M sorbitol, 0.1% (w/v) polyvinylpyrolidone, 3 mM $CaCl_2$, 50 mM NaOH (pH 5.6), 0.5% (w/v) Cellulase Onozuka RS, 0.5% (w/v) Macerozyme R-10, and 0.3% (w/v) BSA. They were kept for 90 min in a shaking water bath, at 28°C, to isolate protoplasts.

Digested leaf cells were filtered through 250 μm stainless steel net and the isolated protoplasts were washed once with the digestion medium without digestion enzymes. The protoplasts thus obtained were precipitated by centrifugation at 20 x g for 5 min, and the supernatant solution was drawn off by a pipette. The protoplast pellet was suspended in a protoplast medium containing 25% (w/v) dextran (M. W. = 60,000 - 90,000), 0.33 M sorbitol, 3 mM $CaCl_2$, and 25 mM Mes/NaOH (pH 5.6). 10 ml of the above protoplast suspension was pipetted into 50 ml glass centrifugal tube. Ten ml of protoplast medium (10% dextran was used in this time), and 8 ml of protoplast medium without dextran were placed successively on the protoplast suspension. After centrifugation at 200 x g for 10 min, the purified protoplasts were found at the interface between 0 and 10% dextran.

The above operations were carried out at about 4°C.

Vacuole Isolation

Vacuoles were prepared from protoplasts by slight modification of the method of Aoki and Nishida (1984). About one ml of purified protoplast suspension was poured into 30 ml of rupturing medium containing 0.13 M sorbitol, 50 mM Tris/HCl (pH 8.0), 2 mM 2-mercaptoethanol, 0.3% (w/v) BSA and 10 mM ethyleneglycolbis-(β-aminoethylether)-N, N, N', N'-tetraacetic acid at 30°C. After 5 min, the above suspension was removed in a 50 ml glass centrifugal tube and 5 ml of vacuole medium containing 5% (w/v) dextran, 0.2 M sorbitol, 25 mM Hepes/Tris (pH 8.0), 1 mM 2-mercaptoethanol, O.1% (w/v) BSA, and 2.5 ml vacuole medium without dextran were placed successively on the vacuole suspension. After centrifugation at 300 x g for 10 min, vacuoles were collected from the interface between 0 and 5% dextran solution. The vacuoles were resuspended in the vacuole medium and used for the following preparation. Vacuole purification was carried out at 4°C.

Small Vacuole Preparation

The purified vacuole suspension was passed through 20 μm nylon mesh fixed to the cut end of a 2 cm^3 syringe. The vacuole was divided in the several small vacuoles by this preparation, and these vacuoles were suspended in the following medium after the centrifugation, and stored on ice; 0.6 M sorbitol, 25 mM Hepes/Tris (pH 8.0), 1 ml 2-mercaptoethanol, 0.1% (w/v) BSA. In this paper, hereafter, we call this divided vacuole as a small vacuole to distinguish from isolated ones. These processes were carried out at about 4°C.

Malate Uptake Experiment

Malate uptake was determined using ^{14}C-malate. The standard reaction mixture contained 0.6 M sorbitol, 30 mM Hepes/Tris (pH 7.3), 1.0 μC_i of ^{14}C-malate (uniformly labeled, specific activity 10 mC$_i$/mmol) and 1 x 10^5 small vacuoles with or without MgATP (Sigma, A-0770) in a total volume of 1 ml. At the end of experimental period 100 μl of this reaction mixture was pipetted into a 400 μl polypropylen microcentrifugal tube which contained previously 50 μl of 0.6 M sorbitol and 25 mM Hepes-Tris (pH 8.0) in the bottom, and 125 μl of silicone oil (Toray silicone, SH-550:SH-556 = 4:1) layer on the sorbitol medium. By the centrifugation for 1 min at x 10,000 g in a microfuge, the small vacuoles were precipitated through silicone oil layer into the sorbitol medium, and then centrifugal tube was immediately put in liquid nitrogen, and the bottom of the tube containing small vacuoles was cut by a knife. ^{14}C-malate was measured with a scintillation counter. Malate uptake was carried out at 20°C.

Marker Enzyme Assay

Enzyme activities were determined in the crude extracts obtained by the sonication of protoplasts and vacuole suspensions, respectively. NADH–cytochrome c reductase and cytochrom c oxidase were assayed according to the procedure of Hodges and Leonard (1974), PEP-carboxylase to Satoh et al. (1980), catalase to Aebi (1965) and chlorophyll to Arnon (1949).

Counting

Number of protoplasts, isolated vacuoles and small vacuoles were counting in a 20 μl aliquot on a haemocytometer without a cover glass under a microscope. Before counting, samples were stained with neutral red to facilitate microscopic counting.

RESULTS AND DISCUSSION

Contamination Check of Isolated Vacuoles

The distribution of marker enzymes of other cell organelles contaminated in isolated vacuoles is shown in Table 1. NADH-cytochrome c reductase and cytochrome c oxidase contaminated in the vacuole preparation by 2.8% and 3.2%, respectively. PEP carboxylase, which is known as a marker enzyme of cytoplasm,

Table 1

Assessment of Contamination of Isolated Vacuoles
by Cytoplasmic Organelles

Enzyme	mU/10^4 protoplasts	mU/10^4 vacuoles	% in vacuoles
NADH–cyt c reductase	1.17	0.033	2.8
Cyt c oxidase	4.18	0.133	3.2
PEP carboxylase	24.10	0.105	0.4
Catalase	1.14	0.015	1.3
Chlorophyll(μg)	3.00	0.062	2.1

Values are given by the mean of two experiments.

catalase and chlorophyll contaminated by 0.4%, 1.3% and 2.1%, respectively. The results indicate that vacuoles isolated in this experiment seem to be purer than that of previous one (Aoki and Nishida, 1984).

Viability of Small Vacuoles

Small vacuoles prepared in this experiment are shown in Fig. 1. We measured the diameter of about 360 small vacuoles under a microscope. Size distributed between 2.5 μm to 42 μm in a diameter, and average diameter was calculated to be about 9.7 μm (Fig. 2). We could not make sure that how many small vacuoles are prepared from one isolated vacuole. Experiment was further confirmed the viability during the experimental period. At 0°C, small vacuoles were almost completely survived without rupturing during 5 h, while at 20°C, about 40 % of total small vacuoles were ruptured. However, during the first 30 min, about 100% viability was observed. Therefore, our experiments were carried out within 30 min from the start of experiment.

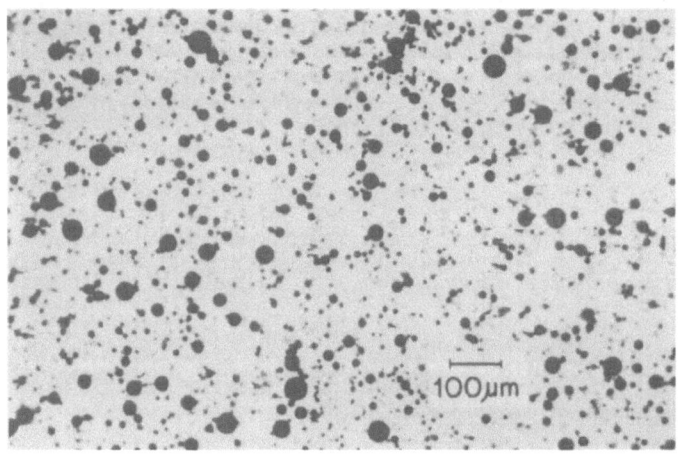

Figure 1

Small vacuoles

Vacuoles were stained with neutral red and photographed under a microscope.

pH Dependence of Malate Uptake

pH dependence of malate uptake carried out by Buser-Suter et al. (1981) showed that optimum pH was 7.5. Our result (Fig. 3) showed that in the presence of 5 mM MgATP, optimum pH was about 7.3. The value was quite similar with that reported by Buser-Suter et al. The pH optimum of ATPase activity found in isolated tonoplast of Kalanchoe daigremontiana was found at pH 8.0 (Aoki and Nishida, 1984; Smith et al., 1984). The reason why pH optimum is different in both phenomenon is not clear at present.

Time Course of ^{14}C-Malate Uptake

Time course of malate uptake with or without 5 mM MgATP is shown in Fig. 4. The uptake was continued without MgATP during 30 min of experimental period. However, uptake was accelerated extremely by the addition of MgATP. Malate uptake without MgATP might be caused by exchange reaction between malate within the vacuoles and ^{14}C-malate contained in the outer reaction medium.

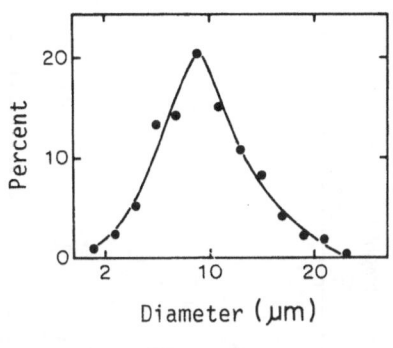

Figure 2

Size distribution of small vacuoles

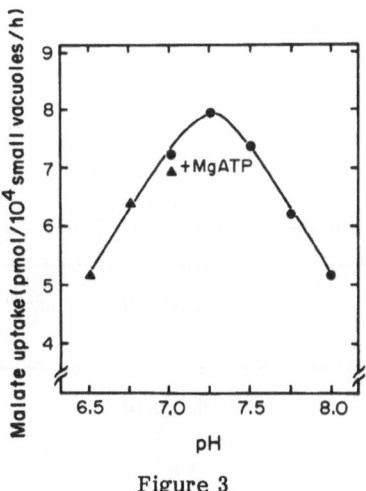

Figure 3

pH dependence of ^{14}C-malate uptake with 5 mM MgATP

Uptake was determined after 10 min incubation.

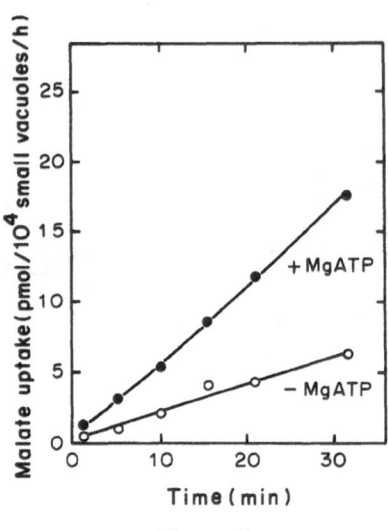

Figure 4

Time-course of ^{14}C-malate uptake
with or without 5 mM MgATP

Table 2

Effect of Inhibitors
on ^{14}-malate Uptake With 5 mM MgATP

Inhibitor	Concentration	Activity (% of control)
Oligomycin	0.14 μg/ml	97
DCCD	50 μM	33
DES	50 μM	21
Quercetin	50 μM	52

All reaction mixtures contained 1% ethanol. The uptake was determined after 5 min incubation.

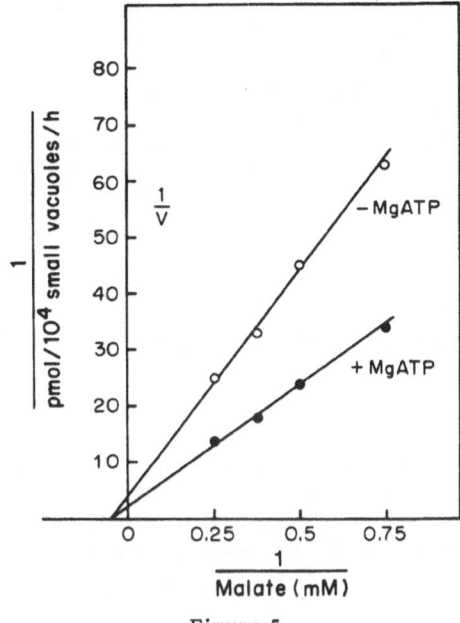

Figure 5

Malate concentration dependence of ^{14}C-malate
with or without 5 mM MgATP

Uptake was determined after 10 min incubation.

ATP in the presence of EDTA did not stimulate malate uptake to any considerable extent indicating that MgATP and not the free ATP was the true substrate for the stimulated malate uptake (data not shown).

Malate Concentration Dependence on ^{14}C-Malate Uptake

In order to make clear the effect of MgATP on the malate uptake, we measured the malate concentration dependence of the uptake in the presence and absence of 5 mM MgATP. A double reciprocal plot of the data (Fig. 5) revealed that MgATP did not affect the K_m of malate uptake (K_m, about 2.5 mM with or without MgATP) toward its substrate. The V_{max} for malate uptake in the presence of MgATP (250 pmol/h/10^4 small vacuoles) was about 2-times higher than that in the absence of MgATP (125 pmol/h/10^4 small vacuoles), which demonstrated that the effect of MgATP on malate uptake was due to an increase of V_{max}.

Effect of Inhibitors

In a previous paper (Aoki and Nishida, 1984) we examined the effect of various inhibitors on the activity of vacuolar membrane ATPase. N, N'-dicyclohexyl-carbodi-imide (DCCD), diethylstilbestrol (DES) and quercetin were found to be powerful inhibitors, so we examined the effect of these inhibitors concomitant with oligomycin which showed no effect on the activity of ATPase. Table 2 showed that 50 μM DCCD, DES and quercetin inhibited malate uptake remarkably, while oligomycin had no effect.

In this experiment we did not determine the malate concentration within the small vacuoles. However, we measured malate concentration of isolated protoplasts, on the assumption that the protoplasts maintain the sphere during the experimental period, and found 0.12 M malate concentration. Lüttge et al. (1982) estimated the malate concentration of Kalanchoe daigremontiana and showed that 22 mM malate concentration was still remained in deacidified leaf. In our reaction medium malate concentration was 0.5 mM, and MgATP accelerated the uptake strongly. Therefore, present experiments indicate that ATP-dependent malate uptake can occur against on transmembrane gradient in CAM cells.

ACKNOWLEGMENT

We thank Prof. Dr. Ulrich Lüttge for his valuable criticisms.

REFERENCES

Aebi, H., 1967, Catalase, in : "Methods of Enzymatic Analysis", Vol. 1, H. U. Bermeyer, ed., Academic Press, New-York.

Aoki, K., and Nishida, K., 1984, ATPase activity associated with vacuoles and tonoplast vesicles isolated from the CAM plant, Kalanchoe daigremontianum, Physiol. Plant., 60:21.

Arnon, D. I., 1949, Copper enzymes in isolated chloroplasts. Polyphenoloxidase in Beta vulgaris, Plant Physiol., 24:1.

Buser, C., and Matile, P., 1977, Malic acid in vacuoles isolated from Bryophyllum leaf cells, Z. Pflanzenphysiol., 82:462.

Buser-Suter, C., Wiemken, A., and Matile, P., 1982, A malic acid permease in isolated vacuoles of a crassulacean acid metabolism plant, Plant Physiol., 69:456.

Hodges, T. K;, and Leonard, R. T., 1974, Purification of a plasma-bound adenosine-triphosphatase from plant roots, in : Methods In Enzymology, 32:392.

Kenyon, W. H., Kringstad, R., and Black, C. C., 1978, Diurnal changes in the malic acid content of vacuoles isolated from leaves of the crassulacean acid metabolism plant, FEBS Letters, 94:281.

Kluge, M., and Tin, I. P., 1978, "Crassulacean Acid Metabolism. Analysis of an Ecological Adaptation", Springer-Verlag, Berlin.

Lüttge, U., Ball, E., and Tromballa, H. W., 1975, Potassium-independence of osmoregulated oscillations of malate^{2-} levels in the cells of CAM leaves, Biochem. Physiol. Pflanzen, 167:267.

Lüttge, U., and Ball, E., 1979, Electrochemical investigation of active malic acid transport at the tonoplast into the vacuoles of the CAM plant Kalanchoe daigremontianum, J. Membrane Biol., 47:401.

Lüttge, U., Smith, J. A. C., and Marigo, G., 1982, Membrane transport, osmoregulation and the control of CAM, in : "Crassulacean Acid Metabolism", I. P. Tin and M. Gibbs, ed., Amer. Soc. Plant Physiol., Rockville, U.S.A.

Nishida, K., and Hayashi, Y., 1979, Deacidification of the leaves of Bryophyllum calycinum under anaerobiosis. Abnormal efflux of malate into the cytoplasm, Plant Cell Physiol, 20, 1209-1215.

Sato, F., Nishida, K., and Yamada, Y., 1980, Activities of carboxylation enzymes and products of $^{14}CO_2$ fixation in photoautotrophically cultured cells, Plant Science Letters, 20:91.

Smith, J. A. C., Uribe, E. G., Ball, E., Heuer, S., and Lüttge, U., 1984, Characterization of the vacuolar ATPase activity of the crassulacean-acid-metabolism plant Kalanchoe daigremontianum, Eur. J. Biochem., 141:415.

PROTON-MOTIVE FORCE AND H$^+$/AMINO-ACID ANTIPORT AT THE TONOPLAST

OF RICCIA FLUITANS RHIZOID CELLS, REVEALED BY DIRECT PROBING

WITH THE pH-SENSITIVE MICROELECTRODE

Eva Johannes and Hubert Felle

Botanisches Institut I, Justus-Liebig-Universität,
Senckenbergstrasse 17-21, D-6300-Giessen, R.F.A.

INTRODUCTION

The electrical properties of H$^+$-amino-acid symport located at the plasmalemma of the freshwater liverwort Riccia fluitans has already been well characterized and described (Felle, 1981, 1983 and 1984; Johannes and Felle, 1985). Tracer compartment analysis proved fast and high accumulation of all amino-acids tested so far, including the nonmetabolizable amino-isobutyric acid (Aib) within the vacuole. It was desirable therefore, to carry out crucial tests, in order to clear up the mode of action of amino-acid transport across the tonoplast of this plant. It is generally assumed that a small electrical potential difference (positive inside the vacuole) and a large pH-gradient exist across tonoplast membranes. The recently developed turgor-resistent pH microelectrodes (Bertl and Felle, 1985; Felle and Bertl, 1986) seemed most promising (a) to measure the proton-motive force directly, and (b) to find out, whether amino-acids are transported along the proton-motive force or possibly in an electrically silent manner.

METHOD

The pH-sensitive (liquid-membrane) microelectrodes have been prepared basically according to Amman et al. (1981). The stabilization of the sensitive tip, necessary for the impalement of plant cells, has been carried out as described before (Bertl and Felle, 1985; Felle and Bertl, 1986). Briefly : The silylated pipettes were submersed with the tip in a mixture of polyvinylchloride and tetrahydrofuran, and a suction applied from the rear. The thus prepared pipettes were then backfilled with the resin (Fluka, No. 82500) first and then with the most appropriate buffer by means of a long glass capillary. These electrodes were calibrated in the test chamber after a stable signal has been recorded. Fig. 1 shows an example of precalibration, impalement into the vacuole, and a short test with the cell. Since the pH electrode measures both, membrane potential ($\Delta\Psi_m$) as well as the pH difference (ΔpH), a high-impedance amplifier (WP Instruments, FD 223) measured and simultaneously subtracted the signal from the pH electrode and from a second, conventional voltage electrode. The difference trace (pH$_o$, pH$_v$, pH$_c$) denotes the actual pH to be tested.

RESULTS AND DISCUSSION

Vacuole or Cytoplasm ?

Measuring internal pH in <u>Riccia</u> rhizoids may give either of two distinct different pH-signals : Is the electrode located within the cytoplasm, the difference trace (pH-electrode minus Ψ_m-electrode) shows pH 7.2 to 7.5, according to the calibration before and after the measurement. But in case the electrode measures within the vacuole, a pH of 4.3 to 5 (mean 4.8) is detected (Fig. 2-A and 2-B). Since this pH-difference of about 2.5 units is roughly equivalent to 150 mV, no error is possible concerning the location of the electrode. This is supported by changing external conditions such as adding weak acids or bases. For instance 1 mM procaine (pK = 9) changes cytoplasmic pH by about 0.8 units, but almost completely abolishes the pH-gradient across the tonoplast (Fig. 2).

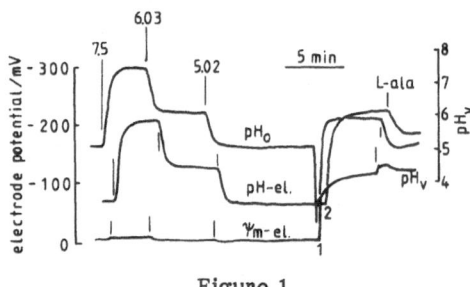

Figure 1

Precalibration of a pH sensitive microelectrode
in 5 mM Tris-Mes buffer (supplemented with 1 mM KCl, NaCl, CaCl$_2$)
at the indicated external pH (pH$_o$)

The trace of the membrane potential (Ψ_m-el.) is substracted from the trace of the pH electrode (pH-el.) yielding the third trace (pH$_o$, pH$_v$). Point 1 denotes the impalement of the cell by the voltage electrode, point 2 the impalement by the pH electrode, with a slow response. The actual internal pH is 4.3, indicating a vacuolar localization of the pH electrode. Addition of 0.1 mM L-alanine (L-ala) depolarizes the cell and alkalizes the vacuole.

Similar results can be obtained with cyanide which stops ATP production of oxidative phosphorylation and thus inhibits the ATPases on both membranes due to the lack of ATP. Fig. 3 shows an acidification of pH$_c$, but an alkalinization of pH$_v$.

Driving Forces Across the Tonoplast

In Fig. 2 and 3 the membrane potential differences between vacuole and cytoplasm are + 20 to + 30 mV, the pH 2.5 to 3 (equivalent to + 150 to + 180 mV). This means, in the resting cell a proton-motive force of roughly + 170 to + 210 mV exists across the tonoplast (compared with the - 300 to - 350 mV at the plasmalemma). Although of different sign, these driving forces are both directed into the cytoplasm for the transport of positive charge. Three important conclusions arise : Firstly, the stoichiometry of the tonoplast ATPase (H$^+$-ATPase) is higher than one. 1 H$^+$/ATP hydrolyzed has been found for the plasmalemma ATPase (Felle, 1981). Secondly, if the amino-acid transport is indeed linked to the translocation of protons, it cannot be a symport. Thirdly, the predominant driving force for substrate (amino-acids, sugars, etc.) is the proton gradient and not the membrane potential as in the plasmalemma.

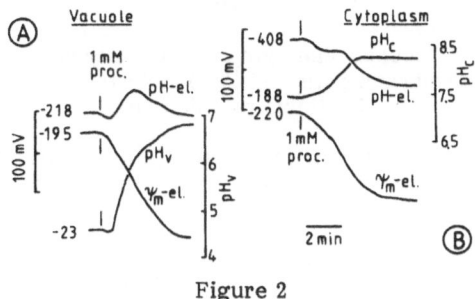

Figure 2

Vacuolar (A) and cytoplasmic (B) recording
of membrane potential (Ψ_m-el.) and pH (pH$_v$, pH$_c$)
in <u>Riccia</u> rhizoid cells

pH-el. denotes the trace of the pH electrode which monitores pH $+\Delta\Psi_m$. The vacuolar pH of 4.5 is changed to 6.8 after the addition of 1 mM procaine, while the cytoplasmic pH is only changed from 7.5 to 8.3. The difference in membrane potential indicates + 25 mV across the tonoplast.

Amino-acid Transport Across the Tonoplast

Fig. 4 shows the reaction of the tonoplast and the vacuolar sap to the addition of several amino-acids. The allover membrane potential (plasmalemma plus tonoplast) depolarizes, with a large portion of the plasmalemma. The vacuolar pH gets less acidic by 0.2 to 0.3 units which has been found for all amino-acids tested so far. The overshoot at the beginning of the reaction is very likely due to the different response time of the two electrodes involved (the pH electrode being slower than the conventional electrode). Since the pH-shift cannot be caused by dissociation/association reactions within the vacuole, and the protonmotive force is directed into the cytoplasm this result is interpreted in terms of a proton/amino-acid antiport.

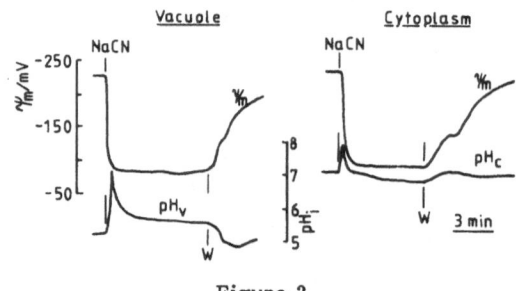

Figure 3

Vacuolar and cytoplasmic recording
of membrane potential (Ψ_m) and pH (pH$_v$, pH$_c$)
in <u>Riccia</u> rhizoid cells

1 mM NaCN causes alkalinization in the vacuole, but acidification in the cytoplasm. W = removal of NaCN. For the sake of clarity the trace of the pH electrode is not shown.

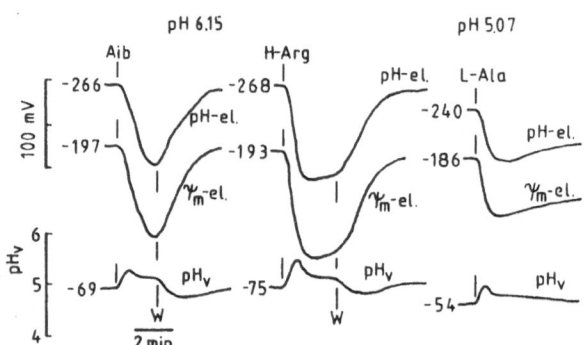

Figure 4

Membrane potential (Ψ_m-el.) and alkalinization of the vacuolar pH (pH$_v$)
in the presence of 0.1 mM of different amino-acids, as indicated

pH-el. = trace of the pH electrode. W = removal of amino-acid.

REFERENCES

Amman, D., Lanter, F., Steiner, R. A., Schulthess, P., Shijo, J., and Simon, W., 1981, Neutral carrier-based hydrogen ion selective microelectrode for extra- and intra-cellular studies, J. Anal. Chem., 53:2267.

Bertl, A., and Felle, H., 1985, Cytoplasmic pH of root hair cells of Sinapis alba resulted by a pH-sensitive microelectrode. Does fusicoccin stimulate the proton pump by cytoplasmic acidification ? J. Exp. Bot., 36:1142.

Felle, H., 1981, A study of the current-voltage relationships of electrogenic active and passive membrane elements in Riccia fluitans, Biochim. Biophys. Acta, 646:151.

Felle, H., 1981, Stereospecificity and electrogenicity of amino-acid transport in Riccia fluitans, Planta, 152:505.

Felle, H., 1983, Driving forces and current-voltage characteristics of amino-acid transport in Riccia fluitans, Biochim. Biophys. Acta, 730:342.

Felle, H., 1984, Steady-state current-voltage characteristics of amino-acid transport in rhizoid cells of Riccia fluitans, Biochim. Biophys. Acta, 772:307.

Felle, H., and Bertl., A., 1986, Light-induced cytoplasmic pH changes and their interrelation to the activity of the electrogenic proton pump in Riccia fluitans, Biochim. Biophys. Acta, 848:176.

Johannes, E., and Felle, H., 1985, Transport of basic amino-acids in Riccia fluitans : Evidence for a second binding site, Planta, 166:244.

SUGAR TRANSPORT ACROSS THE TONOPLAST OF VACUOLES ISOLATED

FROM PROTOPLASTS OF JERUSALEM ARTICHOKE TUBERS

Marco Frehner[*], Felix Keller[**], Philippe Matile[**]
and Andres Wiemken[***]

[*]Department of Biochemistry and Biophysics
University of California, Davis, California 95616, U.S.A.
[**]Department of Plant Biology, University of Zurich
CH-8008-Zurich, and [***]Department of Botany, University
of Basel, CH-4056-Basel, Switzerland

INTRODUCTION

Inulin, a β-(2→1)-linked polyfructosylsucrose (GFF$_n$), is the main carbohydrate of tubers of Jerusalem artichoke (Helianthus tuberosus) where it may constitute up to 80% of the dry weight (Edelman and Jefford, 1968).

By isolating protoplasts and vacuoles from these tubers it was demonstrated recently that all the inulin and all the anabolic and catabolic enzymes of the inulin metabolism are located exclusively in the vacuoles (Frehner et al., 1984). These results provided the basis for the proposal of a model emphasizing the dominant role of the central vacuole of Jerusalem artichoke tubers in inulin metabolism (Fig. 1). Incorporated in this model are also some speculations such as the presence of sugar transport systems across the tonoplast. This is based on the following rationale. In developing tubers transport of sucrose (GF) into the vacuole provides the substrate for the SST (sucrose-sucrose-fructosyltransferase), the starter enzyme of the inulin biosynthesis which combines two sucrose molecules to form a trisaccharide (GFF) and glucose (G). Excess glucose may subsequently be exported out of the vacuole into the cytosol where it may be further metabolized to sucrose whereas the trisaccharide is used as substrate by FFT (fructan-fructan-fructosyl-transferase) to produce inulin in the vacuole. During the resting stage of the tubers (when SST is inactive), the import of sucrose into the vacuole facilitates the formation of low DP inulin by FFT, a process which could be important in regulating the osmotic potential of the cell sap. When the tubers sprout in spring large amounts of free fructose (F) are formed by the action of FEH (fructan-exohydrolase) which need to be transported out of the vacuole again.

In order to test the model for the existence of these sugar transport systems mentioned, sugar uptake experiments were conducted with isolated vacuoles and some preliminary results are now presented.

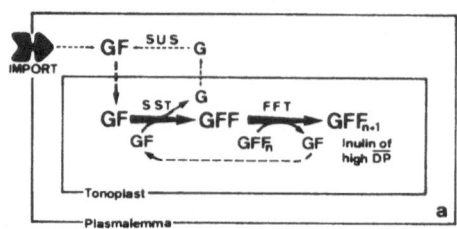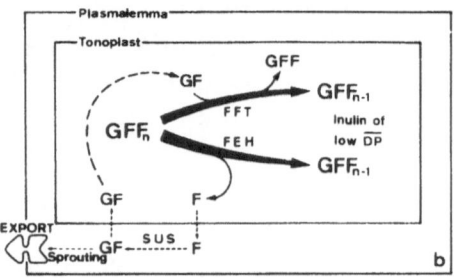

Figure 1

Model proposed
for the compartmentation of inulin metabolism
in Jerusalem artichoke tubers
(from Frehner et al., 1984)

a: Developing stage; b: Resting and sprouting stage; SUS: Sucrose synthesis. Other
abbreviations see Introduction.

MATERIAL AND METHODS

Plant Material and Vacuole Isolation

Vacuoles were isolated from protoplasts of resting tubers of Helianthus
tuberosus as previously described (Frehner et al., 1984). Vacuoles were purified
by allowing them to settle in the protoplast lysate for 10 min and then sucking
off the supernatant. This simplified purification procedure was surprisingly effective
(contamination by MDH < 4% (extravacuolar marker)) and gave higher yields of
vacuoles than our earlier method (Frehner et al., 1984).

Sugar Uptake Measurements

The loose sediment of purified vacuoles was resuspended in 5 volumes of
incubation medium consisting of 1.2 M glycine betaine, 20 mM Hepes (KOH), pH
7.6, 0.1 mM PEG-4000, 2 mM EGTA, 2 mM $CaCl_2$, 2 mM $MgCl_2$, 3 mM KCl.
Unlabelled sugar and other chemicals were added as indicated in the text. Uptake
was initiated by addition of uniformly labelled ^{14}C-sugar (final specific activity
4-10 $\mu C_i/\mu mol$ (sucrose) and 1-2.5 $\mu C_i/\mu mol$ (hexoses), respectively). Incubation
was at room temperature on an orbital shaker (75 rpm). Samples (400 μl) were
withdrawn at the times indicated and loaded onto a washing gradient prepared
in a 3 ml plastic tube and consisting of 1.5 ml 10% Ficoll and 1 ml 6.7% Ficoll in
incubation medium. After 2 min at 1 g all the vacuoles had moved out of the
incubation medium into the 6.7% Ficoll layer. After 20 min at 1 g the uppermost
400 μl of the gradient were withdrawn and the sugar concentrations determined
by GLC (Frehner et al., 1984). This allowed calculation of the specific activities
and monitoring of their change during incubation. After a further 20 min the tubes
were snap-frozen in liquid N_2 and the tips containing the washed vacuoles cut off.
These vacuolar fractions (total of 400 μl) were used to count the radioactivity
taken up by the vacuoles (150 μl) and determine the activities of β-N-acetyl-gluco-
saminidase (3 x 50 μl, marker for the amount of vacuoles recovered) and MDH (50
μl, marker for extravacuolar contamination).

RESULTS

Purity and Stability of Vacuoles

The availability of pure and stable vacuoles is the prerequisite for successful transport studies. The purity of vacuoles isolated from protoplasts of Jerusalem artichoke tubers have been shown to vary considerably depending on the stage of development of the tubers (Frehner et al., 1984). Marker enzyme analyses indicated that the vacuole preparations from growing tubers contained up to 40% extravacuolar contamination whilst those from resting tubers were only contaminated by about 5% and were therefore used throughout our experiments. Furthermore, vacuoles from resting tubers had the additional advantage of being more stable than those from growing tubers (Frehner et al., 1984). In order to quantify the stability of the vacuoles used they were incubated in the incubation medium for up to 40 min on an orbital shaker at 75 rpm. At different time intervals 400 μl of the vacuole suspension was withdrawn and the intact vacuoles were allowed to settle on the bottom of a micro sample tube. Vacuolar markers (β-N-acetylglucosaminidase, glucose plus fructose) were then determined in the supernatant and served as a measure for the rate of bursting of the vacuoles during incubation. The results are shown in Fig. 2. The stability of the vacuoles declined only moderately (by about 10%) during the first 20 min of incubation. From then on the vacuoles became increasingly unstable with a half-life of 30 to 40 min. Therefore incubation times of more than 20 min were avoided in the experiments.

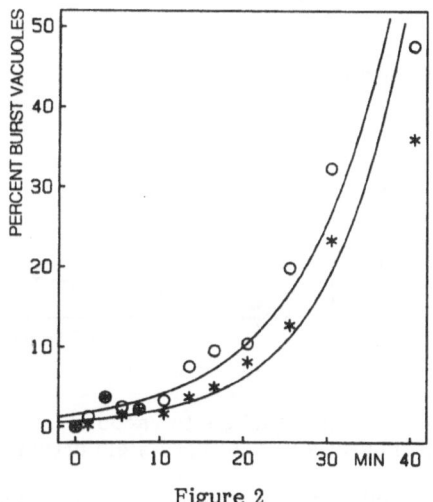

Figure 2

Bursting of vacuoles during incubation
in the incubation medium used for sugar uptake experiments
with vacuoles isolated from protoplasts of Jerusalem artichoke tubers

The percentage of burst vacuoles was derived from the release of the vacuolar markers β-N-acetylglucosaminidase (O) and glucose plus fructose (✳) into the supernatant.

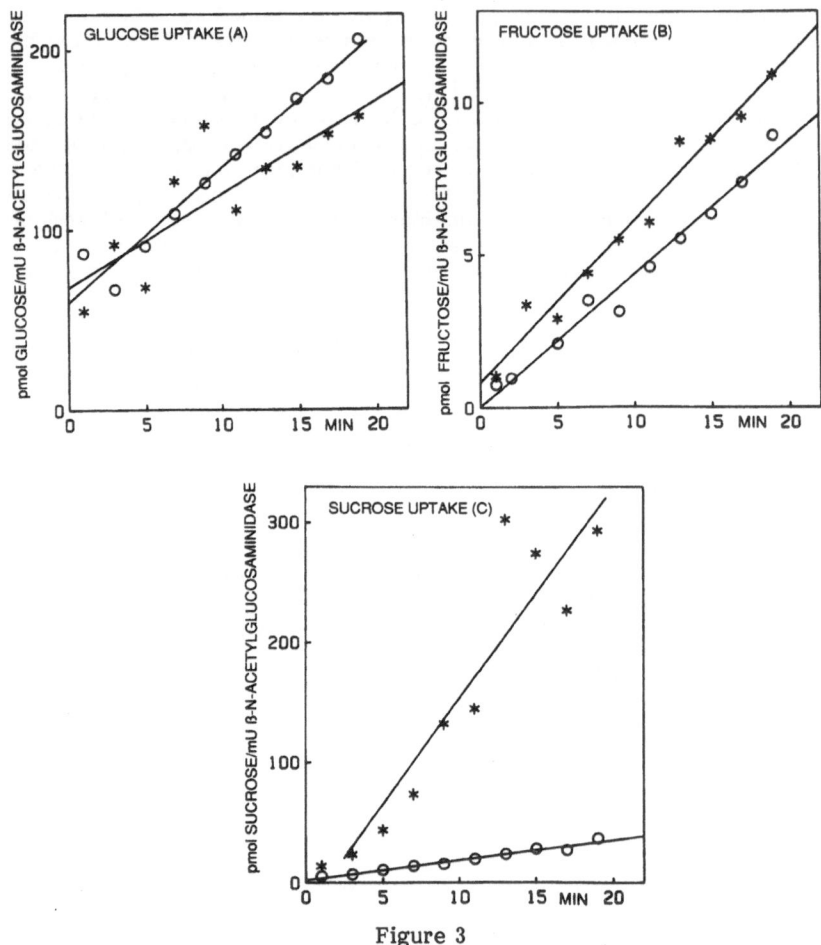

Figure 3

Sugar uptake into vacuoles
isolated from protoplasts of Jerusalem artichoke tubers

The vacuoles were incubated in the incubation medium supplemented with 0.7 mM ^{14}C-glucose (A), 0.03 mM ^{14}C-fructose (B), and 0.5 mM ^{14}C-sucrose (C) in the absence of (O) and presence (*) of 2 mM ATP.

Time Course of Sugar Uptake

Isolated vacuoles were capable of taking up ^{14}C-glucose, ^{14}C-fructose, and ^{14}C-sucrose (Fig. 3). The uptake was linear with time for at least 20 min. A meaningful comparison of the uptake rates of the three sugars is impossible as only one concentration was tested per sugar and these concentrations were different for each sugar due to the release of sugars from burst vacuoles (Fig. 3). As the specific activities decreased during the incubation they had to be determined for each time interval and corrected accordingly.

Specificity of the Hexose Carriers

To examine the recognition characteristics of the transport systems competition between glucose, fructose, and sucrose was tested. After the initial uptake rate has been determined for a ^{14}C-labelled sugar a second, unlabelled sugars was added in excess. The uptake rates before and after the addition of the competitor are shown in Table 1. Competition could only be observed in one instance : Fructose inhibited the uptake of glucose by about 70%.

Table 1

Uptake of ^{14}C-hexoses in the Presence of an Excess
of Potentially Competing Sugars into Vacuoles Isolated
from Protoplasts of Jerusalem Artichoke Tubers

Transport sugar	Competing sugar	Uptake rate (%)
^{14}C-glucose (1-2.6 mM)	none (initial rate)	100
	fructose (30 mM)	30
	sucrose (20 mM)	83
^{14}C-fructose (1-1.8 mM)	none (initial rate)	100
	glucose (27 mM)	84
	sucrose (20 mM)	88

Effect of ATP on Sugar Uptake

When the vacuoles were incubated with labelled sugars in the presence of 2 mM ATP only the uptake of $_{14}$C-sucrose was stimulated (Fig. 3). This stimulation was about ten-fold as compared to the control (no ATP). In a separate experiment a second control was performed by adding 2 mM ADP + P_i to the sucrose incubation medium. The same low rate was observed as in the first control (no ATP) (data not shown). Hexose uptake was unaffected by the addition of 2 mM ATP under the conditions used.

DISCUSSION

The results indicate that vacuoles isolated from protoplasts of Jerusalem artichoke tubers take up sugars. The preliminary nature of this study makes it impossible to draw definite conclusions as to the exact mechanisms of the sugar transport systems involved. However, the different behavior of glucose, fructose, and sucrose in the uptake experiments points to the possibility of at least two different transport systems being involved, one for hexoses and one for sucrose.

The effect of excess sugars as potential competitors on the uptake of glucose and fructose as shown in Table 1 suggests the presence of a hexose permease at the tonoplast with a higher affinity for fructose than for glucose. A glucose permease may be postulated on the grounds on the effect of ATP on sugar uptake. Only the transport of sucrose (and not that of glucose and fructose) was stimulated by 2 mM ATP. As this stimulation was quite pronounced (about ten-fold) an active transport system for sucrose could be suggested similar to that already described (sucrose/H^+ antiport) for vacuoles from red beet (Thom et al., 1986; Doll et al., 1979), cowpea (Knuth et al., 1983), and tonoplast vesicles from sugar beet (Briskin et al., 1985).

Obviously, more evidence is needed to prove the existence of these sugar transport systems unequivocally. Further studies will have to include at least the demonstration of a concentration and pH dependency (evidence for the permease nature of the transport) as well as of the effects of ionophores and the net accumulation of sucrose (evidence for the active nature of the transport).

REFERENCES

Briskin, D. P., Thornley, W. R., and Wyse, R. E., 1985, Membrane transport in isolated vesicles from sugar beet taproot. II. Evidence for a sucrose/H^+ antiport, Plant Physiol., 78:871.
Doll, S., Rodier, F., and Willenbrink, J., 1979, Accumulation of sucrose in vacuoles isolated from red beet tissue, Planta, 144:407.
Edelman, J., and Jefford, T. G., 1968, The mechanism of fructosan metabolism in higher plants as exemplified in Helianthus tuberosus, New Phytologist, 67:517.
Frehner, M., Keller, F., and Wiemken, A., 1984, Localization of fructan metabolism in the vacuoles isolated from protoplasts of Jerusalem artichoke tubers (Helianthus tuberosus L.), J. Plant. Physiol., 116:197.
Knuth, M. E., Keith, B., Clark, C., Garcia-Martinez, J. L., and Rappaport, L., 1983, Stabilization and transport capacity of cowpea and barley vacuoles, Plant Cell Physiol., 24:423.
Thom, M., Leigh, R.A., and Maretzki, A., 1986, Evidence for the involvement of a UDP-glucose-dependent group translocator in sucrose uptake into vacuoles of storage roots of red beet, Planta, 167:410.

Mg^{2+}-ATP-DEPENDENT GLUCOSE TRANSPORT IN TONOPLAST VESICLES

OF NORMAL AND AGROBACTERIUM TUMEFACIENS-TRANSFORMED TOBACCO

CELLS

Thomas Rausch

Botanisches Institut, J. W. Goethe-Universität
Siesmeyerstrasse 70, D-6000-Frankfurt, F.R.G.

INTRODUCTION

Agrobacterium tumefaciens-transformed tobacco cells are indenpendent of exogenous auxin due to the transcription of 2 T-DNA genes, which code for enzymes of auxin biosynthesis (Kemper et al., 1985; Van Onckelen et al., 1986). With primary potato tumors it has been shown that the active glucose uptake by transformed cells is drastically reduced by anti-auxins (Rausch et al., 1984). The inhibitory effect of anti-auxins as well as the modulating effects of chemicals influencing the cytoplasmic Ca^{2+} concentration (A 23187, verapamil) support the hypothesis that auxin modifies indeed active glucose uptake in transformed cells (Rausch, 1986 a and b). Both transformed and non-transformed cells of tobacco have now been further characterized with respect to the tonoplast localized proton/glucose antiport system.

MATERIALS AND METHODS

Non-transformed and Agrobacterium tumefaciens T 37-transformed cells of Nicotiana tabacum L. (cv. White Burley) were grown in Linsmayer-Skoog medium as described by Schäfer et al. (1984). The in vivo 3-O-methyl-D-glucose (OMG) transport assays followed the procedures described elsewhere (Rausch et al., 1984) with some modifications.

Microsomal membranes were prepared according to Rausch et al. (1985). Active OMG transport was assayed by a Millipore filtration technique using ^{14}C-OMG as probe. The 10 K - 50 K microsomal fraction was further separated on a sucrose step gradient according to Chanson et al. (1984).

RESULTS

Both non-transformed (= 2,4-D-dependent) and Agrobacterium tumefaciens-transformed tobacco cells showed a CCCP inhibited active uptake of ^{14}C-OMG which was linear with time for more than 2 hours and was saturated at approximately 200 μM OMG, exhibiting simple Michaelis-Menten kinetics (with a K_m value of 125 μM).

From both cell types crude microsomal membrane fractions were obtained which showed a Mg^{2+}-ATP dependent proton transport with similar characteristics as described for maize microsomal membranes (Mandala et al., 1982; Rausch et al., 1985) : K_m (ATP) = 0.5 mM, K_i (ADP) = 0.2 mM, Ca^{2+}/proton antiport, stimulation by chloride, inhibition by nitrate and DES.

The same microsomal fractions showed a Mg^{2+}-ATP-dependent, CCCP inhibited accumulation of ^{14}C-OMG. The uptake was saturable with an apparent K_m of 110 μM for OMG. Maximum transport activity was found in microsomal membranes isolated during the exponential growth phase of the cell cultures with very low transport activity after the stationary phase was reached. The minor effects of either 50 mM KCl or 50 mM KNO_3 showed that the rate of OMG uptake into the vesicles was not limited by the absolute ΔpH but rather by the number of carrier sites. The maximum OMG accumulation was higher in membranes isolated from transformed cells than from non-transformed cells (approx. 2.6-fold) when calculated per gram fresh weight. All transport assays with crude microsomal membranes were performed in the presence of 100 μM Na-vanadate.

To further characterize the OMG transport system(s) the microsomal fractions were separated on sucrose step gradients (18/26/31/36/45% sucrose w/v; 100 K, 2 hours). The individual fractions of the gradients were assayed for OMG transport activity in the absence of vanadate after they had been equilibrated with ^{14}C-OMG for 20 min. While the fractions at the 18/26% interface showed a Mg^{2+}-ATP-dependent OMG accumulation, membranes at the 31/36% interface (and to a lesser extent at the 36/45% interface) lost OMG upon addition of Mg^{2+}-ATP, indicating the presence of a proton/glucose symport system.

Preliminary experiments with marker enzymes suggest that the low density fraction (proton/glucose antiport) was of tonoplast origin while the higher density fractions (proton/glucose symport) were of plasma membrane origin.

CONCLUSIONS

The experimental approach allowed the comparative in vivo and in vitro characterization of active OMG (and glucose) transport in both transformed and non-transformed cells of tobacco. The apparent similarity between the cellular uptake of OMG and the uptake by tonoplast-enriched membrane fractions suggests that OMG (and glucose) transport in the intact cell may be controlled rather at the tonoplast than at the plasma membrane, at least during the exponential growth phase. However, it can not yet be excluded that the kinetically less defined proton/glucose symport localized on the plasma membrane may have a similar affinity for glucose.

The data provide the basis for a more rigorous analysis of the Ca^{2+} and auxin effects on glucose transport (Rausch, 1986 a and b) in transformed cells with special emphasis on proton transport ATPases and OMG (glucose) carrier sites. In this respect transformed cell clones deficient in one or both auxin genes will be a valuable experimental tool. Studies are under way to further characterize the OMG carrier at the protein level.

REFERENCES

Chanson, A., McNaughton, E., and Taiz, L., 1984, Evidence for a KCl-stimulated, Mg^{2+}-ATPase on the Golgi of corn coleoptiles, Plant Physiol., 76:498.

Kemper, E., Waffenschmidt, S., Weiler, E. W., Rausch, T., and Schröder, J., 1985, Planta, 163:257.

Mandala, S., Mettler, I. J., and Taiz, L., 1982, Localization of the proton pump corn coleoptile microsomal membranes by density gradient centrifugation, Plant Physiol., 70:1743.

Rausch, T., 1986, J. Cellular Biochem., Supplement 10 B.

Rausch, T., 1986, Plant Physiol., Suppl., in press.

Rausch, T. Kahl, G., and Hilgenberg, W., 1984, Primary action of indole-3-acetic acid in crown-gall tumors. Increase of solute uptake, Plant Physiol., 75:354.

Rausch, T., Ziemann-Roth, M., and Hilgenberg, W., 1985, ADP is a competitive inhibitor of ATP-dependent H^+-transport in microsomal membranes from Zea mays L. coleoptiles, Plant Physiol., 77:881.

Schäfer, W., Weising, K., and Kahl, G., 1984, EMBO J., 3:373.

Van Onckelen, H., Prinsen, E., Inze, D., Rüdelsheim, P., Van Lijsebettens, M., Follin, A., Schell, J., Van Montagu, M., and De Greef, J., 1986, F.E.B.S. Letters, in press.

REGULATION OF THE UDP–GLUCOSE GROUP TRANSLOCATOR
BY SOME CYTOPLASMIC COMPONENTS [*]

Margaret Thom and Andrew Maretzki

Hawaiian Sugar Planters' Association
99-193, Aiea Heights Drive
Aiea, Hawaii 96701, U.S.A.

INTRODUCTION

Sucrose is transferred into vacuoles from sugarcane cells via a tonoplast-bound UDP–glucose-dependent group translocation mechanism (Thom and Maretzki, 1985). The group translocator has been postulated to consist of a cluster of at least five enzymes that synthesizes both hexose moieties of sucrose from cytoplasmic UDP–glucose and transfers sucrose-P into the vacuole.

In order to identify the compounds in the cell that regulate sucrose storage in vivo, the effects of some cytoplasmic components, such as sugars, nucleotides, and inorganic cations, on the rate of UDP–glucose uptake into vacuoles were measured. For these experiments, tonoplast vesicles were used so that internal solutes were mostly washed out during vesicle preparation permitting stricter control of the experimental system.

MATERIALS AND METHODS

Sugarcane cell suspensions (a subclone of Saccharum sp. hybrid H50-7209) were grown in White's basal salt mixture supplemented with yeast extract, arginine, sucrose, vitamins, and 2,4-D (Nickell and Maretzki, 1969). Protoplasts were prepared as previously described (Thom et al., 1982) and washed in 25 mM Tris-Mes containing 0.5 M mannitol adjusted to pH 5.6. Vacuoles were isolated from protoplasts as previously described (Thom et al., 1982) except that all solutions were prepared in Tris-Mes containing 0.5 M mannitol adjusted to pH 6.9. Tonoplast vesicles were prepared from isolated vacuoles by a combination of shearing force and osmotic lysis. Vacuoles were suspended in a large volume of Tris-Mes containing 0.25 M mannitol (pH 6.9) and broken by approximately 30 strokes in a glass homogenizer. The tonoplast suspension was layered on a cushion of 10 % (w/v) dextran T-70 and centrifuged at 100,000 g for 1 hr. Tonoplast vesicles were recovered from the interface, diluted, and used for uptake measurements. Uptake was measured as previously described (Maretzki and Thom, 1986).

---

[*]Published as Paper No. 617 in the Journal Series of the Experimental Station, Hawaiian Sugar Planters' Association.

RESULTS AND DISCUSSION

Effects of Cations

Previous experiments on UDP-glucose uptake into tonoplast vesicles from sugarcane vacuoles were performed in White's basal salt mixture (Thom and Maretzki, 1985). When tonoplast vesicles were prepared in Tris-Mes buffer and UDP-glucose uptake was measured in the same buffer, the rate of UDP-glucose uptake was reduced to approximately 10 % of the rate reported for uptake measured in White's basal salt. Addition of a mixture of the major inorganic ions present in White's basal salt restored uptake activity (Table 1). The divalent cations Mg^{2+} and Ca^{2+}, added as Cl^- salts at 10 mM concentration, also restored uptake activity completely; Mn^{2+} partially restored activity, while Zn^{2+} was inhibitory. The monovalent cations Na^+ and K^+ were without effect (Table 1). The mechanism of divalent cation requirement is not clear. These cations may be required for membrane integrity, but most likely they form a complex with UDP-glucose so that the substrate for sucrose group translocation is a Mg- or Ca-UDP-glucose complex.

Table 1

Effects of Cations on UDP-Glucose Uptake

Addition	Ion stimulation nmol . min^{-1} . mg $protein^{-1}$
$MgCl_2$	15.6
$CaCl_2$	13.3
$MnCl_2$	5.8
$ZnCl_2$	− 2.0
NaCl	1.4
KCl	1.4
White's basal salt	9.8

UDP-glucose uptake was measured as described in Materials and Methods. UDP-glucose concentration was 200 μM, salts added were 10 mM. Control value (Tris-Mes buffer) was 2.0 nmol . min^{-1} . mg $protein^{-1}$.

Effects of sugars

Cytoplasmic concentrations of sucrose and glucose in sugarcane cells are both in the range of 10 to 25 mM (Thom et al., 1981). In order to determine if these and other sugars control sucrose storage into vacuoles, the effect of sugars on UDP-glucose uptake was measured. The data in Table 2 showed that the addition of 10 mM glucose stimulated UDP-glucose uptake threefold, while addition of sucrose doubled the uptake rate. Fructose, 3-O-methyl-glucose (an analogue of glucose), 1'-fluorosucrose, and raffinose were witout effect. Both glucose and sucrose increase V_{max} with no change in K_m (unpublished data). The mechanism of activation is not known at this time, but the kinetics suggest an allosteric effect. When 10 mM glucose and $MgCl_2$ were added together to the UDP-glucose uptake medium, the uptake rate was higher not only than the rate obtained with Mg^{2+} or glucose alone, but higher even than the additive stimulation by $MgCl_2$ and glucose (Table 3).

Table 2

Effects of Sugars on UDPG-glucose Uptake

Addition	Relative UDP-glucose uptake (%)
Glucose	152.3
Fructose	0
3-O-methyl-glucose	0
Sucrose	69.2
1'-fluorosucrose	0
Raffinose	0

UDP-glucose uptake was measured as described in Materials and Methods in White's basal salt. UDP-glucose concentration was 200 μM, sugars were 10 mM. Control value was 13.0 nmol . min^{-1} . mg protein^{-1}.

Table 3

Effect of Mg^{2+} and Glucose on UDPG-glucose Uptake

Addition	Stimulation nmol . min^{-1} . mg protein^{-1}
Glucose	3.79
MgCl$_2$	11.18
Glucose + MgCl$_2$	34.48

UDP-glucose uptake was measured as described in Materials and Methods. UDP-glucose concentration was 200 M, glucose 10 mM, and MgCl$_2$ 10 mM. Control value (buffer only) was 1.21 nmol . min^{-1} . mg protein^{-1}.

Effects of Nucleotides

Previous experiments have shown that 2 mM UDP or GDP inhibited UDP-glucose uptake into vacuoles (Thom and Maretzki, 1985). The effect of these and other nucleotides on UDP-glucose uptake by tonoplast vesicles was determined. The results showed that all the nucleotides tested, especially the uridine nucleotides, inhibited UDP-glucose uptake (Table 4). Meyer and Wagner (1985) have reported that in plant cell culture, with the exception of UDP-glucose, the concentration of ATP is higher than that of any other nucleotide. When tonoplast vesicles from sugarcane were incubated in 1 mM ATP and 10 mM MgCl$_2$ with increasing concentrations of UDP-glucose, the inhibition by ATP decreased as the concentration of UDP-glucose was increased (Table 5). This suggests that ATP competes with UDP-glucose for the binding site. One could speculate that the ratio of nucleotides to UDP-glucose plays a regulatory role in sucrose accumulation in vivo. However, Meyer and Wagner (1985) have shown that the ratio of nucleotides to UDP-glucose is close to 1 throughout the grown cycle of suspension culture. This ratio would indicate that nucleotides do not regulate sucrose accumulation in vivo, since there was only approximately a 20% inhibition of UDP-glucose uptake when the nucleotide to UDP-glucose ratio was 1. In addition, while the ratio remained constant throughout the growth cycle, the rate of sucrose accumulation into vacuoles was highest between days 7 and 10 after subculture (Thom et al., 1981).

Table 4

Effect of Nucleotides on UDPG-glucose Uptake

Addition	% of control
$MgCl_2$	100.0
MgADP	28.5
MgATP	19.2
MgUDP	6.9
MgUTP	6.6
MgGDP	15.4
MgGTP	12.6

UDP-glucose uptake was measured as described in Materials and Methods. UDP-glucose concentration was 200 μM, $MgCl_2$ 10 mM, and nucleotides 1 mM. Control value was 20.74 nmol . min^{-1} . mg $protein^{-1}$.

Table 5

Effect of UDP-glucose Concentration
on ATP Inhibition of UDP-glucose Uptake

UDP-glucose (mM)	% inhibition
0.03	90.2
0.10	83.1
0.30	73.4
1.00	66.4
3.00	53.8
10.00	47.7

UDP-glucose uptake was measured as described in Materials and Methods. MgATP concentration was 1 mM.

CONCLUSIONS

The experiments presented in this report show that the group translocator could be stimulated or inhibited by many compounds. The mechanism of regulatory control is not clear but hopefully can be solved when the enzymes are solubilized and purified. Additional experiments on the changes in cytoplasmic concentration of the stimulators or inhibitors, together with the accumulation rate of sucrose into vacuoles over the growth cycle of suspension cultures, would provide further evidence of the in vivo regulatory role of these as well as other cytoplasmic components.

ACKNOWLEDGEMENT

We wish to thank Dr. William D. Hitz of E. I. Du Pont de Nemours and Co. for the gift of 1'-fluoro-sucrose.

REFERENCES

Maretzki, A., and Thom, M., 1986, A group translocator for sucrose assimilation in tonoplast vesicles of sugarcane cells, Plant Physiol., 80:34.

Meyer, R., and Wagner, K. G., 1985, Nucleotide pools in suspension-cultured cells of Datura innoxia. I. Changes during growth in the batch culture, Planta, 166:439.

Nickell, L. G., and Maretzki, A., 1969, Growth of suspension cultures of sugarcane cells in chemically defined media, Physiol. Plant., 22:117.

Thom, M., and Maretzki, A., 1985, Group translocation as a mechanism for sucrose transfer into vacuoles from sugarcane cells, Proc. Natl. Acad. Sci., 82:4697.

Thom, M., Maretzki, A., and Komor, E., 1982, Vacuoles from sugarcane suspension cultures. 1. Isolation and partial characterization, Plant Physiol., 69:1315.

Thom, M., Maretzki, A., Komor, E., and Sakai, W. S., 1981, Nutrient uptake and accumulation by sugarcane cell cultures in relation to the growth cycle, Plant Cell Tissue Organ Culture, 1:3.

PRELIMINARY EVIDENCE FOR SOLUBILIZATION

OF AN ACTIVE TONOPLAST-BOUND VECTORIAL SUCROSE SYNTHESIZING

ENZYME COMPLEX [*]

Margaret Thom and Andrew Maretzki

Hawaiian Sugar Planters' Association
99-193 Aiea Heights Drive
Aiea, Hawaii 96701, U.S.A.

INTRODUCTION

The existence of a UDP-glucose-dependent vectorial mechanism for the formation of vacuolar sucrose on the tonoplast has been demonstrated on a number of different plant species (Thom and Maretzki, 1985; Thom et al., 1986; Oleski et al., 1986). The evidence, so far, indicates that hexose phosphates cannot penetrate the tonoplast membrane any better than can free hexoses, sucrose, or sucrose-P (Thom and Maretzki, 1985; Maretzki and Thom, 1986). This implies that the appearance of labeled sucrose-P and free sucrose inside vacuoles from the external addition of UDPG-glucose labeled in the glucose moiety requires the formation of fructose-6 P from the added UDP-glucose. Indeed, this hypothesis is further strengthened by the fact that neither cold UDP-glucose added simultaneously with labeled fructose-6 P, nor UDP-glucose, labeled in the nucleotide moiety lead to the appearance of label associated with the vacuole (Thom and Maretzki, 1986).

We are exploring means of solubilizing tonoplast proteins to determine the activity of enzymes which can be presumed to participate in the conversion of UDP-glucose and which appear to be intimately associated with the membrane. The present preliminary report establishes that enzymatically active fractions can be recovered from solubilized preparations of tonoplast and thus confirms the precepts for our original hypothesis.

MATERIALS AND METHODS

Plant Material

Sugarcane cell suspensions established from a subclone of Saccharum sp. hybrid H50-7209 were grown in a White's inorganic salt medium supplemented with yeast extract, arginine, sucrose, and 2,4-D (Nickell and Maretzki, 1969). Cells subcultured to mid-to-late log phase (i.e. 8 to 10 days) were used for these experiments.

---

[*]Published as Paper No. 616 in the Journal Series of the Experiment Station, Hawaiian Sugar Planters' Association.

Isolation of Protoplasts and Vacuoles

Cells (12 g fresh weight) were resuspended in 100 ml of White's basal salts, pH 5.6, containing 0.5 M mannitol, 2% cellulysin (Calbiochem), 2% driselase (Plenum Scientific), 0.5% Rhozyme HP-150 (Genencor) and incubated on a rotary shaker at 100 rpm and 30°C for 2 h. At the end of this period the suspension was filtered through a 35 μm nylon mesh cloth to remove intact cells. The resulting protoplast suspension protoplast suspension was sedimented at 50 g and washed three times in 25 mM Tris-Mes buffer, pH 5.6, containing also 0.5 M mannitol (Buffer A). Vacuoles were isolated from protoplasts by shearing the plasma membrane. This was accomplished by layering the protoplast suspension over a 1-ml cushion of 12% Ficoll (Type 400) made up in Buffer A and centrifuging for 60 min at 40,000 rpm (Beckman SW 41 rotor). Vacuoles were recovered at the 0/12% Ficoll interface and were washed three times at 50 g for 5 min in Buffer A at pH 6.9.

Preparation of Tonoplast Membrane

Tonoplast vesicles were obtained from vacuoles via a combination of mechanical disruption and osmotic lysis. Approximately 10^7 vacuoles in a 1.0-ml suspension were added to 10 ml of 25 mM Tris-Mes containing 0.25 M mannitol. Vacuoles were broken with approximately 30 strokes in a glass homogenizer and the suspension was centrifuged at 100,000 g for 1 h after layering it over a 10% dextran T-70 cushion. Tonoplast vesicles banded immediately above the buffer-dextran interface.

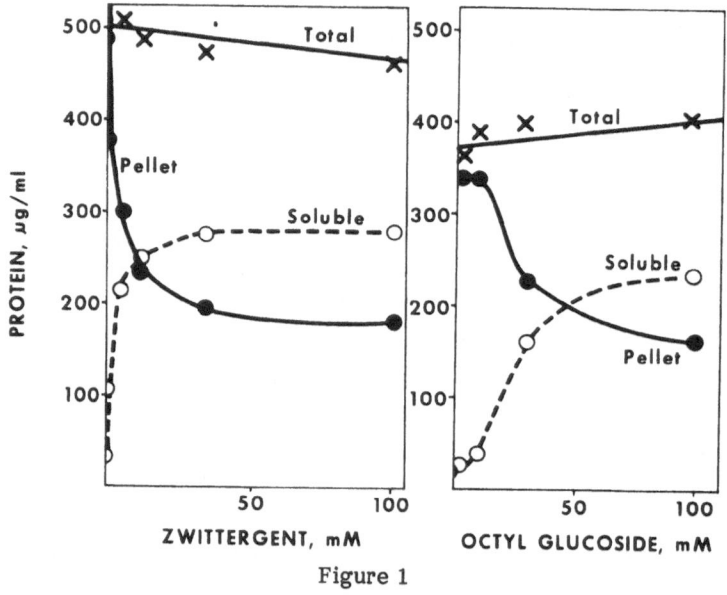

Figure 1

Optimum detergent concentration for solubilization

Tonoplast vesicles were incubated with detergents for 20 min at 4°C. Suspensions were centrifuged at 100,000 g for 60 min. Proteins in the soluble and pellet fractions were determined using Coomassie Blue dye.

RESULTS AND DISCUSSION

Three detergents were chosen in an attempt to achieve satisfactory solubilization of the tonoplast : octyl-glucoside, which has been used extensively for membrane solubilization, and the more recently introduced detergents, Zwittergent 3-12 and Zwittergent 3-14. Fig. 1 shows the results obtained with the two detergents which proved to be superior, Zwittergent 3-14 and octyl-glucoside. Approximately 50% of the total protein was solubilized with 0.3% of Zwittergent 3-14, and this was increased by about another 10% when the concentration was increased to 1%; no further solubilization was achieved above this concentration. With octyl-glucoside a 50% solubilization was also achieved at concentrations between 1% and 2% of the detergent, with little additional solubilization by further increasing the detergent concentration.

The Zwittergent 3-14-solubilized tonoplast preparation and the residual pellet were incubated with the initial substrate UDP-glucose and the suspected intermediates. In each case the subsequent products in the reaction pathway were measured (Table 1). The remaining downstream products in the presumed vectorial pathway were below the limit of detection, probably because the concentration of product (substrate for the next enzyme) was too far below the K_m of the enzyme.

Table 1

Enzyme Activities of the Vectorial Sucrose Synthesis Complex
After Incubation with Zwittergent 3-14

Substrate	Product measured	Enzyme activity	
		Soluble nmol.ml suspension^{-1}	Pellet nmol.ml suspension^{-1}
UDP-glucose	Glucose-1-P	161.1	126.3
Glucose-1-P	Glucose-6-P	15.5	2.7
Glucose-6-P	Fructose-6-P	22.2	22.2
UDP-glucose + Fructose-6-P	Sucrose-P + Sucrose	370.0	145.2

Tonoplast membranes were incubated with Zwittergent 3-14 for 20 min at 4°C, and the suspension was centrifuged for 2 h at 60,000 g. The soluble and pellet fractions were incubated (10 min, 30°C) with 200 μM UDP-glucose or intermediate substrates and the suspected downstream product in the presumed vectorial sucrose synthesis pathway was measured in each case. All other downstream intermediates were tested in each case but activity was too low to detect.

While no conversion of substrates to products could be detected in the intact membrane, phosphoglucose isomerase, phosphoglucomutase, sucrose-phosphate synthase, and a phosphatase activity could all be measured in the Zwittergent 3-14-treated soluble as well as pellet fractions. The detection of enzyme activity in the pellet fraction indicates that the detergent disrupted the normal structure of the membrane so that sites normally shielded from external substrates became accessible to these substrates.

From the present data notthing about the limiting reactions in this multiple enzymic conversion can be concluded. In fact, it is very difficult to draw conclusions about relative enzyme kinetics from such solubilized preparation, since it is likely that the enzymes behave quite differently when oriented in a membrane matrix than they would in a solubilized state. The data show that the enzyme complex is not an inviolate single unit but a complex that can be dissociated into independently functioning parts, whatever its configuration may turn out to be in the intact membrane.

REFERENCES

Maretzki, A., and Thom, M., 1986, A group translocator for sucrose assimilation in tonoplast vesicles of sugarcane cells, Plant Physiol., 80:34.

Nickell, L. G., and Maretzki, A., 1969, Growth of suspension cultures of sugarcane cells in chemically defined media, Physiol. Plant., 22:117.

Oleski, N., Joyce, D., Osteryoung, K., and Bennett, A. B., 1986, Transport properties of tomato fruit protoplast membrane vesicles, Plant Physiol., Suppl., 80:81.

Thom, M., and Maretzki, A., 1985, Group translocation as a mechanism for sucrose transfer into vacuoles from sugarcane cells, Proc. Natl. Acad. Sci. U.S.A., 82:4697.

Thom, M., and Maretzki, A., 1986, Comparison of some characteristics of uridine-diphosphate-glucose-dependent sucrose synthesis in isolated vacuoles and tonoplast vesicles, in : "Phloem Transport", J. Cronshaw, W. J. Lucas, and R. T. Giaquinta, eds., A. R. Liss Inc., New-York.

Thom, M., Leigh, L. A., and Maretzki, A., 1986, Evidence for the involvement of a UDP-glucose-dependent group translocator in sucrose uptake into vacuoles of storage roots of red beet, Planta, 167:410.

SPECIFIC UPTAKE OF THE N-OXIDES OF PYRROLIZIDINE ALKALOIDS

BY CELLS, PROTOPLASTS AND VACUOLES FROM SENECIO CELL CULTURES

Adelheid Ehmke, Kirsten Von Borstel
and Thomas Hartmann

Institut für Pharmazeutische Biologie
der Technischen Universität
D-3300-Braunschweig, F.R.G.

INTRODUCTION

The pyrrolizidine alkaloids (PAs) are secondary plant compounds which are found in several genera of the Asteraceae, Boraginaceae and Fabaceae. In plants they generally occur as mixtures of the tertiary alkaloids and the respective alkaloid N-oxides (Fig. 1). Recent studies in our laboratory with Senecio vulgaris (Asteraceae) revealed the PA-N-oxides not only to be dominating alkaloid form found in the various plant tissues (Hartmann and Zimmer, 1986), but also to be the primary products of PA biosynthesis in root cultures (Von Borstel and Hartmann, 1986).

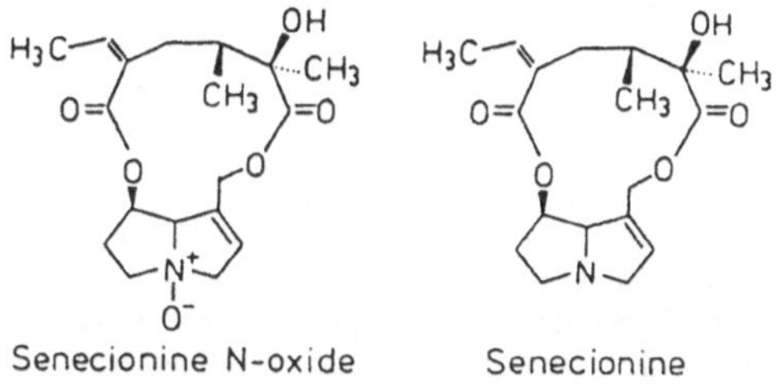

Senecionine N-oxide   Senecionine

Figure 1

Senecionine N-oxide
the predominant pyrrolizidine alkaloid
of Senecio vulgaris,
and its reduced form

N-oxidation alters dramatically the physicochemical properties of the PAs. The N-oxide no longer behaves like a typical alkaloid, but instead becomes a very polar, salt-like compound which is insoluble in most organic solvents and not easily able to nonspecifically permeate biological membranes like an unpolar tertiary alkaloid in its unprotonated form. Thus, a PA-N-oxide could be a molecular species better suited for cellular transport and safe vacuolar accumulation than the respective tertiary alkaloid. To test this hypothesis the studies described here were performed. We demonstrate a specific uptake of the N-oxide into cells and isolated protoplasts derived from alkaloid producing plants and the accumulation of the N-oxide within the vacuoles.

MATERIALS AND METHODS

Cell suspension cultures were established and grown as reported (Hartmann and Toppel, 1986). Monocrotaline (99% pure) was obtained from Aldrich Chem. Comp.; senecionine was isolated from Senecio vulgaris (Hartmann and Zimmer, 1986). The respective N-oxides were prepared according to Craig and Purushothaman (1970). ^{14}C-Senecionine N-oxide (specific activity 82.8 mC$_i$/mmol) was prepared biosynthetically using a Senecio vulgaris root culture (Von Borstel and Hartmann, 1986). 1,4-^{14}C-Putrescine was fed as a precursor. Analysis of nonlabelled alkaloids was performed by means of capillary gas-liquid chromatography (Hartmann and Zimmer, 1986; Von Borstel and Hartmann, 1986; Hartmann and Toppel, 1986).

Protoplasts and vacuoles were isolated from cell suspension cultures of Senecio vulgaris using a slightly modified procedure as given by Kreis and Reinhard (1985). Cells were incubated in a 1:1 mixture of 1 M mannitol and cell culture medium containing 2% cellulase Onuzuka RS (Serva) and 0.3% Rhozyme HP-150 (Rhöm and Haas) at 30°C, pH 4.8. After 2 hr incubation the protoplasts were filtered through a nylon net (100 μm), collected by centrifugation (100 g, 3 min) and washed twice. Lysis of protoplasts was obtained in vacuole buffer (250 mM mannitol, 20 mM Tris, 10 mM Na-EDTA, 2 mM DTT, 8% Ficoll). The vacuoles were filtered through a nylon net (100 μm) and purified by flotation centrifugation : 5 parts of the vacuole suspension were overlayed with 2 parts of vacuole buffer containing 4 % Ficoll, and 1 part of vacuole buffer without Ficoll. After centrifugation at 900 g for 15 min the vacuoles were recovered from the 4%/0% Ficoll interphase. The purity of the preparation was examined microscopically and by marker enzymes. Malate dehydrogenase and glucose-6-phosphate dehydrogenase activities were used as protoplast markers, α-mannosidase activity as vacuolar marker. The purified vacuole preparation contained less than 8% of cytoplasmic contaminations. Quantification of protoplasts and vacuoles were done either by counting in a hemocytometer or by measuring the marker enzyme activities.

RESULTS

The uptake of tertiary PAs and PA-N-oxides by cell suspension cultures is restricted to cells derived from species which produce PAs, as shown for monocrotaline and monocrotaline N-oxide in Table 1. The same results were obtained with senecionine and its N-oxide (not shown). The N-oxide of the quinolizidine alkaloid sparteine, which was chosen as a control, is not taken up by any of the cultures. These results indicate first evidence for the existence of a selective PA uptake (Hartmann and Toppel, 1986).

With the availability of ^{14}C-labelled senecionine and senecionine N-oxide we were able to perform uptake studies in a more physiological concentration range. Comparative uptake experiments with ^{14}C-senecionine and ^{14}C-senecionine N-oxide (10 μM each) revealed two important differences (Table 2) :

   a) Senecionine N-oxide is taken up much more rapidly than the respective tertiary alkaloid.

   b) When protoplasts are prepared from cells located with labelled alkaloid, the N-oxide is completely retained within the protoplast, whereas the tertiary alkaloid is released.

Table 1

Uptake of Monocrotaline (MC), Monocrotaline-N-oxide (MC–ox)
and Sparteine-N-oxide (Sp–ox) by Cell Suspension Cultures

Cell culture	% of alkaloids taken up into the cells		
	MC	MC–ox	Sp–ox
Senecio vulgaris B	28.2	79.0	4.9
Senecio vernalis F	47.3	75.1	2.5
Senecio viscosus	36.8	66.6	n.d.
Symphytum officinale	38.6	34.4	7.5
Daucus carota[a]	2.2	8.6	9.1
Lupinus polyphyllus[a]	6.0	9.9	7.3
Spinacia oleracea[a]	2.2	6.2	8.9

Ca 5 g cells were incubated in 20 ml culture medium containing either
68.5 μM MC, or 120 μM MC–ox, or 430 μM Sp–ox for 5 hr.
[a]Alkaloid background of 2 to 10% found in the cell fractions is due
to alkaloid bound to the apoplast; the cells were not washed.

Table 2

Uptake of [14]-senecionine and [14]-senecionine N-oxide
by Cells of Senecio vulgaris,
and Release of Label from the Cells during Protoplast Preparation

	% Uptake into cells of Senecio vulgaris			
Time of incubation (min)	^{14}C-senecionine (10 μM)		^{14}C-senecionine N-oxide (10 μM)	
60	9	21	21	66
120	16	42	53	83
180	39	46	75	84
	Release of ^{14}C-activity in the medium (%)[a] during isolation of protoplasts			
60	100	110	47	29
120	94	114	51	51

Cells (1 g) were incubated in 5 ml culture medium containing 10 μM labelled alkaloid. After 180 min the cells were transferred into the protoplast isolation medium.

To characterize this differential behavior in more detail protoplasts were allowed to take up the two labelled compounds (10 μM each) for 2 hr. Then vacuoles were isolated from the preloaded protoplasts, purified by flotation centrifugation (9200 g, 15 sec.) and tested for radioactivity. In the feeding experiment with senecionine N–oxide 32% of intact vacuoles could be recovered and these contained 31% of the total N–oxide taken up by the protoplasts. Thus, almost all of the senecionine N–oxide taken up by the protoplasts is localized in the vacuoles. By contrast only trace amounts of label could be recovered from the vacuoles isolated from protoplasts preloaded with senecionine. The vacuolar concentration of senecionine N–oxide was calculated to be 29 μM, which is an approximately 10-fold higher concentration than in the protoplast incubation medium at the end of the incubation period. This clearly shows that senecionine N–oxide is accumulated, and in contrast to its reduced form, retained in the vacuole against a concentration gradient. Preliminary experiments indicate that vacuoles isolated from Senecio vulgaris cells were able to actively take up senecionine N–oxide. The mechanisms involved in this process are presently under investigation in our laboratory.

DISCUSSION

In the past alkaloid uptake and vacuolar accumulation has often been explained by diffusion of the uncharged molecule across membranes and trapping of the protonated molecule within the vacuole (Matile, 1978; Renaudin and Guern, 1982). Recently Deus-Neumann and Zenk (1985 and 1986) demonstrated that alkaloid uptake into isolated vacuoles is an energy-dependent process. Furthermore, the uptake was found to be specific for alkaloids indigenous to the plant from which the vacuoles had been isolated. In this paper we present an other example of specific alkaloid uptake and accumulation, and moreover identified the N–oxide of senecionine as the competent alkaloid form which is translocated into and stored within the vacuoles of Senecio vulgaris cells. Assuming the function of pyrrolizidine alkaloids as powerful components of the plant chemical defense system (Boppré, 1986), a safe vacuolar storage is a prerequisite to maintain a sufficient high protective alkaloid concentration. In this respect, at least within the pyrrolizidine alkaloids, the polar alkaloid N–oxides appear to be superior to the tertiary alkaloids. Alkaloid N–oxides are well known from other alkaloid classes (Phillipson and Handa, 1978). Further studies are needed to decide, whether the storage of alkaloids as N–oxides is of more general importance or unique for the pyrrolizidines.

REFERENCES

Boppré, M., 1986, Insects pharmacophagously utilizing defensive plant chemicals (pyrrolizidine alkaloids), Naturwissenschaften, 73:17.
Craig, J. C., and Purushothaman, K. K., 1970, J. Org. Chem., 35:1721.
Deus-Neumann, B., and Zenk, M. H., 1985, A highly selective alkaloid uptake system in vacuoles of higher plants, Planta, 162:250.
Deus-Neumann, B., and Zenk, M. H., 1986, Accumulation of alkaloids in plant vacuoles does not involve an ion-trap mechanism, Planta, 167:44.
Hartmann, T., and Toppel, G., 1986, Phytochemistry, in press.
Hartmann, T., and Zimmer, M., Organ-specific distribution and accumulation of pyrrolizidine alkaloids during the life-history of two annual Senecio species, J. Plant Physiol., 122:67.
Kreis, W., and Reinhard, E., 1985, Rapid isolation of vacuoles from suspension-cultured Digitalis lanata cells, J. Plant Physiol., 121:385.
Matile, P., 1978, Biochemistry and function of vacuoles, Annu. Rev. Plant Physiol., 29:193.
Phillipson, J. D., and Handa, S. S., 1978, J. Nat. Prod., 41:385.
Renaudin, J. P., and Guern, J., 1982, Compartmentation mechanisms of indole-alkaloids in cell suspensions of Catharanthus roseus, Physiol. Vég., 20:533.
Von Borstel, K., and Hartmann, T., 1986, Plant Cell Rep., in press.

ACCUMULATION OF ORGANIC SOLUTES IN PLANT VACUOLES :

THE INTERPRETATION OF DATA IS NOT SO EASY

Jean Guern, Jean-Pierre Renaudin[*] et Hélène Barbier-Brygoo

Laboratoire de Physiologie Cellulaire Végétale, CNRS / INRA
Boite Postale N°1, F-91.190-Gif-sur-Yvette, and [*]Station de
Physiopathologie Végétale, INRA, B. V. No. 1540, F-21.034-Dijon,
France

INTRODUCTION

Significant progress has been made recently about the mechanisms of transport of solutes across the tonoplast (Alibert and Boudet, 1982; Leigh, 1983; Deus-Neumann and Zenk, 1984; Boller, 1985; Thom and Maretzki, 1985; Deus-Neumann and Zenk, 1986; Thom et al., 1986). The diversity of transport systems tentatively identified at the tonoplast appears rather large (Guern et al., 1987). A few examples illustrate the various types of transmembrane transfer involved in the vacuolar exchanges.

Proton-antiport carriers appear as one of the most important transfer process at the tonoplast. They are considered as responsible for the uptake of molecules as different as O-methylglucose in vacuoles of pea and sugarcane (Guy et al., 1979; Thom and Komor, 1984), sucrose in tonoplast vesicles of sugarbeet (Briskin et al., 1985), citrate in hevea lutoids (Marin and Chréstin, 1985), arginine in yeasts (Ohsumi and Anraku, 1981), esculine in barley (Werner and Matile, 1985), specific alkaloids in <u>Fumaria</u> <u>capreolata</u> (Deus-Neumann and Zenk, 1986). The energy-dependent uptake of malate coupled to the transtonoplast pH or potential gradients, described in barley vacuoles, likely belongs to this category (Martinoia et al., 1985).

One of the most interesting result of the past two years concerning vacuolar uptake has been the identification of a group-transport system at the tonoplast of sugarcane and beet vacuoles, responsible for the synthesis and intravacuolar accumulation of sucrose from external UDPG (Thom and Maretzki, 1985; Thom et al., 1986). The tentative identification of such a system at the tonoplast of grape cells was first reported by Brown and Coombe (1984). A complex interrelation between metabolization, energy dependency and transmembrane transfer is operating in this case.

Other energy-dependent carrier-mediated uptake systems, only partly described, have been reported for sucrose uptake in beet vacuoles (Doll et al., 1979; Willenbrink and Doll, 1979) or specific alkaloids in <u>Catharanthus</u> <u>roseus</u> vacuoles (Deus-Neumann and Zenk, 1984).

Non-concentrative, energy-independant carriers have also been tentatively described for sucrose (Kaiser and Heber, 1984) in barley vacuoles and for malate in <u>Bryophyllum</u> <u>daigremontianum</u> (Buser-Suter et al., 1982). Their biological significance has been assumed to correspond to exchanges between the vacuolar and cytoplasmic compartment without osmotic disturbance.

Aside the recent report of the carrier-mediated uptake of some alkaloids by Zenk's group, a non-catalyzed transmembrane transfer has been described for several molecules of this highly diversified family. The diffusion of these molecules is characterized by a membrane permeability to their ionized form much lower than the one of the neutral form. As a consequence, the transmembrane diffusion is almost restricted to the neutral form whereas the cation is accumulated in the acidic vacuole (Müller, 1976; Renaudin and Guern, 1982; Neumann et al., 1983; Renaudin et al., 1985).

Despite these recent progresses, the study of the mechanisms of solute uptake at the tonoplast is hampered by various problems such as the characterization of the properties of the vacuoles used in the uptake experiments, the manipulation of these fragile objects and also the interpretation of results. Furthermore, only a few data are available on the efflux of solutes from the vacuoles and on the complex interactions between influx and efflux processes which regulate the intravacuolar concentration of solutes. As a consequence, the mechanisms which govern the vacuolar accumulation of solutes, too often implicitly identified to those driven the uptake, are largely unknown. A few examples of the various problems related to the study of the vacuolar uptake and accumulation of solutes in plant vacuoles will be illustrated in this paper with some emphasis on the behavior of alkaloids.

THE CHARACTERIZATION OF TRANSPORT SYSTEM AT THE TONOPLAST: SOME POTENTIAL DIFFICULTIES

Saturation kinetics and specificity of the uptake (often tested by the competition for uptake between analogs) are customary arguments used for the characterization of carrier-mediated transport. Such an approach has been recently illustrated by the elegant work of Deus-Neumann and Zenk showing that (i) the uptake and accumulation of alkaloids only occur to a significant extent in homologous vacuoles i.e. in vacuoles isolated from the plants producing these alkaloids (Deus-Neumann and Zenk, 1984); and (ii) the uptake is very specific of a molecular conformation (Deus-Neumann and Zenk, 1986). These results give strong evidence for the presence of specific alkaloid carriers at the tonoplast. However, methodological problems could be encountered when using the criteria of saturation and specificity ; they will be further discussed later on.

As already said above, the coupling of solute uptake to a transfer of protons through H^+-antiport carriers has been postulated for a variety of molecules. The experimental evidences are in most cases fairly indirect and concern the influence of the external pH and/or the influence of ionophores. Only in a few cases (see for example Thom and Komor, 1984), ΔpH and E_m modifications, corresponding to the antiport of protons, were demonstrated to be associated to solute uptake.

The study of the ATP-dependency of solute uptake is also one of the major elements of the general strategy used to characterize uptake processes. The interpretation of the results is probably not always so straightforward as the apparent ATP-dependency of the carrier activity could in fact indirectly from ATP-induced modifications of other tonoplast properties such as those corresponding to changes in the surface potential (Barbier-Brygoo et al., 1985) or to the phosphorylation of tonoplast proteins (Teulières et al., 1985). Furthermore, the concentration of Mg^{2+} must be controlled carefully when $ATP-Mg^{2+}$ is used in presence of EDTA : a strongly lowered concentration of the ATP complex due to the trapping of Mg^{2+} by EDTA or an excess of Mg^{2+} modifying the tonoplast surface potential could affect the results in an unexpected way.

Ionophores have been extensively used to demonstrate the coupling of solute uptake to ionic fluxes, especially H^+ fluxes. However, in most cases the ionic gradients between the vacuoles and their suspension medium were not known. This could render difficult a proper use of ionophores and a correct interpretation of data. A good example of mastering the use of nigericin through the knowledge of the K^+ gradient in barley vacuoles has been reported by Martinoia et al. (1985). At the opposite, the fact that morphinan uptake by vesicles of Papaver somniferum was inhibited by FCCP but insensitive to NH_4Cl (Homeyer and Roberts, 1984) is rather puzzling. It could suggest the action of FCCP could be more complex than its classical protonophoric activity. As a matter of fact, we have shown that the binding of FCCP to the tonoplast of Acer pseudoplatanus vacuoles does not modify to a significant extent the membrane surface potential but alters the binding of the lipophilic cation TPP^+ to the tonoplast (Barbier-Brygoo et al., 1985). The same result has been obtained recently on vacuoles of Catharanthus roseus where the decrease of the surface charge density induced by 2 mM TPP^+ (from $- 0.63$ e. 1000 A^{o-2} for control vacuoles to $- 0.38$ e. A^{o-2} for treated ones) was reduced by 60% in the presence of 5 μM FCCP (Pradier et al., unpublished results). These results show that, aside its protonophore properties, FCCP inserted into the tonoplast could modify the properties of this membrane. Thus, the straightforward interpretation of the inhibition of solute uptake by FCCP, as evidence for a transport somewhat coupled to the transmembrane proton gradient, has to be considered with more critical attention.

DIVERSITY OF INTRAVACUOLAR EVENTS DISPLACING THE EQUILIBRIUM BETWEEN UPTAKE AND EFFLUX OF SOLUTES. THE "TRAPPING" MECHANISMS

The "trapping" of solutes into vacuoles has attracted some interest. The definition of this phenomenon appears different from one group to another and this is probably at the origin of some confusion in the literature.

A large variety of intravacuolar events, superimposed upon the transmembrane translocation, modify the properties of absorbed solutes, lower the intravacuolar activity of the free solutes and displace the equilibrium between uptake and efflux. Thus, they restrict the tendency of solutes to leak out from the vacuoles and as such they contribute to the vacuolar accumulation. All these reversible events, diagrammatically illustrated in Fig. 1, will be designated under the general term of "trapping".

A first example of such a process is the trapping of the ionized form of weak bases. Under its most simplified form, this trapping is assumed to correspond to the fact that the permeability of membranes to the ionized form of the base is negligible compared to the one of the neutral form. In such a case, the accumulation, linked to ion-trapping, is driven by the transmembrane pH gradient. It is relevant to note that this differential permeability of membranes, including the tonoplast, to the neutral and ionized forms of weak lipophilic bases is at this basis of the use of some of these molecules as pH probes for acidic compartments including vacuoles (see for example Manigault et al., 1983 for 9-aminoacridine and Kurkdjian, 1982 for nicotine). However, such an ideal situation, which cannot be considered as the rule for every weak lipophilic base, has only been checked in a few cases. When the diffusion of the cation cannot be neglected, the intensity of ion-trapping is a function of both the pH and potential gradients (see for example Courtois and Guern, 1980). Here lies the potential for a large variability in the intensity of the vacuolar accumulation of weak bases according to their pK_a, to the relative permeability of the tonoplast to the neutral and cationic forms and to the pH of the external and vacuolar compartments.

More diverse events, such as binding of solutes to various intravacuolar sites or conformation changes, can be also involved in the intravacuolar accumulation of solutes. We have recently shown, for example, (Pradier et al., unpublished results) that lipophilic bases such as serpentine and 9-aminoacridine modify the electropho-

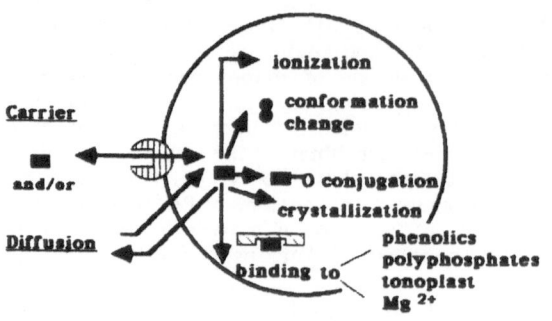

Figure 1

Diagrammatic representation of a variety
of intravacuolar "trapping" events superimposed
to the transmembrane transfer

The ionization of weak lipophilic bases, such as alkaloids or the acridines used as pH probes, gives cations with a much restricted diffusive ability compared to the neutral bases. Conformation shifts of O–coumaric glucosides (Rataboul et al., 1985) or apigenin malonyl–glucoside (Matern et al., 1986) give forms with restricted ability to cross the tonoplast membrane. Metabolization inside the vacuole is another way to restrict the efflux of the absorbed solute or limit the feedback control on the influx carrier (Sharma and Strack, 1985; Strack and Sharma, 1985). An extreme example of modifications of physicochemical properties leading to an increased accumulation is the crystallization of one part of the intravacuolar berberine in Thalictrum minus (Nagakawa et al., 1984). The association of the absorbed metabolites with various intravacuolar binding sites also shifts the equilibrium of uptake towards an increased accumulation. These binding events concern Mg^{2+} for citrate (Marin and Chréstin, 1985), phenolics for alkaloids (Matile, 1976), polyphosphates for arginine and other basic amino–acids (Boller, 1985; Weiss, this issue) or the tonoplast membrane itself (Matile, 1976 and our own unpublished results described in the text).

retic mobility of vacuoles from Catharanthus roseus and Acer pseudoplatanus cells through their binding to the tonoplast, as already demonstrated for TPP^+ (Gibrat et al., 1985).

It is important to realize that (i) these "trapping" events can occur irrespective of the nature of the transmembrane transfer process per se (i.e. the binding to vacuolar sites can be associated with a carrier-mediated uptake as well as with a diffusive entry); (ii) several of them can be combined (i.e. the trapping of the cation of weak bases due to the restricted permeability of the tonoplast can be amplified by the association of the cation to intravacuolar binding sites); and (iii) some of them can be highly specific (i.e. binding, metabolization, conformation changes) and can account for the specificity of the overall accumulation.

This last point, concerning the possible contribution of trapping events to the specificity of vacuolar accumulation, deserves more consideration. It has been implicitely admitted in most studies of vacuolar uptake that the internal concentration measured was the pure reflect of the transmembrane transfer, without influence

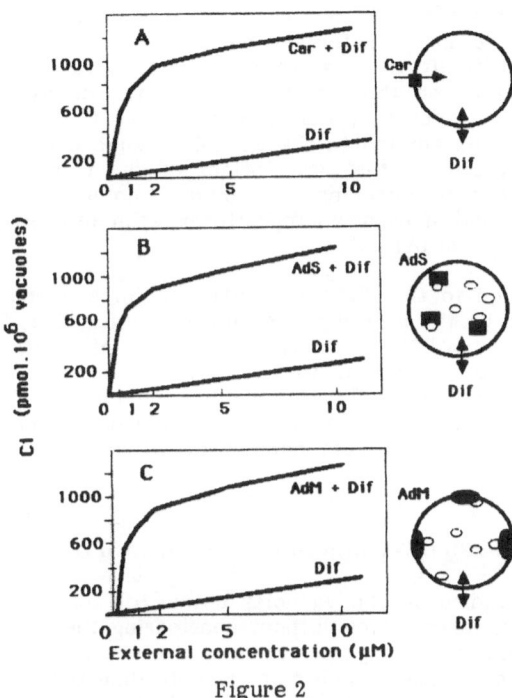

Figure 2

Simulation of the vacuolar accumulation of an alkaloid
driven either by a tonoplastic carrier or by a specific adsorption
on intravacuolar binding sites

The intravacuolar concentrations (C_i) have been calculated at the equilibrium of uptake assuming an external pH of 6.5, an intravacuolar pH of 6.5 and a weakly basic alkaloid with a pK_a of 5.4. It has also been assumed, in agreement with the simplified form of the ion-trap model that, in all cases, the neutral form of the alkaloid diffused through the tonoplast with a permeability coefficient of 0.5 . 10^{-7} cm . s^{-1} whereas the diffusion of the ionized form was neglected.

of subsequent events occurring inside the vacuoles. In fact, the risk that the measurement of vacuolar uptake includes both the transmembrane transfer and surimposed trapping processes has to be considered. This is true if the intravacuolar concentration is measured close to the equilibrium of uptake as illustrated by the simple simulations of Fig. 2. These calculations, starting from rather arbitrary values, have no more ambition than to draw the attention to the fact that the intravacuolar binding to specific sites can, to some extent, mimick saturation kinetics typical of a carrier-mediated transfer. In such a case, the specificity of the overall process of vacuolar accumulation is governed by the specificity of the binding sites. It is more difficult to predict what should be the situation when rates of uptake are measured instead of the accumulation at equilibrium. In this case, the influence of the binding component on the observed kinetics should depend on the relative rates of the transmembrane transfer and binding processes. However for technical reasons it is quite difficult to measure actual initial rates of uptake. When the approach is simply based on the measurement of the intravacuolar concentration after a solute uptake of constant duration, whatever the external solute concentration or the treatments applied to the vacuoles, some caution must be exercized in interpreting the results as the pure reflect of the transmembrane influx. This stresses the importance of checking if a binding component is operating in the accumulation process and of using appropriate experimental procedures to approach the transmembrane transfer per se.

In A, surimposed to the diffusion (Dif), the intravacuolar accumulation was supposed to be driven by the activity of an influx carrier (Car) for the ionized alkaloid with a V_m of 70 pmol h^{-1} 10^6 vacuoles and a K_m of 0.03 μM. In B, the alkaloid cation was assumed to be bound to intravacuolar soluble sites AdS (vacuolar concentration of binding sites : 33 μM, K_d for the ionized alkaloid: 0.03 μM). In C, the influence of the binding of the alkaloid cation to the tonoplast surface (AdM) was simulated assuming a maximum number of sites of 0.35 10^{-3} A$^{\circ 2}$ and a K_d of 0.03 μM.

POSSIBLE INTERACTIONS BETWEEN SEVERAL TRANSPORT SYSTEMS

Most of the studies on the vacuolar uptake of solutes concentrated on the measurement of the inwards fluxes without considering the outwards fluxes limiting the vacuolar accumulation. As a consequence, the characterization of the mechanisms of uptake has progressed much more rapidly than the knowledge of the more complex mechanisms which govern the accumulation. Furthermore, most of the data reported have been interpreted as the result of the activity of only one translocation system.

We consider here the hypothesis that several transmembrane transfer processes coexist for the same molecule, with a special attention to the classical "pump and leak" model where a diffusive exchange and a carrier-mediated influx operate simultaneously. We used this model for a tentative reinterpretation of the data concerning the vacuolar uptake and accumulation of alkaloids. Deus-Neumann and Zenk (1984 and 1986) have recently presented evidence that a highly diversified set of specific carriers were responsible for the uptake of various alkaloids. No evidence for a diffusive uptake and subsequent intravacuolar ion-trapping was obtained for the molecules they studied whereas such events were supported by various data (Müller, 1976; Renaudin and Guern, 1982; Neumann et al., 1983; Renaudin and Guern, 1987). The difference in the results concerning the uptake and accumulation of nicotine was particularly striking, with a specific and rather small accumulation on one hand (Deus-Neumann and Zenk, 1984) and a non-specific (nicotine is used as a vacuolar pH probe) and rather large accumulation on the other hand (Renaudin and Guern, 1987).

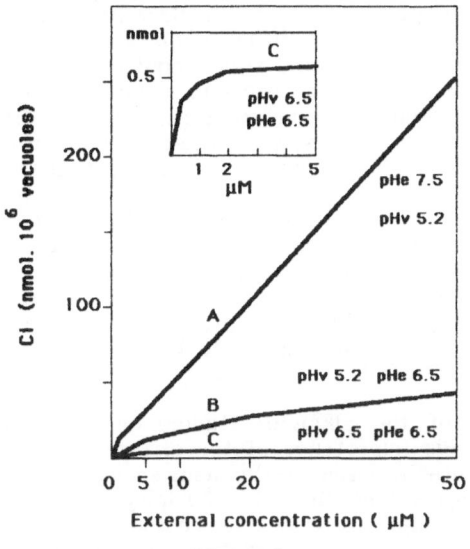

Figure 3

Simulation of the vacuolar accumulation of an alkaloid
as a function of the external concentration
for different values of the transtonoplast pH gradient
assuming that carrier-mediated and diffusive uptake processes coexist
at the tonoplast

The intravacuolar concentrations (C_i) have been calculated at the equilibrium of uptake for an alkaloid with a pK_a = 8.0, crossing the tonoplast through (i) an influx carrier specific for the cationic form (V_m = 12 pmol h^{-1} 10^6 vacuoles - K_m = 0.3 μM; and (ii) a diffusion only concerning the neutral form of the alkaloid (P : 0.5 10^{-6} cm s^{-1}). Curves A, B and C correspond to different values of the external (pH_e) and intravacuolar pH (pH_v). The inset gives some details on curve C for low values of the external concentration.

Our proposal to tentatively interpret such a discrepancy is that specific carriers and diffusive exchanges could coexist for the same molecule with important differences in their relative importance according to (i) the characteristics of the vacuolar preparation used; and (ii) the physicochemical properties and molecular conformation of the solute absorbed.

As a matter of fact, we have recently demonstrated the existence of important differences in the properties of vacuoles according to the type of osmoticum (neutral or saline) used for their isolation (Barbier-Brygoo et al., 1987). Vacuoles isolated in saline medium, as done by Deus-Neumann and Zenk (1984 and

1986), reveal a strongly depressed transtonoplast pH gradient. As a consequence, they likely have a restricted ability to accumulate weak bases under their cationic form. The relative importance of the transtonoplast pH gradient compared to the carrier-mediated component of the vacuolar accumulation was estimated in the case of an alkaloid by simulating different possible situations. This is illustrated by the diagrams of Fig. 3. Curve A, with a nearly linear profile, shows that a high pH strongly favors the diffusive component through cation trapping. At the opposite, curve C displays the appearance of saturation kinetics characteristic of the carrier-mediated uptake, whose function is nearly revealed when the pH gradient is strongly lowered. These simple calculations provide an example of what could be the interaction between three components of a vacuolar accumulation (two membrane transfer processes and an intravacuolar trapping). To what extent curve A should be assumed to reflect a pure diffusive uptake and curve C a pure carrier uptake depend on how precisely are measured the kinetic parameters and on choice of the concentration range studied.

The properties of the alkaloids studied can also potentially amplify the variations in the relative importance of the different components involved in the accumulation. The saturable carrier component should be largely predominant when the uptake concerns an alkaloid with a low basicity accumulated by vacuoles having lost their acidity, an alkaloid diffusing as easily under its ionized form as under the neutral one or an alkaloid with a low diffusion capacity compared to the kinetic parameters of the carrier. Conversely, the uptake of a more basic alkaloid by vacuoles, with a high transtonoplast pH gradient, should be mainly driven by the diffusive and ion-trapping component provided a highly ability of this alkaloid to diffuse under its neutral form through the tonoplast.

Thus if our proposal is correct, one should expect a large variability of situations to occur when studying the mechanisms of uptake and accumulation of members of the highly diversified alkaloid family.

CONCLUSION

Despite recent progress much has to be done to identify the mechanisms of the vacuolar accumulation of solutes with a special attention to the components involved in the transmembrane translocation, to those creating the driving forces and also to the contribution of intravacuolar modifications of the properties of the absorbed solutes.

The strategy of characterization of the uptake processes should be as diversified as possible, looking at different criteria to avoid some of the ambiguities linked to the single and dogmatic use of some of them such as the effects of ATP and protonophores or the study of pH-induced modifications. The evolution towards the isolation, the molecular characterization of the carrier moities and the reconstitution of the uptake systems should avoid the present uncertainties mainly linked to the rather indirect approaches used.

The distinction between uptake and accumulation must finally be stressed. The mechanisms involved in the accumulation are likely more complex than those involved in the uptake as tentatively examplified in this paper. It appears particularly necessary to pay more attention to the mechanisms which insure the reversibility of the accumulation of solutes in the vacuole of plant cells.

REFERENCES

Alibert, G., and Boudet, A. M., 1982, Progrès, problèmes et perspectives dans l'obtention et l'utilisation de vacuoles isolées, Physiol. Vég., 20:289.
Barbier-Brygoo, H., Gibrat, R., Renaudin, J.-P., Brown. S. C., Pradier, J.-M., Gri-

gnon, C., and Guern, J., 1985, Membrane potential difference of isolated plant vacuoles : Positive or negative ? II. Comparison of measurements with microelectrodes and cationic probes, Biochim. Biophys. Acta, 819:215.

Barbier-Brygoo, H., Renaudin, J.-P., Manigault, P., Mathieu, Y., Kurkdjian, A., and Guern, J., 1987, Properties of vacuoles as a function of the isolation procedure, in : "Plant Vacuoles. Their importance in Solute Compartmentation and Their Applications in Biotechnology" B. Marin, ed., Plenum Publishing Corporation, New-York.

Boller, T., 1985, Intracellular transport of metabolites in protoplasts : Transport between cytosol and vacuole, in : "The Physiological Properties of Plant Protoplasts", P. E. Pilet, ed., Springer-Verlag, Berlin.

Briskin, D. P., Thornley, W. R., and Wyse, R. E., 1985, Membrane transport in isolated vesicles from sugarbeet taproot. II. Evidence for a sucrose/H^+ antiport, Plant Physiol., 78:871.

Brown, S. C., and Coombe, B. G., 1984, Proposal for hexose group transport at the tonoplast of grape pericarp cells, Physiol. Vég., 22:231.

Buser-Suter, C., Wiemken, A., and Matile, P., 1982, A malic acid permease in isolated vacuoles of a crassulacean acid metabolism plant, Plant Physiol., 69:456.

Courtois, D., and Guern, J., 1980, Tryptamine uptake and accumulation by Catharanthus roseus cells cultivated in liquid medium, Plant Science Letters, 18:85.

Deus-Neumann, B., and Zenk, M. H., 1984, A highly selective alkaloid uptake system in vacuoles of higher plants, Planta, 162:250.

Deus-Neumann, B., and Zenk, M. H., 1986, Accumulation of alkaloids in plant vacuoles does not involve an ion-trap mechanism, Planta, 167:44.

Doll, S., Rodier, F., and Willenbrink, J., 1979, Accumulation of sucrose in vacuoles isolated from red beet tissue, Planta, 144:407.

Gibrat, R., Barbier-Brygoo, H., Guern, J., and Grignon, C., 1985, Membrane potential difference of isolated plant vacuoles : Positive or negative ? I. Evidence for membrane binding of cationic probes, Biochim. Biophys. Acta, 81:206.

Guern, J., Renaudin, J. P., and Brown, S. C., 1987, The compartmentation of secondary metabolites in plant cell cultures, in : "Cell Culture in Phytochemistry", I. K. Vasil and F. Constabel, eds., Academic Press, in press.

Guy, M., Reinhold, L., and Michaeli, D., 1979, Direct evidence for a sugar transport mechanism in isolated vacuoles, Plant Physiol., 64:61.

Homeyer, B. C., and Roberts, M. F., 1984, Alkaloid sequestration by Papaver somniferum latex, Z. Naturforsch., 39 c:876.

Kaiser, G., and Heber, U., 1984, Sucrose transport into vacuoles isolated from barley mesophyll protoplasts, Planta, 161:562.

Kurkdjian, A. C., 1982, Absorption and accumulation of nicotine by Acer pseudoplatanus and Nicotiana tabaccum cells, Physiol. Vég., 20:73.

Leigh, R. A., 1983, Methods, progress and potential for the use of isolated vacuoles in studies of solute transport in higher plant cells, Physiol. Plant., 57:390.

Manigault, P., Manigault, J., and Kurkdjian, A. C., 1983, A microfluorimetric method for vacuolar pH measurement in plant cells using 9-aminoacridine, Physiol. Vég., 21:129.

Marin, B., and Chréstin, H., 1985, Compartmentation of solutes and the role of tonoplast ATPase in Hevea latex, in : "Biochemistry and Function of Vacuolar Adenosine-triphosphatase in Fungi and Plants", B. P. Marin, ed., Springer-Verlag, Berlin, Heidelberg, New-York and Tokyo.

Martinoia, E., Flügge, I., Kaiser, G., Heber, U., and Heldt, H. W., 1985, Energy-dependent uptake of malate into vacuoles isolated from barley mesophyll protoplasts, Biochim. Biophys. Acta, 806:311.

Matern, U., Reichenbach, C., and Heller, W., 1986, Efficient uptake of flavanoids into parsley (Petroselinum hortense) vacuoles requires acylated glycosides, Planta, 167:183.

Matile, P., 1976, Localization of alkaloids and mechanism of their accumulation in vacuoles of Chelidonium majus laticifers, Nova Acta Leopold., Suppl. 7:139.

Müller, E., 1976, Principles in transport and accumulation of secondary products, Nova Acta Leopold., Suppl. 7:123.

Nagakawa, K., Konagai, A., Fukui, H., and Tabata, M., 1984, Release and crystal-lization of berberine in the liquid medium of Thalictrum minus cell suspension cultures, Plant Cell Reports, 3:254.

Neumann, D., Krauss, G., Hieke, M., and Gröger, D., 1983, Indole alkaloid formation and storage in cell suspension cultures of Catharanthus roseus, Planta Me-dica, 48:20.

Ohsumi, Y., and Anraku, Y., 1981, Active transport of basic amino-acids driven by a proton-motive force in vacuolar membrane vesicles of Saccharomyces cerevisiae, J. Biol. Chem., 256:2079.

Rataboul, P., Alibert, G., Boller, T., and Boudet, A. M., 1985, Intracellular transport and vacuolar accumulation of o-coumaric acid glucoside in Melilotus alba mesophyll cell protoplasts, Biochim. Biophys. Acta, 816:25.

Renaudin, J. P., and Guern, J., 1982, Compartmentation mechanisms of indole alkaloids in cell suspension cultures of Catharanthus roseus, Physiol. Vég., 20:533.

Renaudin, J.-P., Brown, S. C., and Guern, J., 1985, Compartmentation of alkaloids in a cell suspension of Catharanthus roseus : A reappraisal of the role of pH gradients, in : "Primary and Secondary Metabolism of Plant Cell Cultu-res", K. H. Neumann, W. Barz, and E. Reinhard, eds., Springer-Verlag, Ber-lin, Heidelberg, New-York and Tokyo.

Renaudin, J. P., and Guern, J., 1987, Ajmalicine transport into vacuoles isolated from Catharanthus roseus cells, in : "Plant Vacuoles. Their Importance in Solute Compartmentation and Their Applications in Biotechnology", B. Marin, ed., Plenum Publishing Corporation, New-York.

Sharma, V., and Strack, D., 1985, Vacuolar localization of 1-sinapoylglucose:L-mala-te-sinapoyltransferase in protoplasts from cotyledons of Raphanus sativus, Planta, 163:563.

Strack, D., and Sharma, V., 1985, Vacuolar localization of the enzymatic synthesis of hydroxycinnamic acid esters of malic acid in protoplasts from Raphanus sativus leaves, Physiol. Plant., 65:45.

Teulières, C., Alibert, A., and Ranjeva, R., 1985, Reversible phosphorylation of tonoplast proteins involves tonoplast-bound calcium-calmodulin-dependent protein kinase(s) and protein-phosphatase(s), Plant Cell Reports, 4:199.

Thom, M., and Komor, E., 1984, H^+-sugar antiport as the mechanism of sugar uptake by sugarcane vacuoles, FEBS Letters, 173:1.

Thom, M., Leigh, R. A., and Maretzki, A., 1986, Evidence for the involvement of a UDP-glucose-dependent group translocator in sucrose uptake into vacuoles of storage roots of red beet, Planta, 167:410.

Thom, M., and Maretzki, A., 1985, Group translocation as a mechanism for sucrose transfer into vacuoles from sugarcane cells, Proc. Natl. Acad. Sci. U.S.A., 82: 4697.

Werner, C., and Matile, P., 1985, Accumulation of coumarylglucosides in vacuoles of barley mesophyll protoplasts, J. Plant Physiol., 118:237.

Willenbrink, J., and Doll, S., 1979, Characteristics of the sucrose uptake system of vacuoles isolated from red beet tissue. Kinetics and specificity of the sucrose uptake system, Planta, 147:159.

VACUOLE FUNCTION IN NEUROSPORA

Richard L. Weiss

Department of Chemistry and Biochemistry
University of California
Los Angeles, California 90024, USA

INTRODUCTION

The vacuoles of fungi share many properties with the vacuoles of higher plants and the lysosomes of mammalian cells. Vacuoles of higher plants are thought to perform a storage function whereas mammalian lysosomes are the sites of turnover of intracellular macromolecules. Both these functions are performed by fungal vacuoles and both require the movement of molecules between the vacuolar (lysosomal) and cytoplasmic compartments. To understand the physiological functions of vacuoles and the mechanism by which these functions are performed, a simple system amenable to genetic, biochemical and molecular biological manipulation is needed. The vacuoles of the filamentous fungus, Neurospora crassa, appear to satisfy these criteria.

In Neurospora crassa, the vacuoles contain proteases and nucleases as well as a number of basic amino-acids (Vaughn and Davis, 1981). We have chosen to investigate arginine metabolism and its sequestration in the vacuoles as a model for understanding the function of vacuoles. Arginine can be synthesized de novo from glutamate. Under appropriate conditions, arginine can be degraded and used by the organism as a source of nitrogen. Large amounts of arginine are produced and sequestered in the vacuoles (Weiss, 1973). Thus, arginine metabolism in Neurospora crassa involves the movement of arginine across the tonoplast and such movement can affect the pattern of arginine metabolism (Davis, 1986). The mechanism of arginine transport across the tonoplast and the function of vacuolar sequestration of arginine provides an excellent model system for studying the role of the vacuoles in the physiology of eukaryotic organisms.

The relevant enzymatic steps and transmembrane movements of arginine metabolism in Neurospora crassa are shown in Fig. 1. Arginine is synthesized from glutamate by a series of reactions which culminate in the production of arginine in the cytoplasm. The organism normally synthesizes more arginine than is necessary to support optimal growth (protein synthesis) and the excess arginine is sequestered in the vacuoles. This sequestration effectively prevents the catabolism of arginine by maintaining an extremely low concentration of arginine in the cytoplasm – the intramycelial site of arginine catabolism. Arginase, the initial enzyme of the arginine degradation pathway, is present at significant levels in mycelia growing in unsupplemented medium. However, under these conditions, the cytoplasmic arginine concentration is low enough to minimize degradation but sufficient to support maximal rates of protein synthesis (Subramanian et al., 1973). Movement of arginine across the tonoplast can influence arginine metabolism by controlling the cytoplasmic arginine concentration.

315

The sequestration of arginine poses a number of interesting biological questions. First, how is arginine accumulated in the vacuoles ? Second, what role does this sequestration perform in the life cycle of the organism ? In order to address these questions, a variety of experimental approaches have been initiated. Current progress towards answers to these questions is summarized below.

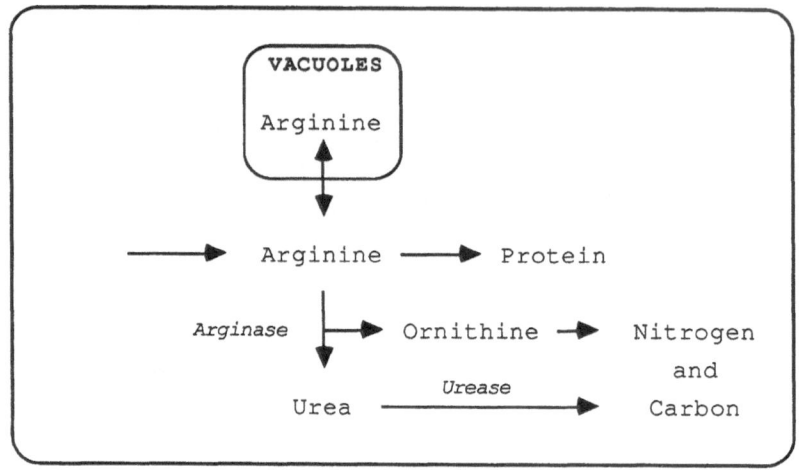

Figure 1

Diagram showing
the relevant biosynthetic and degradative reactions of arginine metabolism,
important enzymes, and transmembrane movement in Neurospora crassa

RESULTS AND DISCUSSION

In order to assess the biological significance of vacuolar compartmentation of arginine, a simple means of measuring the distribution between cytosolic and vacuolar compartments was needed. Upon degradation, arginine yields one molecule of ornithine and one molecule of urea. Mutant strains lacking the enzyme urease (see Fig. 1) accumulate one molecule of urea for each molecule of arginine degraded. Any arginine which accumulates in the cytoplasm will be degraded since arginase is always present at significant levels (Weiss, 1976). Thus, urea accumulation in urease-deficient strains serves as a sensitive measure of cytoplasmic arginine concentration and can be used to detect the accumulation of cytoplasmic arginine, efflux of arginine from the vacuoles, or the failure of cytoplasmic arginine to be taken-up by the vacuoles. This technique was exploited to explore the energetics of vacuolar compartmentation in vivo and the effects of nutritional stress on the sequestration of arginine.

In mycelia growing in the presence of exogenous arginine, approximately 20% of the intramycelial arginine is present in the cytoplasm and degradation is rapid (Weiss, 1976). Upon removal or exhaustion of the exogenous arginine, catabolism ceases abrutly as arginine is taken-up into the vacuoles (Weiss and Davis, 1977). If uptake of arginine into the vacuoles is energy-dependent, then inhibitors or uncouplers of oxidative phosphorylation should prevent such uptake and the cytoplasmic arginine should be degraded. If only uptake is affected, then only the cytoplasmic fraction (20%) of the intramycelial arginine should be degraded. If energy is required for the vacuoles to retain their arginine content, then such inhibitors or uncouplers should cause release of arginine from the vacuoles and result in degradation of both the cytoplasmic and vacuolar arginine pools. Mycelia growing in arginine-supplemented medium were exposed to ^{14}C-arginine to label the cytoplasmic and vacuolar arginine pools and then treated with inhibitors or uncouplers at time zero. Samples were removed at various times and analyzed for radioactive arginine (cytoplasmic plus vacuolar). The results of such experiments are shown in Fig. 2.

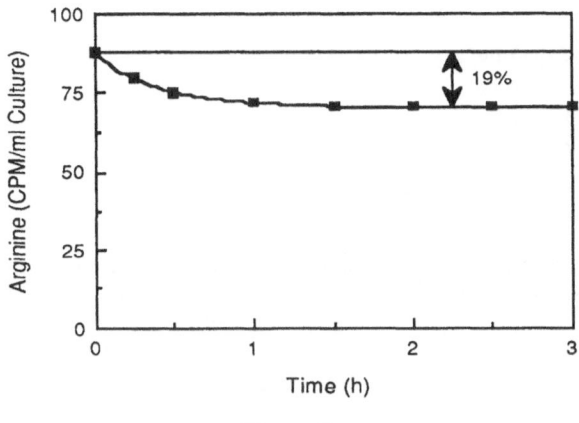

Figure 2

Effect of energy depletion on metabolism
of the intramycelial arginine pools:
cytoplasmic (20%) and vacuolar (80%)

Approximately 19% of the intramycelial arginine was degraded when mycelia growing in arginine supplemented medium were treated with inhibitors or uncouplers of oxidative phosphorylation. This led to the conclusion that energy is required for uptake of arginine into the vacuoles but not for its retention (Drainas and Weiss, 1982 a).

To further characterize the energy-linked uptake of arginine into vacuoles, membrane vesicles derived from isolated vacuoles were used to study arginine transport in vitro. Uptake of arginine was shown to be dependent on ATP hydrolysis. Inhibitors which are specific for the vacuolar membrane ATPase prevented the energy-dependent transport of arginine. Inhibitors of the mitochondrial or plasma membrane ATPases were without effect. These observations are consistent with an energy-linked uptake system mediated by the electrochemical gradient produced in response to the hydrolysis of ATP by the tonoplast ATPase (Zerez et al., 1986; Bowman and Bowman, 1982).

Unlike the arginine transport systems described for membrane vesicles derived from vacuoles of Saccharomyces cerevisiae, the arginine transport system in Neurospora crassa appears to be highly stereospecific. The effect of a variety of compounds on the uptake of radioactive arginine is summarized in Table 1.

Neither lysine nor ornithine inhibited arginine transport suggesting an essential role for the guanido group in recognition by the membrane carrier. D-arginine was slightly inhibitory. Both L-arginine methyl ester and tosyl-L-arginine methyl ester inhibited uptake significantly. These results suggested that the membrane carrier is stereospecific with its primary recognition for the side chain of arginine. A diagrammatic representation of the specificity of the postulated vacuolar arginine carrier is shown in Fig. 3.

317

Table 1

Specificity of Arginine Transport[a]

Inhibitor	Transport (%)
None	100
L-Arginine	26
D-Arginine	57
L-Lysine	89
L-Ornithine	115
L-Arginine methyl ester	29
Tosyl-L-arginine	18

[a]Uptake of arginine into vacuolar membrane vesicles was performed as previously described (Zerez et al., 1986). The concentration of L-^{14}C-arginine was 0.3 mM and the concentration of the inhibitor was 10 mM.

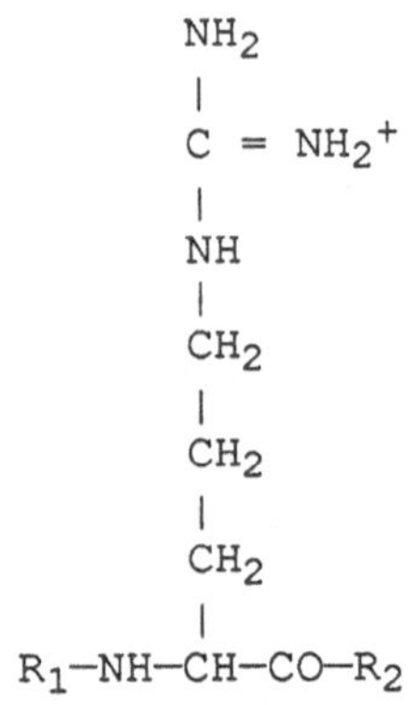

$$
\begin{array}{c}
NH_2 \\
| \\
C = NH_2^+ \\
| \\
NH \\
| \\
CH_2 \\
| \\
CH_2 \\
| \\
CH_2 \\
| \\
R_1-NH-CH-CO-R_2
\end{array}
$$

Figure 3

Binding specificity of arginine carrier

Inhibitors or uncouplers which result in complete loss of intramycelial ATP failed to result in arginine degradation (Drainas and Weiss, 1982 a). This suggested that an energy-independent mechanism must exist to prevent the equilibration of arginine across the tonoplast. We considered a variety of methods to examine the physical state of vacuolar arginine - i.e. its environment and/or binding to macromolecules. In situ nuclear resonance spectroscopy appeared to offer the best means of achieving these goals. Because binding interactions were likely to involve the nitrogen atoms of the arginine, we explored the possibility of utilizing ^{15}N-NMR on intact mycelia of Neurospora crassa. Using a wide-bore spectrometer, we were able to obtain ^{15}N-NMR spectra from intact mycelia of Neurospora crassa which had been labeled with ^{15}NH$_4$Cl (Legerton et al., 1981). Distinct resonances were observed for arginine, both the alpha amino and guanido groups as well as the alpha amino groups of a number of other amino-acids and the amide nitrogen of glutamine.

Proton-coupled spectra were used to determine the pH of various intramycelial compartments (Legerton et al., 1983). A value of pH = 6.1 was obtained for the compartment in which most of the arginine is found (the vacuole). A value of pH = 7.1 was obtained for the environment of intramycelial glutamine and alanine. This is consistent with previous measurements of cytoplasmic and vacuolar pH and the known localization of these amino-acids. These observations confirmed that our spectra were actually observing the appropriate intramycelial environments.

A second nuclear magnetic resonance parameter, the spin lattice relaxation time, $T_{1,dd}$, allowed us to examine the intravacuolar environment and possible binding of arginine to macromolecules. The spin lattice time is inversely proportional to the viscosity (η) of the environment and the cube of the radius (α) of the molecule:

$$1/T_{1,dd} \approx \tau_c = 4\pi\eta\alpha^3/3kT$$

The results of such measurements are summarized in Table 2.

Table 2

Spin-Lattice Relaxation Times for the Nitrogens of Amino-acids
in Aqueous Solution and within Mycelia of <u>Neurospora crassa</u>[a]

	Viscosity (cp)	<u>In vitro</u> T_1 (sec)	<u>In vivo</u> T_1 (sec)
Glutamine N_γ	1.3	4.9	4.1
Arginine $N_{\omega,\omega'}$	1.3	4.6	1.1
Arginine + glycerol	2.8	1.8	
Arginine + glycerol	4.4	1.1	
Arginine + polyphosphate	2.8	1.0	

[a]Data from Kanamori et al. (1982)

The spin lattice relaxation time of intramycelial arginine was considerably shorter than that observed with arginine in aqueous solution. As a control, the spin lattice relaxation time of intramycelial glutamine was also determined. Glutamine is primarily localized in the cytoplasm and therefore should be indicative of the physical properties of this compartment. The results indicate that glutamine exists in an environment similar to that found in a simple aqueous solution. On the other hand, arginine exhibited a considerably shorter relaxation time than that observed in aqueous solution. The smaller value could be mimicked <u>in vitro</u> by the addition of glycerol to the arginine-containing solution or by the addition of polyphosphates. The latter form a complex with arginine (Cramer and Davis, 1984), and the spin lattice relaxation time of the resulting complex is decreased due to an increase in effective molecular size.

A model consistent with the observations described above is shown in Fig. 4. Arginine is taken-up into the vacuoles by a specific carrier in response to the electrochemical gradient produced by the vacuolar ATPase. Once inside the vacuole, arginine is removed from effective equilibrium with "free" arginine, possibly by association with the polyanion, inorganic polyphosphate. In the absence of a membrane potential, arginine will not efflux from the vacuoles because the "free" concentration within this compartment is very low.

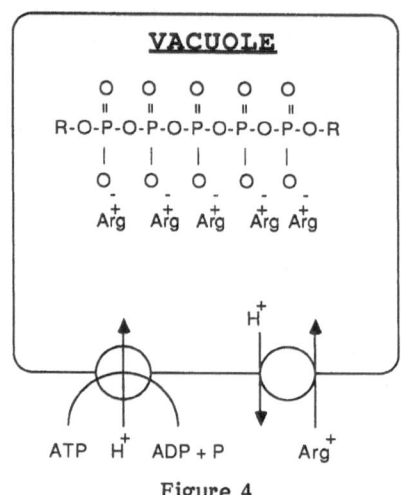

Figure 4

A diagrammatic representation
of the mechanism responsible
for the accumulation and retention of arginine
within the vacuoles of Neurospora crassa

Although consistent with the formation of a macromolecular complex, the spin lattice relaxation data cannot rule-out the possibility that the intravacuolar environment is highly viscous. This would be most consistent with the suggestion that insufficient polyphosphate exists in the vacuole to complex the myriad of cations found in this intramycelial compartment and the failure to detect significant quantities of any other polyanion (Cramer and Davis, 1984).

The results described above fail to provide an answer to the biological question : What is the function of the sequestration of arginine in the vacuoles of Neurospora crassa ? One approach to answering this question is to determine the conditions under which intravacuolar arginine is released from the vacuoles and the metabolic fate of the arginine so released. The metabolism of arginine gives rise, ultimately, to glutamate, ammonia and CO_2. This suggested that mobilization (efflux from the vacuoles) of arginine might provide an alternative source of nitrogen for the organism. To test this possibility, various mutant strains were deprived of essential nutrients - e.g. carbon, nitrogen, amino-acids, etc. and the mobilization of vacuolar arginine was followed by the appearance of urea. (All strains carried the urease-deficient mutation). The results of some of these experiments are summarized in Fig. 5.

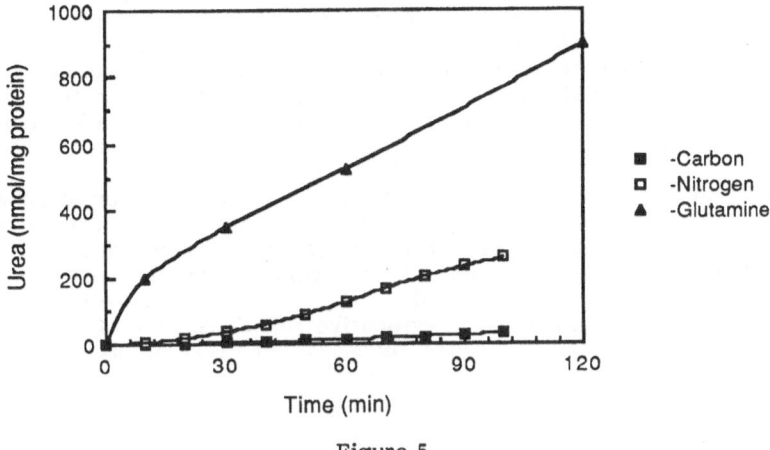

Figure 5

Catabolism of arginine (urea production)
by Neurospora crassa
in response to nutrient deprivation

Urease-deficient strains of Neurospora crassa were starved for carbon, nitrogen or glutamine by transferring appropriate mutant strains to medium lacking carbon, nitrogen or glutamine at zero time.

Nitrogen deprivation resulted in the rapid appearance of urea (arginine catabolism) without concomitant changes in the level of arginase (Legerton and Weiss, 1979). Glutamine starvation resulted in even more rapid and extensive degradation of the arginine pool (98% originally present in the vacuoles). Addition of inhibitors of the vacuolar ATPase to mycelia starving for nitrogen caused an immediate cessation of arginine mobilization (Drainas and Weiss, 1982 b). Degradation was shown to result from redistribution of arginine between the vacuolar and cytoplasmic compartments; the metabolic signal responsible for this mobilization appears to be a decline in the intramycelial pool of glutamine or a metabolite derived from glutamine (Legerton and Weiss, 1984). These results suggest that mobilization or the efflux of arginine from the vacuoles is an energy-dependent process and that the vacuolar sequestration of arginine provides a reserve of nitrogen to be utilized under conditions of nitrogen deprivation.

The experiments described above are consistent with the hypothesis that Neurospora crassa synthesizes excess arginine and sequesters this excess in the vacuoles to provide a reserve of nitrogen for utilization in times of nutritional stress. Interestingly, Neurospora crassa undergoes sporulation in response to nutritional stress – nitrogen deprivation being the most effective stimulus. Differentiation involves the formation of aerial hyphae which must grow and subsequently differentiate into conidia (spores) spatially removed from the normal source of nutrients – the decaying vegetation which supports the growth of vegetative hyphae. Vacuoles can move through the hyphae via cytoplasmic streaming and would provide a convenient mechanism for transporting stored reserves from vegetative hyphae to the tips of growing and differentiating aerial hyphae. Thus vacuoles might serve as a convenient transportation vehicle. This hypothesis requires further investigation.

FUTURE DIRECTIONS

One means of establishing the physiological function of a biological process is to characterize the phenotype of mutants defective in the process. A difficulty in isolating mutants defective in vacuole function is that selection requires the prediction of a mutant phenotype. We are approaching this problem in two ways. First, we have attempted to isolate mutants directly by predicting one possible phenotypic consequence of the inability of vacuoles to accumulate arginine. Second, we hope to identify the arginine carrier, clone its structural gene and create mutants by DNA manipulation and gene replacement. Our progress is summarized below.

Arginine degradation and proline synthesis are interrelated as shown in Fig. 5. Mutants blocked in the synthesis of GSA (glutamate semialdehyde) are proline auxotrophs because arginine and ornithine are not usually degraded; they are efficiently sequestered in the vacuoles. We hypothesized that mutants unable to sequester arginine and ornithine might "suppress" proline auxotrophic mutations by diverting these compounds into degradative reactions. This rationale led to the isolation of mutants (arg-6) in which feedback inhibition of arginine biosynthesis is impaired (Weiss and Lee, 1980).

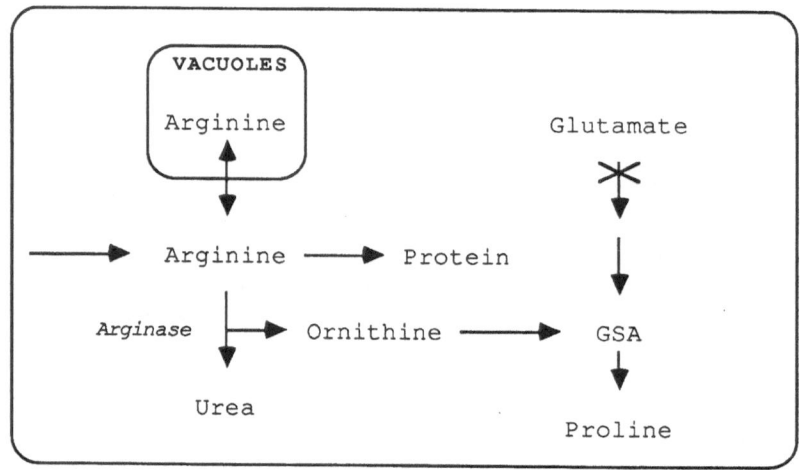

Figure 5

Relationship between the pathways of arginine degradation
and proline biosynthesis
in Neurospora crassa

We have recently isolated 500 additional mutants from a PRO⁻ background by their ability to grow in the absence of proline. To identify strains carrying mutations affecting feedback inhibition, we took advantage of the linkage of this gene (arg-6) to the al-2 mutation which renders the resulting strains white in color. Suppressors unlinked to arg-6 were identified by the absence of linkage to al-2. Sixteen isolates have been obtained and are being characterized both genetically and biochemically for defects in vacuolar compartmentation.

A second method has been designed to avoid the problems inherent in "guessing" possible phenotypes of mutants impaired in vacuolar compartmentation. The general approach is to "construct" the appropriate mutant strain by in vitro mutagenesis and gene replacement. In order to isolate the structural gene for the arginine carrier, we plan to isolate the carrier, prepare antibodies to it, and use the antibodies to identify fusion proteins containing a portion of the carrier's coding region. Normally, we would simply purify the protein using its "activity" to detect it during purification. Because of the low amounts of carrier present and the possibility that its arginine binding capacity might be lost following extraction from the tonoplast, we have attempted to label the protein to facilitate its isolation.

The structural specificity of the postulated arginine carrier has been used to design an active site directed photolabile arginine analog for identification of the specific membrane carrier. This derivative is shown in Fig. 6; it has been shown to competitively inhibit arginine transport into vacuolar membrane vesicles.

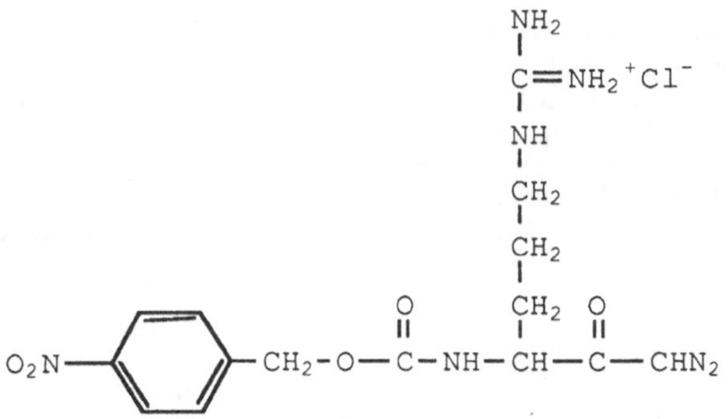

Figure 6

Structure of photosensitive active-site directed inhibitor
of the vacuolar arginine carrier

The derivative has been synthesized in radioactive form from ^{3}H-arginine. Exposure to ultraviolet radiation results in the formation of a free radical which becomes covalently attached to a specific membrane protein with an apparent molecular weight of approximately 40,000 Da. We are confident that this protein is the arginine carrier. We hope to isolate the carrier in sufficiently quantity for sequence analysis and subsequent cloning. Our ultimate aim is to employ in vitro mutagenesis and gene replacement to create mutant strains unable to accumulate arginine in the vacuoles. The phenotype of such mutant organisms should provide valuable insight into the precise physiological role of this process.

REFERENCES

Bowman, E. J., and Bowman, B. J., 1982, Identification and properties of an ATPase in vacuolar membranes of Neurospora crassa, J. Bacteriol., 151:1326.
Cramer, C. L., and Davis, R. H., 1984, Polyphosphate–cation interaction in the amino-acid–containing vacuole of Neurospora crassa, 259:5152.
Davis, R. H., 1986, Compartmental and regulatory mechanisms in the arginine pathways of Neurospora crassa and Saccharomyces cerevisiae, Microbiol. Rev., 50:280.
Davis, R. H., Bowman, B. J., and Weiss, R. L., 1978, Intracellular compartmentation and transport of metabolites, J. Supramol. Struct., 9:473.
Drainas, C., and Weiss, R. L., 1982 a, Energetics of vacuolar compartmentation of arginine in Neurospora crassa, J. Bacteriol., 150:770.
Drainas, C., and Weiss, R. L., 1982 b, Energy requirement for the mobilization of vacuolar arginine in Neurospora crassa, J. Bacteriol., 150:779.

Kanamori, K., Legerton, T. L., Weiss, R. L., and Roberts, J. D., 1982, Nitrogen-15 spin-lattice relation times of amino-acids in Neurospora crassa as a probe of intracallular environment, Biochemistry, 21:4916.

Legerton, T. L., Kanamori, K., Weiss, R. L., and Roberts, J. D., 1981, ^{15}N-NMR studies of nitrogen metabolism in intact mycelia of Neurospora crassa, Proc. Natl. Acad. Sci. U.S.A., 78:1495.

Legerton, T. L., Kanamori, K., Weiss, R. L., and Roberts, J. D., 1983, Measurements of cytoplasmic and vacuolar pH in Neurospora using nitrogen-15 nuclear magnetic resonance spectroscopy, Biochemistry, 22:899.

Legerton, T. L., and Weiss, R. L., 1979, Mobilization of sequestered metabolites into degradative reactions by nutritional stress in Neurospora, J. Bacteriol., 138: 909.

Legerton, T. L., and Weiss, R. L., 1984, Mobilization of vacuolar arginine in Neurospora crassa : Mechanism and role of glutamine, J. Biol. Chem., 259:8875.

Subramanian, K. N., Weiss, R. L., and Davis, R. H., 1973, Use of external, biosynthetic, and organellar arginine by Neurospora, J. Bacteriol., 115:284.

Vaughn, L. E., and Davis, R. H., 1981, Purification of vacuoles from Neurospora crassa, Mol. Cell. Biol., 1:797.

Weiss, R. L., 1973, Intracellular localization of ornithine and arginine pools in Neurospora, J. Biol. Chem., 248:5409.

Weiss, R. L., 1976, Compartmentation and control of arginine metabolism in Neurospora, J. Bacteriol., 126:1173.

Weiss, R. L., and Davis, R. H., 1977, Control of arginine utilization in Neurospora J. Bacteriol., 129:866.

Weiss, R. L., and Lee, C. A., Isolation and characterization of Neurospora crassa mutants impaired in feedback control of ornithine synthesis, J. Bacteriol., J. Bacteriol., 141:1305.

Zerez, C. R., Weiss, R. L., Franklin, C., and Bowman, B. J., 1986, The properties of arginine transport in vacuolar membrane vesicles of Neurospora crassa, J. Biol. Chem., 261:8877.

$\Delta\overline{\mu}_{H}^{+}$-CONTROLLED REVERSIBLE FLUXES OF H$^+$ AND CALCIUM AT THE TONOPLAST BUT QUASI-TOTAL CITRATE SEQUESTRATION WITHIN THE INTACT VACUOLES FROM THE LATEX CELLS OF HEVEA BRASILIENSIS. IMPLICATIONS IN THE PRODUCTION OF NATURAL RUBBER

Hervé Chréstin, Xavier Gidrol, Michel Péan and Bernard Marin

Laboratoire de Physiologie Végétale, Centre ORSTOM
d'Adiopodoumé, Boite Postale N° V-51, Abidjan, Ivory Coast

The latex of Hevea brasiliensis is a fluid cytoplasm which is expelled from wounded latex vessels (articulated, anastomosed cells) (Archer et al., 1963). It contains a vacuolar compartment - the so-called "lutoids" - consisting of micro-vacuoles which can be easily isolated and purified by simple differential centrifugation (Pujarniscle, 1968; Ribaillier et al., 1971; D'Auzac et al., 1982).

Like all plant vacuoles, lutoids exhibit a lower internal pH (about 5.5) than that of their cytosolic environment (about 7.0). They accumulate, in vitro and in vivo, numerous mineral and organic cations such as Mg^{2+}, Ca^{2+}, Cu^{2+}, etc ., and basic amino-acids, as well as anions such as inorganic phosphate and citrate (Ribaillier et al., 1971; D'Auzac and Lioret, 1974; Brzozowska et al., 1974).

Production of latex reflects the intensity of the metabolism within these specialized laticiferous cells. Indeed this regenerative metabolism must be sufficiently active to compensate for the loss of latex (50 to 300 ml or more with a mean value of 35% dry rubber content) at each tapping (generally twice a week).

Rubber production has been shown to be correlated positively with the pH of the cytosol of the latex cells (Table 1), and negatively with the intravacuolar pH (Coupé and Lambert, 1977; Brzozowska-Hanover et al., 1979; Chréstin, 1985). Furthermore, a highly significant inverse relationship was demonstrated between the pH of the cytosolic compartment and the changes in intravacuolar pH (lutoidic pH), strongly suggesting the existence of vectorial H$^+$ fluxes at the level of the lutoidic tonoplast (Table 1) (Brzozowska-Hanower et al., 1979; Chréstin, 1985).

Furthermore, multivariate analysis showed that the latex from high yielding rubber-trees was characterized not only by a slightly alkaline cytosolic pH and a high transtonoplastic H$^+$ gradient, but also by a pronounced accumulation of citrate in the vacuoles, resulting in a high transtonoplastic gradient of citrate, i.e. a low citrate concentration in the cytosol (Table 1).

These relationships were satisfactorily explained by the extreme pH sensitivity (in the physiological pH range) of numerous key enzymes of the cytosolic metabolism

Table 1

Correlation Coefficients
Linking the Latex Production (g dry rubber/tapping/tree),
Cytosolic pH, Transtonoplastic pH Gradient,
Cytosol and Vacuole Citrate Concentrations (mM) in Latex
and the Resulting Transtonoplastic Citrate Gradient

	Latex Production	Vacuolar (Citrate)	Cytosolic (citrate)	(Citrate) Gradient	Cytosolic pH	Transtonoplastic pH Gradient
Latex Production	1	+ 0.562 ***	− 0.768 ***	+ 0.755 ***	+ 0.894 ***	+ 0.822 ***
Vacuolar (Citrate)		1	− 0.369 **	+ 0.778 ***	+ 0.675 ***	+ 0.640 ***
Cytosolic (Citrate)			1	− 0.678 ***	− 0.705 ***	− 0.752 ***
(Citrate) Gradient				1	+ 0.765 ***	+ 0.800 ***
Cytosolic pH					1	+ 0.935 ***
Transtonoplastic pH Gradient						1

Data were obtained from freshly collected latex from 56 rubber trees, (***):
very high significance; (**): high significance.

and their inhibition by certain ions such as Mg^{2+}, Ca^{2+}, citrate, etc., at physiological concentrations (Jacob and D'Auzac, 1967, 1969 and 1972; D'Auzac and Jacob, 1969; Tupy, 1969; Jacob et al., 1979; Chrestin et al., 1984).

All these data led to seeking a mechanism able to control transtonoplastic fluxes of protons and citrate, and therefore wondering about the role of the vacuolar compartment in the control of the cytosolic metabolism within the latex cells of Hevea.

TWO OPPOSING H^+-TRANSLOCATING SYSTEMS AT THE TONOPLAST

H^+ accumulation within intact latex vacuoles was shown to originate from two complementary processes :

- a large, nearly constant pool of protons (accounting for up to 1 pH unit of the transtonoplastic pH gradient) remains sequestered at the thermodynamic equilibrium within the vacuolar compartment, owing to the existence of a transtonoplastic Donnan potential (Crétin, 1982),

- transtonoplastic H^+ fluxes, which determine the cytosolic as well as the vacuolar pH changes (accounting for 0.1 to 1 pH unit of the total transtonoplastic pH gradient), are shown to be under the control of two opposing H^+ translocating systems both located at the level of the tonoplast.

The Inward Proton Pumping Activity of Tonoplast ATPase

The first is the tonoplastic ATPase dependent on Mg^{2+}, revealed and partially described by D'Auzac (1975 and 1977) and more carefully described by Marin and his group (Marin, 1985; Marin et al., 1985 and 1986 a and b). This constitutive membrane ATPase works as a proton pump, catalysing H^+ influx into the vacuole (vacuolar acidification shown by the accumulation of the ΔpH probe ^{14}C-methyl-amine in the vacuolar compartment), causing the alkalinization of the cytosol and an increase in the transtonoplastic pH gradient. As expected, the ATP-dependent transtonoplastic H^+ fluxes were shown to be inhibitable by protonophores such as FCCP (Fig. 1-A and B) (Marin et al., 1981; Crétin, 1982; Crétin et al., 1982; Marin, 1982; Chréstin, 1985; Marin, 1985).

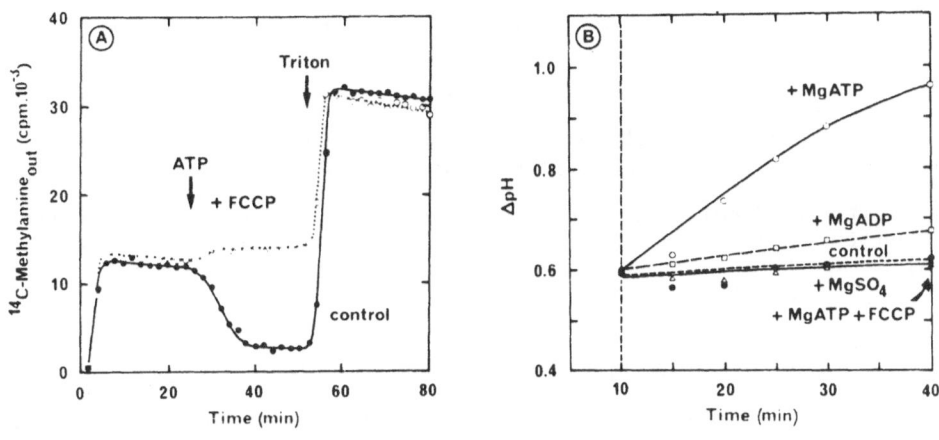

Figure 1

Change in time of ΔpH across the tonoplast (inside acid)
in the presence of ATP and effects of FCCP

ΔpH changes were monitored by following the accumulation of the ΔpH probe ^{14}C-methylamine, using either the flow dialysis technique with intact lutoids (Fig. 1-A) or centrifugation with reconstituted tonoplast vesicles (Fig. 1-B).

Furthermore, in the absence of any energy supply, intact lutoids (Crétin, 1982; Chréstin, 1985) as reconstituted tightly-sealed tonoplast vesicles (Marin, 1982 and 1985 a and b; Marin et al., 1981 and 1985) accumulated $^{86}Rb^+$ (in the presence of valinomycin) and the cationic phosphonium probes, indicating a transmembrane electrical potential difference which was a more negative inside. The addition of MgATP to both materials results in a rapid change in the distribution of the potential probes, corresponding to a transmembrane depolarization (Fig. 2-A and B).

All these results confirmed by the use of fluorescent probes (Marin , unpublished data) led to the conclusion that the Mg–dependent ATPase located on the lutoidic tonoplast works as an electrogenic proton-pump, setting up a high (inside more positive) electrochemical gradient ($\Delta\overline{\mu}_{H}^+$) across the lutoidic tonoplast, then against the thermodynamic equilibrium.

Finally, the proton-pumping ATPase of the lutoidic tonoplast has more recently been fully described and solubilized (Gidrol et al., 1985; Marin et al., 1985 and 1986).

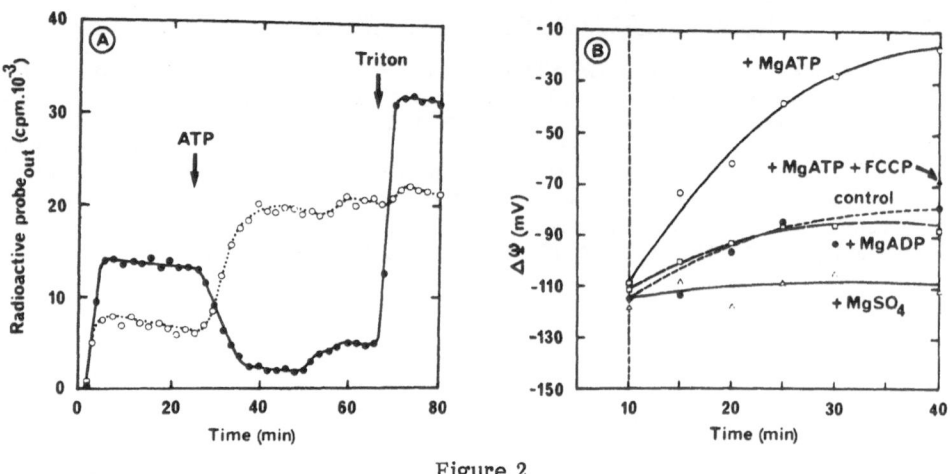

Figure 2

Change in time of the transtonoplastic electrical potential ($\Delta\Psi$)
in the presence of Mg^{2+}-ATP

$\Delta\Psi$ was monitored either following the accumulation of ^{86}Rb in the presence of valinomycin using flow dialysis technique with intact lutoids (open circles, Fig. 2-A) or lipophilic cation probe ^{14}C-TPP^+ with reconstituted tonoplast vesicles using centrifugation (Fig. 2-B).

The Outward H^+-Translocating Activity of a Tonoplastic Redox System

A second H^+-translocating moiety has been revealed on the lutoidic tonoplast. It consists of a NADH-cytochrome $\underline{c}$ (artificial acceptor)-oxido-reductase (Crétin, 1983), perharps the same as the one, including cytochromes b-type, revealed by Moreau et al. (1975). Isopycnic centrifugation experiments confirmed the tonoplastic location of this H^+-pumping redox system, the activity of which closely followed the distribution of the typical lutoidic acid phosphatase and tonoplastic ATPase activities throughout the density gradient profile (Fig. 3) (Chréstin, 1985). The functioning of this redox chain induces H^+ efflux from freshly isolated intact latex vacuoles resulting in acidification of the cytosol and a collapse of transtonoplastic pH gradient (Fig. 4-A) (Crétin, 1983).

The H^+-translocating redox system was shown to work electrogenically and to cause membrane hyperpolarization (inside more negative : Influx of Rb^+) (Fig. 4-B), leading to the collapse of the transtonoplastic electrochemical gradient of protons (Chréstin, 1985).

The partial characterization of this redox H^+-pump showed that it is insensitive towards the classic inhibitors of the cytochromic respiratory chains (KCN, Antimycin A, etc.) and also those of the mitochondrial alternate pathway (Chréstin, 1983 and 1985).

ENERGIZATION OF SOLUTE TRANSPORT AT LATEX CELL TONOPLAST BY THE TWO ELECTROGENIC H^+-PUMPS

The differential functioning of these two opposing H^+-pumps at the latex cells tonoplast is then able to modulate the transtonoplastic electrochemical proton gradient. The latter was shown to energize numerous solute transports through the lutoidic tonoplast (Marin et al., 1982; Marin and Chréstin, 1985). As models, this paper focuses in particular on the processes involved in Ca^{2+} and citrate transport and accumulation within latex cell vacuoles.

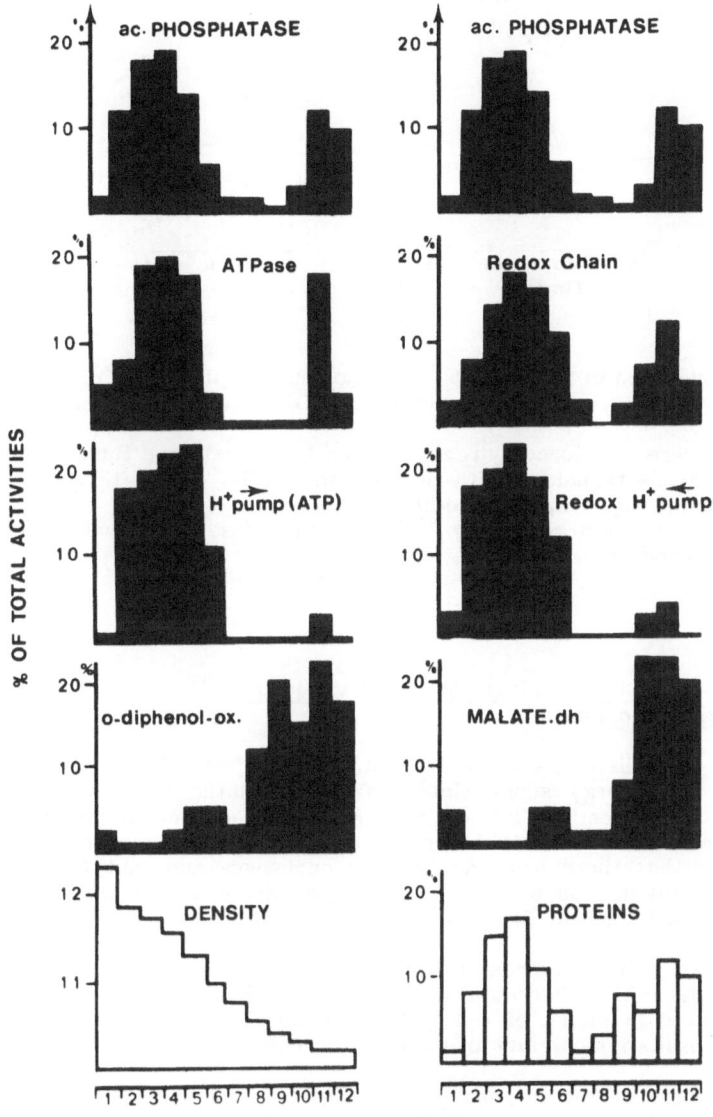

FRACTIONS

Figure 3

Localization of the proton-pumping activities
by density gradient centrifugation
of the latex organelles

An isotonic suspension of the bottom fraction from freshly centrifuged latex was layered on the top of the continuous sucrose gradient (0.6 to 1.8 M sucrose) containing 0.3 M mannitol, 50 mM Hepes-Tris at pH 7.0, and centrifuged 75 min. at 95,000 x g at 8°C. The different fractions were analyzed for their enzymatic and H^+-pumping activities according to Crétin (1982 and 1983).

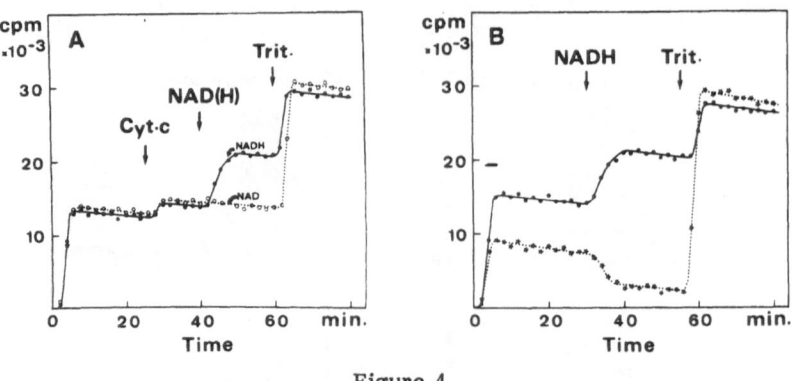

Figure 4

Evolution in time of the transtonoplastic ΔpH and $\Delta\Psi$ changes
during the working of the lutoidic NADH–cytochrome c reductase

ΔpH changes were monitored by recording the transmembrane fluxes of methylamine
using flow dialysis technique with intact lutoids (Fig. 4-A, with NADH: dots, with
NAD: circles; and Fig. 4-B, dots and line).
$\Delta\Psi$ changes were monitored using $^{86}Rb^+$ in the presence of valinomycin (Fig. 4-
B: stars and dotted line).

$\Delta\mu H^+$-controled Reversible Fluxes of "Free" vacuolar Ca^{2+}
At the Lutoidic Tonoplast

Using flow dialysis method (Crétin, 1982), it was observed that even in the
absence of any energy supply intact freshly isolated lutoids could intensively
accumulate Ca^{2+} (freed by lutoid lysis). Since the addition of divalent cations (up
to 2.5 mM Ca^{2+} or 12.5 mM Mg^{2+}) did not lead to release of the accumulated $^{45}Ca^{2+}$,
we concluded that there were no isotopic exchanges and that Ca^{2+} retention by
the lutoids could not be attributed to major adsorption on the external surface
of the membrane (Chréstin, 1985; Marin et al., 1982).

Changes in the external pH induced rapid movement of Ca^{2+} across the
tonoplast (Fig. 5-A) : Decrease in transtonoplastic ΔpH by acidification of the
medium resulted in efflux of calcium while increase in ΔpH by alkalinization of
the medium resulted in calcium influx in the vacuoles.

The supply of the suspension with Mg-ATP, which induces an intravacuolar
acidification, and hence an supplementary increase in the transtonoplastic $\Delta\overline{\mu_H}^+$,
led to simultaneous accumulation of $^{45}Ca^{2+}$ within the lutoids (Fig. 5-A and B)
(Chréstin et al., 1984).

Ionophores such as nigericin or FCCP had no significant effect on Ca^{2+}
fluxes when added to a vacuolar suspension kept in the resting state (no ATP-energized $\Delta\overline{\mu_H}^+$), but when added after the energization of the vacuolar Ca^{2+} uptake
by MgATP, these protonophores were shown to induce a significant efflux of Ca^+
from the vacuoles, equivalent to the level of the ATP-stimulated uptake (Fig. 5-
B) (Marin et al., 1982; Chréstin, 1985).

The addition of the non-electrogenic protonophore NH_4Cl to a non ATP-
energized suspension, which at least in part decreases the non energized ΔpH

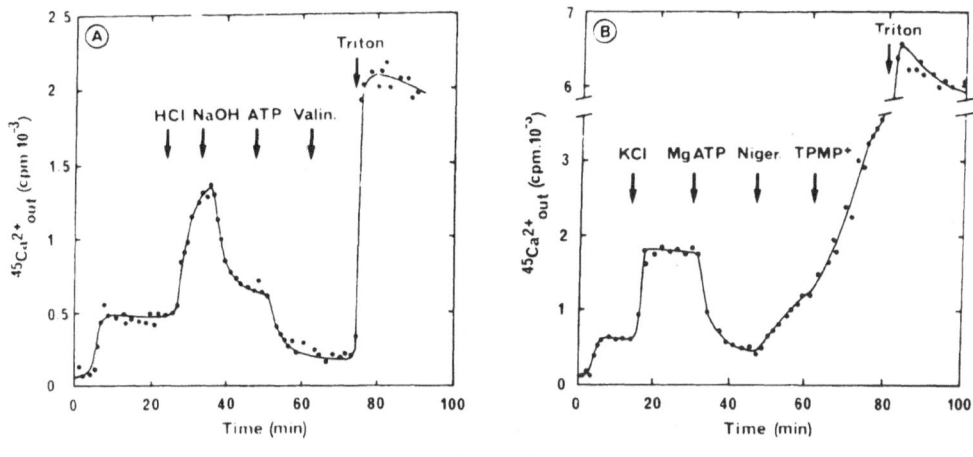

Figure 5

Effect of the changes in tonoplastic proton-motive force
on transtonoplastic calcium fluxes
in a suspension of intact lutoids

Fresh vacuoles were preincubated in the presence of 0.5 mM $CaCl_2$ and $^{45}Ca^{2+}$ at pH 7.5. Additions were as follows in the upper chamber of the flow-dialysis cell: A. HCl (to pH 6.0), NaOH (to pH 7.0), Ca-ATP (2.5 mM), valinomycin (0.8 μg/ml) and Triton (0.1% final concentration). B. Addition of KCl (130 mM plus valinomycin), Mg–ATP (5 mM), nigericin (16 μg/ml), TPMP$^+$ (2.5 mM) and Triton X-100 (0.1%).

(Donnan potential), did not induce significant efflux of Ca^{2+} (Marin et al., 1982). In contrast, the depolarization of the tonoplast by high external concentration of KCl (with valinomycin) or triphenylmethylphosphonium (TPMP$^+$) led to an important efflux of the intravacuolar calcium (Fig. 5-B).

Furthermore, the addition of the ionophore A-23187 (specific for the free divalent cations) was shown to induce only a small efflux of Ca^{2+} when added to resting vacuoles (Chréstin, 1985). In contrast, when added to ATP-energized vacuoles, A-23187 induced a significant efflux of Ca^{2+}, equivalent to the size of the Ca^{2+} pool accumulated in the presence of ATP (like FCCP) (Chréstin et al., 1984). This suggests that the ATP-energized Ca^{2+} transport through the tonoplast leads to the vacuolar accumulation of a pool of free Ca^{2+} against a thermodynamic equilibrium.

Furthermore, the functioning of the tonoplastic H$^+$-pumping system by addition of NADH plus cytochrome c to a suspension of lutoids, preloaded with $^{45}Ca^{2+}$ in the presence of MgATP, leads to an efflux of Ca^{2+} equivalent to the size of the ATP-energized Ca^{2+} pool (Fig. 6), simultaneously with the redox-pump dependent discharge of transmembrane $\Delta\overline{\mu_H}+$ (Chréstin, 1985).

All these data seen as a whole led us to postulate the existence of two kinetic pools of Ca^{2+} within the latex cells vacuoles :

– a major pool sequestrated within the vacuolar compartment, owing probably to adsorption on intravacuolar structures by a Donnan type effect, and which could only be dissipated by high concentrations of KCl (+ valinomycin) or TPMP$^+$;

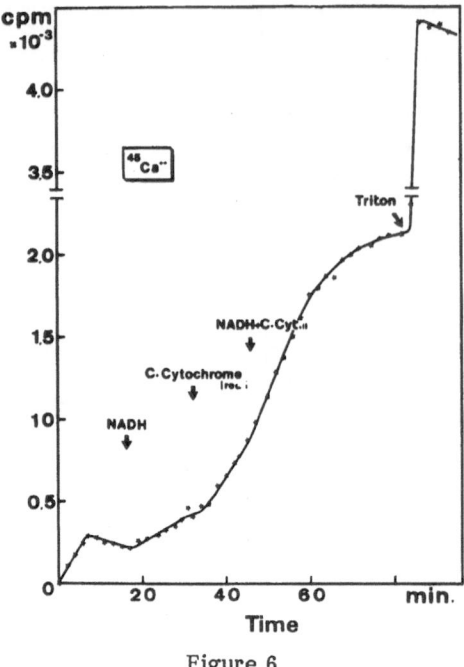

Figure 6

Efflux of calcium from the vacuolar compartment
during the operation of the tonoplastic NADH–cytochrome c–reductase

Intact fresh lutoids were preloaded with $^{45}Ca^{2+}$ (as 0.5 mM $CaCl_2$) in the presence of Mg–ATP (2.5 mM) and then transferred to the flow–dialysis cell. NADH (0.5 mM) and cytochrome c (0.2 mM) were then added first individually and then together. Finally the lutoids were lysed by addition of Triton X–100 (0.1%).

      – an exchangeable pool of "free" Ca^{2+}, accumulated inside the vacuolar compartment against a thermodynamic equilibrium which could be dissipated by ionophores such as FCCP, nigericin + K^+ and ionophore A-23187 and the specific free divalent cations. This pool of free Ca^{2+} accumulated in the vacuolar compartment through ATP-energized $\Delta\bar{\mu}_H+$ could be released into the external medium, both in vivo and in vitro, by the action of the tonoplastic redox protonpump which has been shown to dissipate the transtonoplastic $\Delta\bar{\mu}_H+$.

      Finally, as it could be shown that successive additions of small amounts of Ca^{2+} to an ATP-energized vacuolar suspension led to progressive collapse of ΔpH and even $\Delta\mu_H+$ (Chréstin, 1985), we propose the existence of transport processes corresponding to a Ca^{2+}/H^+ exchange at tonoplast level, and that the adverse transtonoplastic fluxes of free Ca^{2+} remain under the energy-dependent control of both opposing H^+-pumps at the tonoplast.

Sequestration of citrate within the vacuoles of the Latex Cells

      As the latex cell vacuoles accumulate citrate against a steep concentration gradient in vivo (Ribaillier et al., 1971), a lot of work was carried out to characterize the citrate transport processes and energization at latex cell tonoplast in vitro, using either freshly isolated intact lutoids (D'Auzac and Lioret, 1974; Montardy and Lambert, 1977; Chréstin, 1985) or tonoplast vesicles reconstituted from lyophilized lutoids (Marin et al., 1981; Marin, 1982).

The native lutoids, and the tonoplast vesicles as well, were shown to accumulate, in vitro, exogeneous citrate against a steep concentration gradient. The kinetic parameters were clearly defined : Citrate uptake was temperature-dependent and linear for at least 30 min, even in the absence of any metabolic energy supply. Its initial rate as a function of citrate concentration in the medium was shown to display simple Michaelis-Menten kinetics with an apparent K_m value of 7 mM (then in the physiological range), where the three dissociated form predominates (Marin, 1982). The addition of MgATP to a suspension of intact freshly isolated lutoids as well as tonoplast vesicles was shown to generate a large increase in the magnitude of citrate uptake. In the presence of MgATP, the steady state level of citrate uptake and accumulation was shown to be 2 to 5 times higher than the level obtained in the presence of any energy source. Furthermore, the addition of protonophores such as NH_4Cl, FCCP and S-13 caused considerable reduction of citrate uptake in the absence fo energy supply, and completely stopped its activation by Mg-ATP. Finally, it was shown that all the known inhibitors of the lutoidic tonoplast ATPase did inhibit the activation of citrate uptake in the presence of MgATP (Marin et al., 1981; Marin et al., 1982; Marin, 1983 a and b; Chréstin, 1985).

It was then concluded that the energy indispensable for citrate uptake by the lutoids originated from the transtonoplastic gradient of proton, resulting from the functioning of the H^+-pumping ATPase located on the lutoidic tonoplast.

The direct, highly significant relationship linking the amplitude of the transtonoplastic gradient of citrate to the gradient of proton as determined in vivo (Table 1) agrees well with a force derived from the transtonoplastic ΔpH as the energy source for citrate uptake and accumulation in the vacuoles, in vivo, as in vitro.

Moreover, as it was shown that any change in the magnitude of the transtonoplastic gradient of proton (ΔpH) as well as in the transtonoplastic electrical potential gradient ($\Delta \Psi$) induced parallel changes in the magnitude of citrate uptake (Fig. 7) it was definitely concluded that both components of the proton-motive force, then $\Delta \overline{\mu_H}+$ itself, were involved in the energization of citrate uptake. Finally, as uptake of citrate was shown to induce internal alkalinization, Marin (1982) proposed the functioning of a tonoplastic H^+/citrate antiporter.

Since it has been wondered that role of citrate accumulation plays in lutoids as a storage or a detoxifying process, many attempts were made to characterize any mechanism able to control efflux of the citrate accumulated in the lutoids as well as in tonoplast vesicles.

Whatever the technic used : centrifugation and washing (Montardy and Lambert, 1977) or flow dialysis (Chréstin, 1985) and whatever the strategy adopted to try to induce citrate efflux (labile citrate accumulated in vitro, or cold citrate accumulated in vivo) from intact isolated lutoids, such as changes in the transtonoplastic $\overline{H^+}$ gradient by imposed external pH variations, the use of protonophores or diverse ionophores, the functioning of the outwards H^+-pumping tonoplastic redox chain, or changes in the transtonoplastic potential using either KCl + valinomycin or lipophilic cations (TPP^+ or $TPMP^+$), in the presence or absence of various concentrations of exogenous citrate in the medium, and at three temperatures (20, 30 and 40°C), neither of the authors was able to shown any significative efflux of the citrate that had been accumulated in vivo or in vitro by intact vacuoles (Fig. 8-A and B) (Montardy and Lambert, 1977; Chréstin, 1985). It was then concluded that the quasi totality of the citrate accumulated remained definitively entrapped within the native lutoids, constituting in a true detoxification process.

On the contrary, working with tonoplastic vesicles reconstituted from lyophilized lutoids, Marin (1982) did show evidence for the occurrence of massive efflux of citrate : up to 80% of the previously accumulated citrate. This efflux was shown to be temperature-dependent, and to increase with the concentration

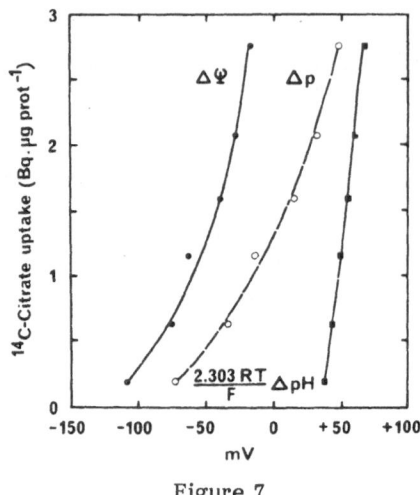

Figure 7

Relation between the components of the proton-motive force (Δp) and citrate incorporation by tonoplast vesicles

of citrate in the medium. Yet, upon the uptake of labelled citrate by these vesicles, the isotopic enrichment of the internal compartment did not exceed 8% of the external specific activity. As a result, isotopic equilibrium was never reached.

From these data obtained on native vacuoles and tonoplastic vesicles, it was finally concluded that there might be exist an "internal compartmentation" of the vacuolar citrate into two distinct kinetic pools: a minor, directly exchangeable pool, and a major one (more than 97% of the total vacuolar citrate) assumed to remain sequestrated within the native vacuoles. Insofar as it was shown that citrate and Mg^{2+} are present in quasi stoichiometric concentrations in latex, and in particular in the lutoids (Coupé, 1977; Jacob, unpublished data), it was proposed that the vacuolar citrate might be sequestrated in the complex form citrate^{3-}-Mg^{2+} in intact native lutoids (Marin, 1982; Marin et al., 1982). It was then suggested that the massive efflux of citrate observed in tonoplast vesicles was somewhat artifactual (i.e. non physiological), and might be due to the loss of intralutoidic sequestrating factors, including Mg^{2+}, during tonoplast vesiculation from lyophilized lutoidic membrane in vitro. It was then assumed that in vivo the triacid is accumulated and sequestrated inside the vacuolar compartment in the complex form citrate^{3-}-Mg^{2+}, considered as the impermeant form which represents by far the largest pool of the vacuolar citrate (more than 97%). A very minor pool (less than 3%) of free citrate^{3-}, might exist as a possible exchangeable pool. Computer kinetic simulations based on this model were shown to be quite reconciliable with these experimental data (see Marin, 1982). Such models involved the working of citrate translocator antiporterly with a proton, one molecule of Mg^{2+} being simultaneously transferred from the cytosol to the intra-lutoidic space, then being sequestrated in this complex non-permeant form (Marin, 1982). Fig. 8-A agrees

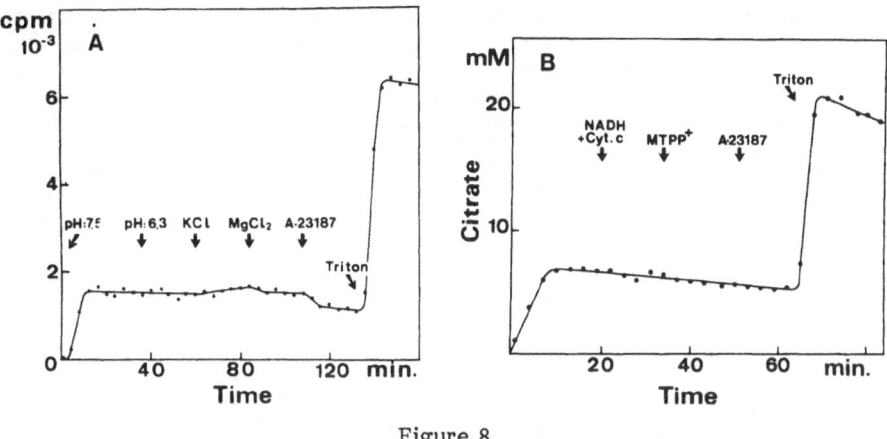

Figure 8

Unfruitful attempts to provoque efflux of the intravacuolar citrate
with intact lutoids

A. Intact fresh lutoids were preloaded with ^{14}C-citrate at pH 7.5 (external citrate: 5.5 mM), and transferred to the flow dialysis cell. The medium was then acidified by addition of HCl. Further additions were as follows : KCl (130 mM + valinomycin), MgCl$_2$ (2.5 mM), A-23187 (0.2 μg/ml) and then Triton X-100 0.1%.
B. Intact lutoids were incubated in unlabelled citrate (7 mM) for 30 min and then transferred to the flow-dialysis cell. Additions were as follows: NADH (1 mM) plus cytochrome c (0.5 mM), then TPMP$^+$ (2 mM) and then A-23187 (0.2 μg/ml) without Mg^{2+}. Finally, the lutoids were lysed with Triton X-100 (0.1%). Citrate was determined enzymatically in the flow-dialysis effluent.

with the suspected important role of Mg^{2+} in citrate uptake and accumulation, as far as in the presence of the divalent cation ionophore A-23187 and the addition of excess MgCl$_2$ to the suspension induces a significant uptake of citrate by intact lutoids.

Taking into consideration the fact that citrate (a potent inhibitor of some cytosolic key enzymes) remains mainly entrapped within the native lutoids, it is proposed that this vacuolar compartment essentially plays a role of detoxifying trap, ensuring detoxification of the cytosolic metabolism against any excessive accumulation of this triacid in the cytosol. This proposal fully explains the direct highly significant relationships linking the production of latex (rubber) with high transtonoplastic gradient of citrate in the latex, i.e. low concentration of citrate in the cytosol but high accumulation in the vacuolar compartment in vivo (Table 1).

CONCLUSION

Control of metabolism within the latex cells by intracellular pH and by ionic composition of the cytosol has attracted more and more interest during the past few years, because of their major impact on natural biosynthesis of rubber, and hence on production. When present in excess, H$^+$, Ca^{2+} and citrate which are potent inhibitors of some key enzymes of the cytosolic metabolism, are removed from the latex cell cytosol and accumulate in the vacuoles.

As far as protons are concerned, all the data reported here demonstrate the existence of two H^+-pumping systems, definitely located on the lutoidic tonoplast, able to control opposite transtonoplastic fluxes of protons. Their respective sensitivity towards the pH of the medium (see Chréstin et al., in this issue) is in good agreement with their functioning as a true biophysical pH-stat, controlling adverse fluxes of H^+ across the lutoidic tonoplast using an energy-consuming system, thus regulating the cytoplasmic pH. It was shown that there are two kinetic pools of accumulated protons inside intact vacuoles. A more or less constant one, equivalent to 1 pH unit, accumulated at the thermodynamic equilibrium in the vacuolar compartment owing to some Donnan potential. This immobilized pool could only be released by the artificial neutralization of the Donnan potential through additions of permeant cations. A second pool of H^+ with highly variable size (accounting for 0.1 to 1 pH unit of the total transtonoplastic ΔpH gradient), forms an exchangeable pool of free protons accumulated in the vacuolar space against the thermodynamic equilibrium through the functioning of the tonoplastic H^+-pumping ATPase. This exchangeable pool of H^+ can be reinjected in the cytosolic compartment through the working of the outwards tonoplastic H^+-pumping redox system.

The two opposing H^+-pumps were shown to work electrogenically, and then to modulate the amplitude of the transtonoplastic electrochemical proton gradient ($\Delta\overline{\mu_H}^+$), which has been shown to energize transports of various solutes across the membrane (Marin, 1982; Marin and Chréstin, 1985), and in particular citrate and calcium.

Each of the citrate and the calcium pools accumulated in the vacuolar space could be subdivided into two distinct kinetic pools. The first and major pool corresponds to a non exchangeable pool where solutes remain entrapped within the vacuolar compartment according to some Donnan potential (this is the case of H^+ and probably Ca^{2+}) or other trapping processes such as immobilization of citrate in the impermeant complex form $citrate^{3-}$-Mg^{2+}. The second, minor pool, whose size varies considerably, reversibly accumulated in the vacuolar compartment. These exchangeable pools can be placed again at the disposal of the cytosolic metabolism through the functioning of outward H^+-pumping redox system. The exchangeable pool can be relatively important as far as Ca^{2+} is concerned (up to 25% of the total vacuolar calcium), whereas it is reduced (3 to 5% to non-existent) for citrate.

Consistently with the functioning of the two opposing H^+ pumps at the lutoidic tonoplast, we propose that the lutoids, the vacuolar compartment of the latex cells, play a triple role as a "biophysical pH-stat", a detoxifying trap (citrate, etc.) and a storage compartment (Ca^{2+}, H^+, ..), thus controlling homeostasis in the cytosol and favouring active metabolism within the cells, thus resulting in high latex (natural rubber) production.

REFERENCES

Archer, B. L., Barnard, D., Cockbain, E. G., Dickenson, P. B., and Mac Mullen, A. I., 1963, Structure, composition and biochemistry of Hevea latex, in : "The Biochemistry and Physics of Rubber-like Substances", L. Bateman, ed., Mac Laren and Sons Ltd., London.

D'Auzac, J., Crétin, H., Marin, B., and Lioret, C., 1982, A plant vacuolar system : the lutoids from Hevea brasiliensis, Physiol. Vég., 20:311.

D'Auzac, J., and Jacob, J. L., 1969, Regulation of glycolysis in latex of Hevea brasiliensis, J. Rubb. Res. Inst. Malaya, 21:417.

D'Auzac, J., and Lioret, C., 1974, Mise en évidence d'un mécanisme d'accumulation du citrate dans les lutoides du latex d'Hevea brasiliensis, Physiol. Vég., 12: 617.

D'Auzac, J., 1975, Caractérisation d'une ATPase membranaire en présence d'une phosphatase acide dans les lutoides d'Hevea brasiliensis, Phytochemistry, 14:671.

D'Auzac, J., 1977, ATPase membranaire de vacuoles lysosomales : les lutoides d'Hevea brasiliensis, Phytochemistry, 16:1881.

Brzozowska-Hanower, J., Crétin, H., Hanower, P., and Michel, M., 1979, Variations de pH entre compartiments vacuolaire et cytoplasmique au sein du latex d'Hevea brasiliensis. Influence saisonnière et action du traitement par l'Ethrel, générateur d'éthylène. Répercussion sur la production et l'apparition d'encoches sèches, Physiol. Vég., 17:851.

Brzozowska, J., Hanower, P., and Chézeau, R., 1974, Free amino-acids of Hevea brasiliensis latex, Experientia, 30:894.

Chréstin, H., 1985, "La Vacuole dans l'Homéostasie et la Sénescence des Cellules Laticifères d'Hevea", Etudes et Thèses, ORSTOM, Paris.

Chréstin, H., Gidrol, X., Marin, B., Jacob, J. L., and D'Auzac, J., 1984, Role of the lutoidic tonoplast in the control of the cytosolic homeostasis within the laticiferous cells of Hevea, Z. Pflanzenphysiol., Bd. 114 S:269.

Coupé, M., 1977, Etudes Physiologiques sur le Renouvellement du Latex d'Hevea brasiliensis : Action de l'Ethylène. Importance des Polyribosomes. Thèse Doc. Etat Sci. Nat., Montpellier.

Coupé, M., and Lambert, C., 1977, Absorption of citrate by the lutoids from the latex vessels, and rubber production by Hevea, Phytochemistry, 16:45.

Crétin, H., 1982, The proton gradient across the vacuo-lysosomal membrane of lutoids from the latex of Hevea brasiliensis. I. Further evidence for a proton-translocating ATPase on the vacuo-lysosomal membrane of intact lutoids, J. Membrane Biol., 65:175.

Crétin, H., 1983, Efflux transtonoplastique de protons lors du fonctionnement d'un système transporteur d'électrons (la NADH-cytochrome c-réductase) membranaire des vacuo-lysosomes du latex d'Hevea brasiliensis, C. R. Acad. Sci, Paris, Série III, 296:101.

Crétin, H., Marin, B., and D'Auzac, J., 1982, Characterization of a magnesium-dependent proton translocating ATPase on the Hevea latex tonoplast, in : "Plasmalemma and Tonoplast, Their Functions in the Plant Cells", D. Marmé, E. Marre, and R. Hertel, eds., Elsevier/ North Holland Biomedical Press, Amsterdam.

Gidrol, X., Marin, B., Chréstin, H., and D'Auzac, J., 1985, The functioning of the tonoplast H^+-translocating ATPase from Hevea latex in physiological conditions, in : "Biochemistry and Function of Vacuolar Adenosine-triphosphatase in Fungi and Plants", B. P. Marin, ed., Springer-Verlag, Berlin, Heidelberg, New-York and Tokyo.

Jacob, J. L., and D'Auzac, J., 1967, Sur l'existence conjointe d'une hexokinase et d'une fructokinase au sein du latex d'Hevea brasiliensis, C. R. Acad. Sci., Paris, Série D, 265:260.

Jacob, J. L., and D'Auzac, J., 1969, Sur quelques caractéristiques originales de la pyruvate-kinase du latex d'Hevea brasiliensis, Bull. Soc. Chim. Biol., 51:511.

Jacob, J. L., and D'Auzac, J., 1972, La glyceraldehyde-3-phosphate deshydrogénase du latex d'Hevea brasiliensis. Comparaison avac son homologue phosphorylante, Eur. J. Biochem., 31:255.

Jacob, J. L., Primot, L., and Prévot, J. C., 1979, Purification et étude de la phospho-énol-pyruvate carboxylase du latex d'Hevea brasiliensis, Physiol. Vég., 17:501.

Marin, B., 1982, "Le Fonctionnement du Transporteur Tonoplastique du Citrate du Latex d'Hevea brasiliensis", Trav. Doc. ORSTOM, Vol. 144, ORSTOM, Paris.

Marin, B., 1983 a, Sensitivity of tonoplast-bound adenosine-triphosphatase from Hevea to inhibitors, Plant Physiol., 73:973.

Marin, B., 1983 b, Evidence for an electrogenic adenosine-triphosphatase in Hevea tonoplast vesicles, Planta, 157:324.

Marin, B., 1985 a, The control by $\Delta\overline{\mu_H}^+$ of the tonoplast-bound H^+-translocating adenosine-triphosphatase from rubber-tree (<u>Hevea</u> <u>brasiliensis</u>) latex, <u>Biochem. J.</u>, 229:459.

Marin, B., 1985 b, "Biochemistry and Function of Adenosine-triphosphatase from Fungi and Plants", Springer-Verlag, Berlin, Heidelberg, New-York and Tokyo.

Marin, B., Crétin, H., and D'Auzac, J., 1982, Energization of solute transport and accumulation at the tonoplast in <u>Hevea</u> latex, <u>Physiol. Vég.</u>, 20:233.

Marin, B., and Gidrol, X., 1985, Chloride-ion stimulation of the tonoplast H^+-translocating ATPase from <u>Hevea</u> <u>brasiliensis</u> (rubber ree) latex. A dual mechanism, <u>Biochem. J.</u>, 226: 85.

Marin, B., Gidrol, X., Chrestin, H., and D'Auzac, J., 1986, The tonoplast proton-translocating ATPase of higher plants as third class of proton-pumps, <u>Biochimie</u>, 68:1263.

Marin, B., Marin-Lanza, M., and Komor, E., 1981, The proton-motive potential difference across the vacuo-lysosomal membrane of <u>Hevea</u> <u>brasiliensis</u> (rubber tree) and its modification by a membrane-bound adenosine-triphosphatase, <u>Biochem. J.</u>, 198:365.

Marin, B., Preisser, J., and Komor, B., 1985, Solubilization and purification of the ATPase from the tonoplast of <u>Hevea</u>, <u>Eur. J. Biochem.</u>, 151:131.

Montardy, M. C., and Lambert, C., 1977, Diverses propriétés de l'absorption du citrate, du malate et du succinate par les lutoides du latex d'<u>Hevea</u> <u>brasiliensis</u>, <u>Phytochemistry</u>, 16:677.

Moreau, F., Jacob, J. L., Dupont, J., and Lance, C., 1975, Electron transport in the membrane of the lutoids from the latex of <u>Hevea</u> <u>brasiliensis</u>, <u>Biochim. Biophys. Acta</u>, 396:116.

Pujarniscle, S., 1968, Caractère lysosomal des lutoides du latex d'<u>Hevea</u> <u>brasiliensis</u>, <u>Physiol. Vég.</u>, 6:27.

Ribaillier, D., Jacob, J. L., and D'Auzac, J., 1971, Sur certains caractères vacuolaires des lutoides du latex d'<u>Hevea</u> <u>brasiliensis</u>, <u>Physiol. Vég.</u>, 9:423.

Tupy, J., 1969, Stimulatory effects of 2,4-dichlorophenoxyacetic and 1-naphtylacetic acids on sucrose level, invertase activity, and sucrose utilization in the latex of <u>Hevea</u> <u>brasiliensis</u>, <u>Planta</u>, 88:144.

AJMALICINE TRANSPORT INTO VACUOLES ISOLATED

FROM CATHARANTHUS ROSEUS CELLS

Jean-Pierre Renaudin[*] and Jean Guern

Laboratoire de Physiologie Cellulaire Végétale, CNRS/INRA, Boite Postale N°1, F-91.190-Gif-sur-Yvette, and [*]Station de Physiopathologie Végétale, INRA, B. V. No. 1540, F-21.034-Dijon, France

INTRODUCTION

We have studied previously the compartmentation of ajmalicine in Catharanthus roseus cells (Renaudin and Guern, 1982; Renaudin et al., 1985) and of nicotine in Nicotiana tabacum and Acer pseudoplatanus cells (Kurkdjian, 1982). The behavior of the ^{14}C-labelled alkaloids added in the cell suspensions was in good agreement with the predictions of the so-called ion-trapping model : These alkaloids apparently diffused passively through the plasmalemma and the tonoplast, mainly under their neutral form, and did accordingly distribute between the cells and the medium. They were accumulated within cells, likely in vacuoles, against a concentration gradient, because of the relative acidity of the vacuolar compartment and of the low capacity for diffusion of the alkaloid cation. The distribution of alkaloids between the cells (i.e. the vacuoles) and the medium was dynamic and it was a function of 1) the acidity constant of the alkaloid, and 2) the pH difference between the vacuole and the medium. Due to the high pH (ca. 7.5) and low relative volume (less than 20%) of the cytoplasm, the accumulation of this compartment was likely neglectable. It was assumed that accumulation in whole cells merely reflected the accumulation within vacuoles, thus allowing the indirect approach of vacuolar properties.

The present report aims to evaluate this major assumption by studying directly the accumulation of ajmalicine by vacuoles isolated from Catharanthus roseus cells. A further reason to undertake this study is issued from recent reports about the translocation of alkaloids across the tonoplast of isolated vacuoles (Deus-Neumann and Zenk, 1984; Deus-Neumann and Zenk, 1986). These authors concluded that a very specific active transport of alkaloids was the only mechanism for the translocation and accumulation of vindoline in vacuoles from Catharanthus roseus cells (Deus-Neumann and Zenk, 1984) and S-reticuline and S-scoulerine in vacuoles from Fumaria capreolata cells (Deus-Neumann and Zenk, 1986). This conclusion was extrapolated to the general case of the alkaloids, for which a highly diversified set of specific carriers at the tonoplast was hypothesized, and the ion-trapping model was negated (Deus-Neumann and Zenk, 1986).

CHARACTERIZATION OF THE VACUOLES

All experiments were performed with 5/6-day-old cell suspensions, i.e. at the late exponential growth phase. Culture conditions, protoplast and vacuole preparation are described elsewhere (Barbier-Brygoo et al., 1987; Renaudin et al., 1986). The vacuoles were recovered in a medium comprising 550 mM sorbitol and 10 mM Hepes-KOH, pH 7.3, after osmotic lysis of the protoplasts.

A quantitative characterization of the content of the cells, protoplasts, and vacuoles has been reported elsewhere (Renaudin et al., 1986). The isolated vacuoles were reasonably stable as to their number per ml, since 75% and 60% remained after 5 h and 24 h, respectively. One major feature of these vacuoles was their high acidity corresponding to a transtonoplast ΔpH of ca. 2.0 pH units when incubated at pH 7.3 (Barbier-Brygoo et al., 1987). The contamination rates from enzymatic non-vacuolar markers ranged from 3 and 13% (Renaudin et al., 1986) and the activity per mg protein of α-mannosidase, a vacuolar marker, was enriched more than 10 times in vacuole suspensions compared to protoplasts (Table 1). The two main indole alkaloids synthesized by the cells, ajmalicine and serpentine, were present in isolated vacuoles to very different extent (Table 1). By using -mannosidase as a vacuolar marker, we found that only 41% ajmalicine but up to 215% serpentine was present in isolated vacuoles compared to protoplasts. From previous experience with ajmalicine, which is a lipophilic alkaloid with a low acidity constant (pK_a = 6.3), it was likely that a large part of ajmalicine initially present in the vacuoles was lost by diffusion during their isolation. As to serpentine, which is a highly polar high pK_a (= 10.8) alkaloid, the rather surprising value of 215% present in the vacuoles has been interpreted as due to the selection of a special class of vacuoles with a larger content in serpentine than the average population (Renaudin et al., 1986). This hypothesis lay upon the previously observed very large cell-to-cell dispersion, over a 20-fold range, of the concentrations of serpentine in cell populations (Brown et al., 1984). A significant selection could thus occur even in the present case where the recovery of the vacuoles has a rather good yield estimated to 20% of the initial protoplast population based on α-mannosidase (Renaudin et al., 1986).

Table 1

Content[a] of Protoplasts and Isolated Vacuoles

Compounds	Content in 10^6		% in vacuoles[b]
	Protoplasts	Vacuoles	
Protein (g[c])	464	23	9
α-Mannosidase (pkat)	217	106	100
Ajmalicine (pmol)	155	38	41
Serpentine (pmol)	71	91	215

[a]The data are average values from several experiments.
[b]Calculated by assuming 100 % α-mannosidase to be vacuolar.
[c]bovine serumalbumin equivalents.

TRANSPORT OF AJMALICINE INTO THE VACUOLES

Measurement of The Uptake

A detailed description of the procedure to measure the uptake will be presented since several pitfalls have to be avoided in such experiments. Moreover, a complete description of the methodologies is necessary to compare the results with other possibly contradictory results.

The uptake into vacuoles of ($^{14}CH_3COO$)-ajmalicine (1.6 MBq μmol^{-1}, CEA) and (pyrrolidine-2-^{14}C)-nicotine (1.9 MBq μmol^{-1}, NEN Research Products) was measured at room temperature by silicon layer filtering flotation according to a procedure slightly modified from Martinoia et al. (1985). To 500 μl of vacuole suspension (2-3 10^6 vacuoles ml^{-1}, internal volume 2.8%) 300 μl were added of an osmotically adjusted solution comprising 2.5 mg ml^{-1} bovine serumalbumin, 15% (w/v) Nycodenz (a neutral triiodinated derivative of benzoic acid, Nyegaard and Co), ^3H_2O (600 kBq ml^{-1}, CEA) and the ^{14}C-alkaloid at various concentrations. Bovine serumalbumin (0.9 mg ml^{-1} final concentration) was added in order to prevent the quantitatively large adsorption of ajmalicine on plastic walls. The presence of Nycodenz (5.6% w/v final concentration) avoided phase inversion with the silicon layer during further centrifugation. ^3H_2O was systematically present in order to estimate the internal volume of vacuoles actually recovered after silicon filtration. In parallel experiments, ^{14}C-mannitol (2.0 MBq ml^{-1}, NEN Research Products), a not permeating substance, was added instead of the alkaloid at a final concentration 10-20 kBq ml^{-1}, in order to correct for the small amount of medium adhering to the vacuoles as they migrate through silicon.

At various times, 100 μl aliquots of the incubation solution were put into a 400 μl thin polypropylene microcentrifugation tube and centrifuged about 3 s at 10,000 g in a microfuge (Eppendorf). This was necessary in order to eliminate all the incubation solution from the upper walls of the tube. The sample was then overlaid with 200 μl phenylmethyl silicon oil AP 100/AR 200 (100/83, v/v, Wacker Chemie, München, FRG) and 100 μl of an osmotically adjusted solution containing 1 mg ml^{-1} bovine serumalbumin was added on the top of this. The incubation was terminated by centrifugation (10,000 g, 30 s). The upper phase, sometimes including a small amount of silicon, was collected and the radioactivity of ^3H and ^{14}C counted by liquid scintillation.

Data were analyzed according to Heldt (1980). In classical experiments, the intravacuolar volume actually recovered above silicon amounted to 1.5 - 2.0% of the volume of the incubation solution, which means that nearly all the vacuoles were recovered. The extravacuolar residual medium recovered above silicon represented 13% of the corresponding intravacuolar volume. It has to be pointed out that the use of tritiated water allows to take into account the possible lysis of the vacuoles during the incubation or the centrifugation steps, which is not the case when vacuoles are only counted at the beginning of the incubation.

Kinetics of Transport

Fig. 1-A shows a typical kinetics of uptake of ajmalicine by isolated vacuoles. ^{14}C-ajmalicine was absorbed very quickly by the vacuoles so that, as soon as after 3 min, a stable distribution of it between the vacuoles and the medium occurred. At that time, the concentration of ^{14}C-ajmalicine in the vacuoles was higher that in the medium, which correspond to a concentration ratio C_i/C_e (intravacuolar versus extravacuolar concentration) ranging from 5 to 10.

The uptake of ^{14}C-ajmalicine by the vacuoles did not simply corresponds to an isotopic exchange of the labelled alkaloid with intravacuolar endogenous ajmalicine. The concentration of the latter was calculated to be 4 μM in isolated vacuoles (Renaudin et al., 1986), and the accumulation of exogenous ajmalicine in vacuoles was observed to concentrations up to 350 μM (Fig. 2).

The addition of 0.5 - 2 mM MgATP to the incubation medium did not change the pattern of the kinetics neither the intensity of the accumulation. Such a result cannot be used to argue for a diffusive uptake or against an energy-dependent carrier mediated uptake. The isolated vacuoles kept an acidic pH near 5.3 - 5.6 (Barbier-Brygoo et al., 1987), and were strongly buffered, as seen from their ionic content (Renaudin et al., 1986). Then, a positive effect of MgATP on proton pumping by the tonoplast ATPase could not be expected to result in a significant increase of

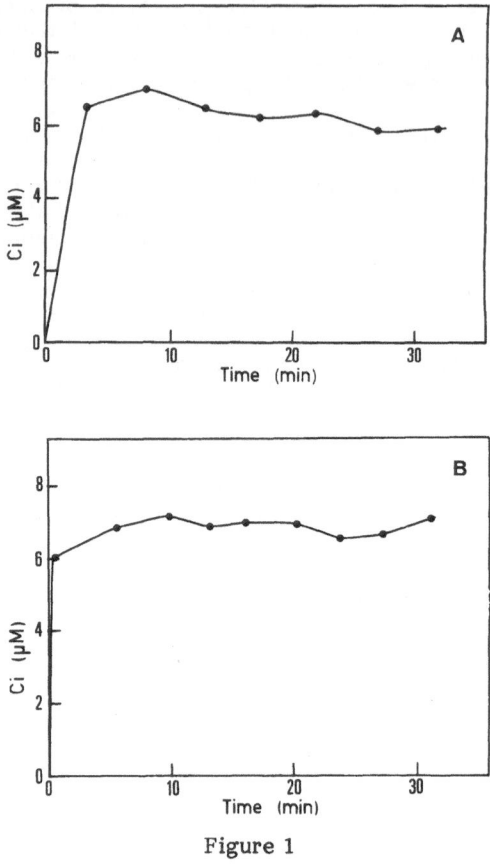

Figure 1

Kinetics of uptake of (^{14}C)-ajmalicine
by isolated vacuoles (A)
and by the cells from which the vacuoles derived (B)

^{14}C-ajmalicine (1.6 MBq μmol^{-1}) was added at time zero at a final concentration 1.1 μM and 0.79 μM in vacuole and cell suspensions, respectively. Incubation conditions are described in the text. The experiment was designed in order to have roughly the same number per ml of cells present as vacuoles and cells, i.e. 0.84 10^6 cells present as vacuoles in the vacuole suspension (corresponding to 1.68 10^6 vacuoles ml^{-1}, Renaudin et al., 1986), and 1.3 10^6 cells ml^{-1} in the cell suspension. But the cumulated internal volume was much lower in the vacuole suspension (= 1.6%) than in the cell suspension (= 8.2%). At the equilibrium, the concentration ratio C_i/C_e of ^{14}C-ajmalicine is 5.5 in the vacuole suspension and 37.4 in the cell suspension.

the already large transtonoplast ΔpH. Some caution must thus be exercised in interpreting the apparent absence of effect of MgATP on the uptake of a metabolite, when the energy necessary to its intravacuolar accumulation comes from the pH component of the proton electrochemical potential gradient through the tonoplast.

The kinetics of uptake of ^{14}C-ajmalicine by vacuoles was compared to that by the cells from which the vacuoles derived (Fig. 1-B). In these experiments, the cells were kept in their culture medium, buffered with Hepes-KOH to the same pH as the incubation medium of the vacuoles (pH 7.3). The kinetics of uptake by the cells looked exactly similar to that by the vacuoles (Fig. 1-A), which eliminated artefactual adsorption of ajmalicine onto the vacuoles as the reason for the very quick uptake. The accumulation capacity of the cells, reflected by the concentration ratio C_i/C_e at the equilibrium, and their cumulated relative volume (ca. 8%) were high enough to induce a decrease of the extracellular concentration much higher than the one corresponding to the vacuolar suspension. Thus, whereas the internal concentration at equilibrium was about the same for cells and vacuoles, the concentration ratios were very different, ^{14}C-ajmalicine being near 7-times more concentrated in the cells relative to the medium than in the vacuoles. According to the ion-trap model, such a difference can be due to 1) a shift of vacuolar pH towards more ajmalicine values during vacuole isolation, 2) a modification of the tonoplast properties during the isolation with an increasing permeability to the alkaloid cation, and 3) a not neglectable extravacuolar accumulation in cells. It is difficult up to now to evaluate the points 2) and 3). As to 1), measurements of the vacuolar pH with the ^{31}P-NMR technique on cell suspensions (pH = 4.91 $\pm$ 0.03) and corresponding vacuolar suspensions (pH = 5.29 $\pm$ 0.05) have shown that some alkalinization of the vacuoles occur during their isolation and can account for at least one part of the difference observed in the accumulation capacity.

Evidence for Translocation by Diffusion and Accumulation by Ion-Trapping

Uptake as a Function of the Initial Concentration of Ajmalicine. Previous experiments with whole cells have enabled to measure the initial rate of ajmalicine by the cells in the first 1.5 min after addition of the alkaloid (Renaudin and Guern, 1982). The initial rate was a linear function of the initial concentration of ajmalicine in the medium, in the tested range 0.02 - 6.9 μM, and it was affected by the external pH value in a way compatible with the preferential diffusion of the neutral form of ajmalicine. Moreover, the efflux of ajmalicine from the cells proceeded with the same kinetics as the influx (Renaudin et al., 1985). This strongly indicated that the translocation of ajmalicine across plasmalemma and tonoplast occurred by simple diffusion mainly of the neutral form of the molecule.

For technical reasons, it was not possible to measure the initial rate of ajmalicine by isolated vacuoles, since the first measurement was obtained only after 3 min, once the equilibrium was reached. Fig. 2 shows that the overall accumulation of ajmalicine by cells and by isolated vacuoles was also a linear function of its initial concentration in the medium, between 0.3 and 40 μM. The pattern of the kinetics remained unchanged whatever the concentration. Such results are in favor of the translocation of ajmalicine across the tonoplast by simple diffusion. For comparison, the K_m values reported for the active transport of alkaloids across the tonoplast were near 0.4 μM (Deus-Neumann and Zenk, 1984 and 1986).

Uptake as a Function of the ΔpH across the Tonoplast. Vacuoles were purified in a medium buffered either at pH 7.3 with 10 mM Hepes-KOH or at pH 6.4 with 10 mM Mes-KOH. The intravacuolar pH was measured by microfluorimetry with 9-aminoacridine as a vacuolar pH probe (Manigault et al., 1983). The vacuolar pH did not change, being 5.3, with the external pH (Table 2). Ajmalicine was more concentrated in the vacuoles suspended at pH 7.3 (C_i/C_e = 14.8) than in the vacuoles at pH 6.4 (C_i/C_e = 6.2). Theoretical concentration ratio calculated from the pK_a of ajmalicine and the values of the pH in the medium and in the vacuoles were

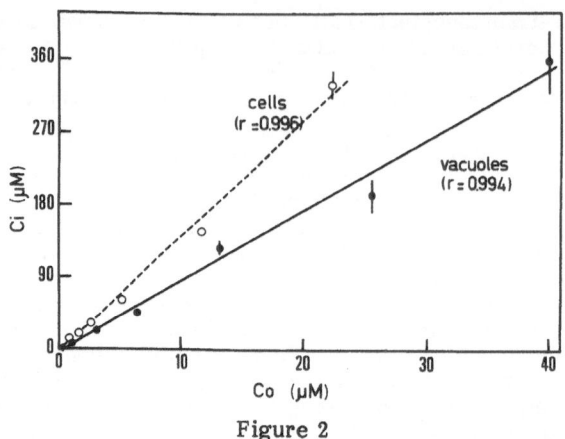

Figure 2

Uptake as a function of the initial concentration of ajmalicine
in the incubation medium of vacuoles and cells

The conditions have been described in Fig. 1. C_O is the initial concentration of
^{14}C-ajmalicine in the medium, at specific radioactivity constant in the case of
vacuoles (1.6 MBqμmol^{-1}) or varying from 0.060 to 1.6 MBqμmol^{-1} in the case of
cells. C_i is the intravacuolar or intracellular concentration of ^{14}C-ajmalicine at
the equilibrium of the uptake, calculated as the mean $\pm$ SE of 6 measurements
between 3 and 30 min.

rather close to the experimental ones. This corroborates our previous results with
whole cells showing that the higher the pH gradient between the vacuole and the
medium, the more ajmalicine was concentrated within cells (Renaudin and Guern,
1982).

One conclusion of such results is that very careful checking of the pH of
the incubation medium has to be made in the case of different treatments or during
the aging and possible lysis of the vacuoles, as it is a major parameter determining
the accumulation of alkaloids within vacuoles. The interpretation of the action
of effectors should also integrate possible effects on the intravacuolar pH.

Uptake as a Function of the pK_a of the Alkaloid. When the diffusion of
the alkaloid cation can be assessed to be nul, the only specificity of the ion-trapping
model with regards to the alkaloid is related to its pK_a (see the formula in Table
2). The higher the pK_a, the more the alkaloid will be accumulated in vacuoles.
A more realistic view has however to consider also that the diffusion of the cation
cannot be quite neglected, which is especially important at the quantitative level
for high pK_a alkaloids (see Renaudin and Guern, 1982, for discussion).

As a matter of fact, nicotine (pK_a = 8.0), which is not produced by
Catharanthus plants, was 12 times more concentrated in the vacuoles than ajmalicine

Influence of the pH Gradient Across the Tonoplast
on the Accumulation of ^{14}C-ajmalicine by Vacuoles

pH_e	pH_i	C_i/C_e	
		experimental[a]	theoretical[c]
6.4	5.3	6.2 ± 0.5	6.1
7.3	5.3	14.8 ± 0.5	9.6

[a]Measured by microfluorimetry with 9-aminoacridine.
[b]Mean ± SE of 5 determinations.
[c]Calculated by the formula :

$$C_i/C_e = (1 + 10^{pKa - pH_i})/(1 + 10^{pKa - pH_e}).$$

(pK_a = 6.3) (Table 3). Due to the high concentration ratio of nicotine (= 91.4), its extravacuolar concentration at the equilibrium had to be calculated after correction for the presence within vacuoles of relatively large amount of nicotine. The kinetics of uptake of nicotine was very similar to that of ajmalicine. The larger accumulation of nicotine correspond to rather satisfactorily with the theoretical values predicted from the ion-trapping model (Table 3). Such results are in agreement with the reported use of nicotine as a vacuolar pH probe (Kurkdjian, 1982; Kurkdjian et al., 1985). A similar effect of the pK_a value has been already reported for the accumulation in Catharanthus roseus cells of ajmalicine (pK_a = 6.3), tabernanthine, a non-Catharanthus indole alkaloid (pK_a = 8.1), and tryptamine (pK_a = 10.2) (Renaudin, 1981).

Table 3

Influence of the pK_a of the Alkaloid
on its Accumulation by Vacuoles

Alkaloid	pKa	C_i/C_e		
		experimental[a]	theoretical[b]	
			pH_i = 5.3	pH_i = 5.5
Ajmalicine	6.3	7.5 ± 0.6	10.2	6.8
Nicotine	8.0	91.4 ± 8.1	88.5	55.9

[a]Mean ± SE of 5 determinations.
[b]Calculated by the formula in Table 2 considering two values of the vacuolar pH (pH_e = 7.3).

Such results are however contradictory to another report (Deus-Neumann and Zenk, 1984) indicating that vacuoles could only absorb the alkaloids which the corresponding plant biosynthesizes. As an example, vacuoles of Catharanthus roseus were unable to absorb nicotine. We are currently looking for possible explanations to such a discrepancy (Guern et al., 1987). One first point is that these authors (Deus-Neumann and Zenk, 1984) isolated the vacuoles in 500 mM NaCl. We have observed that this leads to a large alkalinization of the vacuolar sap compared to vacuoles isolated in sorbitol (Barbier-Brygoo et al., 1987). Ion-trapping would then become strongly depressed due to the low pH gradient across the tonoplast.

CONCLUSION

These results agree satisfactorily to a model in which ajmalicine crosses the tonoplast by passive diffusion and is accumulated within vacuoles against a concentration gradient because of the pH gradient across the tonoplast and the low diffusion capacity of the ajmalicine cation. Moreover the accumulation within vacuoles is not specific since nicotine, not produced by the Catharanthus genus, can be also accumulated in vacuoles. No evidence for a specific carrier has been obtained. However, the possibility that a carrier-mediated uptake could coexist with diffusion cannot be excluded (Guern et al., 1987). One must then admit that in our conditions diffusion was by far the most important transfer process. Two kinds of factors could be at the origin of a favoured diffusive transfer compared to the situation encountered by others (Deus-Neumann and Zenk, 1984 and 1986): 1) the high transtonoplast ΔpH of the vacuoles we have isolated, and 2) the lipophilicity coupled to the very weak basicity of the alkaloid studied here. The variety of vacuolar characteristics (Barbier-Brygoo et al., 1987) and the large diversity of physicochemical properties in the alkaloid family allow many different situations in term of uptake, which must be further explored before building any generalized model.

The above results concern in fact only short-term uptake of ajmalicine by the vacuoles. From the analysis of the isotopic dilution of exogenous labelled ajmalicine by endogenous ajmalicine, we have previously shown that a true equilibrium of the exogenous ajmalicine between the cells and the medium, identical to that of the endogenous ajmalicine, was only reached after more 5 h (Renaudin et al., 1985). This reveals in fact a second mechanism of accumulation, complementary to the ion-trap process. Preliminary results indicate that ajmalicine is in fact present in vacuoles both under a quickly exchangeable form, in equilibrium with extracellular ajmalicine according to the ion-trapping model, and under a much more slowly exchangeable form we suggest to consist in bound molecules.

REFERENCES

Barbier-Brygoo, H., Renaudin, J.-P., Manigault, P., Mathieu, Y., Kurkdjian, A., and Guern, J., 1987, Properties of vacuoles as a function of the isolation procedure, in : "Plant Vacuoles. Their Importance in Solute Compartmentation and Their Applications in Biotechnology", B. Marin, ed., Plenum Publishing Corporation, New-York.

Brown, S. C., Renaudin, J. P., Prévot, C., and Guern, J., 1984, Flow cytometry and sorting of plant protoplasts : Technical problems and physiological results from a study of pH and alkaloids in Catharanthus cells, Physiol. Vég., 22:541.

Deus-Neumann, B., and Zenk, M. H., 1984, A highly selective alkaloid uptake system in vacuoles of higher plants, Planta, 162:250.

Deus-Neumann, B., and Zenk, M. H., 1986, Accumulation of alkaloids in plant vacuoles does not involve an ion-trap mechanism, Planta, 167:44.

Guern, J., Renaudin, J. P., and Barbier-Brygoo, H., 1987, Accumulation of solutes in plant vacuoles : Interpretation of data is not so easy, in : "Plant Vacuoles. Their importance in Solute Compartmentation and their Applications in Biotechnology", B. Marin, ed., Plenum Publishing Corporation, New-York.

Heldt, H. W., 1980, Measurement of metabolite movement across the envelope and of the pH in the stroma and the thylakoid space in intact chloroplasts, Methods in Enzymology, 69:604.

Kurkdjian, A. C., 1982, Absorption and accumulation of nicotine by Acer pseudoplatanus and Nicotiana tabaccum cells, Physiol. Vég., 20:73.

Kurkdjian, A. C., Quiquampoix, H., Barbier-Brygoo, H., Péan, M., Manigault, P., and Guern, J., 1985, Critical evaluation of methods for estimating the vacuolar pH of plant cells, in : "Biochemistry and Function of Vacuolar Adenosine-triphosphatase in Fungi and Plants", B. P. Marin, ed., Springer-Verlag, Berlin, Heidelberg, New-York and Tokyo.

Manigault, P., Manigault, J., and Kurkdjian, A. C., 1983, A microfluorimetric method for vacuolar pH measurement in plant cells using 9-aminoacridine, Physiol. Vég., 21:129.

Martinoia, E., Flügge, I., Kaiser, G., Heber, U., and Heldt, H. W., 1985, Energy-dependent uptake of malate into vacuoles isolated from barlkey mesophyll protoplasts, Biochim. Biophys. Acta, 806:311.

Renaudin, J. P., 1981, Uptake and accumulation of an indole alkaloid, ^{14}C-tabernanthine, by cell suspension cultures of Catharanthus roseus (L.) G. Don and Acer pseudoplatanus L., Plant Science Letters, 22:59.

Renaudin, J. P., and Guern, J., 1982, Compartmentation mechanisms of indole alkaloids in cell suspension cultures of Catharanthus roseus, Physiol. Vég., 20:533.

Renaudin, J.-P., Brown, S. C., and Guern, J., 1985, Compartmentation of alkaloids in a cell suspension of Catharanthus roseus : A reappraisal of the role of pH gradients, in : "Primary and Secondary Metabolism of Plant Cell Cultures", K. H. Neumann, W. Barz, and E. Reinhard, eds., Springer-Verlag, Berlin.

Renaudin, J. P., Brown, S. C., Barbier-Brygoo, H., and Guern, J., 1986, Quantitative characterization of protoplasts and vacuoles from suspension cultured cells of Catharanthus roseus, Physiol. Plant., in press.

ROLE OF THE VACUOLE IN METAL ION TRAPPING

AS STUDIED BY IN VIVO ^{31}P-NMR SPECTROSCOPY

Philip R. Pfeffer, Shu-I Tu, Walter V. Gerasimowicz
and Richard T. Boswell

United States Department of Agriculture, Agricultural Research
Service, Eastern Regional Research Center
600, East Mermaid Lane, Philadelphia, Pennsylvania 19118, USA

INTRODUCTION

^{31}P-NMR spectroscopy has been extensively used to study the energetic profiles in intact plant tissues (Roberts and Jardetzsky, 1981; Roberts, 1984; Loughman and Ratcliffe, 1984). These reports clearly demonstrate that this technique can be used to measure energy status, ATP/ADP ratio (Roberts et al., 1985), changes in intracellular pH (Roberts et al., 1981), effects of hypoxia (Roberts et al., 1984), phosphate movement (Rebeille et al., 1983), and aluminium ion toxicity (Pfeffer et al., 1986).

In a number of cases Mn^{2+} has been demonstrated to produce toxic effects in growing plants (Pfeffer et al., 1986). For example, in cotton Mn^{2+} lowers ATP concentrations and respiration rates as well as altering the activity of certain enzymes and hormones (Sirkar and Armin, 1979). In some cases healthy roots have been shown to reduce Mn^{2+} toxicity by precipitating oxidized Mn as MnO_2 on root surfaces (Foy, 1984). Plants such as maize may protect itself from such toxicity by entrapment of relatively high concentrations of the metal ion in the vacuole (Foy, 1973). A similar observation has been proposed for Cu^{2+} as well (Woodhouse and Walter, 1981).

Paramagnetic Mn^{2+} is a useful probe for examining the movement of a divalent cation via the ^{31}P-NMR spectrum of intact plant tissue (Loughman and Ratcliffe, 1984; Pfeffer et al., 1986). This is accomplished because the metal ion induces paramagnetic broadening on the ^{31}P resonances and this can be used to monitor the location of the metal as it migrates from one intracellular compartment to another (Loughman and Ratcliffe, 1984; Pfeffer et al., 1986).

In this report, we present our observations of 161.7 MHz ^{31}P-NMR studies of corn root tip tissue in which we have examined the movement of the divalent paramagnetic cation Mn^{2+} in and out the meristem tissue under different energetic conditions.

MATERIAL AND METHODS

Plant Tissue

Maize (Zea mays L. var FRB-73, Illinois Foundation Seeds) were germinated in a growth chamber at 28°C for 72 h as previously described (Pfeffer et al., 1986). Each experiment required approximately 700-900, 3-5 mm root tips which were generally examined by ^{31}P-NMR in the perfused state within one hour of excision.

Experimental Solutions

All solutions at pH 4 were unbuffered and contained from 0.1 to 10 mM calcium sulfate, 50 mM glucose, 25 mM sucrose or 5 mM 2-deoxyglucose with the addition of 1 mM $MnCl_2$. The pH of these perfusate solutions were monitored and adjusted throughout the experiments. Solutions at pH 6.0 were buffered with 10 mM Mes. All solutions were continually saturated with either, O_2 or N_2 gas and circulated through the root tissue at a rate > 45 ml/min. Prior to the connection between the perfusion tubes and the NMR tube, a 30 sec vacuum evacuation of the tube was carried out to remove trapped gas bubbles from between the roots. This procedure was important to prevent heterogeneous broadening of the spectra. A change from one perfusate solution to another was accomplished by means of a two way double stopcock assembly that connected two separate reservoirs to the peristaltic pump and NMR tube assembly. Prior to changing from one solution to another a 100 ml flush of the new perfusate through the system was carried out to minimize contamination.

NMR Experiments

A narrow bore (54 mm) JEOL GX-400 NMR spectrometer operating at 21-22°C was used to obtain the 161.7 MHz ^{31}P spectra of 700-900 excised (3-5mm) root tips as described previously (Pfeffer et al., 1986). In order to safely carry out perfusion/NMR experiments (without leakage hazards) for extended periods of time ($\sim$ 48 h) inside the magnet, an additional external suction tube was set into the reservoir of the NMR tube spinner housing to prevent overflow of liquid (in the event of a leak at the NMR cap) onto the probe and shim coil insert. In all other respects, the perfusion system design and operation was as previously described (Pfeffer et al., 1986). A reference capillary containing 120 mM HMPA (hexamethylphosphoramide) was used to give a satisfactory size reference peak for each spectrum. HPMA exhibited a resonance at 13.78 δ downfield from MDP (methylene diphosphoric acid). All chemical shifts were referenced relative to MDP which was assigned a value of 0.0 δ.

The concentrations of mobile phosphorus compounds in the root tissue samples, averaged over the total sample volume within the detector coils were determined by comparing the area of the signal from the HMPA resonance (observed in the tissue spectra) to the area response given by the phosphorus resonances for 1 mM standard solutions of ATP, glucose-6-phosphate and P_i in the same sample volume. Both the tissue and the standard solution spectra were acquired under quantitative conditions. Adjustements were then made for differences in signal responses (see above) to establish a direct relationship between the area of the resonances obtained under the fast and slow acquisition regions. A typical spectrum obtained from approximately 800 tips exhibited a concentration profile within the coil volume of the probe (1$\sim$2 ml) of approximately 1.0 - 2.0 mM sugar phosphates, 0.6 - 0.7 mM cytoplasmic P_i, 1.0 - 3.0 mM vacuolar P_i, 0.3 - 0.5 mM NTP, and 0.3 - 0.5 mM UDPG + NAD depending upon the age and size of the tips used. Since the data collection process for the experiment often lasted overnight, it was important to access the possible cellular aging effect on NMR signal characters. Spin lattice relaxation times (T_1) were determined by the inversion recovery method (180 - 90°) with 16 s repetition times for the non-nucleotide resonances and 4 s for the nucleotide resonances between scans, respectively. Relaxation values were calculated using a two-parameter exponential fit. The T_1 values, as reported earlier (Pfeffer et al., 1986) for both nucleotide and non-nucleotide resonances at the initiation and completion of a 34 h sequence of experiments were found to be identical within experimental error.

Estimation of pH

Cytoplasmic and vacuolar pH were estimated from standard calibration curves for P_i that have been described earlier (Pfeffer et al., 1986). Only very small changes, < 0.1 pH unit, were observed in the cytoplasmic pH over the duration of experiments up to 34 h. Because of the insensitivity of the chemical shift of the vacuole P_i at pH 5.5 no measurable change in the vacuolar pH could be detected throughout these experiments.

RESULTS AND DISCUSSION

Extensive use of ^{31}P-NMR in plant tissue studies has made it possible to evaluate the pH values, P_i content and NTP/ADP ratios of the cytosol and vacuole and to some extent pH changes in the vacuole. Fig. 1 shows the standard ^{31}P-NMR profiles of maize root tips (~ 800) under perfusion conditions using rapid (A) and short (B) instrument recycling times. The assignment of each well established resonance (Roberts, 1984; Loughman and Ratcliffe, 1984) is given in the figure. Although the rapid pulsing-technique (A) gives rise to distortion in the ratio of nucleotides to the other phosphates, it can still be used to enhance the former

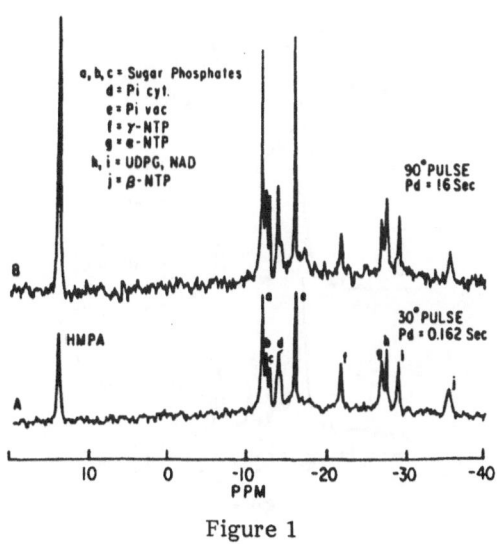

Figure 1

Intensity distortion due to rapid aquisition
A 161.7 MHz ^{31}P spectrum of approximately 800 excised corn root tips (3-5 mm)
taken with (A) rapid acquisition parameters
and with (B) slow acquisition parameters as indicated above

Each spectrum was obtained with a spectral width of 16,000 Hz and 2,000 data points zero filled to 16,000.

so that better quantitation of relative changes can be assessed. In order to evaluate the real concentrations of the various components, conversion for relaxation differences has been established to standardize area of resonance peaks. Typically, the concentration of compounds represented in the spectra approximately 0.60.7 mM cytoplasmic P_i, 1.0 - 3.0 mM vacuolar P_i, 0.3 - 0.5 mM NTP and 0.3 - 0.5 mM UDPG and NAD based on the resonance areas compared with the standard HPMA (hexamethylphosphoramide) at 13.78 δ and standard solutions of the corresponding compounds.

Effects of Mn^{2+}

Introduction of 1 mM Mn^{2+} into the perfusion medium containing a sufficient supply of glucose (50 mM) under aerobic conditions caused a rapid broadening of the NTP resonances (Fig. 2). This broadening is due to the paramagnetic properties of Mn^{2+} and its subsequent shortening of the relaxation properties of the ^{31}P nuclei of the strongly binding compounds. After 90 min of perfusion we also observed that P_i and sugar phosphates in the cytoplasm had undergone strong broadening. By 150 min, significant broadening of the vacuolar P_1 had occurred indicating that

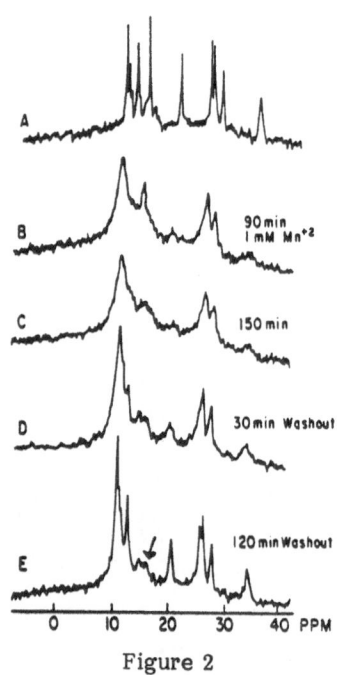

Figure 2

Mn^{2+} uptake in vacuole

A. ^{31}P spectra of approximately 800 excised corn root tips after perfusion for 2 hr with 0.1 mM $CaSO_4$, 50 mM glucose and 10 mM Mes buffer at pH 6.0 with O_2.
B. Same as in A after the addition of 1 mM $MnCl_2$ for period indicated.
C. Same as B after period indicated.
D. Washout following C for 30 minutes; washout after 120 minutes.

migration of Mn^{2+} across the tonoplast and into the vacuole became prominent. Exchange of the perfusate to one not containing Mn^{2+} caused a partial reversal of the broadening effects on all resonances representing components within the cytoplasm, however continual perfusion up to 5 hr (not shown) gave no regeneration of the vacuole P_i signal as in Fig. 2-E. Based on binding studies of Mn^{2+}, ATP and P_i as given in Fig. 3 we noted that the full broadening of the P_i resonance by Mn^{2+} required 6.4 times more Mn^{2+} at pH 5.5 than at pH 7.5. Consequently, in order to effect comparable line broadening of the vacuolar P_i resonance, the vacuolar (Mn^{2+}) must be 6.4 times higher that of cytoplasm. On a per phosphorus basis, ATP requires about 5.7 times the concentration of Mn^{2+} to undergo comparable linebroadening as cytoplasmic P_i. However, the ATP-Mn^{2+} complex is formed preferentially because of its tighter binding constant defined as :

$$\frac{ATP \; M_n \; K_{stab}}{P_i \; M_n \; K_{stab}} = 158$$

Considering the effective concentration of mobile phosphorus compounds observed in the spectra we estimate that there is approximately 0.015 mM Mn^{2+} associated with the P_i representing the vacuolar compartment when broadening is complete and 0.03 mM Mn^{2+} associated with the P_i and nucleotide in the cytoplasm.

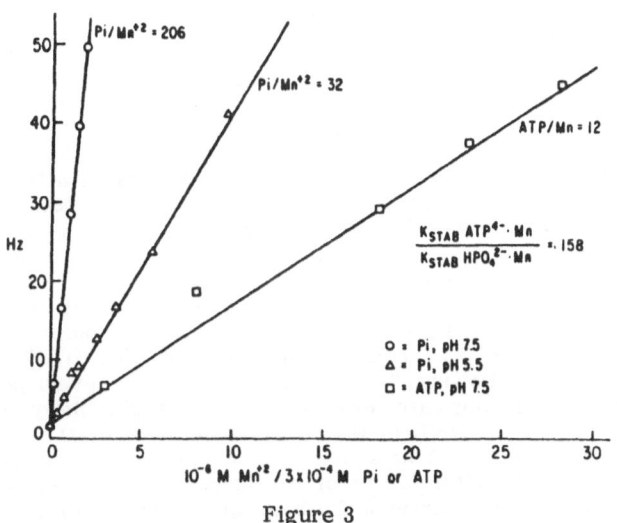

Figure 3

Phosphate and ATP linebroadening with Mn^{2+}

Mn^{2+} Migration as a Function of Carbohydrate Supply and Metabolism

In order to understand the effects of the carbohydrate supply on the influx of Mn^{2+} we examined its relative rate of uptake in the absence of exogenous glucose. Omission of glucose in the perfusion medium followed by the introduction of Mn^{2+} showed a marked slowing of the migration and subsequent broadening of the vacuolar P_i resonance (90 min with glucose, 135 min without glucose). These results indicate that at a somewhat lowered energy state, the Mn^{2+} migration across the tonoplast membrane is inhibited. Movement, however, into the cytoplasm as evidenced by

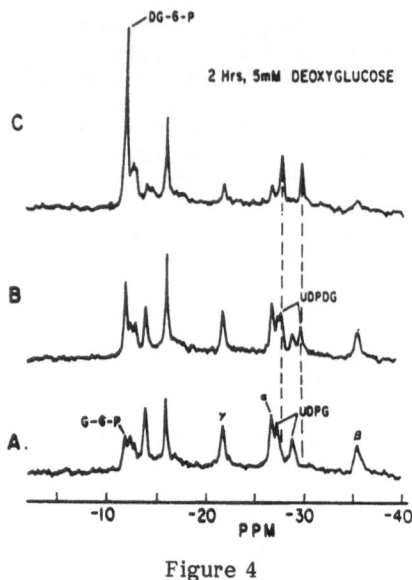

Figure 4

Incubation with 2–deoxy–glucose

[31]P spectra of corn root tips (A) before addition of 5 mM 2–deoxy–glucose, (B) after 30 minutes treatment with 5 mM 2–deoxy–glucose, (C) after 2 hr treatment with 5 mM 2–deoxy–glucose.

rapid cytoplasmic resonance linebroadening was insignificantly affected. To evaluate this phenomenon more directly we decided to look at the response of the metal ion movement by inhibiting the glycolytic pathway directly. Fig. 4 shows the effect of the initial incubation of excised root tips with 5 mM 2–deoxyglucose in the absence of any other exogenous carbohydrate source. After 0.5 hr we noted the buildup of the 2–deoxyglucose–6–phosphate in the cytoplasm due to its lack of conversion to fructose–6–phosphate·in the glycolytic pathway. Also we observed the production of two new resonances at ~ 27 and 29 ppm corresponding to the 2–deoxyglucose derivatives of UDPG. Following 2 hr of perfusion, NTP levels decreased by 2/3 (due to the decrease of substrate–level and oxidative phosphorylation) and UDPG was replaced by UDP-2-deoxy-G. In this state, treatment of the tips in the normal manner with 1 mM Mn^{2+} for 2 hr showed diminished Mn^{2+} broadening or invasion into the vacuole (Fig. 5). After 5.5 hr of washings, little if any, broadening of the vacuolar P_i resonance was observed. Thus in the presence of a minimal level of NTP only a marginal amount of metal ion could be moved effectively and trapped in the vacuole.

Effects of Hypoxia (N_2) on Metal Ion Trapping

It has been well established that during hypoxia, maize root tips can be sustained by anaerobic metabolism with much reduced levels of NTP and production of lactate and ethanol (Roberts et al., 1984). In this altered state the pH of the cytoplasm is maintained at pH 6.7-6.9 as opposed to 7.5 to 7.6 in the aerobic state (Fig. 6). Also note (in Fig. 6-B) that under hypoxia we observed the presence of NDP, an increased concentration of cytoplasmic P_i and only 35 - 40% of the original NTP levels. Following 108 min of exposure to Mn^{2+} we observed the complete loss

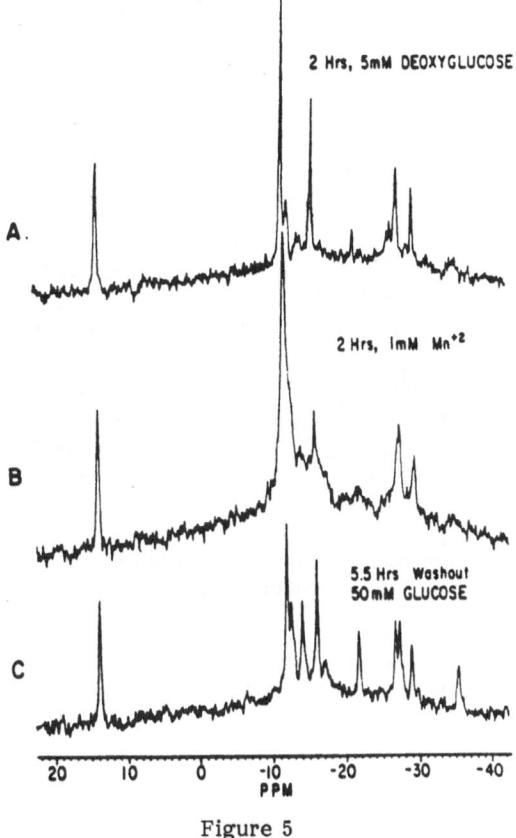

Figure 5

Suppression of Mn^{2+} migration with 2–deoxy–glucose

^{31}P spectra showing suppression of Mn^{2+} migration into the vacuole following a 2 hr treatment with 5 mM 2-deoxy-glucose. A. Spectrum prior to treatment with 1 mM $MnCl_2$; B. Spectrum following 2 hr treatment with 1 mM $MnCl_2$; C. Spectrum resulting from 5.5 hr wash-out with 10 mM Mes, pH 6.0, 50 mM glucose and 0.1 mM $CaSO_4$ perfusate.

of the ATP resonances, however no change in the observed line widths corresponding to the other cytoplasmic compounds. This suggests that only a very minimal amount of Mn^{2+} migration across the plasmalemma has taken place. Efforts to wash out this small amount of Mn^{2+} from the cytoplasm under an N_2 atmosphere was unsuccessful. With the resumption of an O_2 atmosphere, however, Mn^{2+} washout was complete and full generation of the narrow line width spectrum was complete.

Competition of Ca^{2+} with Mn^{2+} for Migration in Root Tips

It is well established that the application of Ca^{2+} to the soil decreases the effect of toxic metal ions (Lund, 1970). Yet, it is not known whether this is a direct consequence of Ca^{2+} competing more favorably than other metal ions for movement

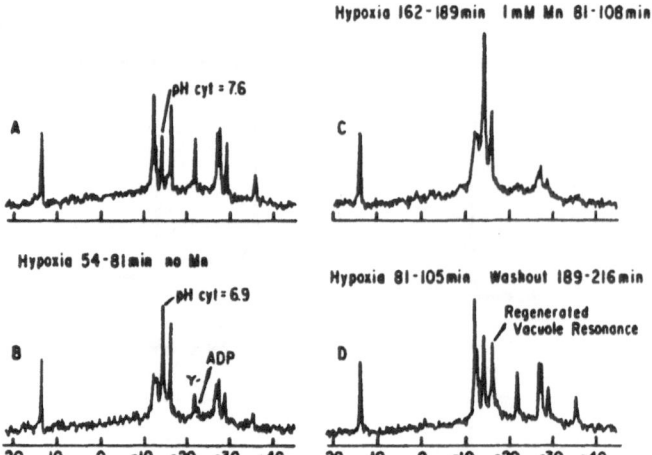

Figure 6

Suppression of Mn^{2+} migration to vacuole
under hypoxia

^{31}P spectrum of corn root tips at pH 6.0, 10 mM Mes, 0.1 mM $CaSO_4$ and 50 mM glucose. A. With O_2; B. Following 54–81 minutes of hypoxia (N_2); C. Spectrum during 162–189 minutes of hypoxia and 81–108 minutes perfusion with medium containing 1 mM $MnCl_2$; D. Final washout under O_2 after 189–216 minutes with medium that does not contain Mn^{2+}.

across the plasmalemma. Treatment of maize root tips with a perfusate containing 10.0 mM Ca^{2+}, 5.0 mM glucose at pH 6.0 produced a significantly less broadened spectrum following 150 min of exposure to Mn^{2+} (Fig. 7). Subsequent washout of the Mn^{2+} containing perfusate gave a spectrum which showed a slight indication of the presence of Mn^{2+} in the vacuole with no indication of its presence in the cytoplasm. Clearly, under the employed conditions, Mn^{2+} movement across the plasmalemma in both directions is free of restrictions. However, in the presence of high concentration of Ca^{2+}, less Mn^{2+} can reach the cytoplasm within the limited 150 min of perfusion. Presumably, Mn^{2+} has to compete with Ca^{2+} for the same transport mechanism associated with the plasma membrane. Consequently, the secondary migration of a diminished concentration of Mn^{2+} across the tonoplast is significantly suppressed. We have also observed a similar effect with Cd^{2+}, except in this instance Ca^{2+} was capable of preventing the cytoplasmic "poisoning" induced by Cd^{2+} invasion (unpublished results).

Divalent Cation Transport Mechanism

The data obtained in this study demonstrate that the uptake of metal ions, e.g. Mn^{2+}, is regulated by the energetic status (O_2 + exogenous glucose > O_2 > O_2 + 2-deoxy-glucose > N_2 + glucose) of the root cells. Furthermore, the uptake of different divalent cations may compete for the same membrane apparatus. Based on the accepted general concepts (Racker, 1979) of ion transport developed mainly from research in non-plant systems, a mechanism, as depicted in Fig. 8, is proposed to account for the observations mentioned in this report.

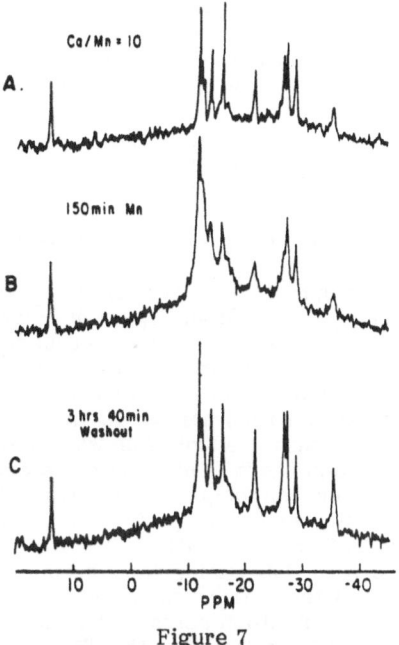

Figure 7

Ca suppression of Mn passage

A. ^{31}P spectrum of corn root tips at pH 6.0, 10 mM Mes, 10.0 mM $MnSO_4$, 50 mM glucose, O_2 atmosphere; B. Following 150 min treatment with same medium containing 1 mM $MnCl_2$; C. Following 3 hr. and 40 minutes of washout with medium as in A except containing 0.1 mM $CaSO_4$.

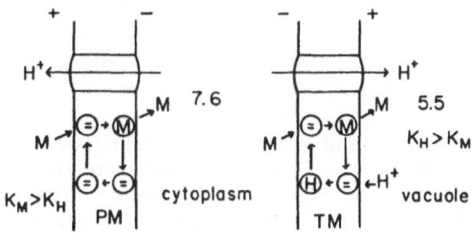

Figure 8

Proposal for transport mechanism of M(2+)

Our basic assumptions are the following :

- (1) The proton pumping processes associated with the plasma membrane (PM) and the tonoplast membrane (TM) generate transmembranous proton electrochemical potentials in the directions as shown in Fig. 8 (Sze, 1985) (active H^+-pumping outward from cytoplasm, membrane potential negative in cytoplasm).

- (2) The actinal passage of metals (M) through the membranes is facilitated by certain negatively-charged membrane carriers which can bind with either protons or metal ions.

- (3) The affinities of carriers to protons (K_H) or metal ions (K_M) are regulated by pH.

Thus, under the conditions of O_2 + exogenous glucose, the added metal ion, e.g. Mn^{2+}, binds with the PM carrier to form a natural complex which diffuses through PM and releases the metal. The negatively charged carrier (the alkaline cytoplasmic pH $\sim$ 7.8 precludes the protonotion of PM carrier) is then driven back to the exterior surface of the membrane for the next loading of metal ion. Although the transport is diffusion in nature (along a concentration gradient), the process is facilitated by the energy imput in the form of a membrane potential to enhance the returning of the negatively charged carrier to the exterior surface of PM. The transport of M from the cytoplasm to vacuole follows a similar mechanism. However, the low pH of the vacuole region tends to protonate the carrier on the inner surface of TM. The protonated carrier can only slowly diffuse back to the cytoplasmic side (no facilitated movement) when the protons are released. Since the motion of the TM carrier is not facilitated in either direction, the uptake across TM is slower than that associated with PM. During the washout process, the concentration gradient of M is reversed and the carrier mediated back diffusion through the PM takes place. However, the back diffusion across TM is hindered because the vacuolar metal ion can not compete with protons for the TM carrier. Thus, metal ions are trapped.

Under hypoxia conditions, the cytoplasmic pH decreases to $\sim$ 6.9 or lower. This acidic condition promote the protonation of both PM and TM carriers at cytoplasmic surfaces. Consequently, the facilitated motion of the PM carrier is hindered and the uptake across PM is by diffusion only. The transport across TM becomes unnoticeable because the protonated TM carrier cannot release its protons. The inefficient washout of M from cytoplasm under hypoxia is expected because the relatively acidic cytoplasm may trap M by the similar mechanism for vacuolar M trapping.

The experimental results obtained under the conditions of either the omission of exogeneous glucose or the use of 2-deoxy-glucose are also expected since the cells are in intermetant energy states between O_2 + exogeneous glucose and N_2 + glucose.

REFERENCES

Foy, C. D., 1973, Manganese and plants, in : "Manganese", National Academy of Sciences, National Research Council, Washington, D. C.

Foy, C. D., 1984, Physiological effects of hydrogen, aluminium and toxicities in acid soil, in : "Soil Acidity and Liming", American Society of Agronomy, Inc., Madison, Wisconsin, USA.

Loughman, B. C., and Ratcliffe, R. G., 1984, Nuclear magnetic resonance and the study of plants, in : Advances in Plant Nutrition", Vol. 1, Praeger Publishers, New-York.

Lund, Z. F., 1970, The effect of calcium and its relation to some other cations on soybean root growth, Soil Sci. Soc. Am. Proc., 34:456.

Pfeffer, P. E., Tu, S.-I., Gerasimowicz, W. U., and Cavanaugh, J. R., 1986, In vivo ^{31}P-NMR studies of corn root tissue and its uptake of toxic metals, Plant Physiol., 80:77.

Racker, E., 1979, Transport of ions, Act. Chem. Res., 12:338.

Rebeille, F., Bligny, R., Martin, J.-B., and Douce, R., 1983, Relationship between the cytoplasm and the vacuole phosphate pool in Acer pseudoplatanus cells, Arch. Biochem. Biophys., 225:143.

Roberts, J. K. M., 1984, Study of plant metabolism in vivo using NMR spectroscopy, Ann. Rev. Plant Physiol., 35:375.

Roberts, J. K. M., Callas, J., Wemmer, D., Walbot, V., and Jardetzky, O., 1984, The mechanism of cytoplasmic pH regulation in hypoxic maize root tips and its role in survival and hypoxia, Proc. Natl. Acad. U. S. A., 81:3379.

Roberts, J. K. M., and Jardetzky, O., 1981, Monitoring of cellular metabolism by NMR, Biochim. Biophys. Acta, 639:53.

Roberts, J. K. M., Jardetzsky, N. M., and Jardetzsky, O., 1981, Intracellular pH measurements by ^{31}P-NMR. Influence of factors other than pH on ^{31}P chemical shifts, Biochemistry, 20:5389.

Roberts, J. K., Lane, A. N., Clark, R. A., and Nieman, R. H., 1985, Relationship between the rate of synthesis of ATP and the concentrations of reactants and products of ATP hydrolysis in maize root tips, determined by ^{31}P-NMR, Arch. Biochem. Biophys., 240:712.

Sirkar, S., and Amin, J.V., 1979, Influence of auxin on respiration of manganese toxic cotton plants, Ind. J. Exp. Biol., 17:618.

Sze, H., 1985, H$^+$-translocating ATPases : Advances using membranes, Annu. Rev. Plant Physiol., 36:175.

Woodhouse, H. W., and Walter, S., 1981, The physiological basis of copper toxicity and copper tolerance in higher plants, in : "Copper in Soils and Plants", Academic Press, Sydney, Australia.

DYNAMICS OF LYSOSOMAL FUNCTIONS

IN PLANT VACUOLES

Thomas Boller and Andres Wiemken

Botanisches Institut, Abteilung Pflanzenphysiologie
Universität Basel, Hebelstrasse 1
CH-4056-Basel, Switzerland

INTRODUCTION

We have recently reviewed the dynamics of vacuolar compartmentation in plants (Boller and Wiemken, 1986). Here, we discuss one particular aspect, the lysosomal functions of vacuoles, in more detail.

Vacuoles of yeast (Matile and Wiemken, 1967; Wiemken et al., 1979) and of higher plants (reviewed in Boller, 1982; Boller and Wiemken, 1986) contain most of the intracellular activity of many typical " lysosomal" acid hydrolases. For example, α-mannosidase, considered to be a typical lysosomal enzyme in animal tissues, is exclusively vacuolar in protoplasts isolated from yeast (Wiemken et al., 1979) and from various plants (Boller and Kende, 1979). While the enzyme is bound to the vacuolar membrane in yeast (Van der Wilden et al., 1973), it is soluble in higher plants (Boller and Kende, 1979). In higher plants, α-mannosidase can therefore be used as a marker for the vacuolar contents. A number of other glycosidases and phosphodiesterases, including RNase (Abel and Glund, 1986), are also primarily vacuolar.

The localization of proteases has been studied in a variety of plants. In most cases, proteolytic activity has been found to be located primarily in the vacuole (Nishimura and Beevers, 1978; Boller and Kende, 1979; Heck et al., 1981; Lin and Wittenbach et al., 1982; Thayer and Huffaker, 1984; Canut et al., 1985 a; Bhalla and Dalling, 1986).

Thus, in general, the enzymatic equipment of plant vacuoles is homologous to that of the animal lysosomes. It should be noted that many of these hydrolases are present, in addition, in the apoplast (Matile, 1975). In bean leaves, even the proteolytic activity is largely extracellular (Van der Wilden et al., 1983). In cell suspension cultures, a number of hydrolytic activities are found in the culture medium (Wink, 1985). Therefore, as pointed out by Matile (1975), the apoplast, like the vacuole, may be considered part of the "lytic compartment" of plants.

What is the functional significance of the "lysosomal enzymes" in plant vacuoles ? Below, we want to discuss three possible functions.

Autophagy, the hydrolysis and degradation of cytoplasmic constituents, is a function classically attributed to the lysosome. Ultrastructural studies indicate that autophagic processes play a role in the ontogeny of vacuoles. However, it remains an open question whether or not mature plant vacuoles participate in autophagy.

The mobilization of specific vacuolar reserves by hydrolases is a slightly different matter. In this case, specific reserve proteins or carbohydrates accumulate in the vacuole and can later be mobilized by hydrolases. Several well-documented examples implicate vacuolar hydrolases in the mobilization of storage molecules.

Heterophagy, the engulfment and degradation of foreign material, is another of the classical functions of lysosomes. In this regard, it is interesting that processes similar to heterophagy occur in many mutualistic and antagonistic interactions between plants and microorganisms.

AUTOPHAGY : DEGRADATION OF CYTOPLASMIC COMPONENTS

Autophagy in the early development of vacuoles

The ontogeny of vacuoles has been much studied by electron microscopy. In thin sections of developing meristems, membrane-bound hollow rings have been observed which encircle portions of the cytoplasm (see e.g. Buvat and Robert, 1979). Cytochemical tests have shown these rings to contain acid hydrolases (Buvat and Robert, 1979).

Marty (1978) has interpreted such structures on the basis of three-dimensional pictures of thick sections stains with zinc iodide-osmium, obtained with the high-voltage electron microscope. He observed, in material from early developmental steps, a heavily stained tubular membrane network at the trans face of the Golgi apparatus, called GERL (Golgi-ER-lysosome system). During cell expansion, the GERL network appeared to enclose portions of the cytoplasm by lateral expansion and fusion of membrane tubules. According to Marty (1978), the sequestered cytoplasm is finally sealed off by a double layer of GERL-derived membrane sheets, and the narrow space between the membranes is filled with lysosomal enzymes which, after rupture of the inner membrane, digested the enclosed cytoplasm.

While Marty's (1978) pictures are very suggestive, it should be noted that the final phase of the process, i.e. the total enclosure and dissolution of the cytoplasm, could not actually be seen in his pictures. Membrane configurations similar to the early phases of ontogeny have also been observed by in vivo fluorescence microscopy of developing Allium stomatal cells (Palewitz et al., 1981). Here, there is no evidence for autophagic processes. Hilling and Ameluxen (1985) have suggested that in various tissues, the vacuole originates from circularly arranged ER by dilatation of the ER cavern and displacement of the encircled cytoplasm rather than by autophagic processes.

It would be interesting to complement the ultrastructural studies with a biochemical analysis of protein turnover. This might lead to a more quantitative assessment of the importance of lytic processes in the early development of vacuoles.

Autophagy in mature vacuoles ?

As pointed out earlier by Leigh (1979), the localization of hydrolases in mature vacuoles does not prove that they have any function there. They could be present in vacuoles as remains of the autophagic processes that took place early in the ontogeny of the vacuole.

In recent years, a number of investigators examined a possible role of vacuoles in autophagy, particularly in the degradation of chloroplast proteins during senescence. Wittenbach et al. (1982) studied senescing wheat leaves. They found that chloroplast number and the main chloroplast protein, RuBPCase, decreased concomitantly. This, in conjunction with electron microscopic pictures of chloroplasts engulfed in the vacuole, lead them to conclude that the whole chloroplasts are transported into the vacuoles and then degraded.

Martinoia et al. (1983) studied senescing barley leaves and obtained different results. They found that the chloroplast number remained approximately constant during senescence. Later, Wardley et al. (1984) reexamined senescing wheat leaves. They observed that chloroplast stability decreased in the course of senescence. Using special precautions to preserve the structural integrity of chloroplasts after disruption of the protoplasts, they found that the number of chloroplasts was reduced by less than 20% during senescence, while RuBPCase content diminished 90%.

This suggests that chloroplasts are not taken up as whole structures and then digested. Rather, RuBPCase disappears from the chloroplasts, indicating that it is either digested within the chloroplast (Schmidt and Mishkind, 1983) or that it is exported specifically to the vacuole (Vaughn and Duke, 1981).

Perharps, it is only in terminal senescence, when the tonoplast disintegrates, that indiscriminate degradation of cytoplasmic components takes place.

HYDROLYTIC MOBILIZATION OF RESERVE SUBSTANCES

Mobilization of reserve proteins

Storage proteins are deposited in protein bodies, an organelle considered homologous to the vacuole. In developing legume cotyledons, protein bodies are in fact generated by multiple division of the central vacuole (see Vitale and Chrispeels, 1984). In cereal endosperm, protein bodies appear to be formed from the ER directly (Higgins, 1984).

During germination, the proteins stored in the protein bodies are mobilized by hydrolysis. Immunocytochemical ultrastructural studies in mung bean have shown that newly formed protease first appears in small vesicles and is later transferred into the protein bodies (Baumgartner et al., 1978). This protease degrades the main storage protein, vicilin. Ultrastructural evidence indicates that, at this stage, protein bodies function also as autophagic organelles which take up and digest portions of the cytoplasm (Van der Wilden et al., 1980). In Ricinus endosperm, the protein bodies fuse during germination and form large vacuoles. These vacuoles can be isolated and continue to degrade the storage proteins in vitro (Nishimura and Beevers, 1979). The protein bodies in barley aleurone cells can similarly form vacuoles under appropriate conditions (Jacobsen et al., 1985).

In general terms, the accumulation and mobilization of storage proteins in seeds is a partcularly clear example of the dynamics of vacuolar function. During legume seed development, for example, the vacuole is filled with specific storage proteins via the ER–Golgi pathway (Chrispeels, 1984); at the same time, the vacuole fragments into protein bodies which are dehydrated. During germination, protease is imported, again via a vesicular pathway, and the stored proteins are degraded.

Accumulation and mobilization of specific storage proteins in vacuoles may occur also in vegetative organs. In the so-called paraveinal mesophyll of soybean leaves, specific glycoproteins accumulate in the vacuoles at anthesis and disappear again during pod filling (Franceshi and Giaquinta, 1983 a and b).

A rapid degradation of proteins has been found in vacuoles isolated from Acer cell cultures (Canut et al., 1985 b). At present, it is not yet clear whether this represents turnover of cytoplasmic proteins taken up unspecifically by the vacuoles, or of specific proteins sequestered in the vacuole for temporary storage. Interestingly, abnormal proteins, containing fluorophenylalanine instead of phenylalanine, accumulate in the vacuole to a larger extent than normal proteins; both the normal and abnormal proteins are similarly degraded (Canut et al., 1986).

Little is known about the mobilization of storage proteins in roots and tubers. It is interesting, however, that the predominant storage protein of potato, patatin, has an enzymatic activity : It is a typical lysosomal hydrolase, lipid acyl hydrolase (Racusen, 1984). This example illustrates that acid hydrolases can have the function of storage proteins.

Mobilization of Polysaccharides

Sucrose is stored in the vacuoles of some storage organs, e.g. in beet root. Leigh et al. (1979) have shown that mobilization of vacuolar sucrose is correlated with an increase in activity of vacuolar invertase. Similarly, in Stachys tubers which store stachyose in the vacuoles, the first enzyme used for mobilization of this oligosaccharide, α-galactosidase, occurs in the vacuole (Keller and Matile, 1985).

Some plants store fructans in the vacuole. The vacuoles isolated from such plants, e.g. from barley leaves (Wagner et al., 1983; Wagner and Wiemken, 1986) or from Jerusalem artichoke tubers (Frehner et al., 1984), contain both the fructosyl-transferase necessary for fructan formation and the fructan hydrolase necessary for its breakdown. Here, accumulation and mobilization are probably regulated by increases and decreases of enzyme activities as well as by changes in substrate and effector concentrations (Wagner et al., 1986).

HETEROPHAGY : INTERACTIONS WITH MICROORGANISMS

Little is known about heterophagic functions of plant vacuoles. Animal cells generally take up the foreign material to be degraded by endocytosis. Ultrastructural work indicates that plant protoplasts can take up extracellular material by endocytosis (see e.g. Tanchak et al., 1984; Joachim and Robinson, 1984). More surprisingly, it has recently been found that intact yeast cells (Riezman, 1985; Makarow, 1985) have an endocytotic pathway also, despite of the turgor pressure and the rigid cell wall which, according to common sense, render endocytosis impossible. Endocytosis may also exist in intact cells of higher plants (Hübner et al., 1985; Romanenko et al., 1986).

In interactions of plants with microrganisms, processes reminiscent of heterophagy come into play. Some examples are discussed below.

Rhizobia

A long-recognized special case of endocytosis occurs in legume nodules. Rhizobia are taken up into membrane-bound vesicles within the cytoplasm (see e.g. Werner and Mörschel, 1978). While the bacteria multiply intracellularly, they remain enclosed within a plant membrane, the so-called peribacteroid membrane. The peribacteroid membrane contains an ATPase (Blumwald et al., 1985) and antigenic determinants (Brewin et al., 1985) in common with the plant plasmalemma and the Golgi system. However, the peribacteroid space also contains α-mannosidase, a marker for plant vacuoles and the lytic compartment (Mellor et al., 1984), indicating that the engulfed bacterium is in a space resembling a vacuole. Nevertheless, in normal, functional nodules, endocytosis is not followed by lysis of the bacteria. The process of "heterophagy" is arrested after endocytosis. In senescing pea nodules, bacteroids degenerate and disintegrate within the peribacteroid membrane (Truchet and Coulomb, 1973; Kijne, 1975). Here, the heterophagic process appears to continue, although its is not clear whether plant hydrolases or bacterial autolytic enzymes cause the observed degeneration. Cases of complete heterophagy emerge also from ultrastructural work with certain ineffective soybean nodules (Mellor et al., 1984; Werner et al., 1984). In these examples, the peribacteroid membranes fuse and form vacuoles in which the bacteroids are lysed.

Biotrophic Fungi

In interactions of plants with pathogenic biotrophic fungi, the fungal haustoria protrude deeply into the plant cell. The haustoria remain surrounded by a plant membrane, the so-called extrahaustorial membrane, which is continuous with the plasma membrane. However, ultrastructural work shows that the extrahaustorial membrane differs from the plasma membrane with regard to staining and histochemical properties (Manners and Gay, 1983; Harder and Chong, 1984). In interactions of plant roots with symbiotic fungi forming vesicular-arbuscular mycorrhiza, the intracellular "arbuscules" are surrounded by a similar type of membrane (Scannerini and Bonfante-Fasolo, 1983). Mature arbuscules occupy almost half of the cell volume, at the expense of vacuolar volume; later, arbuscules degenerate, and the vacuolar volume increases again (Toth and Miller, 1984). It is tempting to speculate that the plant membrane surrounding the fungal organ can develop into a tonoplast-like membrane, and that the fungus is thereby enclosed in a kind of vacuole within the plant cell. As in the case of rhizobia, the heterophagic process is temporarily arrested in mid-term.

However, some heterotrophic plants, for example the Monotropaceae and most orchid seedlings, live on heterophagy. They also form mycorrhiza but appear to digest the intracellular fungal hyphae for nutrition. Ultrastructural studies of mycorrhiza of some Monotropaceae (Duddridge and Read, 1982; Robertson and Robertson, 1982) have revealed that, during this process, each intracellular fungal hypha remains enclosed in a membranous sac which is continuous with the plant plasmalemma. The fungal hypha appears to rupture at its tip and to release material into this vacuole-like membranous sac. Unfortunately, the biochemistry of these interactions have not yet been studied in detail. It remains to be seen if plant hydrolases take part in the digestion of the released fungal material.

Hypersensitive Reaction

Lytic processes are also important in the plant's defense against pathogens. One of the most frequently observed defense reactions of plants is the co-called hypersensitive reaction, in which a few cells around an invading pathogen die and release the toxic chemicals and the lytic enzymes stored in their vacuoles. In this context, it is interesting that plant vacuoles can contain large amounts of chitinase (Boller and Vögeli, 1984). Chitinase has no substrate in the plant itself but readlily attacks and partially digests fungal cell walls (Boller et al., 1983).

CONCLUSION

The lytic functions of plant vacuoles are similar to those of the animal lysosomes but differ in detail. Autophagy of cytoplasmic material, an important function of the animal lysosomes (Dean, 1984), may play some role in the ontogeny of vacuoles, but there is no clear-cut evidence, at present, that this type of autophagy occurs in mature vacuoles. However, plant vacuoles frequently accumulate and mobilize specific reserve substances. This dynamic process can be considered a specialized form of autophagy. The uptake of extracellular material by endocytosis, an important mode of access to the lysosome in animal cells (Dean, 1984), has only recently been discovered in fungal and plant cells. However, processes resembling heterophagy occur in the interactions of plants with microorganisms. Both mutualistic and pathogenic microorganisms can be taken up into the interior of plant cells by invaginations of the plasma membrane. Frequently, the heterophagic process appears to be arrested in mid-term. However, in some cases, the microorganisms appear to be "digested" within the plant cell. It remains to be demonstrated that the lytic enzymes of the plant play a role in these interactions.

REFERENCES

Abel, S., and Glund, K., 1986, Localization of RNA-degrading enzyme activity within vacuoles of cultured tomato cells, Physiol. Plant., 66: 79.

Baumgartner, B., Tokuyasu, K. T., and Chrispeels, M. J., 1978, Localization of vicilin peptidohydrolase in the cotyledons of mung bean seedlings by immunofluorescence microscopy, J. Cell. Biol., 79:10.

Bhalla, P. L., and Dalling, M. J., Endopeptidases and carboxypeptidase enzymes of vacuoles from mesophyll protoplasts of the primary leaf of wheat seedlings, J. Plant Physiol., 122:289.

Blumwald, E., Fortin, M. G., Rea, P. A., Verma, D. P. S., and Poole, R. J., 1985, Presence of host-plasma membrane type H^+-ATPase in the membrane envelope enclosing the bacteroids in soybean root nodule, Plant Physiol., 78:665.

Boller, T., 1982, Enzymatic equipment of plant vacuoles, Physiol. Vég., 20:247.

Boller, T., and Kende, H., 1979, Hydrolytic enzymes in the central vacuole of plant cells, Plant Physiol., 63:1123.

Boller, T., Gehri, A., Mauch, F., and Vögeli, U., 1983, Chitinase in bean leaves : Induction by ethylene, purification, properties, and possible function, Planta, 157:22.

Boller, T., and Vögeli, U., 1984, Vacuolar localization of ethylene-induced chitinase in bean leaves, Plant Physiol., 74:442.

Boller, T., and Wiemken, A., 1986,, Dynamics of vacuolar compartmentation, Annu. Rev. Plant Physiol., 37:137.

Brewin, N. J., Robertson, J.G., Wood, E.A., Welles, B., Larkins, A. P., Galfre, G., and Butcher, G. W., 1985, Monoclonal antibodies to antigens on the peribacteroid membrane from Rhizobium-induced root nodules of pea cross-react with plasma membranes and Golgi bodies, EMBO J., 4:605.

Buvat, R., and Robert, G., 1979, Vacuole formation in the actively growing root meristem of barley (Hordeum sativum), Amer. J. Bot., 66:1219.

Chrispeels, M. J., 1984, Biosynthesis, processing and transport of storage proteins and lectins in cotyledons of developing legume seeds, Phil. Trans. R. Soc. London B, 304:309.

Canut, H., Alibert, G., and Boudet, A. M., 1985 a, Proteases of Melilotus alba mesophyll protoplasts. I. Intracellular localization, Plant Physiol., 39:163.

Canut, H., Alibert, G., and Boudet, A. M., 1985 b, Hydrolysis of intracellular proteins in vacuoles isolated from Acer pseudoplatanus cells, Plant Physiol., 79:1111.

Canut, H., Alibert, G., Carrasco, A., and Boudet, A. M., 1986, Rapid degradation of abnormal proteins in vacuoles from Acer pseudoplatanus L. cells, Plant Physiol., 81:460.

Dean, R. T., 1984, Modes of access of macromolecules to the lysosomal interior, Biochem. Soc. Trans., 12:911.

Duddridge, J. A., and Read, D. J., 1982, An ultrastructural analysis of the development of mycorrhizas in Monotropa hypopitys L., New Phytol., 92:203.

Franceschi, V. R., and Giaquinta, R.T., 1983 a, The paraveinal mesophyll of soybean leaves in relation to assimilate transfer and compartmentation. I. Ultrastructure and histochemistry during vegetative development, Planta, 157:411.

Franceschi, V. R., and Giaquinta, R. T., 1983 b, The paraveinal mesophyll of soybean leaves in relation to assimilate transfer and compartmentation. II. Structural, metabolic and compartmental changes during reproductive growth, Planta, 157:422.

Frehner, M., Keller, F., and Wiemken, A., 1984, Localization of fructan metabolism in the vacuoles isolated from protoplasts of Jerusalem artichoke tubers (Helianthus tuberosus L.), J. Plant Physiol., 116:197.

Harder, D. E., and Chong, J., 1984, Structure and Physiologia of Haustoria, in: "The Cereal Rusts", Vol. 1, W. R. Bushnell and A. P. Roelfs, eds., Academic Press, Orlando.

Heck, U., Martinoia, E., and Matile, P., 1981, Subcellular localization of acid proteinase in barley mesophyll protoplasts, Planta, 151:198.

Higgins, T. J. V., 1984, Synthesis and regulation of major proteins in seeds, <u>Annu.</u> <u>Rev. Plant Physiol.</u>, 35:191.

Hilling, B., and Amelunxen, F., 1985, On the development of the vacuole. II. Further evidence for endoplasmic reticulum origin, <u>Eur. J. Cell Biol.</u>, 38:195.

Hübner, R., Depta, H., and Robinson, D. G., 1985, Endocytosis in maize root cap cells. Evidence obtained using heavy metal solutions, <u>Protoplasma</u>, 129: 214.

Jacobsen, J. V., Zwar, J. A., and Chandler, P. M., 1985, Gibberellic acid-responsive protoplasts from mature aleurone of Himalaya barley, <u>Planta</u>, 163:430.

Joachim, S., and Robinson, D. G., 1984, Endocytosis of cationized ferritin by bean leaf protoplasts, <u>Eur. J. Cell Biol.</u>, 34:212.

Keller, F., and Matile, P., 1985, The role of the vacuole in storage and mobilization of stachyose in tubers of <u>Stachys sieboldii</u>, <u>J. Plant Physiol.</u>, 119:369.

Kijne, J. W., 1975, The fine structure of pea root nodules. 2. Senescence and disintegration of the bacteroid tissue, <u>Physiol. Plant Pathol.</u>, 7:17.

Leigh, R. A., 1979, Do plant vacuoles degrade cytoplasmic components ? <u>Trends Biochem. Sci.</u>, 4:37.

Leigh, R. A., ap Rees, T., Fuller, W. A., and Banfield, J., 1979, The location of acid invertase activity and sucrose in the vacuoles of storage roots of beet root (<u>Beta vulgaris</u>), <u>Biochem. J.</u>, 178:539.

Lin, W., and Wittenbach, V. A., 1981, Subcellular localization of proteases in wheat and corn mesophyll protoplasts, <u>Plant Physiol.</u>, 67:969.

Makarow, M., 1985, Endocytosis in <u>Saccharomyces cerevisiae</u>. Internalization of amylase and fluorescent dextran into cells, <u>EMBO J.</u>, 4:1861.

Manners, J. M., and Gay, J. L., 1983, The host-parasite interface and nutrient transfer in biotrophic parasitism, <u>in</u> : "Biochemical Plant Pathology", J. A. Callow, ed., Wiley, Chichester.

Martinoia, E., Heck, U., Dalling, M. J., and Matile, P., 1983, Changes in chloroplast number and chloroplast constituents in senescing barley leaves, <u>Biochem. Physiol. Pflanzen</u>, 178:147.

Marty, F., 1978, Cytochemical studies on GERL, provacuoles, and vacuoles in root meristematic cells of <u>Euphorbia</u>, <u>Proc. Natl. Acad. Sci. USA</u>, 75:852.

Matile, P., 1975, "The lytic compartment of plant cells", Springer, Berlin.

Mellor, R. B., Dittrich, W., and Werner, D., 1984, Soybean root response to infection by <u>Rhizobium japonicum</u> : Mannoconjugate turn-over in effective and ineffective nodules, <u>Physiol. Plant Pathol.</u>, 24:61.

Nishimura, M., and Beevers, H., 1978, Hydrolases in vacuoles from castor bean endosperm, <u>Plant Physiol.</u>, 62:44.

Nishimura, M., and Beevers, H., 1979, Hydrolysis of protein in vacuoles isolated from higher plant tissue, <u>Nature</u>, 277:412.

Palevitz, B. A., O'Kane, D. J., Kobres, R. E., and Raikhel, N. V., 1981, The vacuole system in stomatal cells of <u>Allium</u>. Vacuole movements and changes in morphology in differentiating cells as revealed by epifluorescence, video- and electron microscopy, <u>Protoplasma</u>, 109:23.

Racusen, D., 1984, Lipid acyl hydrolase of patatin, <u>Can. J. Bot.</u>, 62:1640.

Riezman, H., 1985, Endocytosis in yeast : Several of the yeast secretory mutants are defective in endocytosis, <u>Cell</u>, 40:1001.

Robertson, D. C., and Robertson, J. A., 1982, Ultrastructure of <u>Pterospora andromedea</u> Nuttall and <u>Sarcodes sanguinea</u> Torrey mycorrhizas, <u>New Phytol.</u>, 92:539.

Romanenko, A. S., Kovtun, G. YU., and Salyev, R. K., 1986, Effect of metabolic inhibitors on pinocytosis of uranyl ions by radish cells : Probable mechanisms of pinocytosis, <u>Ann. Bot.</u>, 57:1.

Scannerini, S., and Bonfante-Fasolo, P., 1983, Comparative ultrastructural analysis of mycorrhizal associations, <u>Can. J. Bot.</u>, 61:917.

Schmidt, G. W., and Mishkind, M. L., 1983, Rapid degradation of unassembled ribulose-1,5-biphosphate carboxylase small subunits in chloroplasts, <u>Proc. Natl. Acad. Sci. USA</u>, 80:2632.

Tanchak, M. A., Griffing, L.R., Mersey, B. G., and Fowke, L. C., 1984, Endocytosis of cationized ferritin by coated vesicles of soybean protoplasts, Planta, 162:481.

Thayer, S. S., and Huffaker, R. C., 1984, Vacuolar localization of endoproteinases EP_1 and EP_2 in barley mesophyll cells, Plant Physiol., 75:70.

Toth, R., and Miller, R. M., 1984, Dynamics of arbuscule development and degeneration of a Zea mays mycorrhiza, Amer. J. Bot., 71:449.

Truchet, G., and Coulomb, P., 1973, Mise en évidence et évolution du système phytolysosomal dans les cellules des différentes zones de nodules radiculaires de pois (Pisum sativum L.). Notion d'hétérophagie, J. Ultrastruct. Res., 43:36.

Van der Wilden, W., Matile, P., Schellenberg, M., Meyer, J., and Wiemken, A., 1973, Vacuolar membranes : Isolation from yeast cells, Z. Naturforsch. Teil C, 28:416.

Van der Wilden, W., Herman, E. M., and Chrispeels, M. J., 1980, Protein bodies of mung bean cotyledons as autophagic organelles, Proc. Natl. Acad. Sci. USA, 77:428.

Van der Wilden, W., Seghers, J. H. L., and Chrispeels, M. J., 1983, Cell walls of Phaseolus vulgaris contain the azocoll digesting proteinase, Plant Physiol., 73:576.

Vaughn, K. C., and Duke, S. O., 1981, Evaginations from the plastid envelope : A method for transfer of substances from plastid to vacuole, Cytobios, 32:89.

Vitale, A., and Chrispeels, M. J., 1984, Transient N-acetylglucosamine in the biosynthesis of phytohemagglutinin : Attachment in the Golgi apparatus and removal in the protein bodies, J. Cell Biol., 99:133.

Wagner, W., Keller, F., and Wiemken, A., 1983, Fructan metabolism in cereals : Induction in leaves and compartmentation in protoplasts and vacuoles, Z. Pflanzenphysiol., 112:359.

Wagner, W., and Wiemken, A., 1986, Properties and subcellular localization of fructan hydrolase in the leaves of barley (Hordeum vulgare L. cv. Gerbel), J. Plant Physiol., 123:429.

Wagner, W., Wiemken, A., and Matile, P., 1986, Regulation of fructan metabolism in leaves of barley (Hordeum vulgare L. cv. Gerbel), Plant Physiol., 81:444.

Wardley, T. M., Bhalla, P. L., and Dalling, M. J., 1984, Changes in the number and composition of chloroplasts during senescence of mesophyll cells of attached and detached primary leaves of wheat (Triticum aestivum L.) leaves, Plant Physiol., 75:421.

Werner, D., Mörschel, E., Kort, R., Mellor, R. B., and Bassarab, S., 1984, Lysis of bacteroids in the vicinity of the host cell nucleus in an ineffective (fix) root nodule of soybean (Glycine max), Planta, 162:8.

Wiemken, A., Schellenberg, M., and Urech, K., 1979, Vacuoles : The sole compartments of digestive enzymes in yeast (Saccharomyces cerevisiae), Arch. Microbiol., 123:23.

Wink, M., 1984, Evidence for an extracellular lytic compartment of plant cell suspension cultures : The cell culture medium, Naturwissenschaften, 71:635.

Wittenbach, V. A., Lin, W., and Hebert, R. R., 1982, Vacuolar localization of proteases and degradation of chloroplasts in mesophyll protoplasts from senescing primary wheat leaves, Plant Physiol., 69:98.

PROTEIN DEGRADATION IN VACUOLES

FROM ACER PSEUDOPLATANUS L. CELLS

Hervé Canut, Gilbert Alibert, Antoine Carrasco
and Alain M. Boudet

Centre de Physiologie Végétale, Université Paul Sabatier
Unité Associée au C.N.R.S. n° 241
118, Route de Narbonne, F-31.062-Toulouse-Cedex, France

INTRODUCTION

In higher plants, intracellular protein breakdown is a fundamental process which has important implications in biochemical regulation, protein turnover (Davies, 1982) and physiological events such as germination (Mikola, 1983) or senescence (Thimann, 1980; Laurière, 1983). However, the mechanisms that control the degradation of proteins are still obscure even though the characterization of plant proteolytic enzymes, and particularly their intracellular location has been extensively studied.

The vacuole is the main site of protease deposition (Boller, 1982; Canut et al., 1985 a). So, this conspicuous organelle was generally considered as a lytic compartment of plant cells (Matile, 1975) and indeed was shown to be directly involved in the breakdown of storage proteins during germination in seeds (Nishimura and Beevers, 1979) as well as in lysis during senescence (Matile and Wikenbach, 1971). In this way, the vacuole appears similar to the lysosome system in animal cells (Matile, 1978). However, no firm experimental evidence supports the involvment of vacuolar proteases in the turnover of proteins occurring in growing or mature cells. During this continuous breakdown of proteins, the different potential substrates of proteases are degraded at various rates leading to various half-lives. Thus, proteolysis must be selective and the factors which determine this selectivity have been suggested to be closely dependent upon the physical properties of the protein (Davies, 1982). For example, the charge and the molecular weight of proteins have been shown to be correlated with relative rates of degradation (Coates and Davies, 1983). In animal cells, the glycan moieties of glycoproteins seem to play an important role in the stabilization of the polypeptides against the proteolysis (Olden et al., 1982), but in contrast abnormal proteins (amino-acid analog containing proteins) tend to have a higher rate of degradation than normal proteins (Goldberg and St John, 1976).

In order to answer some of the previous questions concerning both the sub-cellular location of protein turnover-associated hydrolytic events and the mechanisms involved in the selectivity of protein breakdown, we designed experiments on Acer pseudoplatanus cell suspension cultures. The basic procedure consisted in following, in protoplasts and the corresponding vacuoles, the concentration changes in different proteins including glycoproteins and abnormal proteins.

EXPERIMENTAL TOOLS

Detailed procedures for all the experimental approaches have been already reported (Canut et al., 1985 b; Canut et al., 1986; Canut et al., submitted), and just a brief survey will be provided.

Protoplasts and vacuoles from Acer pseudoplatanus cell suspension cultures were obtained and purified at different times of the culture cycle as reported by Alibert et al. (1982). The procedure for preparation of vacuoles allows, in one-step, the lysis of the protoplast plasmalemma and the rapid separation (less than 1 min) of the released vacuoles from the other compartments of the protoplast.

For the labelling of cell proteins, radioactive precursors (^{3}H-Leu, ^{14}C-Phe) were injected aseptically through the culture vial cotton wool plug of 7-d-old cells, i.e. during the exponential phase of growth. The glycoproteins were labelled both on their polypeptide chains (^{3}H-Leu) and on their carbohydrate moieties (^{14}C-mannose), the abnormal proteins were obtained in the presence of an amino-acid analog (^{14}C p-fluoroPhe).

Usually, the incorporation of precursors proceeded for 18 h, then the cells were harvested, protoplasts and vacuoles prepared, and used for protein determinations (quantity and radioactivity). Radioactivity in the TCA-precipitable material from protoplasts and the corresponding vacuoles maintained in various conditions was followed with time.

Table 1

Protein Contents of Protoplasts and Vacuoles
at Different Times of the Culture Cycle

| Culture Age (Days) | Number of Vacuoles Per Protoplast[a] | μg protein in | | % of Protein in Vacuoles[c] |
		10^6 Protoplasts	Vacuoles Corresponding to 10^6 Protoplasts[b]	
4	2.04	589	159	26.9
8	1.64	505	188	37.2
10	1.38	409	141	34.4
14	1.03	493	78	15.8

[a], calculated using α-mannosidase as vacuolar marker.
[b], each value is corrected by the factor corresponding to the number of vacuoles per protoplast.
[c], similar results were obtained from three independent experiments.

HYDROLYSIS OF INTRACELLULAR PROTEINS IN VACUOLES

We investigated the protein content of protoplasts and vacuoles from Acer pseudoplatanus cells at different times in their culture cycle (Table 1). The highest amount of protein per protoplast was obtained in 4-d-old-cells. Thereafter, the protein content decreased but increased again in 14-d-old-cells. An unusual high

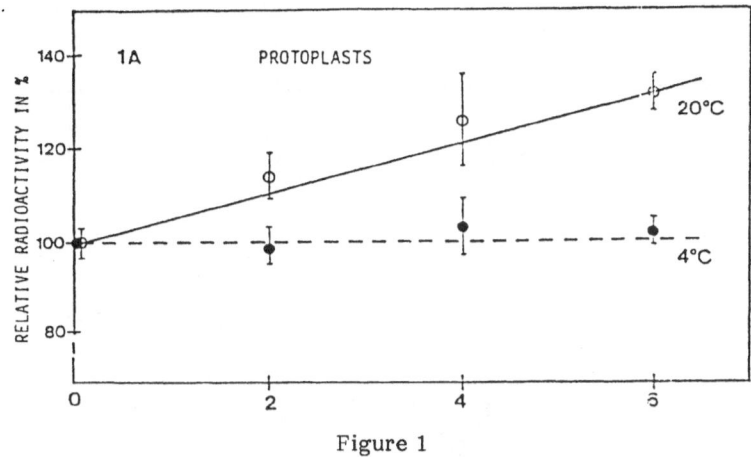

Figure 1

Time-course changes of TCA-insoluble radioactivity
in protoplasts obtained from Acer cells
prelabelled with ^{3}H-leucine for 18 h

Protoplasts were incubated either at 20°C (o) or at 4°C (•). At time zero, 100% corresponds to 19,740 dpm for 10^6 protoplasts.

proportion of these proteins was found in the vacuoles, i.e. respectively 37 and 34% for 8-d- and 10-d-old-cells. However, in 14-d-old-cells, the vacuolar proteins represented only 15% of the total proteins.

The concentration changes observed either in protoplasts or in vacuoles indicate that the intracellular proteins are in dynamic state, and that synthesis, degradation and transport of proteins into the vacuole occur during the culture. For the study of vacuolar proteolysis reported here, only 8-d-old-cells were used since : (a) the level of vacuolar proteins was maximal (37%) whereas at that time, the protein content of protoplast decreased; and (b) the study of protein breakdown during the exponential phase of growth avoided processes of generalized protein breakdown that occur in senescent cells during the stationary phase of growth.

Incorporation of ^{3}H-leucine into TCA-precipitable material occurs when protoplasts are maintained at 20°C showing that isolated protoplasts are able to sustain protein synthesis from endogenous amino-acids (Fig. 1). In contrast, no protein synthesis occurs at 4°C.

Vacuoles, isolated from protoplasts, contain a large part of the labelled proteins (30%), indicating the rapid transport of some of the newly synthesized proteins from the cytoplasm to the vacuole. Following incubation of isolated vacuoles, the TCA-insoluble radioactivity decreases with time (Fig. 2). Half of the vacuolar labelled proteins were hydrolyzed after 6 h incubation at 20°C. Similar results were obtained when the total vacuolar proteins were measured (inset, Fig. 2) since 42% of the proteins were hydrolyzed at 20°C after 6 h incubation. This decrease is temperature- (Fig. 2) and pH-dependent (Fig. 3) as are enzymatic processes. As a consequence, it appears very likely that Acer vacuole which house proteolytic enzymes (data not shown) are able to efficiently hydrolyze their endogenous proteins.

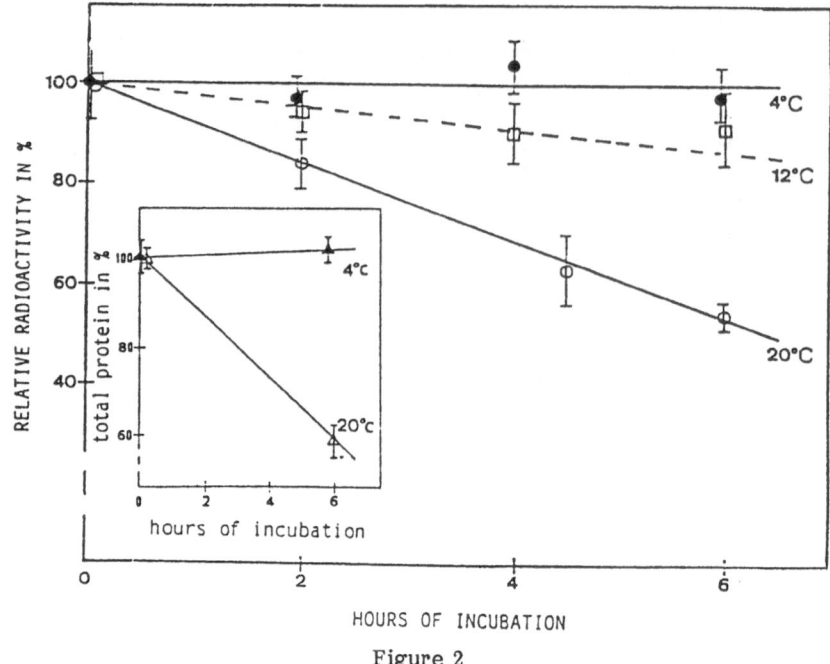

Figure 2

Decrease of TCA-insoluble radioactivity in vacuoles
incubated at 20°C (○), 12°C (□) and 4°C (●)

Vacuoles were isolated from cells prelabelled with ^{3}H-leucine for 18 h. At time zero, the TCA-insoluble radioactivity was 6,280 dpm for 10^6 protoplasts and 1,880 dpm for vacuoles corresponding to 10^6 protoplasts. For the decrease of total proteins in vacuoles (inset) incubated at 20°C (△) and 4°C (▲), at time zero the total proteins content was 106 μg equivalent BSA for 10^6 vacuoles.

Such a demonstration of vacuolar proteolysis, in the case of storage protein hydrolysis, has already been reported in protein bodies isolated from Ricinus communis (Nishimura and Beevers, 1979). However, some differences between their and our work can be underlined. In Ricinus, the vacuolar apparatus is quite particular, since the protein bodies coalesce early in germination to form the large central vacuole (Van der Wilden et al., 1980). In addition, only one vacuolar polypeptide, that was deposited inside the vacuole before the germinating hydrolytic events, is extensively degraded in these isolated organelles. In contrast, the fluorographs of the proteins (Fig. 4-A) obtained after electrophoretic analysis of the vacuolar fraction indicated that, in Acer cells, a large number of polypeptides were present in the vacuole and undergo proteolysis. However, a few bands were recovered with the same intensity (arrows). Such proteins resistant to proteolysis might represent intrinsic proteins of the vacuolar apparatus, either hydrolytic enzymes in the sap or constitutive proteins of the tonoplast. Simultaneously, a diffuse radioactive band appeared at the bottom of the gel corresponding to small polypeptides probably arising from the hydrolyzed proteins. The same observations were made for the silver-stained protein patterns (Fig. 4-B). Thus, the proteolytic capacity of the Acer vacuole was not limited to a few specific proteins but may be connected with the turnover of many intracellular proteins.

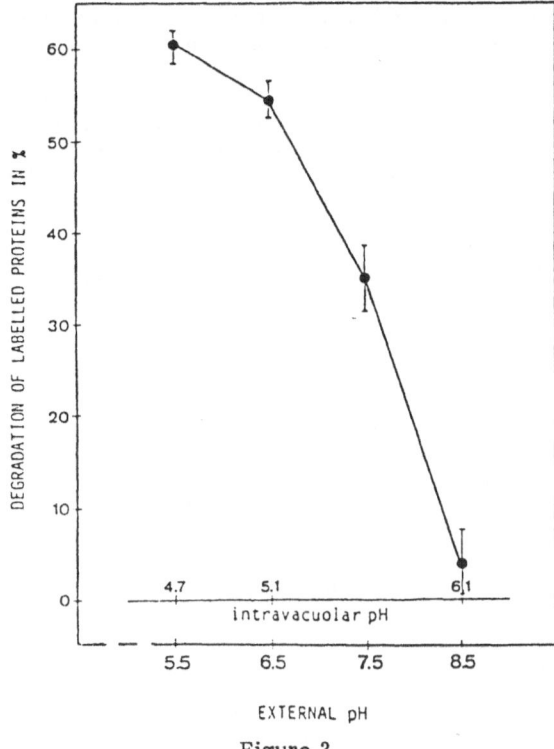

Figure 3

pH profile of proteolytic rates in isolated vacuoles

Vacuoles were isolated from cells prelabelled with ^{3}H-leucine as in Fig. 2. Intravacuolar pH was determined as described by Kurkdjian and Guern (1981).

Such an assumption necessarily implies a selectivity of the degradation processes since proteins are degraded at various rates. This essential aspect of intracellular protein breakdown was studied through the hydrolysis of particular proteins. We have thus followed the behaviour of two sets of proteins : glycoproteins and abnormal proteins.

HYDROLYSIS OF GLYCOPROTEINS

The idea that carbohydrates are involved in the protection of glycoproteins against proteolytic attack was introduced by Ashwell and Morell (1974) and revived by recent studies including the use of inhibitors of protein glycosylation (Olden et al., 1985).

In Acer pseudoplatanus, two vacuoles-located hydrolytic enzymes, mannosidase and β-N-acetylglucosaminidase, are glycoproteins. These two enzymes were demonstrated to exhibit a great stability of their activities, during autolysis experiments performed by incubation of Acer protoplast extracts at 30°C for 4 h (data not shown). Thus, the possibility that, in plants, the carbohydrate moiety of glycoproteins is a general factor involved in their resistance to proteolytic attack cannot be ruled out.

Experiments involving ^{3}H-leucine and ^{14}C-mannose-fed cells were performed to label the newly-synthesized proteins both on their polypeptide chains and on their carbohydrate groups.

The distribution of the TCA-soluble and insoluble radiolabelled materials was determined between the vacuolar and extravacuolar compartments of the protoplasts (Table 2) : 60% of the free ^{3}H-leucine and 80% of the free ^{14}C-mannose were recovered inside the vacuoles. In the TCA-precipitable fraction from vacuoles, the ^{3}H and ^{14}C radioactivity levels were very similar (i.e. 30%). This result was confirmed by the comparison of the ^{14}C/^{3}H ratios : Identical values were also obtained in protoplasts, vacuolar and extravacuolar compartments showing that the intracellular distribution of glycoproteins was very close to that of total proteins. So, glycoproteins appear to be transferred into the vacuole at the same rate as the other proteins.

Table 2

Distribution of TCA-precipitable and TCA-soluble Materials
in Protoplasts and Vacuoles Isolated from <u>Acer</u> Cells

	^{14}C-mannose radioactivity dpm	^{3}H-leucine radioactivity dpm	Ratio ^{14}C/^{3}H
A- TCA-precipitable Materials :			
10^6 protoplasts	30372	75177	0.40
Vacuoles[a] corresponding to 10^6 protoplasts	9273	26079	0.36
Extravacuolar compartments	21099	29079	0.43
% in Vacuole	31	35	
B- TCA-soluble Materials :			
10^6 protoplasts	67232	91104	0.74
Vacuoles[a] corresponding to 10^6 protoplasts	55662	56747	0.98
Extravacuolar compartments	11570	34357	0.34
% in Vacuole	83	62	

[a], each value is corrected according to the number of vacuoles per protoplast (i.e. 1.94) : this factor was calculated using α-mannosidase as vacuolar marker.

After their isolation, vacuoles were able to extensively degrade their protein content. Since the decreases of ^{3}H and ^{14}C radioactivities in the TCA-precipitable material were identical (Fig. 5), the glycoproteins and the bulk of the radiolabelled proteins appeared to be broken down in a similar extent. Half of the labelled proteins were hydrolyzed after 6 h incubation.

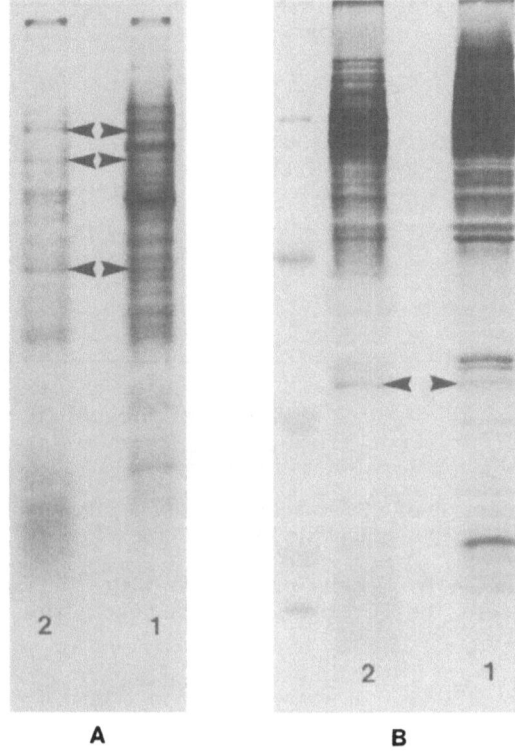

Figure 4

SDS–polyacrylamide gel electrophoresis patterns
of vacuole polypeptides
A, fluorograph; B, silver–staining

The amount of protein loaded in each well was derived from equal amounts of vacuoles : (1), time zero of an incubation period; (2), time 6 h. The apparent molecular weights were estimated by comparison with standards (92, 43, 24, 18 and 14 KD). Vacuoles were isolated from cells prelabelled with ^{3}H–leucine for 18 h.

So, as demonstrated for the leucine tritiated labelled proteins, the vacuolar compartment appears to be the major site of glycoprotein degradation in the cell. This confirms again the prominent role of this organelle in basal protein turnover. In addition, identical rates of transfer into, and degradation inside, the vacuoles were found both for glycoproteins and total newly–synthesized proteins, indicating no particular properties of the bulk of glycoproteins as far as breakdown is concerned. Thus, the presence of covalently bound carbohydrates does not seem to be a determining factor in the selectivity of protein degradation in Acer cells.

HYDROLYSIS OF ABNORMAL PROTEINS

Abnormal proteins can result from mutations, biosynthetic errors, or spontaneous denaturations and represent potentially harmful polypeptides. In animal cells, both lysosomal and cytosolic pathways are involved in the breakdown of

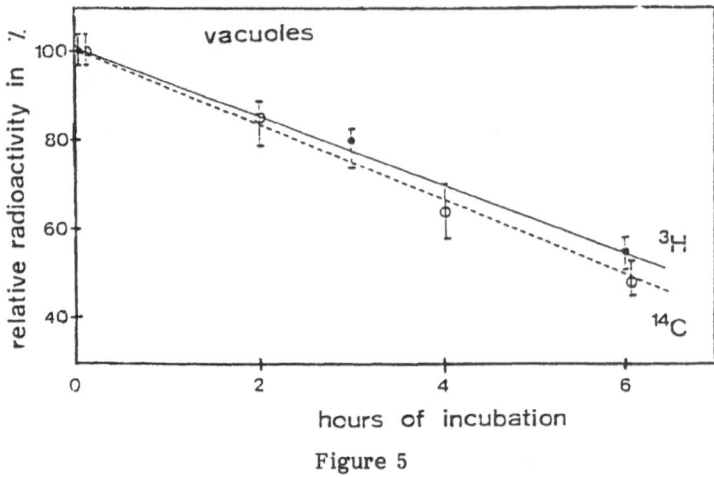

Figure 5

Decrease of TCA-insoluble radioactivity in vacuoles
incubated at 20°C for 6 h

Vacuoles were isolated from cells prelabelled with ^{3}H-leucine and ^{14}C-mannose for 18 h. At time zero, the TCA-precipitable radioactivities of vacuoles were those indicated in Table 2.

endogeneous proteins, but the cytosolic pathway alone is implied in the degradation of abnormal proteins. This allows rapid degradation of these deleterious molecules (Etlinger and Goldberg, 1977; Hershko et al., 1982; Wharton and Hipkiss, 1984). In plants, nothing is known about the abnormal protein degradation systems.

In Acer cells, labelling experiments involving ^{14}C-phenylalanine (reference culture) or ^{14}C p-fluorophenylalanine (test culture) were performed for comparative studies on the degradation of newly-synthesized normal and abnormal proteins (proteins containing the amino-acid analog).

During a 6 h incubation period, protoplasts isolated from phenylalanine fed cells actively incorporate the labelled amino-acid into the proteins, while protoplasts isolated from p-fluorophenylalanine fed cells do not (Fig. 6). As the uptake and intracellular compartmentation of the labelled precursors were comparable in reference and test experiments, the lack of a net gain in labelled proteins containing the amino-acid analog may be due either to the absence of protein synthesis or to a high rate of degradation of the abnormal synthesized proteins.

To determine whether protein synthesis or degradation was responsible for the observed results, isolated protoplasts were incubated with benzylamine. This weakly basic amine has been reported by us (Rataboul et al., 1985) and others (Kurkdjian, 1982) to specifically accumulate in the plant vacuole and to cause at high concentration a sharp rise of the intravacuolar pH. Due to the pH properties of vacuolar proteases, such a pH rise is supposed to inhibit vacuolar protein degradation (Fig. 3). Indeed, protein radiolabelling (^{14}C-phenylalanine) was effectively enhanced in the amine-treated protoplasts compared to non treated ones (Fig. 6). Moreover, the net gain in labelled proteins resulting from the benzylamine treatment was slightly higher for the abnormal proteins (^{14}C p-fluorophenylalanine) than for the normal ones, suggesting a greater rate of degradation of the abnormal proteins in the vacuolar compartment.

However, vacuoles isolated from the protoplasts degrade their endogenous proteins actively and at the same rate, whatever the precursor fed to the cells. Thus, a specific vacuolar proteolytic system acting on the abnormal proteins does not seem responsible for their greater rate of degradation. We then determined the distribution of normal and abnormal proteins between the vacuolar and extravacuolar compartments of protoplasts (Fig. 7). In reference culture cells, about 30% of the newly synthesized proteins (^{14}C-phenylalanine containing proteins) were recovered in the vacuolar fraction. In contrast, in test culture cells, a higher proportion of abnormal proteins (60%) was present in the vacuolar fraction compared to the reference experiment. So, in comparison to the normal proteins, the abnormal proteins seem to be preferentially directed towards the vacuole. Since abnormal and normal proteins are degraded at the same rate inside the vacuole, the abnormal proteins will be more rapidly eliminated.

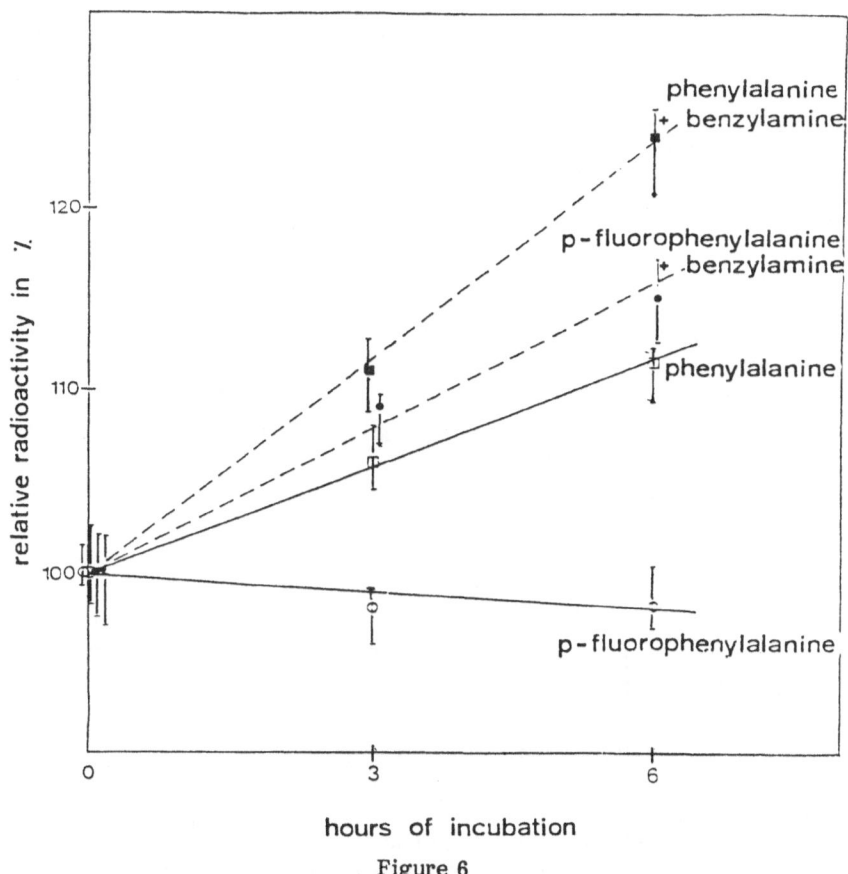

Figure 6

Variations of TCA-precipitable radioactivity in protoplasts :
Effect of benzylamine

Cells were fed with 25 μC$_i$ of ^{14}C-phenylalanine for 18 h in the reference culture, with 25 μC$_i$ of ^{14}C-phenylalanine for 18 h in the reference culture, with 25 μC$_i$ of ^{14}C-p-fluorophenylalanine for 18 h in the test culture. Protoplasts from reference cells (□) or from test cells (○) and benzylamine-treated protoplasts (10 mM for 10 min) isolated from reference cells (■) or from test cells (●) were incubated at 20°C for 6 h.

CONCLUSION

Taken together, the results presented here clearly demonstrate the involvement of vacuole in intracellular protein degradation in growing cell suspension cultures. However, the concomitant occurrence of a cytosolic degradation pathway involving more discrete or more specific enzymes cannot be ruled out. Vacuolar protein breakdown per se appears to be mainly non specific, in accordance with the broad specificity described for vacuolar proteases (Canut et al., 1985). We can therefore assume that the control of protein breakdown (with the exception of the intrinsic vacuolar proteins) is more related to transport of proteins from the cytoplasm to the vacuole, than to a specificity of the vacuolar proteases towards a given class of substrate.

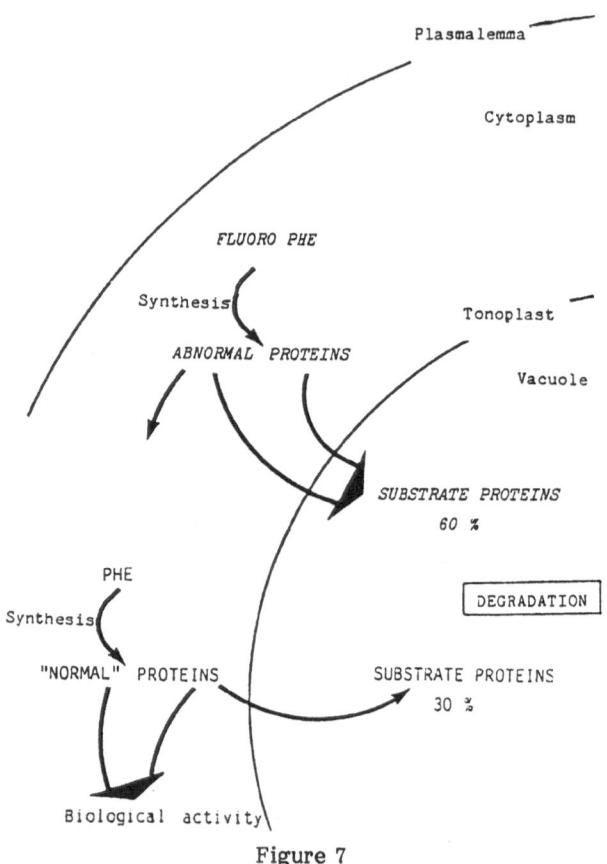

Figure 7

Distribution of normal and abnormal proteins
between the vacuolar and extravacuolar compartments
of Acer protoplasts

Cells were prelabelled as in Fig. 6

ACKNOWLEDGEMENT

This work was supported by grants from the Centre National de la Recherche Scientifique and from the Université Paul Sabatier of Toulouse.

REFERENCES

Alibert, G., Carrasco, A., and Boudet, A. M., 1982, Changes in biochemical composition of vacuoles isolated from Acer pseudoplatanus L. during cell culture, Biochim. Biophys. Acta, 721:22.

Ashwell, G., and Morell, A. G., 1974, The role of surface carbohydrates in the hepatic recognition and transport of circulating glycoproteins, Adv. Enzymol., 41:99.

Boller, T., 1982, Enzymatic equipment of plant vacuoles, Physiol. Vég., 20:247.

Canut, H., Alibert, G., and Boudet, A. M., 1985, Proteases of Melilotus alba mesophyll protoplasts : I. Intracellular localization. Plant Science Letters, 39:163.

Canut, H., Alibert, G., and Boudet, A. M., 1985 b, Hydrolysis of intracellular proteins in vacuoles isolated from Acer pseudoplatanus L. cells, Plant Physiol., 79:1090.

Canut, H., Alibert, G., Carrasco, A., and Boudet, A. M., 1986, Rapid degradation of abnormal proteins in vacuoles from Acer pseudoplatanus L. cells, Plant Physiol, 81, in press.

Canut, H., Alibert, G. and Carrasco, A., 1986, Intracellular degradation of glycoproteins in Acer pseudoplatanus L. cells, submitted.

Coates, J. B., and Davies, D. D., 1983, The molecular basis of the selectivity of protein degradation in stressed senescent barley (Hordeum vulgare cv. Proctor) leaves, Planta, 158:550.

Davies, D. D., 1982, Physiological aspects of protein turnover, in : "Encyclopedia of Plant Physiology, New Series, Vol. 14 A : Nucleic Acids and Proteins in Plants", D. Boulder and B. Parthier, eds., Springer-Verlag, Berlin.

Etlinger, J. D., and Goldberg, A. L., 1977, A soluble ATP-dependent proteolytic system responsible for the degradation of abnormal proteins in reticulocytes, Proc. Natl. Acad. Sci. USA, 74:54.

Goldberg, A. L., and St-John, A. C., 1976, Intracellular protein degradation in mammalian and bacterial cells : Part 2, Annu. Rev. Bochem., 45:747.

Hershko, A., Eytan, E., Ciechanover, A., and Haas, A. L., 1982, Immunochemical analysis of the turnover of ubiquitin protein conjugates in intact cells. Relationship to the breakdown of abnormal proteins, J. Biol. Chem., 257: 13964.

Kurkdjian, A., 1982, Absorption and accumulation of nicotine by Acer pseudoplatanus and Nicotiana tabacum cells, Physiol. Vég., 20:73.

Kurkdjian, A., and Guern, J., 1981, Vacuolar pH measurement in higher plant cells. I. Evaluation of the methylamine method, Plant Physiol., 67:953.

Laurière, C., 1983, Enzymes and leaf senescence, Physiol. Vég., 21:1159.

Matile, P., 1975, The lytic compartment of plant cells, M. Alfert, W. Beerman, G. Rudkin, W. Saudritter and P. Stittle, eds., Vol. 1, Springer-Verlag, Wien, New-York.

Matile, P., 1978, Biochemistry and function of vacuoles, Annu. Rev. Plant Physiol., 29:193.

Matile, P. and Wikenbach, F., 1971, Function of lysosomes and lysosomal enzymes in the senescing corolla of the morning glory (Ipomea purpurea), J. Expt. Bot., 22:759.

Mikola, J., 1983, Proteinases, peptidases, and inhibitors of endogenous proteinases in germinating seeds, in : "Seed proteins", J. Daussant, J. Mossé, and J. Vaughan, eds., Ann. Proc. The Phytochemical Society of Europe, Vol. 20, Academic Press, London.

Nishimura, M., and Beevers, H., 1979, Hydrolysis of protein in vacuoles isolated from higher plant tissue, Nature, 277:412.

Olden, K., Parent, J. B., and White, S. L., 1982, Carbohydrate moieties of glycoproteins. A reevaluation of their function, Biochim. Biophys. Acta, 650:209.

Olden, K., Bernard, B. A., Humphries, M. J., Yeo, T. K., White, S. L., Newton, S. A., Bauer, H. C., and Parent, J. B., 1985, Function of glycoprotein glycans, Trends Biochem. Sci., 10:78.

Rataboul, P., Alibert, G., Boller, T., and Boudet, A. M., 1985, Intracellular transport and vacuolar accumulation of o-coumaric acid glucoside in Melilotus alba mesophyll cell protoplasts, Biochim. Biophys. Acta, 816:25.

Thimann, K. V., 1980, The senescence of leaves, in : "Senescence in Plants", R.

C. Aldeman and G. S. Roth, eds., CRC Series in Ageing, CRC Press, Boca Raton, Florida.

Van der Wilden, W., Herman, E. M. and Chrispeels, M. J., 1980, Protein bodies of mung bean cotyledons as autophagic organelles. Proc. Natl. Acad. Sci. USA, 77:428.

Wharton, S. A., and Hipkiss, A. R., 1984, Abnormal proteins of shortened length are preferentially degraded in the cytosol of cultured MRC5 fibroblasts, FEBS Letters, 168:134.

PROTEASE ACTIVITY IN OAT LEAF VACUOLES

Henry Van der Walk and Leendert Van Loon

Department of Plant Physiology, Agricultural University
Arboretumlaan 4
NL-6703-BD-Wageningen, The Netherlands

Senescence of leaves is characterized by a gradual decline in protein-content. At the initial stage of protein breakdown, oat leaves, attached to the plant, do not exhibit an increase in protease activity, indicating that synthesis of additional protease is not required for protein degradation (Van Loon et al., 1986).

Compartmentation and changes in the subcellular localization of enzymes and substrates might be the mechanism regulating the rate of protein loss. Previously it was found that part of the protease activity in oat leaves is located extracellularly, associated with the cell walls (Van Loon et al., 1980). The subcellular localization of the intracellular protease activity was further investigated by using protoplasts.

Oat leaves contain two types of proteases : an acidic one with a pH optimum of 4.5 and an neutral one with an optimum at pH 7.5. Both types of proteases are found in protoplasts, isolated from mesophyll cells of the leaves (Van der Walk et al., 1984).

To further elucidate the subcellular localization of these proteases, vacuoles were isolated from the protoplasts. Protoplasts were lysed in 0.10 M K_2HPO_4, 1 mM DTT, 5 mM EDTA at pH 8.0. After 2 - 5 minutes at 28°C, vacuoles were

Table 1

Protein Content, Glucose-6-phosphate Dehydrogenase
α-mannosidase and Protease Activity in Absorbances/ml

	protein	glucose 6-phosphate dehydrogenase	α-mannosidase	protease	
				pH 4.5	pH 7.5
5% Ficoll	0.828	0.039	0.149	0.164	0.121
3% Ficoll	0.068	0.008	0.034	0.027	0.038
vacuoles	not determined	0.007	0.292	0.262	0.007

released and after centrifugation on a 5%/3% Ficoll gradient, protease activity in the vacuole–containing upper layer was determined. Activities in both Ficoll layers were also measured. Vacuoles were further purified by repeated centrifugation in 3% Ficoll. Both activities and the activities of a cytoplasmic marker enzyme (glucose-6-phosphate dehydrogenase) and a vacuolar marker enzyme (α-mannosidase) are presented in Table 1.

Thus, acidic protease activity is compartmentalized in the vacuole, whereas the neutral protease appeared to be located exclusively in the cytoplasm. The acidic protease may function in rapid general protein breakdown, particularly in the late stages of senescence. In contrast, the neutral proteases may play a role both in normal protein turnover and in the acceleration of protein breakdown at the onset of senescence. In how far vacuolar enzymes are involved in cellular processes at an early stage of leaf senescence is still unclear.

REFERENCES

Van Loon, L. C., Van der Walk, H. C. P. M., Haverkort, G. J., and Lokhorst, G. I., 1986, J. Plant Physiol., in press.
Van Loon, L. C., Van der Lubbe, J. L. M., and Van der Walk, H. C. P. M., 1980, Abstr. FESPP II, Congress Santiago di Compostella, p. 285-A,
Van der Walk, H. C. P. M., and Van Loon, L. C., 1984, Vacuolar localization of acidic protease in oat leaves, in : Abstracts 4th Congress of The Federation of European Societies of Plant Physiology, Strasbourg, France, July 29–August 3, 1984, p. 335-336.

COMPARTMENTATION OF GLYCOSIDASES IN A LIGHT VACUOLE FRACTION

FROM THE LATEX OF <u>LACTUCA SATIVA</u> L.

Roger Giordani[*], Georges Noat[*]
and Francis Marty[**]

[*]Centre de Biochimie et de Biologie Moléculaire, C.N.R.S.
31, Chemin Joseph Aiguier, F-13.402-Marseille Cedex 9
and [**]Laboratoire de Biologie de la Différenciation Cellulaire
Faculté des Sciences de Luminy, F-13.288-Marseille-Cedex 9
France

INTRODUCTION

Articulated laticifers in young growing roots of <u>Lactuca sativa</u> L. arise from aligned sets of cells, the end walls as well as local areas of transverse walls of which disintegrate when branched tubes are formed with adjacent cells (Giordani, 1977 and 1979). Before being completely perforated cell walls become locally thinner. Simultaneously, numerous membrane vesicles originating from the plasmalemma and loaded with polysaccharides are present in the neighboring cytoplasm and some have been observed inside the vacuoles (Giordani, 1980). Because of the general lytic function of vacuoles (Marty et al., 1980; Matile, 1978) the results of our previous experiments suggest that cell wall components carried by endocytic vesicles might be digested in the vacuoles of the latex in differentiating laticifers of <u>Lactuca sativa</u>. Electron microscope cytochemistry has previously shown a marked increase in cellulase and acid hydrolase activities in cell wall areas while these are being remodeled (Giordani, 1980 and 1981). Cellulase and acid phosphatase activities have also been detected in small membrane vesicles (Giordani, 1980).

The possible involvement of vacuole in the intracellular digestion of membrane vesicles was investigated by the measure of glycosidic activities corresponding to different polysaccharidic cell wall components, cellulose, pectin and hemicelluloses. The latex flow of articulated laticifer is a good source of vacuoles because without the disadvantage of protoplasts that the properties of vacuoles may be changed upon exposition of tissues to wall–degrading enzymes (1978). The development of mechanical method for the isolation of vacuoles (Leigh and Branton, 1976) from intact tissues enable a biochemical approach of intravacuolar cell wall degradation process. For that the vacuoles have been isolated from crude latex of <u>Lactuca sativa</u> by flotation in a discontinuous step gradient of metrizamide (15% – 7.5% and 0%). Two vacuoles fractions have been collected at the 15% 7.5% and 7.5% – 0% interfaces (Giordani and Marty, 1983). Glycosidic assays in vacuoles fractions are then possible. This study was undertaken to examine the glycosidic activities of the vacuoles from the latex of <u>Lactuca sativa</u> and their putative role in the digestion of cell wall components during the differentiation of laticifers.

MATERIAL AND METHODS

Isolation of Vacuoles

Fresh latex was tapped from prefloral stems of <u>Lactuca sativa</u> grown under greenhouse conditions. The latex, diluted with 19% metrizamide in buffer (1.5 M sorbitol, 1 mm EDTA, 10 Tris-HCl pH 7.6) was fractionated as previously described (Giordani and Marty, 1983). Fractions were recovered from the pellet (fraction F_1), from the loading zone (fraction F_2), from the 15/7.5% interface (fraction F_3), from the 7.5% layer (fraction F_4), from the 7.5/0% interface (fraction F_5) and from the 0% layer (fraction F_6).

Glycosidases assays

The following glycosidic activities were determined by measuring the release of p-nitrophenol from the substrates shown in parentheses : α-glucosidase (PNP-α-D-glucopyranoside), β-glucosidase (PNP-β-D-glucopyranoside), α-galactosidase (PNP-α-D-galactopyranoside), β-galactosidase (PNP-β-D-galactopyranoside), α-mannosidase (PNP-α-D-mannopyranoside), β-N-acetylglucosaminidase (PNP-N-acetyl-β-D-glucosaminide), β-xylosidase (PNP-β-D-xylopyranoside), β-fucosidase (PNP-D-fucopyranoside), α-arabinosidase (PNP-α-L-arabinofuranoside). The assay mixture contained in 2 ml of 1 mM substrate in 0.1 M pH 5.0 succinate buffer, 30 - 100 μl of crude latex or subfraction. After 4 hr of incubation at 30°C the reaction was stopped by addition of 1 ml of 0.5 M Na_2CO_3 and the absorbance of p-nitrophenol formed was recorded at 400 nm on Philips DU 10 spectrophotometer.

Cellulase and Pectinase Assays

Cellulase and pectinase were measured viscometrically using respectively 4 ml of 1% (w/v) carboxymethylcellulose (CMC) solution and 0.3% (w/v) polygalacturonic acid methyl ester (fruits pectin) solution in 0.1 M pH 5.0 succinate buffer at 30°C. The reaction is started by addition of 1 ml of crude latex or vacuolar fractions, F_3 and F_5 (enzyme solution). Specific viscosity (η sp) determination were effected with a Hewlett-Packard 5901 B viscosimeter. Readings were recorded at 30 min intervals for 5 hr. A reaction mixture containing 1 ml of boiled enzyme solution was used as a blank.

Protein Determination

Protein determinations were carried out according to Bradford (1976) with BSA as a protein standard.

Reagents and Chemicals

p-nitrophenyl-glycosides, p-nitrophenol were obtained from Sigma Chemical Co.; carboxymethyl-cellulose and pectin were purchased from B.D.H. Chemicals Ltd., Poole, Dorset, U.K.; metrizamide is obtained from Nyegaard & Co. As. Oslo, Sweden; all other reagent grade chemicals were purchased from Merkx, Darmstadt, Federal Republic of Germany.

RESULTS

Fractions collected from the discontinuous metrizamide gradient were examined to determine whether p-nitrophenyl-glycosidase, carboxymethyl-cellulase and pectinase activities were present and the specific activities of each fraction were compared with those found in the total latex (Pl. I, Pl. II). Significant glycosidase activities were recovered in fraction F_3 and fraction F_5 (Table 1). It is noteworthy that the lightest of those two fractions showed the highest recovered

Table 1

Substrate Specificity for Glycosidase Activities at pH 5.0
in Crude Latex of <u>Lactuca</u> <u>sativa</u> L. and Various Fractions

Substrates	Crude Latex	Gradient Fractions					
		F_1	F_2	F_3	F_4	F_5	F_6
α-D-glucoside	0.46	1.69	0.94	1.76	5.52	46	0
β-D-glucoside	1.55	5.64	0	14.1	4.74	204	0
α-D-galactoside	36.5	40.6	6.5	4.58	6.7	39.5	15.6
β-D-galactoside	3.88	3.95	0.94	4.4	3.35	36	0
α-D-mannoside	4.35	7.9	1.139	3.52	0	6.6	6.16
β-N-acetyl-glucosaminide	0.77	0	1.23	5.3	8.09	20	9.05
β-D-xyloside	3.18	1.69	0	0	5.52	304	66.60
β-D-fucoside	10.88	15.8	2.8	0	6.12	674	0
α-D-arabinoside	3.88	4.5	1.66	0	8.9	214	9.05
Total protein (mg)	0.508	0.035	0.312	0.14	0.045	0.012	0.0045

Using p-nitrophenyl-substitued glycosides as substrates, the enzymic activity is expressed in μM min^{-1} mg protein^{-1} based on p-nitrophenol standard. The total protein content of the crude latex and gradient fractions is given at the bottom of the table.

at the interface between the 7.5% metrizamide and 0% metrizamide layers displayed specific activities which were 9 to 130 times higher than that of the original sample, but only 3% of the protein. The fraction banding at the 15/7.5% metrizamide interface contained α- and β-glucosidases, β-N-acetylglucosaminidase with relative specific activities of about 4 to 9. For comparison, the same glycosidic activities were sought in latex sap from non-articulated laticifers of <u>Asclepias</u> <u>curassavica</u> L. Except for α-mannosidase activity, the specific activities of all the other enzymes were lower in the sap of <u>Asclepias</u> <u>curassavica</u> latex than in the light fraction 5 from the latex of <u>Lactuca</u> <u>sativa</u> (Table 2).

The specific activities of cellulase and pectinase were also found to have increased in fractions F_3 and F_5 from the latex of <u>Lactuca</u> <u>sativa</u> (Table 3). These fractions contained 16% and 76% of the cellulase activity, and 9% and 96% of the pectinase activity of the crude latex, respectively. The specific activities of cellulase and pectinase were also found to have increased 2.5 to 4-fold in fraction F_3 and more than a hundred-fold in fraction F_5. There was no measurable viscosity change when the CMC or pectin mixtures are incubated without latex, fraction F_3 or fraction F_5; also no change in viscosity when the CMC or pectin solutions are incubated with fraction F_3 or fraction F_5 preliminary boiled for 10 min (not shown). A blank with boiled latex is not possible as regards latex granules coagulation.

On electron microscope examination, fractions F_3 (Fig. 1) and F_5 (Fig. 2) were found to be abundant in membrane vesicles, as expected. But surprisingly, their morphologies did not reflect the differences found in the specific activities

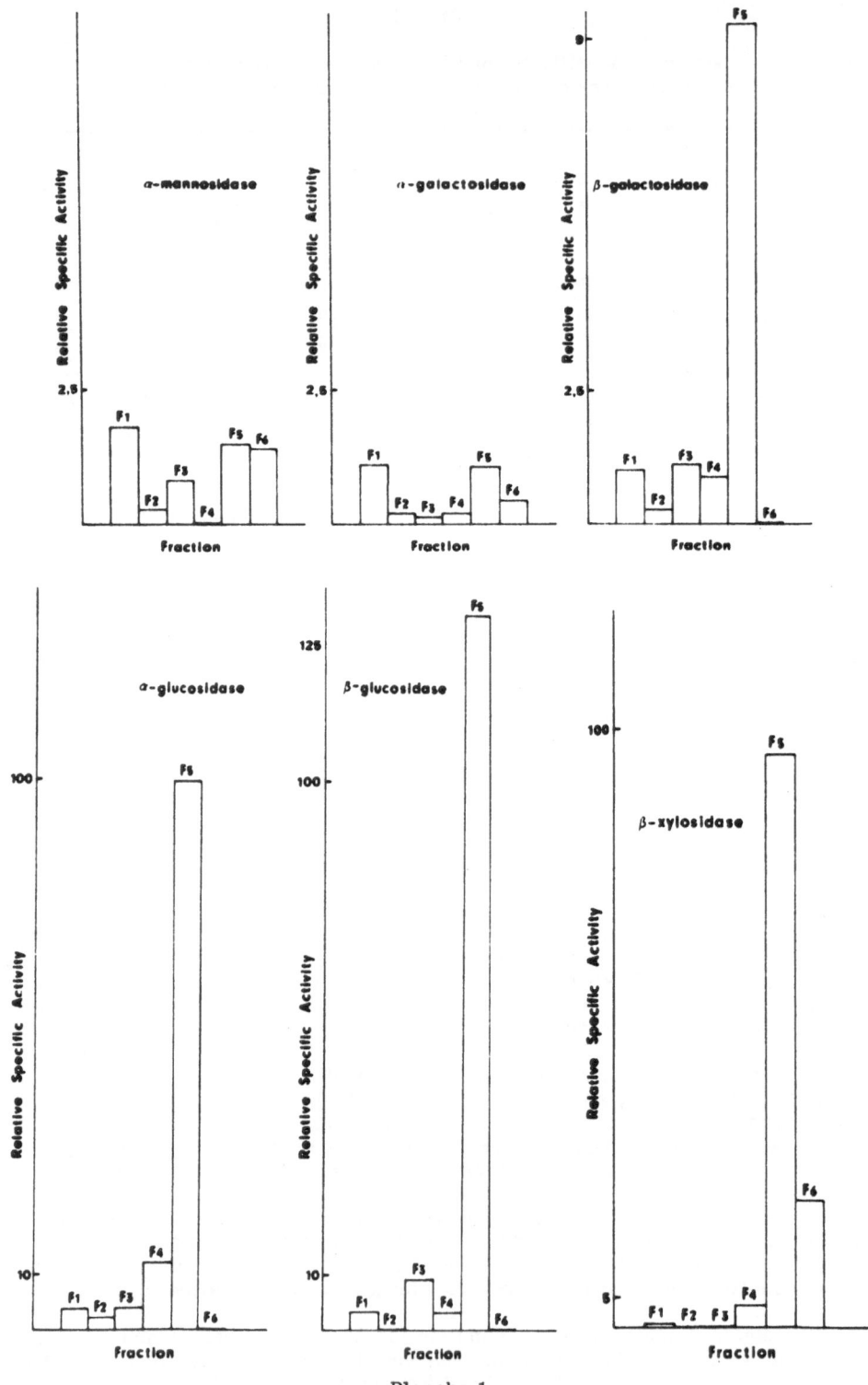

Planche 1

Glycosidases Activities
(α-mannosidase, α-galactosidase, β-galactosidase,
α-glucosidase, β-glucosidase and β-xylosidase)

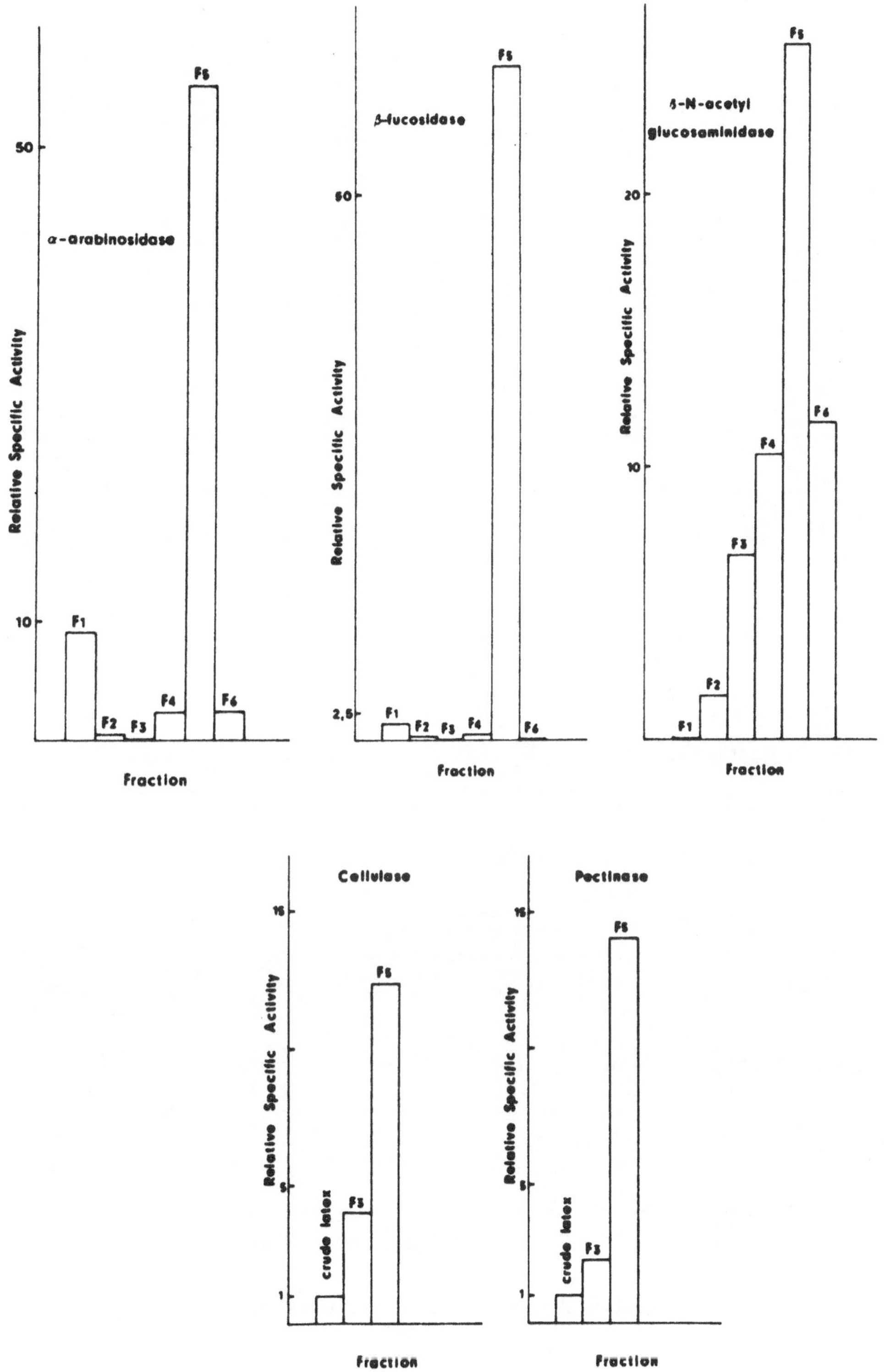

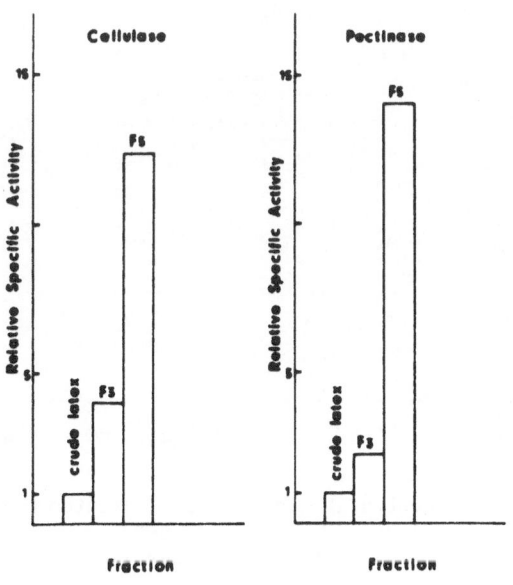

Planche 2

Glycosidases Activities
(α-arabinosidase, β-fucosidase, β-N-acetyl–glucosaminidase,
cellulase and pectinase)

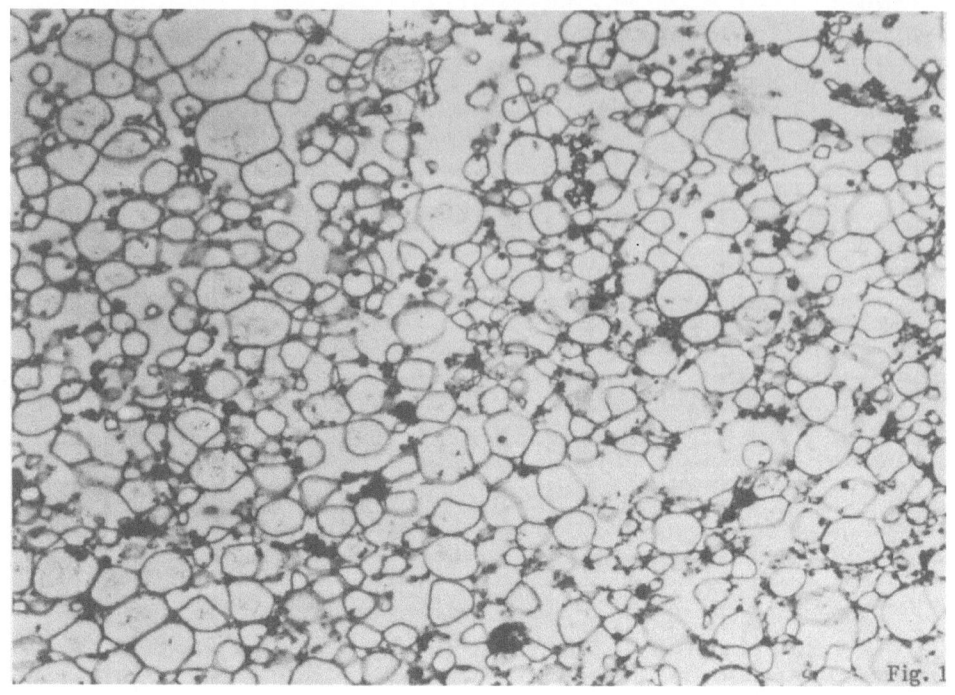

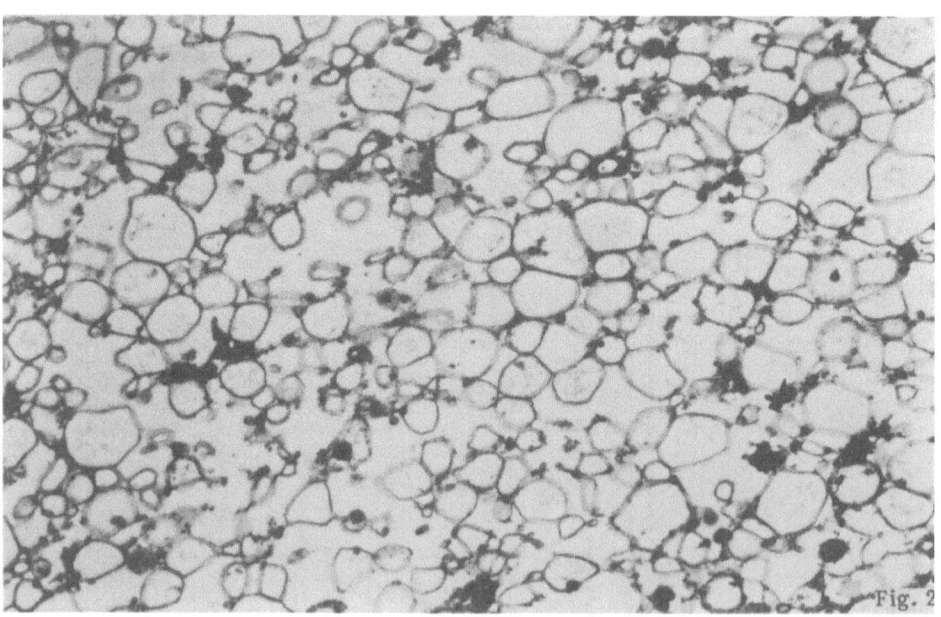

Section through pellets of isolated vacuoles
(G x 12,000)
Fig. 1. Fraction F_3

Fig. 2. Fraction F_5

of the glycosidases. Both fractions consisted predominantly of empty membranous bags 0.7 μm in size, very similar to the small vacuoles of the latex. The most obvious contaminants of the light vacuole fraction were lipophilic particles trapped in between the empty vacuoles.

Table 2

Substrate Specificity for Glycosidase Activities at pH 5.0
in Vacuolar Sap for <u>Asclepias</u> Latex
by Comparison with that of F$_3$ and F$_5$ Vacuolar Fractions From <u>Lactuca</u> Latex

Substrates	Vacuolar Sap From Asclepias Latex	F$_3$ Heavy Vacuole Fraction From Lactuca Latex	F$_5$ Light Vacuole Fraction From Lactuca Latex
α-D-glucoside	1.1	1.76	46
β-D-glucoside	6.5	14.1	204
α-D-galactoside	8.2	4.58	39.5
β-D-galactoside	8.3	4.4	36
α-D-mannoside	27.4	3.52	6.6
β-N-acetyl-glucosaminide	14.6	5.3	20
β-D-xyloside	1.1	0	304
β-D-fucoside	11.1	0	674
α-L-arabinoside	1	0	214

Enzyme activity is expressed as the change in μM min^{-1} mg protein^{-1}.

Table 3

Cellulase and Pectinase Activities at pH 5.0
in Crude Latex and Vacuolar Fractions F$_3$ and F$_5$

Fraction	Cellulase		Pectinase		Total Protein (mg)
	Total Activity	Specific Activity	Total Activity	Specific Activity	
Crude latex	44	3.07	18	1.25	14.3
F$_3$	6.8	12.3	1.6	2.9	0.55
F$_5$	33.5	380	15.5	176	0.088

The enzyme activity is given as a percentage of the decrease in specific viscosity per hour (total activity) and per mg protein (specific activity).

DISCUSSION

The light vacuole fraction (F_5) is active towards the artificial substrates commonly utilized instead of cellulose, pectins and hemicelluloses which are occuring naturally in plant cell walls. From these results we suggest that fraction F_5 is potentially able to degrade cell wall fragments.

The specific activities of α- and β-glucosidases, β-galactosidase, β-N-acetyl-glucosaminidase, β-xylosidase, β-fucosidase and α-arabinosidase are much higher in the light vacuole fraction from the articulated laticifers of Lactuca sativa than in the vacuoles isolated from parenchyma cells (Butcher et al., 1977; Boller and Kende, 1979; Saunders and Gillepsie, 1984). The differential pattern of specific activities between the light vacuole fraction from Lactuca latex and the vacuole sap from Asclepias indicates that the vacuolar glycosidases in the latex of Lactuca probably have a tissue-specific function. A cellulase activity has been localized at the ultrastructural level in the articulated laticifers of Papaver somniferum (Nessler and Mahlberg, 1981). A cellulase activity has been also demonstrated in the latex from a variety of articulated laticifers - Carica papaya, Musa textilis, Achras sapota and various species of Hevea - (Sheldrake, 1969; Sheldrake and Moir, 1970); conversely no activity is detectable in the latex from non articulated laticifers - Dycra costulata, Euphorbia pulcherima, Ficus elastica, Ficus indica (Sheldrake, 1969). It has been suggested that cellulase is involved in the removal of end walls during the differentiation of articulated laticifers (Sheldrake, 1969; Sheldrake and Moir, 1970). Our results with the latex of Lactuca sativa are in agreement with this interpretation. Pectinase activity has been reported in the latex of Asclepias syriaca (Wilson et al., 1976) and by cytochemistry in non-articulated laticifers of Nerium oleander (Allen and Nessler, 1984). The pectinase would be involved in the intrusive tip growth of this type of laticifers (Wilson et al., 1976; Allen and Nessler, 1984). These results present the first evidence for the involvement of pectinase in the cell wall perforation process in articulated laticifers. The role of cellulase in the degradation of the articulated laticifer walls of Papaver somniferum was pointed out (Nessler and Mahlberg, 1981). In the latex of Lactuca sativa cellulase and pectinase compartmentation in the vacuole makes their involvement in intrusive growth unlikely. Their role in the breakdown of cell wall fragments internalized in the vacuole are more likely.

It should be observed that preliminary hydrolytic attack on the walls during degradation (Giordani, 1981) could lead to a higher accessibility of the parietal substrates for the vacuolar hydrolases, particularly for the cellulase. Then a delayed but more important degradation should be possible. The detection of several hydrolases, including cellulase, on the parietal degradation sites (Giordani, 1981) suggests the involvement of a synergetic complex of enzymes reminding to us those previously observed in fungus culture filtrates (Gascoigne and Gascoigne, 1960; Wood, 1968) or in vascular tissue homogenates of superior plants (Sheldrake, 1970). In the case of Lactuca sativa latex we cannot completely exclude the possible contamination of vacuolar enzymatic activities by cytoplasmic glycosidases adsorbed on the external face of the tonoplast during bleeding. To confirm the possible degradation of wall compounds by the vacuoles the action of vacuolar fractions on natural substrates should be proved. Isolation of membrane vesicles, containing wall fragments, from latex being difficult, owing to their precocity and transitory existence, we will try in the first time to test the hydrolytic activities on polysaccharidic fractions obtained from wall preparations.

REFERENCES

Allen, R., and Nessler, C., 1984, Cytochemical localization of pectinase activity in laticifers of Nerium oleander L., Protoplasma, 119:74.
Boller, T., and Kende, H., 1979, Hydrolytic enzymes in the central vacuole of plant cells, Plant Physiol., 63:1123.

Bradford, M., 1976, A rapid and sensitive method for the quantitation of microgram quantities of protein utilizing the principle of protein-dye binding, Anal. Biochem. 72:248.

Butcher, H., Wagner, G., and Siegelman, H., 1977, Localization of acid hydrolases in protoplasts, Plant Physiol., 59:1098.

Gascoigne, J., and Gascoigne, M., 1960, "Biological Degradation of Cellulose", Butterworths, London.

Giordani, R., 1977, Degradation des parois terminales durant la différenciation des laticifères articulés anastomosés de Lactuca sativa L., C. R. Acad. Sci., Paris, Ser. D, 284:569.

Giordani, R., 1979, Ultrastructure des laticifères articulés de la Laitue, C. R. Acad. Sci., Paris, Ser. D, 288:615.

Giordani, R., 1980, Dislocation du plasmalemme et libération des vésicules pariétales lors de la dégradation des parois terminales durant la différenciation des laticifères articulés, Biol. Cell., 38:231.

Giordani, R., 1981, Activités hydrolasiques impliquées dans le processus de dégradation pariétal durant la différenciation des laticifères articulés, Biol. Cell., 40:217.

Giordani, R., and Marty, F., 1983, Isolement et purification des vacuoles des laticifères articulés de Lactuca sativa L., Ann. Sc. Nat., Bot., Paris, 13ème Série, 5:229.

Leigh, R., and Branton, D., 1976, Isolation of vacuoles from root storage tissue of Beta vulgaris L., Plant Physiol., 58:656.

Marty, F., Branton, D., and Leigh, R., 1980, Plant vacuoles, in : "The Biochemistry of Plants", Vol. 1, P. K. Stumpf and E. E. Conn, eds., p. 625, Academic Press, New-York.

Matile, P., 1978, Biochemistry and function of vacuoles, Ann. Rev. Plant Physiol., 29:193.

Nessler, C., and Mahlberg, P., 1981, Cytochemical localization of cellulase activity in articulated, anastomosing laticifers of Papaver somniferum L., Amer. J. Bot., 68:730.

Saunders, J., and Gillespie, J., 1984, Localization and substrate specificity of glycosidases in vacuoles of Nicotiana rustica, Plant Physiol., 76:885.

Sheldrake, A., 1969, Cellulase in latex and its possible significance in cell differentiation, Planta, 89:82.

Sheldrake, A., 1970, Cellulase and cell differentiation in Acer pseudoplatanus, Planta, 95:167.

Sheldrake, A., and Moir, G., 1970, A cellulase in Hevea latex, Physiol. Plant, 23:267.

Wilson, K., Nessler, C., and Mahlberg, P., 1976, Pectinase in Asclepias latex and its possible role in laticifer growth and development, Amer. J. Bot.,63:1140.

Wood, T., 1968, Cellulolytic enzyme system of Trichoderma koningii. Separation of components attacking native cotton, Biochem. J., 109:217.

THE POSSIBLE ROLE OF ENDOPLASMIC RETICULUM

IN THE BIOSYNTHESIS AND TRANSPORT OF ANTHOCYANIN PIGMENTS

George J. Wagner

Agronomy Department, University of Lexington
Lexington, Kentucky 40,546-0091, U.S.A.

INTRODUCTION

The availability of methods for isolating higher plant vacuoles from a variety of plant tissues has led to recent studies of the compartmentation and mechanisms of accumulation of various plant secondary metabolites. Long suspected and inferred vacuolar location of many secondary metabolites has been confirmed in studies of vacuoles/extravacuole compartmentation using protoplasts and vacuoles isolation after slicing of root tissues (Matile, 1984; Ryan and Walker-Simmons; Wagner, 1985). More recently, a number of investigators have asked questions about which secondary products are taken up by isolated vacuoles prepared from which species, and what chemical and stereochemical-conformational properties are required for metabolite uptake into isolated vacuoles in vitro ((Deus-Neumann and Zenk, 1986; Matern et al., 1986; Werner and Matile, 1985; Rataboul et al., 1985). The results of these studies are quite interesting, but at present, it is not clear if these experiments elucidate the in vivo mechanism(s) for vacuolar accumulation of secondary metabolites. Much additional work is needed and is not doubt forthcoming.

Less uncertain perharps, but still unresolved, is the question of the localization of the biosynthetic machinery for producing secondary products in the cell. Mounting evidence suggests that multienzyme complexes or clusters may function in a number of secondary product pathways (Stafford, 1981; Hrazdina and Wagner, 1985). This notion principally comes from evidence for channeling between certain enzymes of certain pathways, ulstrastructural studies, and from cytochemical evidence for product-containing vesicles being associated with endoplasmic reticulum on one hand and fusing with the vacuoles or plasma membrane on the other. It is not illogical that secondary product pathways should exist as enzyme clusters since these pathways are multienzyme, are often branched to yield different products in different plants (i.e. phenyl-propanoid pathway) and therefore require complex regulation, usually contain one or more oxido-reductases (often integral membrane proteins - membrane anchoring), and because reactive intermediates are not uncommun in these pathways.

Also, the concept of "soluble", free-in-the cytosol proteins and enzymes is being generally displaced by a view in which enzymes and pathways of both procaryotes and eucaryotes are located as clusters with related proteins on membranes or cytoskeletal elements perharps in specific proximity to related processes and organelles/membranes (Wombacher, 1983). This notion was evidenced early by studies which showed that centrifugation of elongated algal cells and euglena resulted in concentration of the bulk of the cellular protein at the centripital end of these cells, suggesting sedimentation of most cellular protein with cytomatrix and organellar components (Guilliermond, 1941). The recent work of Paine, Masters and their colleagues are very telling in this regard. Paine et al. (1984) injected frog oocytes with a gelatin matrix which could be permeabilized by proteins of specific mol. wt. After a period of time the cells were frozen, cryodissected and the isolated gel matrix was analyzed to determine which "soluble" proteins were free to penetrate the matrix in vivo. Results showed that perharps 80% of proteins previously characterized as soluble were fixed within the cell and were not normally soluble in the cytosol (Paine, 1984). The work of Masters et al. and others has established the specificity and regulatory associations of enzymes with membranes (Wombacher, 1983; Guilliermond, 1941; Paine, 1984; Masters, 1984).

Thus, our concept of the biosynthesis and accumulation of plant secondary metabolites is beginning to mature to a view in which biosynthesis occurs in enzyme clusters (perharps often associated with the endoplasmic reticulum) and products are subsequently transported to and accumulated within the cell (vacuoles) or outside the cell into cell wall or outside the cell wall sites (exudate droplets). One might expect that transport also occurs by non-random processes.

A very much open question is what mechanism(s) or modes of transport are associated with metabolite accumulation in the vacuole or secretion outside the cell. For many secondary metabolites a part of this question which is seldom addressed is the question of whether or not the cytosol-cytomatrix can tolerate the presence of secondary metabolites which are potentially reactive to proteins-enzymes. There is in vitro data which suggest that phenols, flavanoids and other secondary products at low concentration may be inhibitory to critical metabolic processes (Van Sumere, 1975). However, it is possible that microcompartmentation protects cytoskeletal proteins and enzymes or that toxic effects seen in vitro do not occur in vivo because of some unknown protective mechanisms.

POSSIBLE MODES OF BIOSYNTHESIS AND ACCUMULATION
OF SECONDARY METABOLITES

Several modes of biosynthesis and accumulation of secondary metabolites destined for vacuolar accumulation are presented in Table 1.

Table 1

Possible Modes of Biosynthesis and Vacuolar Accumulation
of Secondary Metabolites

1)	·Formation on/in ER or elsewhere (chloroplasts), release into cytosoil and uptake by tonoplast - "cytosolic" enzymes protected by microenvironments ?
2)	Formation on/in ER or elsewhere, vesicle packaging and targeting for vacuole ?
3)	Mode 1 or 2, depending on product.
4)	Close proximity of site of biosynthesis and tonoplast.

In mode 1, products are formed in/on the endoplasmic reticulum or elsewhere and are subsequently released and diluted into the cytosol. When these products contact a tonoplast transporter, they are taken up and sequestered-stored within the vacuole. A modified version of mode 1 might include complexation of a potentially toxic metabolite at the site of synthesis - where complexation permits movement through the cytosol as an inert species. This author knows of no direct experimental evidence which supports this mode 1. Suggestive evidence to be discussed below comes from studies using isolated vacuoles. In mode 2, secondary metabolites produced on/in membranes and organelles, are packaged in product-impermeable vesicles, and by some means vesicles are targeted to and fuse with tonoplast resulting in vacuolar deposition. It is possible that both modes 1 and 2 occur depending on the secondary product involved.

In mode 4, close proximity between the site of biosynthesis and the tonoplast may create a microenvironment in which products can enter the vacuole without dilution into the cytosol or packaging within vesicles.

Evidence relating to possible modes of accumulation of various secondary metabolites in vacuoles and those excreted are listed in Table 2.

Table 2

Evidence Relating to Possible Modes of Accumulation
of Secondary Metabolites

Tannins	via vesicles from RER; C_4-OH ER bound	Parham and Kaustinen (1977); Diers and others; Czichi and Kindl and others (1981)
Naphtoquinone pigments (excreted)	via vesicles from ER	Tsukuda and Tabata (1984)
Terpenoids (excreted)	via vesicles from ER, chloroplasts ?; HMG–Co A reductase ER bound	Carde, Vemer and Peterson (see West et al., 1979); Brooker and Russell and others (1985)
o-coumaric glycoside	cis-form trapping	Rataboul et al. (1985)
Alkaloids	1) via vesicles from ER	Nessler and Mahlberg (1977)
	2) selective, diffusion weak base trapping (?)	Renaudin and Guern (1982); Deus-Neumann and Zenk (1986) and others
	3) berberine alkaloids via vesicles	Amann et al. (1986)
Cyanogenic glycosides	most enzymes ER-bound; chloroplast glycosylation ?	Conn et al. (1984)
Flavanoids	Selective, acetylated glycosides ?	Matern et al. (1986)

There is good evidence for the occurrence of mode 2 in the case of tannins. The work of Parham and Kaustinen (1977) and others (see Wagner, 1981) provide strong cytochemical-ultrastructural evidence for synthesis of tannins on/in the endoplasmic reticulum and their transport to the vacuole via endoplasmic reticulum-derived vesicles. Similarly, recent work of Tsukuda and Tabata suggests the endo-plasmic reticulum origin of naphtoquinone pigment containing vesicles which appear to fuse with the plasma membrane and deposit these pigments outside the cell. Vesicular routes of transport have also been suggested for lipids and alkaloids (see Tsukuda and Tabata, 1984). Recently Amann et al. (1986) reported the occurrence of vesicles containing the isoquinoline alkaloid berberine and two enzymes involved in biosynthesis of this product. It was concluded that vesicles were derived from Golgi, however this was not established. For terpenoids (lipophilic and potentially reactive) ultrastructural evidence suggests secretion outside the cell by a vesicular route (see West et al., 1979). As already mentioned, less compelling evidence supporting a vesicular route is the argument and evidence that various critical metabolic processes are inhibited in vitro by low levels of phenols, flavonoids, terpenes (Van Sumere et al., 1975).

Support for mode 1 comes from a number of investigators who have recently presented very interesting results showing that isolated vacuoles can be selective in accumulating secondary metabolites (Matern et al., 1986). However, one may argue that these results point to a vacuolar trapping mechanisms. Boudet's group was the first to show trapping of o-coumaric but not p-coumaric glycoside in isolated vacuoles of Meliloites suggesting a mechanism at the tonoplast which could result in accumulation of a specifically substituted metabolite (Rataboul et al., 1985). It has been known for some time that weak bases such as alkaloids can accumulate in vacuoles via an ion trap mechanism whereby unprotonated species penetrate the vacuole but once within protonation prevents exit. Thus a downhill chemical gradient is maintained allowing accumulation. The question of the selectivity in accumulation of alkaloids and the importance of an ion trap mechanism versus one in which specific tonoplast porters function in alkaloid transport and accumulation is one of much current debate (see Guern et al., this volume).

It is perharps too early to evaluate in a uptake-mechanism-sense results which demonstrate selective uptake of secondary metabolites by isolated vacuoles. Questions remain, such as how vacuoles may have been altered by isolation, the consequences of disruption of the probable cytosol-vacuole interface, and others. Testing systems using vacuoles isolated by different basic approaches (Wagner, this volume) may be useful in resolving these questions, just as this approach is perharps beginning to explain variation in proton pump and ATPase activities of vacuoles isolated by various methods (see Pugin, and Columbo et al., this volume). Nevertheless, the results using isolated vacuoles are most interesting.

EXPERIMENTS ON ANTHOCYANIN ACCUMULATION

We (G. Hrazdina and G. Wagner) have been investigating the subcellular site of biosynthesis and mechanism of vacuolar accumulation of anthocyanin pigments and have, with others, provided evidence for association of enzymes involved in this synthesis with endoplasmic reticulum (Hrazdina and Wagner, 1985; Wagner, 1981). We have also reported evidence which is consistent with the view that biosynthesis occurs via multienzyme complexes associated with the endoplasmic reticulum (Hrazdina and Wagner, 1985). Here we report results of preliminary experiments which suggest that anthocyanin produced on/in the endoplasmic reticulum is packaged in vesicles for transport to the vacuole.

Hippeastrum petals from intermediate stage buds active in anthocyanin biosynthesis were sliced into 1 mM strips to allow penetration of aqueous solution and were floated on B5 culture medium containing hormones and 1 μC_i ^{14}C-cinnamic acid (56 mC_i/mmole - Amersham). After various periods of time, pieces (0.15 mg fresh weight) were homogenized in 150 mM Tricine-KOH, pH 7.5, containing 10 mM KCl, 1 mM MgCl$_2$, 1 mM EDTA, 2 mM DTT, 3% sucrose, 1.5 mg PVP and homogenates were centrifuged at 1000 g and the supernatants separated by isopycnic centrifugation as previously described (Wagner and Hrazdina, 1984). Fractions corresponding to rough endoplasmic reticulum (RER) which were identified using NADPH cytochrome c̲ reductase as an ER marker and EDTA shift experiments (Fig. 1) were recovered. Pelargonidin-3-rutinoside (principal anthocyanin pigment in this tissue) was separated by TLC on microcrystalline cellulose using BuOH:HOAc:H_2O, 4:1:5 (v/v) as solvent. Label in pigment was observed to increase with time up to 180 minutes (not shown) after which time it decreased. When tissue was transferred to fresh medium after 60 minutes of labeling (endogenous chase) or to medium containing 0.05 mM unlabeled cinnamic acid (external chase), label in RER-associated pigment was observed to decrease (Fig. 1). These results suggest that labeled anthocyanin formed from ^{14}C-cinnamic acid can occur in association with RER. Dilution and lysis of RER fractions and separation of membrane and supernatant by centrifugation at 100,000xg, 1 hr suggests that labeled anthocyanin is associated with the RER lumen (not shown). However, it is recognized that pigment loosely associated with the cytosol face of the ER might also be solubilized under such conditions.

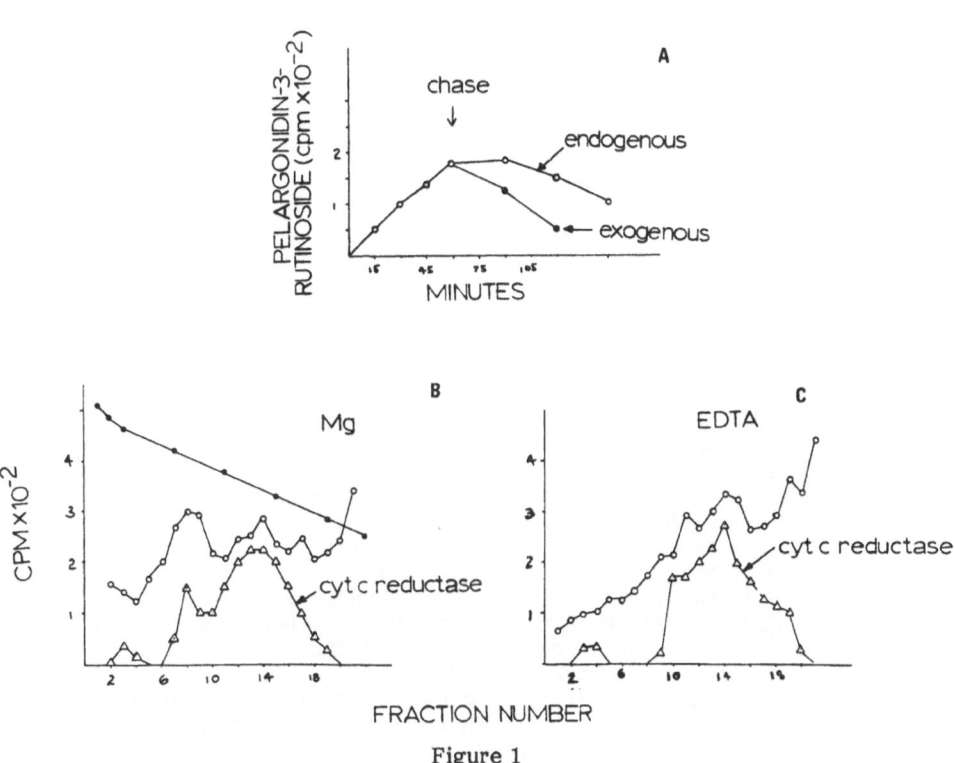

Figure 1

Formation of pelargonidin-3-rutinoside
in Hippeastrum petal slices

A: time course labeling (see text for explanation), B: isopycnic separation of labeled membranes using Mg-containing sucrose gradient, and C: as in B but using EDTA-containing gradient.

Therefore, while these data support earlier studies suggesting biosynthesis on the ER and packaging in the lumen, they do not allow distinction between such packaging and release into the cytosol (mode 1, Table 1). Efforts to isolate anthocyanin-containing vesicles have led to inconclusive results. Vesicles (1.04 g/cc density) containing labeled product after incubation of Hippeastrum bud petals in ^{14}C-cinnamic acid, ^{14}C-phenylalanine, ^{14}C-UDP-glucose and also ^{3}H-choline have been isolated by isopycnic separation (Wagner, unpublished data). However, labeling of these vesicles is very low and products have not been identified as anthocyanin (data not shown).

Fig. 2 describes a modification of a working model presented earlier to summarize data concerning biosynthesis of anthocyanin on/in the RER and subsequent transport to the vacuole.

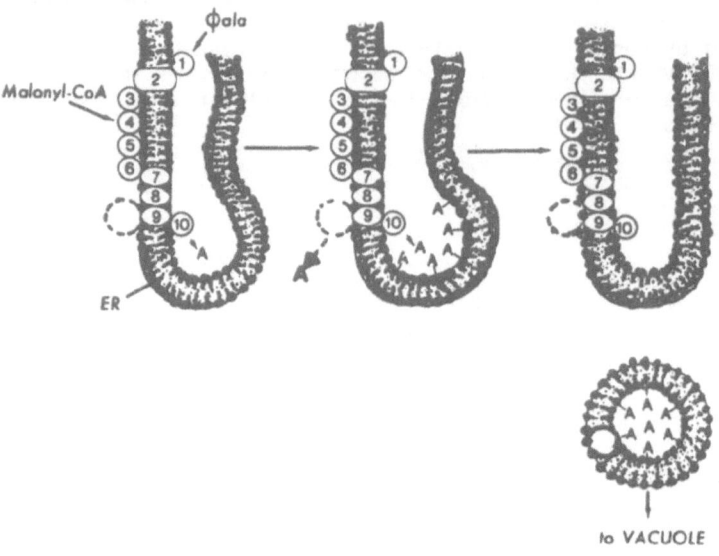

Figure 2

Working model explaining available data
on phenyl-propanoid and flavonoid enzyme localization

In the synthesis of cyanidin-3-glucoside an endoplasmic reticulum (ER) portion bears internal phenylalanine ammonia-lyase 1, that is associated with transmembrane cinnamate 4-hydroxylase 2, that channels its product, p-coumarate, to the cytoplasmic face of the ER where the four consecutive enzymes of the pathway, p-coumarate/CoA ligase 3, chalcone synthase 4, chalcone isomerase 5, and flavonone 3-hydroxylase 6, are located. Further transformation of flavan 3-ol to anthocyanidin anhydrobase takes place on the three membrane embedded oxidoreductases (putative) flavanol 3'-hydroxylase 7, flavanol-4-keto-reductase 8, and flavan 3,4-diol dehydratase 9. Cyanidin anhydrobase is glucosylated by a glucosyltransferase 10, that resides on the lumen face of the ER-membrane. In this model, products are sequestered in specific regions of the ER that is destined for vesiculation to form transport vesicles. Glycosylated flavonoids are transported to the central vacuole where they accumulate. Glycosylated phenyl-propanoids may be transported by a similar mechanism to the plasmalemma and excreted into cell wall regions for lignification. Alternatively, if last enzyme of the pathway is on the cytoplasmic face, products may be released to the cytosol (dashed circle). Model is modified from Hrazdina and Wagner (1985).

The dashed circle represents the possible occurrence of glucosyltransferase on the cytoplasmic face of the RER. If this enzyme was the last enzyme of the biosynthetic pathway and it was localized on the cytoplasmic face of the RER (both questions unresolved), pigment glycosides might be released into the cytosol and possibly accumulate in the vacuole via modes 1 or 4. If this enzyme is located on lumen side, anthocyanin might be transported to the vacuole packaged in vesicles. More studies are required to resolved these possibilities.

SUMMARY

Current interest in determing the subcellular sites of biosynthesis of plant secondary products and the modes of their accumulation in vacuoles or secretion outside cells has moved us closer to elucidating the biochemistry and cell biology of secondary product formation and accumulation. Understanding these processes is essential if future attempts to modify and amplify secondary product production are to be maximally efficient and successful. Application of biotechnology approaches to plant modification not only require that specific targets (enzymes) and their genes be identified, but also that regulatory aspects of pathways be understood and that the cell's capacity to accomodate amplification be known.

REFERENCES

Amann, M., Wanner, G., and Zenk, M. H., 1986, Intracellular compartmentation of two enzymes of berberine biosynthesis in plant cell cultures, Planta, 167:310.

Conn, E. E., 1984, Compartmentation of secondary compounds, in : "Proceedings of the Phytochemical Society of Europe", A. M. Boudet, G. Alibert, G. Marigo, and P. J. Rea, eds., Vol. 24, 1, Clarendon Press, Oxford.

Deus-Neumann, B., and Zenk, M. H., 1986, Accumulation of alkaloids in plant vacuoles does not involve an ion-trap mechanism, Planta, 167:44.

Guilliermond, A., 1941, "The Cytoplasm of the Plant Cell", Chronica Botanica, Waltham, Massachussetts.

Hrazdina, G., and Wagner, G. J., 1985, Compartmentation of plant phenolic compounds : Sites of synthesis and accumulation, in : "Proceedings of the Phytochemical Society of Europe", C. F. Sumere, and P. J. Lea, eds., Vol. 25, p. 119, Clarendon Press, Oxford.

Matern, V., Reichenbach, C., and Heller, W., 1986, Efficient uptake of flavanoids into parsley vacuoles requires acetylated glycosides, Planta, 167:183.

Masters, C., 1984, Interactions between glycolytic enzymes and components of cytomatrix, J. Cell. Biol., 99:222.

Matile, P., 1984, Das toxische kompartiment der pflanzell, Naturwissen., 71:18.

Nessler, C. L., and Mahlberg, P. G., 1977, Ontogeny and cytochemistry of alkaloidal vesicles in laticifers of Papaver somniferum L., Amer. J. Bot., 64:541.

Paine, P. L., 1984, Diffusive and non-diffusive proteins, in vivo, J. Cell. Biol., 99:219.

Parham, R. A., and H. M. Kaustinen, H. M., 1977, On the site of tannin synthesis in plant cells, Bot. Gaz., 138:465.

Rataboul, P., Alibert, G., Boller, T., and Boudet, A. M., 1985, Intracellular transport and vacuolar accumulation of o-coumaric acid glucoside in Melilotus alba mesophyll cell protoplasts, Biochim. Biophys. Acta, 816:25.

Renaudin, J. P., and Guern, J., 1982, Compartmentation mechanisms of indole alkaloids in cell suspension cultures of Catharanthus roseus, Physiol. Vég., 20:533.

Russell, D. W., 3-hydroxy-3-methylglutaryl-CoA reductases from pea seedlings, Methods in Enzymol., 110:26.

Ryan, C. A., and Walker-Simmons, M., 1983, Methods in Enzymol., 96:580.

Stafford, H. A., 1981, Compartmentation in natural product biosynthesis by multienzyme complexes, in : "The Biochemistry of Plants", E. E. Conn, ed., Vol. 7, p. 117, Academic Press, New-York.

Tsukada, M., and Tabata, M., 1984, Intracellular localization and secretion of naphtoquinone pigments in cell cultures of Lithospermum erythrorhizon, Planta Medica, 51:338.

Van Sumere, C. F., Albrecht, J., Dedonder, A., DePooter H., and I. Pe', 1975, in

: "The Chemistry and Biochemistry of Plant Proteins", J. Harborne, and
C. F., Van Sumere, eds., p. 211, Academic Press, New-York and London.

Wagner, G. J., 1981, Compartmentation in plant cells : The role of the vacuole,
in : "Recent Advances in Phytochemistry", L. L. Creasy, and G. Hrazdina,
eds., Vol. 16, p. 1, Plenum Press, New-York and London.

Wagner, G., 1985, in : "Modern Methods of Plant Analysis", H. F. Liskins, and J.
F. Jackson, eds., Vol. 1, p. 105, Springer-Verlag, Berlin, Heidelberg, New-
York, and Tokyo.

Wagner, G. J., and Hrazdina, G., 1984, Endoplasmic reticulum as a site of phenyl-
propanoid and flavanoid metabolism in Hippeastrum, Plant Physiol., 74:901.

Werner, C., and Matile, P., 1985, Accumulation of coumaryl-glucosides in vacuoles
of barley mesophyll protoplasts, J. Plant. Physiol., 118:237.

West, C. A., Dudley, M. W., and Dueber, M. T., 1979, Regulation of terpenoid biosynthesis in higher plants, in : "Recent Advances in Phytochemistry", T. Swain
and G. R. Waller, eds., Vol. 13, p. 163, Plenum Press, New-York and London.

Wombacher, H., 1983, Molecular compartmentation by enzyme cluster formation,
Mol. Cell Biochem., 56:155.

PATCH CLAMP STUDIES ON THE TRANSPORT OF IONS

ACROSS THE MEMBRANE OF BARLEY VACUOLES

Ulrich I. Flügge[*], Rainer Hedrich and J. M. Fernandez

Max-Planck Institut für Biophysikalische Chemie
and [*]Institut für Biochemie der Pflanze
3400-Gottingen, Federal Republic of Germany

INTRODUCTION

In photosynthesis of C_3 plants such as wheat, spinach and barley, the main part of the fixed carbon is converted to sucrose (Giersch et al., 1980; Stitt et al., 1980). However, significant amounts of the assimilated carbon are also found in malate, which exhibits pronounced diurnal concentration (Gerhardt et al., 1986). In assimilating barley mesophyll protoplasts it could be shown that malate is rapidly transported in the vacuoles (Kaiser et al., 1982). In spinach leaves, the vacuolar malate concentration at the end of the day can increase up to 50 mM whereas the cytosolic malate concentration remains at about 5 mM (Gerhardt et al., 1986). This suggested that the malate transport into the vacuoles is an energy driven process. During the following night period malate is released into the cytosol where it presumably serves as a substrate for the mitochondrial respiration. In addition to malate, H^+ and K^+ ions also cross the vacuolar membrane in order to maintain electroneutrality and isoosmolarity between the cytosol and the vacuole. Using vacuoles isolated from barley mesophyll protoplasts we have recently shown that malate can be transported into the vacuoles against its concentration gradient (Martinoia et al., 1985). The driving force of the active transport is provided by an H^+-translocating ATPase located in the vacuolar membrane (Martinoia et al., 1985). Like other H^+-translocating ATPases, this ATPase can be inhibited by vanadate, olygomycin and azide but it is stimulated in the presence of anions (in particular chloride) and inhibited by nitrate.

In this work, we study the transport of malate, H^+ and K^+ across the vacuolar membrane using the patch-clamp technique.

RESULTS AND DISCUSSION

Vacuoles were isolated as described elsewhere (Martinoia et al., 1985) and resuspended in a medium containing 50 mM KCl, 1 mM $CaCl_2$, 50 mM Hepes-imidazol pH 7.3 adjusted of a final osmolarity of 600 mosmol (bath solution). The recording pipette was filled with 50 mM KCl, 1 mM $CaCl_2$, 50 mM Hepes-Mes pH 5.5 adjusted with sorbitol to 550 mosmol. High resistance seals between the patch pipette and the vacuolar membrane were easily achieved when the pipette was pressed against the vacuole and suction to the pipette interior was applied ("vacuole attached" mode analogous to "whole cell", Hamil et al., 1981). The "whole vacuole" configuration was formed when the membrane patch underlying the pipette tip

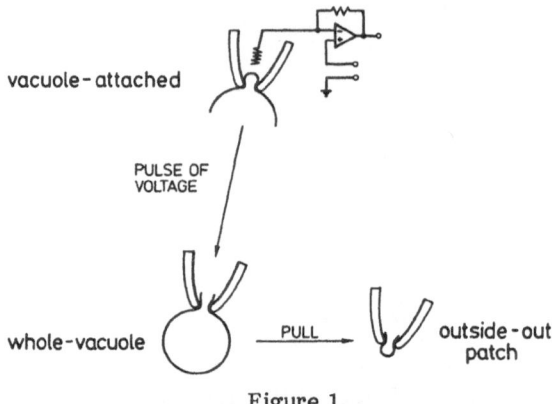

Figure 1

Patch–clamp technique applied to plant vacuoles
(according to Hamill et al., 1981)

was broken by a pulse of hyperpolarizing voltage. Withdrawal of the pipette forms a membrane patch with the cytoplasmic side facing the bath solution (outside–out patch, Fig. 1).

In "vacuole attached" patches we have observed single channel currents. Fig. 2 shows that discrete step–like transitions of the current of about 1 pA can be seen when a channel opens or closes. The single channel conductance was found to be 60-80 pS. The opening probability of the single channel is voltage dependent with increasing opening at negative membrane potentials. By changing the ion composition of both sides of the membrane the channels proved to be non–selective and were found to be permeable to K^+ as well as to malate as outlined below in more detail.

2 pA

20 ms

Figure 2

Single–channel openings from an outside–out patch
recorded in 50 mM KCl and a membrane potential of – 50 mV
(inside negative)

Fig. 3–A shows typical ionic currents recorded in a "whole vacuole" configuration. In symmetric 50 mM KCl solutions (holding potential 0 mV) hyperpolarizing (inside negative) voltage steps evoked large inward currents but no current at depolarizing potentials, indicating the presence of inwardly directed rectifying channels through which cations can pass into or anions can be released from the vacuole. To determine the specificity of these channels, experiments were performed with different ion composition in the pipette and the bath solution, respectively. Fig. 3–B (closed symbols) shows current–voltage (I/V) relationships

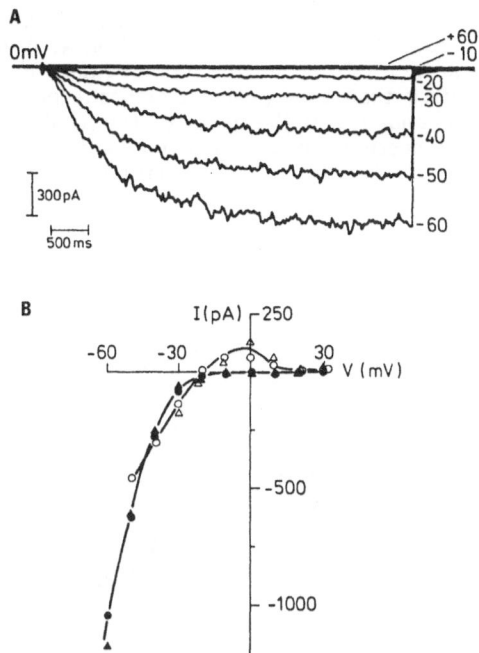

Figure 3

A. "Whole-vacuole" currents in response to a series of voltage steps from a holding from a holding potential of 0 mV in symmetric 50 mM K_2malate.

B. Current-voltage relationship of the vacuolar membrane measured in symmetric 50 mM K_2malate (closed symbols) and after changing external solution to 20 mM K^+ and 5 mM malate (open symbols).Zero current potentials were determined by measuring the reversal of the tail currents.

obtained with 50 mM K_2malate (pH 7.3) in the bath and 50 mM K_2malate (pH 5.5) in the pipette. Under these conditions only inwardly rectifying currents with a zero-current potential of 0 mV could be observed. When the external bath solution was changed to 20 mM K^+ and 5 mM malate^{2-} the zero current potential shifted to - 17 mV (Fig. 3-B, open symbols). This shift in the reversal potential could be attributed neither to K^+ (E_{K^+} = - 40 mV) nor to malate ($E_{mal}2-$ = + 20 mV, $E_{mal}1-$ = + 150 mV) alone but rather indicates that both ions can pass through these channels.

To determine whether the current channel measured in the "whole vacuole" configuration (Fig. 3-A) is due to single channel currents we used outside-out patches prepared as described above. Holding the excised patches at a hyperpolarizing potential (- 40 mV) current fluctuations were observed indicating activation of many channels (Fig. 4-A). When the membrane potential was stepped to a depolarizing value (+ 60 mV) the closing of individual channels could be observed. A similar experiment was performed measuring currents in a "whole vacuole" configuration (Fig. 4-B). The membrane potential was stepped from a holding potential of 0 mV to a hyperpolarizing value (- 80 mV) and subsequently to different depolarizing levels. During the second voltage pulse, the currents ("tails") decreased with a voltage dependent time course reversing their direction at 0 mV. As expected from the strong voltage dependence of the membrane currents, the decline of the current was much faster at positive than at negative membrane potentials. Since the time course and the voltage dependence of the decay of the whole cell current is identical to that of single channel currents as shown in Fig. 4-A, it can be concluded, that the current measured in the "whole vacuole" configuration is generated by inward rectifying channels.

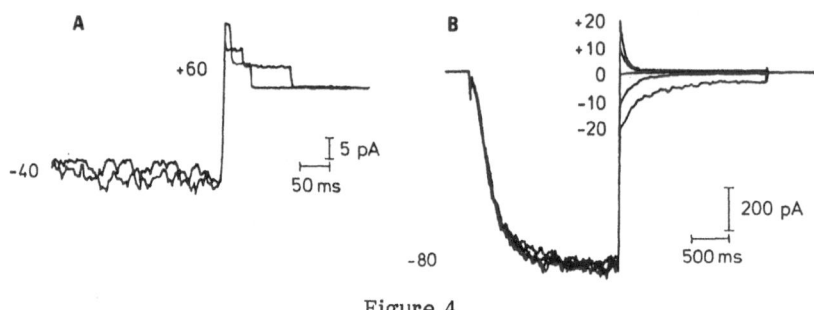

Figure 4

A. Membrane currents recorded from an outside-out patch. The membrane potential was held at a hyperpolarizing value (- 40 mV) and then stepped to a depolarizing value of + 60 mV (symmetric 50 mM K_2malate solutions).

B. "Whole-vacuole" currents during a double-pulse experiment. The membrane potential was stepped from a holding potential of 0 mV with a first hyperpolarizing pulse to - 80 mV and then to various depolarizing levels (symmetric 50 mM K_2malate solutions).

In further experiments we studied that ATP-dependent H^+-translocation into barley mesophyll vacuoles, which is suggested to be involved in anion accumulation. Using N-methylglucamine–glutamate (NMG) solutions (pH 7.3) the vacuolar resting potential was zero and no currents could be measured under voltage clamp at both depolarizing and hyperpolarizing potentials, indicating that the channels were blocked.

Externally added MgATP resulted in a depolarization of the membrane potential to + 50 mV (Fig. 5-A, current clamp) indicating the presence of an electrogenic H^+ pump energized by MgATP. Proton pumping could be totally inhibited by low concentrations of the H^+-ATPase inhibitor tributyltin. When the membrane potential was held at 0 mV (voltage clamp) the pump current generated by the H^+-ATPase could be measured directly. Increasing concentration of MgATP in the extravacuolar medium resulted in an increase of the inward directed pump current to a maximal value of about 70 pA (Fig. 5-B). The double-reciprocal plot yields a linear function, indicating ATP saturation of H^+ pumping (Fig. 5-C). In various experiments, mean values for K_M(ATP) and I_{max} were found to be 0.8 mM and 50-85 pA, respectively. The estimated K_M value of 0.8 mM is in the range of the cytoplasmic ATP concentration as measured in isolated C_3 protoplasts (Stitt et al., 1982). Under physiological conditions the cytoplasmic ATP concentration is about 1-2 mM (Stitt et al., 1982). The pH in the cytoplasm is about 7-7.8 and 5-6 in the vacuole. Our results show that these conditions will produce a pump current of 50 pA per vacuole. Assuming that malate short circuits the H^+ current by following H^+ passively, vacuoles could accumulate malate under these conditions

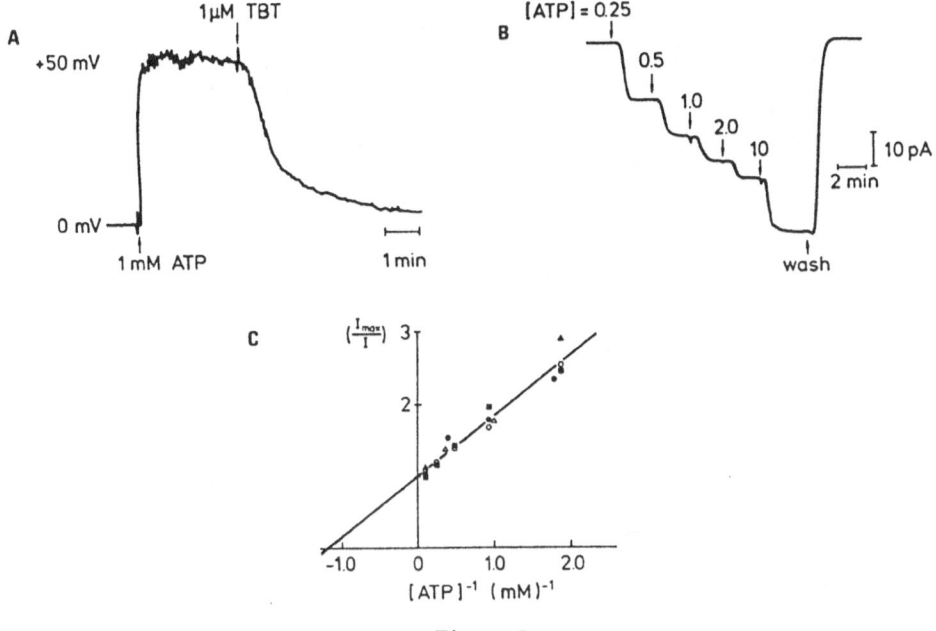

Figure 5

A. Current clamp recording of the vacuolar membrane potential. The arrows indicate the application of 1 mM MgATP or 1 M tributyltin (TBT), respectively.
B. ATP-induced pump currents. The membrane potential was held at 0 mV. At the times indicated, increasing concentrations of MgATP were added.
C. Concentration dependence of the pump current on external MgATP. Data derived from four different experiments similar to that shown in Fig. 5-B.

at a rate of 12.5 fmole per vacuole per second. If the pump current is entirely due to H^+, then two H^+ ions are pumped into the vacuole per each malate^{2-} accumulated.

In summary we have presented direct evidence for the presence of an electrogenic H^+ pump in the vacuolar membrane which provides enough current to drive the accumulation of malate against its concentration gradient as observed in whole leaves (Gerhardt et al., 1986) or intact isolated vacuoles (Martinoia et al., 1985). Furthermore we have described a pathway for the flux of malate and K^+ through the inwardly rectifying channels across the vacuolar membrane.

REFERENCES

Gerhardt, R., Stitt, M., and Heldt, H. W., 1986, Plant Physiol., in press.
Giersch, C., Heber, U., Kaiser, G., Walker, D. A., and Robinson, S. P., 1980, Intra-cellular metabolite gradients and flow of carbon during photosynthesis of leaf protoplasts, Arch. Biochem. Biophys., 205:246.
Hamill, O. P., Marty, A., Neher, E., Sakmann, B., and Sigworth, F. J., 1981, Pflügers Arch. Ges. Physiol., 39J:85-100.
Kaiser, G., Martinoia, E., and Wiemken, A., 1982, Rapid appearence of photosynthetic products in the vacuoles isolated from barley mesophyll protoplasts by a new fast method, Z. Pflanzenphysiol., 107:103.
Martinoia, E., Flügge, U. I., Kaiser, G., Heber, U., and Heldt, H. W., 1985, Energy-

dependent uptake of malate into vacuoles isolated from barley mesophyll protoplasts, Biochim. Biophys. Acta, 806:311.

Stitt, M., McLilley, R. and Heldt, H. W., 1982, Adenine nucleotide levels in the cytosol, chloroplasts and mitochondria of wheat leaf protoplasts, Plant Physiol., 70:971-977.

Stitt, M., Wirtz, W., and Heldt, H. W., 1980, Metabolite levels during induction in the chloroplast and extra-chloroplast compartments of spinach protoplasts, Biochim. Biophys. Acta, 593:85-102.

INTRACELLULAR DISTRIBUTION OF ORGANIC AND INORGANIC ANIONS

IN MESOPHYLL CELLS : TRANSPORT MECHANISMS IN THE TONOPLAST

Enrico Martinoia, Michael J. Schramm, Ulf-Ingo Flügge[*]
and Georg Kaiser

Lehrstuhl Botanik I der Universität,
Mittlerer Dallenbergweg 64, D-8700-Würzburg
and *Lehrstuhl für Biochemie der Pflanzen der Universität
Untere Karspüle 2, D-3400-Göttingen, FRG

INTRODUCTION

Plants contain organic as well as inorganic ions. Both play an important role in metabolism. Malic acid accumulates in the vacuoles of CAM plants during the night, and this process is reversed during the daytime (Osmond and Holtum, 1981). In contrast, in leaves of C_3 plants, vacuolar levels of malate are high at the end of the day and low in the morning (Gerhardt and Heldt, 1984). Both examples reflect the dynamics of vacuolar compartmentation.

Uptake and vacuolar storage of inorganic ions enable the plant to maintain the necessary difference in water potential between soil and leaves without expanding energy for the production of organic osmotica. High vacuolar contents of inorganic salts which may be toxic when accumulated in the cytoplasm are not restricted to plants adapted to saline environments. Even in plants grown under normal conditions, inorganic salts account for the major part of the osmotically active substances.

Vacuolar storage of anions must be under complex regulation. Some anions are remobilized and others remain in the vacuole once deposited there (Jeschke, 1979; Christensen et al., 1981). The knowledge of anion compartmentation and their transport properties are first steps toward the understanding of this complex phenomena.

MATERIAL AND METHODS

Methods used in this report have been described before (Martinoia et al., 1981; Kaiser et al., 1982; Kaiser and Heber, 1984; Martinoia et al., 1985 and 1986).

RESULTS AND DISCUSSION

Compartmentation of Anions

Distribution of anions in leaves and protoplasts. Green mesophyll protoplasts from barley primary leaves contained only about 55 % of the total leaf anions when compared on chlorophyll basis (Table I).

Table 1

Content of Anions and Anion Ratios
in Primary Leaves and Protoplasts of Barley

	Leaves	Mesophyll Protoplasts	% in Mesophyll
	μmol mg^{-1} Chl		
Measured Inorganic Anions	121.1	67.3	55.6
Cl^{-1}	70.5	30.5	43.3
NO_3^-	32.7	21.1	64.5
P_i	16.2	14.2	87.7
SO_4^{2-}	1.7	1.5	88.0
Cl^-/NO_3^-	2.16	1.45	
Cl^-/P_i	4.35	2.15	
Cl^-/SO_4^{2-}	41.5	20.3	

At first sight, this suggests significant loss of anions during protoplast isolation. However, microscopic examination revealed that about 40% of the barley leaf cells were free of chloroplasts. Moreover, protoplasts with only a few chloroplasts, which had a lower specific density than protoplasts rich in chloroplasts were lost during the first part of the isolation procedure.

Results of an efflux study revealed little loss of anions from floating leaves whose lower epidermis had been removed to facilitate diffusion between tissue

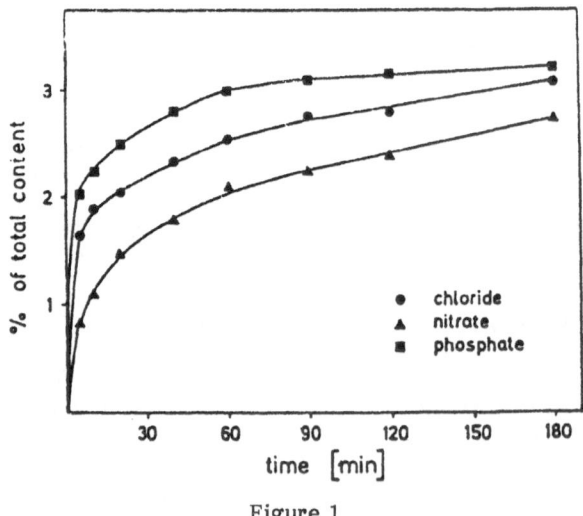

Figure 1

Loss of anions from floating leaves

Lower epidermis had been stripped to facilitate exchange between mesophyll and bathing solution.

and solution (Fig. 1). About 2% of the leaf anions were lost during the first 10 minutes. This loss probably reflects diffusion of apoplastic anions in the medium. In the subsequent 3 h leakage of anions was only about 1%.

The ratios of chloride to phosphate, chloride to nitrate and chloride to sulfate were significantly higher in leaf extracts than in extracts from protoplasts (Table I). This is not due to a preferential loss of chloride during the protoplast isolation procedure but to an accumulation of chloride in the epidermis (Table 2). In contrast, phosphate is accumulated in the mesophyll and largely excluded from the epidermis.

Table 2

Content of Anions
in Homogenates of Primary Leaves and Isolated Epidermis

	Leaf Homogenate mM	Epidermis mM
Cl^-	18.2	37
NO_3^-	159	209
P_i	21.2	0.8

From the slow leakage of ions from leaves (Fig. 1) and protoplasts (not shown) it can be concluded that mesophyll protoplasts from barley primary leaves are suited for investigating the localization of anions.

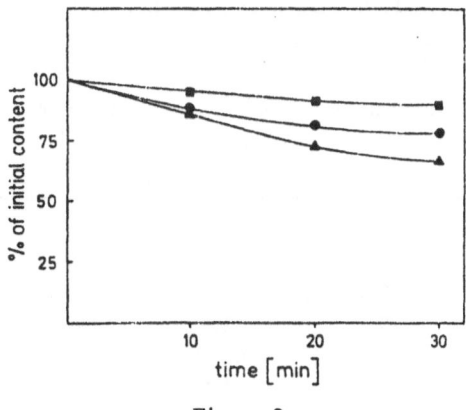

Figure 2

Time-dependent release of chloride (●), nitrate (▲) and phosphate (■)
from isolated vacuoles

Intracellular Distribution of Anions in Mesophyll Protoplasts. Depending on the method, vacuole isolation and purification is rather time consuming and solutes may leak out from the vacuole. It is therefore important to ensure, that leakage of the solutes to be investigated is negligible within the time scale of vacuole isolation. As shown in Fig. 2, anions are released from isolated vacuoles. Loss of nitrate was faster than that of chloride, whereas phosphate efflux was very slow.

Using the silicone oil technique (Kaiser et al., 1982), vacuoles were isolated at 4°C within 30 s. Under these conditions, efflux of anions is negligible. Table 3 shows data on the intracellular distribution of anions between the vacuole and the extravacuolar space of mesophyll protoplasts. Data are based on the assumption that on average the vacuole of mesophyll cells occupies 80% of the cellular volume. As expected, the bulk of anions is localized in the vacuolar compartment. Nevertheless, there are important differences between the different species. The chloride-, sulfate- and potassium-gradients across the tonoplast are very small, whereas phosphate and to a greater extent nitrate and sodium are accumulated in the vacuole. Since homeostatic mechanisms are thought to maintain an ionic composition of the cytoplasm which facilitates functioning of metabolism (Kaiser et al., 1983), the vacuole may be viewed as a storage compartment for ions. Interest should therefore be focused on cytoplasmic rather on vacuolar anion concentrations. Sodium and nitrate concentrations are very low in the extravacuolar space (4-6 mM).

However, it should be noted, that for the determination of the extravacuolar concentrations the difference of the ion content between the protoplast and the vacuole, which occupies 80% of the cellular volume has to be calculated. It is obvious that the error of such a calculation increases with an increasing accumulation of the ion in the vacuolar space. The extravacuolar phosphate concentration of almost 34 mM is significantly higher than concentrations reported for darkened chloroplasts (McLilley et al., 1977). However our data were obtained from protoplasts which had been suspended in a medium of higher osmolarity than that used for the chloroplast experiments described by McLilley et al. (1977). In the presence of sorbitol both chloroplasts and protoplasts respond to modest changes in osmolarity as near-perfect osmometers (Kaiser et al., 1983). Correction for osmolarity results in a phosphate concentration comparable to that reported for isolated darkened chloroplasts.

Table 3

Concentrations of Anions in Protoplasts and Vacuoles
Isolated from Barley Leaves

	Protoplasts	Vacuoles mM	Extra-vacuolar Space	Proportion %
Measured Inorganic Anions	175.6	188.6	122.8	
Cl^-	56.0	56.1	55.6	80.3
NO_3^-	35.1	42.9	4.1	97.7
P_i	58.4	64.5	33.6	88.5
SO_4^{2-}	26.1	25.1	30.0	77.0
K^+	134.1	142.2	101.4	84.9
Na^+	35.0	42.2	5.3	96.5

In the experiments of Table 3 chloride seems to be distributed equally between vacuolar and extravacuolar space, which results in a cytoplasmic concentration of about 56 mM. However, if barley seedlings are stressed by the addition of sodium chloride, they take up sodium chloride to overcome osmotic stress. In the experiments of Table 4 the chloride concentration in protoplasts was increased to 153 mM, the nitrate concentration to 157.5 mM. Nevertheless, the cytoplasmic chloride and nitrate concentrations remained at the levels shown for unstressed plants (Tables 3 and 4), indicating that homeostasis is maintained in the extravacuolar space. These observations are in agreement with the results recently published by Steingröver et al. (1986). Using the nitrate reductase method for the estimation of the nitrate pools, they have shown that the cytoplasmic nitrate concentration is held constant at 5 - 8 mM, whereas vacuolar nitrate is increased severalfold during the night.

Table 4

Concentration of Anions in Protoplasts and Vacuoles
Isolated from Salt-stressed Barley

	Protoplasts	Vacuoles mM	Extra-vacuolar Space	Vacuolar Proportion %
Cl^-	153.3	180.1	47.1	93.9
NO_3^{2-}	157.5	195.2	6.8	99.1

Transport of Anions

Transport of Organic Anions   From the results shown above, we conclude that homeostatic mechanisms maintain an ionic composition of the cytoplasm which is optimal for the functioning of metabolism. Therefore, transport of anions plays a central role in the metabolism of plants. There is the question of how ions are transported. L-malate uptake follows saturation kinetics (Fig. 3).

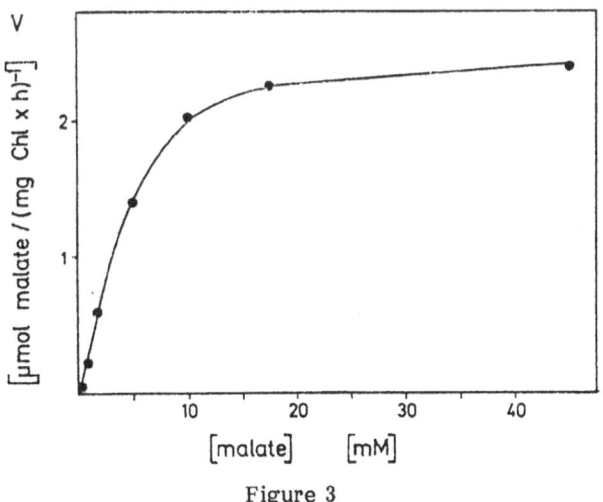

Figure 3

Concentration-dependent uptake of ^{14}C-malate
by isolated vacuoles

The transport mechanism is not specific for L-malate, since other di- and tricarboxylic acids act as competitive inhibitors (Table 5). Citrate acts slightly overcompetitively. Even malonylaminocyclopropane carboxylic acid (malonyl ACC) which is produced during stress in many plant species acts as competitive inhibitor.

Table 5

Competitive Inhibition of ^{14}C-Malate Transport into Isolated Vacuoles by Carboxylates

The average value for the K_m of L-malate was 2.5 mM.

Treatment	K_i (mM)
D-malate	2.0 – 3.2
Citrate	1.7 – 2.7
Oxaloacetate	3.6 – 4.8
Tartronate	4 – 7
Malonyl–ACC	5 – 7
Malonate	9 – 12
Succinate	9 –12

Uptake of malate (Fig. 4), citrate and malonylaminocyclopropane carboxylic acid (data not shown) into isolated vacuoles is ATP-dependent. ATP is required as the Mg^{2+}-complex. ATP stimulation of ^{14}C-malate uptake is not due to an enhanced malate exchange. Addition of MgATP only affects V_{max} but not the apparent K_m

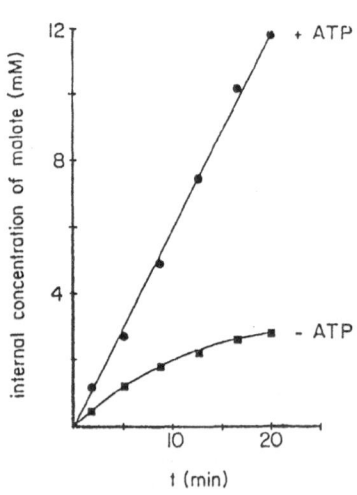

Figure 4

Time-course of ^{14}C-malate uptake in presence or absence of MgATP

Vacuoles were incubated in a medium containing 1.1 mM malate.

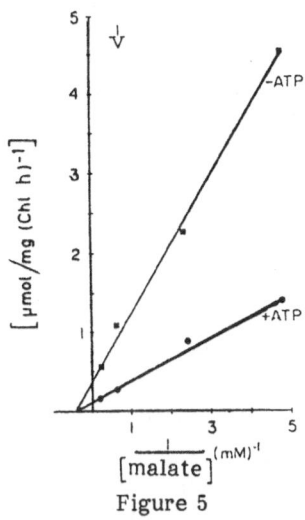

Figure 5

Concentration–dependent uptake of ^{14}C-malate
in presence or absence of MgATP

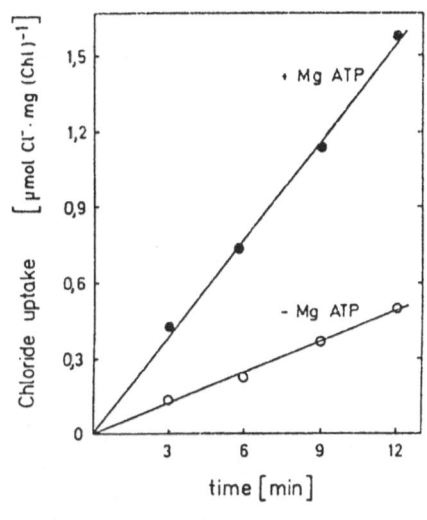

Figure 6

Time–course of ^{36}Cl$^-$–uptake
by isolated barley vacuoles

Vacuoles were incubated in a medium containing 10 mol m^{-3} Cl$^-$ and 10 mol m^{-3} MgATP (●) or 10 mol m^{-3} MgSO$_4$ (○) respectively.

for malate (about 2.5 mM) of the malate transporter (Fig. 5). Inhibitor studies suggest energization of malate uptake by the proton-translocating ATPase, which has been described for tonoplast vesicles (Sze, 1985) (Table 6). Like this ATPase, malate transport is inhibited by DCCD (N,N'-dicyclohexylcarbodiimide) and DES (diethyl-stilbestrol). Vanadate, which inhibits the plasmalemma ATPase has no influence

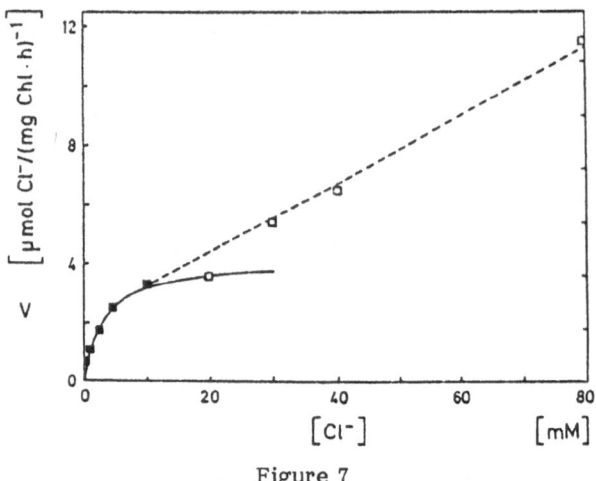

Figure 7

Concentration–dependence of $^{36}Cl^-$–uptake

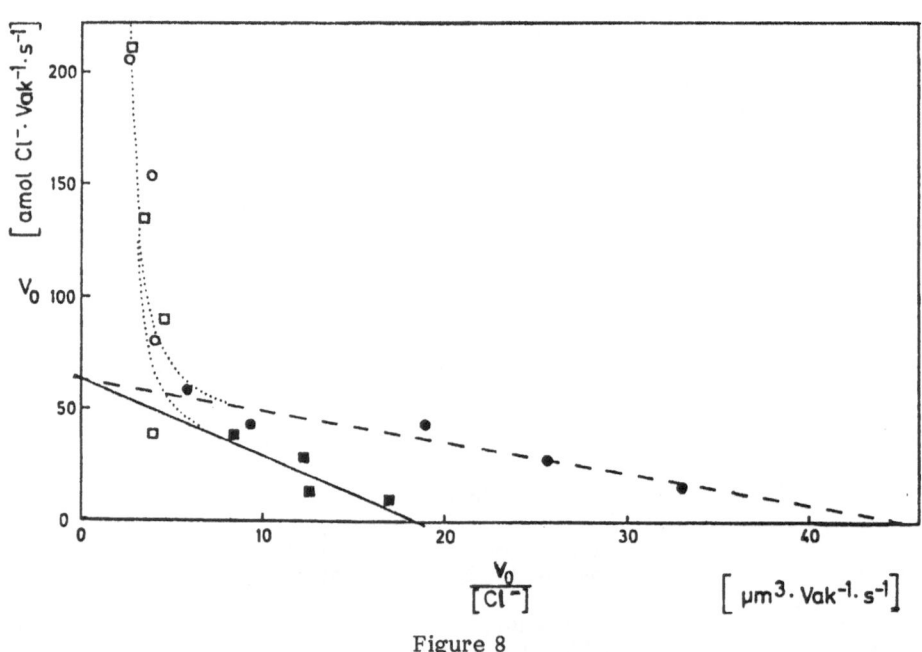

Figure 8

Concentration–dependence of $^{36}Cl^-$–uptake
by isolated barley vacuoles
in the absence (●) or presence (■) of 5 mol m^{-3} NO$_3^-$

Data are shown in a Woolf–Augustinsson–Hofstee–Plot.

Table 6

Effects of Inhibitors on the Rate of MgATP-dependent [14]-Malate Uptake
into Isolated Vacuoles

Experiment	Treatment	Malate Transport (% of Control)
I	Control	100
	+ DCCD (100 μM)	37
	+ DES (250 μM)	38
	+ Vanadate (200 μM)	98
	+ Azide (1 mM)	98
	+ Oligomycin (40 μg/ml)	93
	without MgATP	34
II	+ NO_3^- (40 mM)	20
	+ Cl^- (40 mM)	50

The uptake rate for the control was 1.96 μmol malate 10^{-7} (vacuoles) h^{-1} in experiment I, respectively 2.5 μmol malate 10^{-7} (vacuoles) h^{-1} in experiment II.

The higher inhibition of the malate transport system in presence of nitrate than in presence of chloride may not only due to a partial inhibition of the tonoplast ATPase. As we have seen in Fig. 2 nitrate crosses the tonoplast faster than chloride. Since the non-energized uptake of malate is only slightly inhibited by this inorganic anions (not shown), we suggest that the mechanism of energization is the same for all these anions.

Table 7

Inhibition of $^{36}Cl^-$-Uptake
by 4,4'-diisothiocyano-2,2'-stilbene disulfonic acid (DIDS)

	Chloride Uptake mol Cl^- 10^{-7} (Vacuoles) h^{-1}	
	2.2 mol m^{-3} Cl^-	20 mol m^{-3} Cl^-
Control	0.78	2.36
50 mmol m^{-3} DIDS	–	1.75
500 mmol m^{-3} DIDS	0.15	1.77

Vacuoles were preincubated for 15 min at 273 K in media containing bovine serumalbumin (1.5 Kg m^{-3}).

on malate transfer. Azide and oligomycin, both inhibitors of the mitochondrial ATPase have no effect on MgATP-stimulated malate transport. The tonoplast ATPase is known to be inhibited by nitrate. We have observed that 40 mM nitrate reduces the uptake of malate to 20% of the control rate. However, chloride which is known to activate the tonoplast ATPase inhibits the MgATP-activated malate transport, too. Inhibition may be due to competition for energy.

Transport of Inorganic Anions. Fig. 6 shows the ATP dependence of chloride uptake. In contrast to the saturation kinetics of malate uptake, chloride uptake shows two clear discernible kinetic phases (Fig. 7). At low concentrations, a saturable component is involved with an apparent K_m of about 2 mM, whereas at higher concentrations chloride uptake is proportional to the external chloride concentrations. The involvement of two kinetic components is also illustrated in Fig. 8. Addition of nitrate changes the slope of the saturable component of the chloride uptake system, indicating that nitrate acts as competitive inhibitor of chloride uptake. The calculated K_i is 3 mM. As demonstrated for other chloride transport systems, chloride transfer across the tonoplast can be inhibited by DIDS 4,4'-diisothiocyano-2,2'-stilbenedisulfonic acid) (Table 7). Inhibition is very strong at low chloride concentrations, whereas at higher external concentrations inhibition is less pronounced.

CONCLUSIONS

The tonoplast of barley mesophyll vacuoles contains transport systems which facilitate the transfer of anions between the cytoplasm and the vacuolar space, contributing to the maintenance of cytoplasmic homeostasis. Uptake of several anions is increased in the presence of MgATP. We suggest that the energized uptake of different anions is supported by only one proton-translocating ATPase.

REFERENCES

Christensen, L. E., Below, F. E., and Hageman, R. H., 1981, The effect of ear removal on senescence and metabolism of maize, Plant Physiol., 68:1180.

Gerhardt, R., and Heldt, H. W., 1984, Measurement of subcellular metabolite levels in leaves by fractionation of freeze-stopped material in non-aqueous media, Plant Physiol., 75:542.

Jeschke, W. D., 1979, in : "Recent Advances in the Biochemistry of Cereals", D. L. Laidman and R. G. Wyn-Jones, eds., Academic Press, London.

Kaiser, G., and Heber, U., 1984, Sucrose transport into vacuoles isolated from barley mesophyll protoplasts, Planta, 161:562.

Kaiser, G., Martinoia, E., and Wiemken, A., 1982, Rapid appearence of photosynthetic products in the vacuoles isolated from barley mesophyll protoplasts by a new fast method, Z. Pflanzenphysiol., 107:103.

Kaiser, W. M., Weber, H., and Sauer, M., 1983, Photosynthetic capacity, osmotic response and solute content of leaves and chloroplasts from Spinacea oleracea under salt stress, Z. Pflanzenphysiol., 113:15.

Martinoia, E., Flügge, U. I., Kaiser, G., Heber, U., and Heldt, H. W., 1985, Energy-dependent uptake of malate into vacuoles isolated from barley mesophyll protoplasts, Biochim. Biophys. Acta, 806:311.

Martinoia, E., Heck, U., and Wiemken, A., 1981, Vacuoles as storage compartments for nitrate in barley leaves, Nature, 289:292.

Martinoia, E., Schramm, M. J., Kaiser, G., Kaiser, W. M., and Heber, U., 1986, Transport of anions in isolated barley vacuoles. 1. Permeability to anions and evidence for a Cl⁻-uptake system, Plant Physiol., 80:895.

McLilley, C. R., Chon, C. J., Mosbach, A., and Heldt, H. W., 1977, The distribution of metabolites between spinach chloroplasts and medium during photosynthesis in vitro, Biochim. Biophys. Acta, 460:259.

Osmond, C. B., and Holtum, J. A. M., 1981, in : "The Biochemistry of Plants", M. D. Hatch and N. K. Boardman, eds., Vol. 8, Academic Press, New-York.

Steingröver, E., Ratering, P., and Siesling, J., 1986, Daily changes in uptake, reduction and storage of nitrate in spinach grown at low light intensity, Physiol. Plant., 66:550.

Sze, H., 1985, H⁺-translocating ATPases : Advances using membrane vesicles, Annu. Rev. Plant Physiol., 36:175.

VACUOLAR AND CYTOSOLIC METABOLITE POOLS

BY COMPARATIVE FRACTIONATION

OF VACUOLATE AND EVACUOLATE PROTOPLASTS

Manfred Steingraber and Rüdiger Hampp

Universität Tubingen, Lehrstuhl Biologie I
Auf der Morgenstelle, 1
D-7400-Tübingen 1, Federal Republic of Germany

INTRODUCTION

Knowledge about the distribution of metabolites between subcellular compartments is important for the understanding of regulatory events. Orthophosphate (P_i), e.g., is involved in most biochemical processes of energy transduction within the cell, thereby affecting important metabolic routes such as respiration, glycolysis or photosynthesis. Especially partioning of photosynthetically fixed CO_2 between starch and sucrose appears to be regulated by the stromal and cytosolic pool sizes of P_i.

While levels of metabolites contained in chloroplasts and mitochondria of mesophyll protoplasts can be determined by recent techniques (Hampp, 1980; Goller et al., 1982; McLilley et al., 1982), cytosolic and vacuolar pools can only be resolved by ^{31}P-NMR spectroscopy (Foyer et al., 1982; Rebeille et al., 1983; Foyer and Spencer, 1986).

In this paper we report on a completely different and more widely applicable approach. Evacuolate protoplasts were isolated from ordinary oat mesophyll protoplasts and were subjected to a rapid fractionation technique. By comparison with vacuolate protoplasts, pool sizes of cytosolic and vacuolar metabolites can be determined.

MATERIALS AND METHODS

Isolation and Evacuolation of Protoplasts

Seedlings of Avena sativa L. (cv. Arnold) were grown in hydroponic culture for 7 d. Illumination (about 9 W m^{-2}) was started after 3 d of germination in the dark at 26°C and about 80 % relative humidity. Enzymatic isolation from 0.5 to 1 mm wide leaf segments and purification of protoplasts was according to Hampp and Ziegler (1980).

Evacuolation of protoplasts was carried out in principal as reported by Griesbach and Sink (1983). Purified protoplasts (0.5 ml, about 3 10^6) were carefully layered on top of 3.2 ml of a preformed percoll gradient (0.5 M mannitol, 100 mM

CaCl$_2$, 5 mM 3-(N-morpholino)propane-sulfonic acid (Mops) pH 7.0, dissolved in Percoll) and centrifuged (34,000 rpm, 25 min, 23°C in a Beckman SW 60 rotor). They were centrifuged for additional 15 min. This resulted in three bands which consisted of plasma-surrounded vacuoles (top), aggregations of broken protoplasts, and evacuolated protoplasts (lowermost band; Fig. 1). The evacuolated protoplasts were carefully removed by suction, washed twice with 0.5 M mannitol, and stored on ice for at least 60 min. This period was necessary to obtain maximal rates of light-dependent O$_2$ evolution.

The integrity of the protoplasts (80 % on average) was assayed enzymatically (glycolate oxidase; Nishimura et al., 1985).

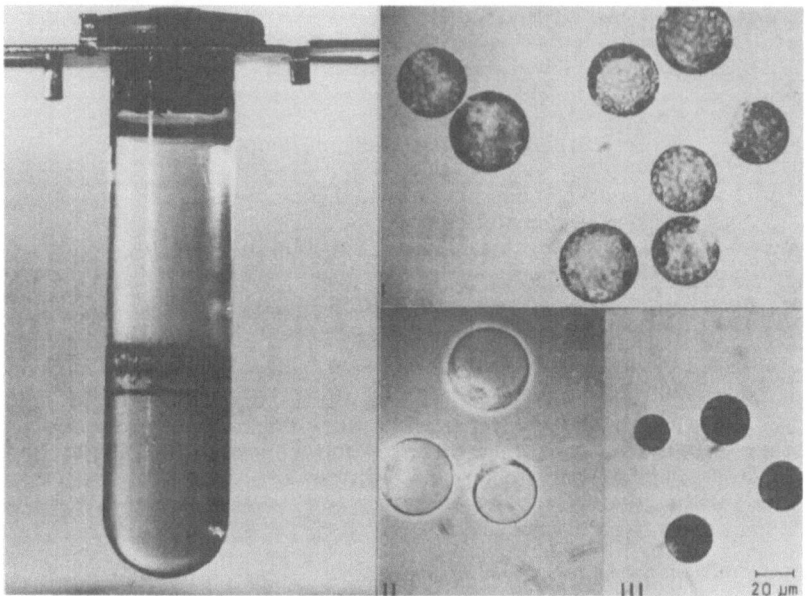

Figure 1

Fractions after centrifugation of oat mesophyll protoplasts
on a self-generating percoll gradient

(I) vacuolate protoplasts; (II) "pseudo-vacuoles" : vacuole with cytoplasmic remainder and plasma membranes; (III) evacuolate protoplasts.

Incubation of Protoplasts

Vacuolate protoplasts (about 10^6 ml^{-1}) were suspended in 0.5 M sorbitol, 5 mM KHCO$_3$, 7.5 mM CaCl$_2$, 25 mM Tricine (pH 7.6) and incubated in a Hansatech oxygen electrode at 20°C in a volume of about 1.5 ml. With evacuolate protoplasts, 0.5 M raffinose was used instead of sorbitol. This prevented settling of the protoplast population (isopycnic density in Percoll : about 1.15) and subsequent damage by the stirring bar.

Fractionation of Protoplasts

Fractionation by oil filtration (Goller et al., 1982) of evacuolate protoplasts did not result in a satisfactory separation of mitochondria from cytosolic markers. We therefore used the membrane filtration technique, developed by McLilley et al. (1982). Protoplasts were suspended, shortly before the experiment was started, in a medium, containing 0.3 M raffinose, 0.05 M Hepes (pH 7.6), 0.05 M KCl, 1 mM $MgCl_2$, 1 mM $NaHCO_3$, 0.2% soluble PVP and 0.06 %BSA. Because of the missing vacuole the reduction in osmotic pressure (0.3 instead of 0.5 M osmoticum) did not break the protoplasts, but increased the quality of the fractionation. The protoplast suspension (1.5 ml, 0.6 10^6 protoplasts) was transferred into 2.5 ml syringes, attached to Sartorius membrane filter holders (Fig. 2).

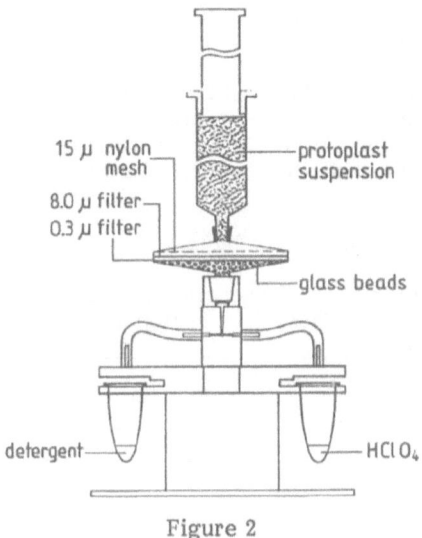

Figure 2

Experimental set up for the rapid fractionation
of evacuolate protoplasts

The respective filtrates are evenly distributed between receptacles containing either detergent (enzymes) or acid/base (metabolites). For additional details see McLilley et al. (1982).

The complete set of filtration units (3) was thermostatted (20°C). Fractionation was initiated by forcing the suspension through a 12 μm nylon mesh (Züricher Beuteltuchfabrik, Switzerland) and different sets of membrane filters. With the nylon mesh only, a protoplast homogenate resulted. From this homogenate chloroplasts could be withdrawn by an additional 8 μm-membrane filter (cellulose nitrate, Sartorius; homogenate minus chloroplasts). By the insertion of a second membrane filter (0.45 μm; cellulose acetate, Sartorius), a protoplast homogenate resulted which had a considerably decreased level of mitochondrial markers (Table 1). By calculation, metabolite levels associated with chloroplasts, mitochondria and the cytosol could be obtained from quenched (Eppendorf vial containing 100 μl of 0.5 N NaOH or 0.5 N HCl, in addition to 500 μl filtrate) or solubilized filtrates (Eppendorf vial containing 0.06 mM Zwittergent Z-12, Calbiochem). For further details, see McLilley et al. (1982). Background determinations were made with aliquots of the respective protoplast suspensions which were pelleted (300 g; 2 min), and the supernatant treated (addition of acid or base) as described above.

Determination of Metabolites

Adenine nucleotides were determined by bioluminescence (Hampp et al., 1982; Hampp, 1985). Oxidized and reduced pyridine nucleotides by enzymatic cycling (Lowry and Passonneau, 1972; Kato et al., 1973). For the determination of all other metabolites reactions were used, which were selectively coupled to pyridine nucleotides (Lowry and Passonneau, 1972; "specific step"). These were followed by enzymatic cycling of the pyridine nucleotides formed in each specific reaction. Detailed information on all these reactions is given elsewhere (Hampp et al., 1984).

RESULTS

Properties of evacuolate mesophyll protoplasts

In Table 1 some characteristics of vacuolate and evacuolate oat mesophyll protoplasts are compared. With regards to the content of marker enzymes, there are specific differences. While acid phosphatase, a reliable marker for the vacuolar compartment in oat, is completely absent from evacuolate protoplasts, chloroplasts (NAD:triose-phosphate dehydrogenase) are retained. Of the cytosolic compartment (PEP carboxylase) more than 50% are lost during evacuolation (see also "pseudo vacuoles" in Fig. 1). Mitochondria, however, are largely retained.

Table 1

Characteristics of Vacuolate and Evacuolate Oat
Mesophyll Protoplasts

	Vacuolate Protoplasts	Evacuolate Protoplasts
Light-dependent O_2 evolution (μmol/mg chlorophyll x h)	65 ± 10	40 ± 15
Time needed for photo-synthetic induction (min)	2	2
Acid phosphatase (vacuole)	100	0
Phosphoenolpyruvate carboxylate (cytosol)	100	45 ± 10
NADP-triosephosphate-dehydrogenase (chloroplasts)	100	95 ± 5
Fumarase (mitochondria)	100	80 ± 15

Enzyme activities in % of control. Values $\pm$ S.D.; n =4.

Inspite the severe interference with cellular functions, due to evacuolation, a considerable part of the capacity for light-dependent oxygen evolution is preserved. This can be taken as an indicator for physiological integrity. Additional support for the assumption of physiological integrity comes from regeneration experiments. Evacuolate mesophyll protoplasts are able to develop a vacuole within 2 d and can be regenerated to intact plants (Griesbach and Sink, 1983; Burgess and Lawrence, 1985; Naton and Hampp, unpublished).

Cytosolic and Vacuolar Metabolite Pools

Cytosolic and vacuolar concentrations of a range of metabolites are given in Table 2, in addition to those within chloroplasts and mitochondria. Vacuolar contents were calculated on the basis of protoplast number (10^6):

$$Vac = P(+) - (a*Cyt + Chl + b*Mit.)_{P(-)}$$

where the symbols represent nmol metabolite / 10^6 protoplasts, contained in vacuolate protoplasts (P(+)), and in the cytosol (Cyt), chloroplasts (Chl) and mitochondria (Mit) of evacuolate protoplasts (P(-)). The factors (a, b) correct for the loss of the respective compartment during evacuolation (see PEP carboxylase and fumarase in Table 1). Concentrations were calculated from volume determinations with vacuolate protoplasts (μl/10^6 protoplasts : chloroplasts, 2.5; mitochondria, 1.0; cytosol, 2.0; vacuole, 18; Hampp et al., 1982) and the respective volume of evacuolate protoplasts after incubation in 0.3 M raffinose (total volume : 9.1 pl/protoplast). Under the assumption that only swelling of the cytosol occurs (no changes in organellar volumes), its volume was taken as 5.8 pl/protoplast (corrected for a 20% loss of mitochondrial space).

Table 2

Compartmentation of Metabolites
between Chloroplasts, Mitochondria, Cytosol and Vacuole
of Oat Mesophyll Protoplasts

Metabolite	Cytosol	Vacuole	Chloroplasts	Mitochondria
Inorganic Phosphate	7.1	2.6	7.1	8.3
Malate	2.4	3.7	0.8	2.4
ATP + ADP	0.3	0	0.6	0
NADP + NADPH	0.053	0	0.079	0
NAD + NADH	0.138	0	0.114	0.230
Triose-phosphates	0.15	–	0.65	0.69
3-P-glyceric acid	0.37	0	0.85	1.25

The association of the different metabolites with chloroplasts and mitochondria largely agrees with recent determinations obtained with ordinary oat mesophyll protoplasts (Hampp et al., 1984 and 1985). The additional resolution of cytosolic and vacuolar pool sizes shows that ATP, ADP, reduced and oxidized di- and triphosphopyridine nucleotides, as well as 3-phosphoglyceric acid are virtually absent from the vacuolar space of dark treated protoplasts. The vacuolar concentration of malate was higher, that of inorganic phosphate lower compared to the cytosol.

DISCUSSION

This paper introduces a new approach for the determination of cytosolic pool sizes of metabolites. By comparison of vacuolate and evacuolate protoplasts the vacuolar space can be assessed as well. A first application of this technique for a compartmentational analysis of a range of metabolites indicates distinct differences in cytosolic and vacuolar pools. In the case of P_i it is possible to compare our data with recent ^{31}P-NMR measurements. Rebeille et al. (1983), e.g., reported cytosolic and vacuolar concentrations of 6 and 1.5 mM respectively (Acer pseudoplatanus cells) which is close to our values in oat mesophyll protoplasts (7.1 and 2.6; Table 2). It should be stated, however, that P_i levels can show considerable variations due to plant nutrition (Foyer and Spencer, 1986).

For the other metabolites comparable data are only available from non-aqueous fractionations. Dietz and Foyer (1986) observed levels of 3-phosphoglyceric acid in the chloroplast of barley leaves which were more than twice as high as those associated with the cytosol (compare our Table 2). Similarly the concentration of malate calculated for the cytosol of oat mesophyll protoplasts (2.4 mM) is in the range reported by Gerhardt and Heldt (1984) for nonaqueously fractionated spinach leaves (about 1 mM). Due to the acidic pH and its hydrolytic properties (Matile, 1984) the absence from the vacuolar compartment of pyridine and adenine nucleotides is as expected.

It can be objected that evacuolate protoplasts as used in this communication constitute an even more artificial system than an ordinary (vacuolate) protoplast already is. However, inspite of the loss of one compartment (vacuole) and the considerable reduction in size of another (cytosol) evacuolate protoplasts are physiologically functional. This is shown by their ability to perform unimpaired photosynthesis (starch and sucrose formation, not shown), regenerate a vacuole and even complete plants (Griesbach and Sink, 1983; Burgess and Lawrence, 1985). With respect to photosynthesis the cytosolic P_i pool by itself is obviously sufficient to sustain CO_2 fixation. This should be due to the recycling of P_i during formation of starch or sucrose and is supported by the observation that the transport of P_i across the tonoplast membrane is rather slow (Woodrow et al., 1984).

Taken together the approach suggested in this paper offers some new possibilities for assessing vacuolar and cytosolic functions :
- generally, the association of a given metabolite or enzyme with the respective compartment can easily determined;
- cytosolic concentrations of metabolites can be measured which, to a considerable extent, are also localized in vacuoles (due to problems with cross contamination (e.g. P_i) this is not possible with non-aqueous techniques);
- compartmented processes (e.g., photosynthesis, photorespiration) which are not dependent on the presence of the vacuole can be studied without the interference of lytic constituents during cell homogenation and fractionation;
- tonoplast formation can be studied in detail.

ACKNOWLEDGMENT

This study was supported by a grant from the Deutsche Forschungsgemeinschaft (Ha 970/6-8).

REFERENCES

Burgess, J., and Lawrence, W., 1985, Studies of the recovery of tobacco mesophyll protoplasts from an evacuolation treatment, Protoplasma, 126:140.

Dietz, K.-J., and Foyer, C., 1986, The relationship between phosphate status and photosynthesis in leaves. Reversibility of the effects of phosphate deficiency on photosynthesis, Planta, 167:376.

Foyer, C., and Spencer, C., 1986, The relationship between phosphate status and photosynthesis in leaves. Effects on intracellular orthophosphate distribution, photosynthesis and assimilate partitioning, Planta, 167:369.

Foyer, C., Walker, D. A., Spencer, C., and Mann, B., 1982, Observations on the phosphate status and intacellular pH of intact cells, protoplasts and chloroplasts from photosynthetic tissue using phosphorus-31 nuclear magnetic resonance, Biochem. J., 202:429.

Gerhardt, R., and Heldt, H. W., 1984, Measurement of subcellular metabolite levels in leaves by fractionation of freeze-stopped material in non-aqueous media, Plant Physiol., 75:542.

Goller, M., Hampp, R., and Ziegler, H., 1982, Regulation of the cytosolic adenylate ratio as determined by rapid fractionation of mesophyll protoplasts of oat. Effect of electron transfer inhibitors and uncouplers, Planta, 156:255.

Griesbach, R. J., and Sink, K. C., 1983, Evacuolation of mesophyll protoplasts, Plant Science Letters, 30:297.

Hampp, R., 1980, Rapid separation of the plastid, mitochondrial and cytoplasmic fractions from intact leaf protoplasts of Avena. Determination of an in vivo ATP pool sizes during greening, Planta, 150:291.

Hampp, R., 1985, ADP and AMP, Luminometric method, in : "Methods in Enzymatic Analysis", Vol. 7, J. Bergmeyer and M. Grassl, eds., Verlag Chemie, Weinheim.

Hampp, R., Goller, M., and Füllgraf, H., 1984, Determination of compartmented metabolite pools by a combination of rapid fractionation of oat mesophyll protoplasts and enzymic cycling, Plant Physiol., 75:1017.

Hampp, R., Goller, M., Füllgraf, H., and Eberle, I., 1985, Pyridine and adenine nucleotide status, and pool sizes of a range of metabolites in chloroplasts, mitochondria and the cytosol/vacuole of Avena mesophyll protoplasts during dark/light transition : Effect of pyridoxal phosphate, Plant Cell Physiol., 26:99.

Hampp, R., Goller, M., and Ziegler, H., 1982, Adenylate levels, energy charge and phosphorylation potential during dark/light and light/dark transition in chloroplasts, mitochondria and cytosol of mesophyll protoplasts from Avena sativa L., Planta, 62:448.

Hampp, R., and Ziegler, H., 1980, On the use of Avena protoplasts to study chloroplast development, Planta, 147:485.

Kato, T., Berger, S. J., Carter, J. A., and Lowry, O. H., 1973, An enzymatic cycling method for nicotine adenine dinucleotide with malate and alcohol dehydrogenase, Anal. Biochem., 53:86.

Lowry, O. H., and Passonneau, J. V., 1972, "A flexible system of enzymatic analysis", Academic Press, New-York.

Matile, P., 1978, Biochemistry and function of vacuoles, Annu. Rev. Plant Physiol., 29:193.

McLilley, C. R., Stitt, M., Mader, G., and Heldt, H. W., 1982, Rapid fractionation of wheat leaf protoplasts using membrane filtration. The determination of metabolite levels in chloroplasts, cytosol and mitochondria, Plant Physiol., 70:965.

Nishimura, M., Douce, R., and Akazawa, T., 1985, A simple method for estimating intactness of spinach leaf protoplasts by glycolate oxidase assay, Plant Physiol., 78:343.

Rebeille, F., Bligny, R., Martin, J.-B., and Douce, R., 1983, Relationship between the cytoplasm and the vacuole phosphate pool in Acer pseudoplatanus cells, Arch. Biochem. Biophys., 225:143.

Woodrow, I. E., Ellis, J. R., Jellings, A., and Foyer, C. H., 1984, Compartmentation and fluxes of inorganic phosphate in photosynthetic cells, Planta, 161:525.

SUBCELLULAR LOCALIZATION

OF PRIMARY AND SECOND CARDIAC GLYCOSIDES,

AND THE RELATED 16'-O-GLUCOSYL-TRANSFERASE

IN CELL SUSPENSION CULTURES OF DIGITALIS LANATA

Wolfgang Kreis, Ursula May and Ernst Reinhard

Pharmazeutisches Institut, Universität Tübingen
D-7400-Tübingen, F.R.G.

INTRODUCTION

Plant cell cultures, like microorganisms, are able to effect structural modifications on exogenously supplied steroids (Reinhard, 1974; Stochs, 1980). In this context the biotransformation of cardiac glycosides by suspension-cultured Digitalis lanata cells (Reinhard et al., 1975) is of special interest because of the world-wide use of these compounds in the treatment of heart diseases.

Upon the addition of various cardenolides to Digitalis lanata suspension cultures different reactions occur, including 16'-O-glucosylation, 15'-O-acetylation, and 12 β-hydroxylation (Fig. 1). Among these reactions the hydroxylation step is of some pharmaceutical importance, since 12 β-hydroxylation of digitoxin (produced during the technical isolation of digoxin from Digitalis lanata plants) would lead to digoxin which, due to its pharmacological properties is the cardenolide most in demand. However, attempts to use digitoxin a precursor for the production of digoxin derivatives with cultured foxglove cells have failed so far, so β-methyl-digitoxin – prepared from digitoxin by chemical methylation – is used as the substrate of choice, leading to β-methyldigoxin as the main product (for a recent review see Alfermann et al., 1983).

Within our general effort to optimize the cardenolide biotransformation process with cultured Digitalis cells we decided to study, on the cellular level, the accumulation of cardiac glycosides. Our investigations involved both uptake and wash-out studies with intact cells, the analysis of isolated vacuoles, and the cellular localization of a glucosyltransferase catalyzing the formation of primary cardiac glycosides from secondary ones.

MATERIALS AND METHODS

Suspension cultures were established from Digitalis lanata callus (Heins, 1978) and routinely subcultured every 10.5 d in a modified Murashige and Skoog (1962) medium as described (Kreis and Reinhard, 1985).

To isolate the protoplasts cultured cells were incubated for 1.5 h in a modified

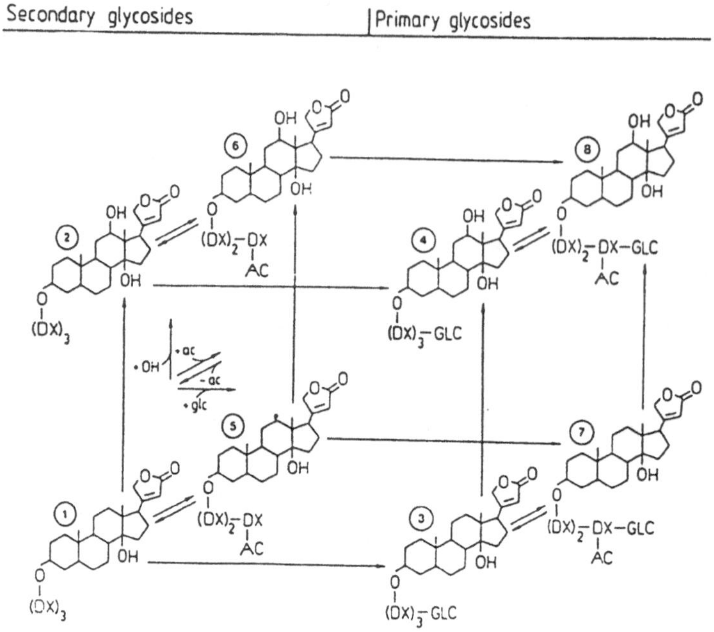

Figure 1

Proposed interrelationships of cardenolide biotransformation
in <u>Digitalis</u> <u>lanata</u> cell cultures

DX, digitoxosyl; GLC, glucosyl; AC, acetyl; 1, digitoxin; 2, digoxin; 3, purpurea-glycoside A; 4, deacetyl-lanatoside C; 5, α-acetyldigitoxin; 6, α-acetyldigoxin; 7, lanatoside A; 8, lanatoside C.

Murashige and Skoog medium containing 0.5 M mannitol, 1% purified Cellulase Onozuka RS (Yakult Honsha Co. Ltd., Japan), and 0.3% Rhozyme HP 150 (Rhom and Haas, U.S.A.). After incubation the protoplast suspension was filtered through two layers of nylon cloth (mesh size : 135 μm and 60 μm, respectively). The protoplasts were pelleted by centrifugation, the supernatant discarded and the pellets washed twice. Finally the protoplasts were purified by floatation through 22% w/v sucrose.

Vacuoles were released from the protoplasts by osmotic shock treatment at pH 8 in the presence of EDTA. The protoplast lysate containing intact vacuoles was filtered through nylon cloth (100 μm) and the vacuoles then purified by floatation through a three-step Ficoll 400 (Sigma, F.R.G.) gradient (Fig. 2). Details of protoplast and vacuole isolation have been described in a previous publication (Kreis and Reinhard, 1985).

The purity of isolated vacuoles was tested by marker enzyme analysis. Glucose6-phosphate dehydrogenase and malate dehydrogenase (Bergmeyer, 1974) served as the markers for extravacuolar contamination. α-mannosidase was used as a reference for comparing protoplasts and vacuoles on a quantitative basis (Boller, 1982). Protein was determined with Bradford's (1976) method.

UDP-glucose:digitoxin-16'-O-glucosyl-transferase was assayed according to Kreis et al. (1986). Cells, protoplasts, vacuoles, cell-free incubation mixtures, or culture medium were analysed by HPLC for their cardenolide content (Kreis et al., 1986).

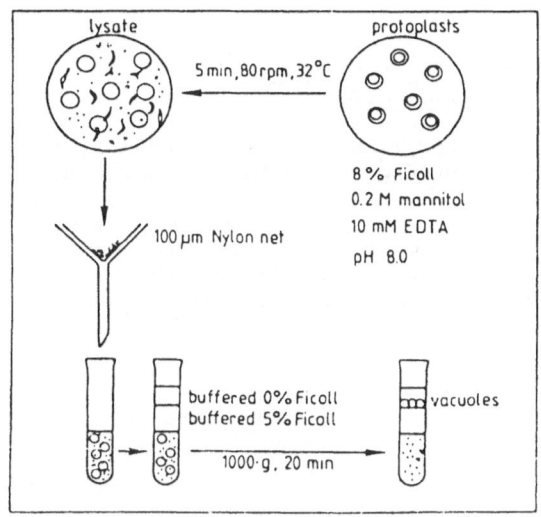

Figure 2

Outline of the vacuole isolation procedure
(Kreis and Reinhard, 1985)

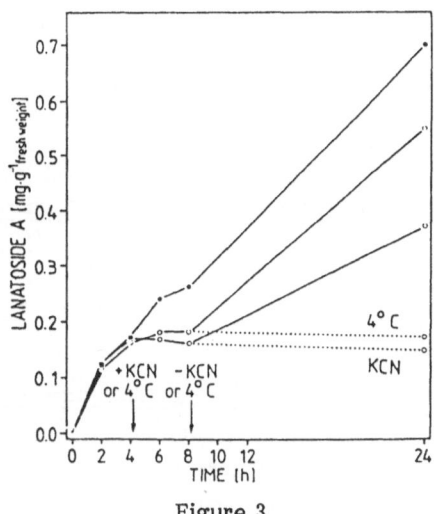

Figure 3

Uptake of primary cardiac glycosides
by cultured <u>Digitalis</u> <u>lanata</u> cells

The inhibitory effects of 2 mM potassium cyanide (empty circles, KCN) or low temperatures (empty circles, 4°C) were examined. Cells were incubated with lanatoside A (100 mg l^{-1}) under standard conditions (24°C, rpm, 2000 lx) for 4 h and then cultivated under the respective inhibitory conditions for either 24 h (dotted lines) or 4 h (full lines). In the latter case the cells were washed free from the inhibitor and then transferred into fresh culture medium containing 100 mg l^{-1} lanatoside A (or transferred from 4°C conditions back to 24°C conditions). Filled circles show the time course of lanatoside A uptake of a control (kept for 24 h under standard conditions).

RESULTS AND DISCUSSION

Specific Uptake of Primary Cardiac Glycosides by Suspension-cultured Digitalis Cells

We started our investigations into the cellular organization of cardenolide biotransformation by establishing the uptake kinetics of various cardiac glycosides. We found that the uptake of primary glycosides, such as lanatoside A or purpurea-glycoside A, was strongly inhibited by cyanide, dinitrophenol, or low temperatures. If the cells were washed free from the respective inhibitor and transferred into fresh medium (or transferred from 4°C conditions to 24°C conditions) the accumulation of primary glycosides was reactivated, indicating that the inhibition could be reversed (Fig. 3).

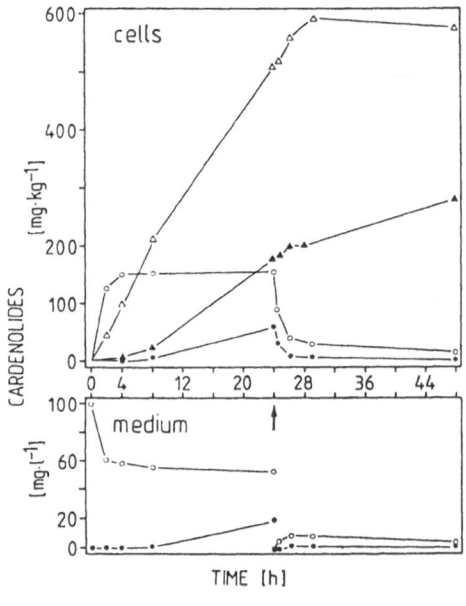

Figure 4

Retention of primary cardiac glycosides
by cultured <u>Digitalis</u> <u>lanata</u> cells
Time-course of product formation in a biotransformation experiment
with digitoxin (initial concentration 100 mg l^{-1}) as substrate

Circles indicate secondary glycosides : digitoxin (empty) and digoxin (filled); triangles indicate primary glycosides : purpurea-glycoside A (empty) and deacetyl-lanatoside C (filled). After 24 h of incubation with digitoxin cultured cells were transferred into fresh medium without cardenolides (indicated by an arrow). Primary glycosides were retained by the cells, while secondary glycosides passed into the culture medium.

The uptake of secondary glycosides, on the other hand, was not affected by cyanide or dinitrophenol. Obviously, secondary cardiac glycosides entered the cells by diffusion , whereas primary glycosides were taken up actively. The ability to accumulate the latter type of cardenolides against a concentration gradient seemed to be restricted to the genus Digitalis. Other cell cultures examined were not able to accumulate these glycosides (data not shown).

Upon incubation of Digitalis cells with digitoxin a series of biotransformation products appeared in the cells. Of these products only digoxin, a secondary cardiac glycoside, could be detected on the surrounding medium. When cells containing cardenolides were transferred into fresh culture medium, the cellular secondary glycosides passed into the culture medium down a different gradient. The primary glycosides, on the other hand, remained stored (Fig. 4). Obviously, as in the experiments with exogenous cardenolides, the secondary glycosides formed by biotransformation have free access across the plasmalemma, while the accumulation of primary glycosides involves a specific uptake mechanism.

The Vacuole as the Storage Site for Primary Cardiac Glycosides

During the past decade, techniques for isolating vacuoles on a large scale have been developed and refined (see Wagner, 1983 for a review). As a consequence the important role of this compartment in the metabolism and storage of secondary plant products has been realized (see Matile, 1984 and references therein).

The primary glycosides produced via the biotransformation of digitoxin accumulate in suspension-cultured foxglove in fairly high concentrations (approx. 2% of dry weight) and were supposed to be located in the vacuolar sap. Using the α-mannosidase content of protoplasts and vacuoles as the basis for our calculations, we found that purpurea-glycoside A, either supplied exogenously or produced by biotransformation, was located almost exclusively in the vacuoles isolated from cultured cells (94% $\pm$ 17%, n = 9, in typical biotransformation experiments with an initial digitoxin concentration of 40 mg l^{-1} and vacuoles prepared after 2 to 5 days of incubation). Secondary glycosides could never be detected in isolated vacuoles.

Further studies showed that the stored cardenolides were retained by the tonoplast under in vitro conditions. This perfect retention inside the vacuole is in agreement with several reports on other systems used for investigating the mechanisms of the vacuolar storage of glycosides (e.g. Alibert et al., 1982; Werner and Matile, 1985).

In summary, our results indicate that primary but not secondary cardiac glycosides are stored in the vacuoles of suspension-cultured Digitalis cells.

Subcellular Localization
of a UDP-glucose:digitoxin 16'-O-glucosyltransferase

Although cardenolide biotransformation by Digitalis lanata cells has been investigated very extensively, nothing was known about the enzymes involved in these reactions. Recently, Petersen and Seitz (1985) reported on the isolation of a digitoxin 12 β-hydroxylase from cell cultures of Digitalis lanata. From their results the authors concluded that this enzyme is a cytochrome P-450-dependent mixed-function monooxygenase.

A common principle of product accumulation by plant cells features the glycosylation of a given compound followed by the transtonoplast import of the product into the vacuolar space (see Matile, 1984 for a recent review). In several cases both the glycosylation of exogenously supplied compounds and the vacuolar storage of the corresponding products have been demonstrated (e.g. Ohlrogge et al., 1980; Alibert et al., 1982; Schmitt and Sandermann, 1982; Werner and Matile,

1985). Obviously, a similar mechanism is responsible for the selective storage of primary glycosides in the vacuoles of suspension-cultured foxglove cells. Therefore, we felt that investigations into the properties and localization of the enzyme catalyzing the formation of primary glycosides from secondary ones would lead to a better understanding of the cellular mechanism involved in the biotransformation of cardenolides by Digitalis lanata cells.

We found that the enzymatic glucosylation of secondary glycosides to their corresponding primary glycoside is performed by a UDP-glucose:digitoxin 16'-O-glucosyltransferase (Kreis et al., 1986). Some of the properties of this enzyme are summarized in Table 1.

Table 1

Some Properties of the UDP-glucose:digitoxin 16'-O-glucosyltransferase
from Digitalis lanata cells (Kreis et al., 1986)

Parameter examined		Relative DGT Activity (%)
pH dependence	pH 7.4	100
	pH 7.1	50
	pH 7.7	50
Cofactor requirement	18 mM UDP-glucose (K_m = 5 mM)	100
	0 mM UDP-glucose	0
	18 m other sugar nucleotides	0
Activation	Control	100
	10 mM Dithiothreitol	264
	10 mM Mercaptoethanol	295
	10 mM Ascorbate	455
Substrate	Digitoxin (K_m = 110 μM)	100
	Digoxin (K_m = 410 μM)	71
	α-Acetyldigitoxin (K_m = 540 μM)	21
	α-Acetyldigoxin (K_m = 810 μM)	18

High glucosyl-transferase activity was found in cells, protoplasts, and crude protoplast lysates containing intact vacuoles, but never in isolated purified vacuoles. On the other hand, most of the transferase activity was found in the soluble protein fraction, whereas no activity could be detected in microsomal preparations. Although the possibility of dissociation from membranes during the isolation procedure cannot be ruled out, we inferred that the digitoxin glucosyl-transferase is located in the cytosol rather than associated with the tonoplast. In this context it has to be mentioned, however, that Löffelhardt and Kopp (1981), when investigating the glucosylation of convallatoxin to convalloside in Convallaria majalis found the glucosyl-transferase activity associated with a very light membrane fraction which was thought to represent vacuolar membranes. As a consequence the authors proposed that the cardenolide glucosylation in the cells of Convallaria leaves takes place at the tonoplast.

Subcellular Organization of Cardenolide Biotransformation
in Digitalis lanata Cell Cultures

    Although the elucidation of uptake, transformation, storage, and excretion
of cardenolides by suspension-cultured foxglove cells is far from being complete,
a model featuring the probable mechanisms involved in these processes may be
proposed (Fig. 5).

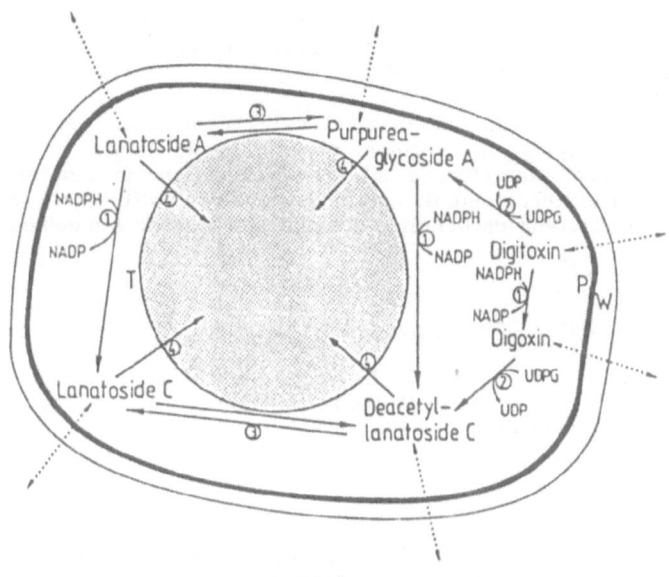

Figure 5

Model proposed for the cellular organization
of cardenolide biotransformation
in cultured Digitalis lanata cells

1, digitoxin 12 β-hydroxylase; 2, UDP-glucose:digitoxin 16'-O-glucosyl-transferase;
3, acetyl-transferase; 4, cardenolide carrier; W, cell wall; P, plasmalemma; T,
tonoplast.

    According to this model both primary and secondary glycosides enter the
Digitalis cell by simple diffusion. The primary glycosides are then accumulated
in the vacuolar sap by an energy requiring process. The enzymes catalyzing 12
β-hydroxylation, 15'-O-acetylation/deacetylation, or 16'-O-glucosylation are located
in the cytoplasm as either soluble or membrane-bound proteins. Digoxin produced
upon incubation with digitoxin (12 β -hydroxylation) may appear in the surrounding
medium, but cellular glucosylation and the subsequent vacuolar storage of the
primary glycoside formed (deacetyl-lanatoside C) forces digoxin to re-enter the
cell down a diffusion gradient.

    Now, the particular case of the 12 -hydroxylation of β-methyldigitoxin and
the appearance of the product (β-methyldigoxin) in the culture medium is easily
understood : β-methyldigitoxin is not a substrate for the glucosyltransferase, thus
no vacuolar storage form is available. The same restrictions hold true for the
hydroxylated product, namely β-methyldigoxin. As a consequence it diffuses into
the culture medium, from which it can be isolated.

CONCLUSIONS

The understanding of biosynthetic mechanisms is a key element in the development of productive cell culture processes. A knowledge of the properties and the localization of the enzymes catalyzing the various steps of secondary product synthesis or biotransformations may result in the establishment of cell culture processes with increased productivity. As emphasized by Deus-Neumann and Zenk (1984) the transport systems involved in the uptake and storage of natural products may be a limiting factor in the production of valuable metabolites by cell cultures. Future uptake studies with isolated vacuoles may help to elucidate the mechanisms involved in the storage of secondary metabolites, and this understanding may help us increase product accumulation, or to develop techniques for recovering stored metabolites without destroying the cells.

With this in mind, cultured foxglove cells seem to be a suitable system for studying the metabolism of cardiac glycosides in terms of their storage, remetabolization from the storage site, and excretion. In addition, the results obtained from cell cultures and the methods developed with them may help us to investigate and understand cardenolide storage and transport mechanisms of intact Digitalis plants.

ACKNOWLEDGEMENT

This work was supported by the Deutsche Forschungsgemeinschaft (Re 131/11-1). The authors would like to thank H. Wolter and Ch. Waiblinger for their assistance in various phases of this project.

REFERENCES

Alfermann, A. W., Bergmann, W., Figur, C., Helmbold, U., Schwantag, D., Schuller, I., and Reinhard, E., 1983, Biotransformation of β-methyldigitoxin to β-methyldigoxin by cell cultures of Digitalis lanata, in : "Plant Biotechnology", S. H. Mantell and H. Smith, eds., Cambridge University Press, Cambridge.

Alibert, G., Boudet, A. M., and Rataboul, P., 1982, Transport of o-coumaric acid glucoside in isolated vacuoles of sweet cover, in : "Plasmalemma and Tonoplast : Their Functions in the Plant Cell", D. Marmé, E. Marré, and R. Hertel, eds., Elsevier, Amsterdam.

Bergmeyer, H. U., 1974, "Methoden der Enzymatischen Analyse", 3. Aufl., Verlag Chemie, Weinheim.

Boller, T., 1982, Enzymatic equipment of plant vacuoles, Physiol. Vég., 20:247.

Bradford, M. M., 1976, A rapid and sensitive method for the quantitation of microgram quantities of protein utilizing the principle of protein-dye binding, Anal. Biochem., 72:248.

Deus-Neumann, B., and Zenk, M., 1984, A highly selective alkaloid uptake system in vacuoles of higher plants, Planta, 162:250.

Heins, M., 1978, Screening of Digitalis lanata plants and cell cultures for hydroxylation capacity, in : "Production of Natural Compounds by Cell Culture Methods", A. W. Alfermann and E. Reinhard, eds., Gesellschaft für Strahlen und Umweltforschung mbH, München.

Kreis, W., and Reinhard, E., 1985, Rapid isolation of vacuoles from suspension of cultured Digitalis lanata cells, J. Plant Physiol., 121:385.

Kreis, W., May, U., and Reinhard, E., 1986, UDP-glucose:digitoxin 16'-O-glucosyltransferase from suspension-cultured Digitalis lanata cells, Plant Cell Reports, submitted.

Löffelhardt, W., and Kopp, B., 1981, Subcellular localization of glucosyltransferases involved in cardiac glycoside glucosylation in leaves of Convallaria majalis, Phytochemistry, 20:1219.

Matile, P., 1984, Das toxische Kompartment der Pflanzenzelle, Naturwiss., 71:18.

Murashige, T., and Skoog, F., 1962, A revised medium for rapid growth and bioassays with tobacco tissue cultures, Physiol. Plant., 15:473.

Ohlrogge, J. B., Garcia-Martinez, J. L., Adams, D., and Rappaport, L., 1980, Uptake and subcellular compartmentation of gibberellin A_1 applied to leaves of barley and cowpea, Plant Physiol., 66:422.

Petersen, M., and Seitz, H. U., 1985, Cytochrome P-450-dependent digitoxin 12-β-hydroxylase from cell cultures of Digitalis lanata, FEBS Letters, 188:11.

Reinhard, E., 1974, Biotransformations by plant tissue cultures, in : "Tissue culture and Plant Science 1974", H. E. Street, ed., Academic Press, London.

Reinhard, E., Boy, H. M., and Kaiser, F., 1975, Umwandlungen von Digitalis Glykosiden durch Zellsuspensionskulturen, Planta Med. Suppl., 163.

Stohs, S. J., 1980, Metabolism of steroids in plant tissue cultures, in : "Advances in Biochemical Engineering", Vol. 16, Plant cell cultures I, A. Fiechter, ed., Springer-Verlag, Berlin.

Schmitt, R., and Sandermann, H., 1982, Specific localization of β-D-glucoside conjugates of 2,4-dichlorophenoxyacetic acid in soybean vacuoles, Z. Naturforsch., 37c:772.

Wagner, G., 1983, Higher plant vacuoles and tonoplasts, in : "Isolation of Membranes and Organelles from Plant Cells", J. L. Hall and A. L. Moore, eds., Academic Press, London.

Werner, C., and Matile, P., 1984, Accumulation of coumarylglucosides in vacuoles of barley mesophyll protoplasts, J. Plant Physiol., 118:237.

TONOPLAST : THE CONTROLLING SITE FOR THE IN VITRO SOLUTE TRANSPORT IN DEVELOPING SOYBEAN COTYLEDONS

Willy Lin

Central Research and Development Department
Experimental Station E. I. du Pont de Nemours and Co.
Wilmington, Delaware 19898, U.S.A.

INTRODUCTION

In vitro accumulation into developing soybean cotyledons have been shown to be energy dependent, and is an active uptake process (Lichtner and Spanswick, a and b; Thorne, 1982; Lin et al., 1984; Schmitt et al., 1984; Lin, 1985 b). Studies on the concentration dependency, interference of external pH, and effect of several metabolic inhibitors on both sugar and amino-acid uptake into excised intact cotyledons and isolated protoplasts have provided circumstantial evidences for the involvement of a proton cotransport mechanism (Lichtner and Spanswick, 1981 a and b; Thorne, 1982, Lin et al., 1984; Schmitt et al., 1984; Lin, 1985 b).

Recent studies with excised intact soybean cotyledons (Lichtner and Spanswick, 1981 a and b; Thorne, 1980 and 1982) and isolated cotyledonary protoplasts (Lin et al., 1984; Schmitt et al., 1984) have shown a biphasic sucrose concentration dependent uptake kinetic in soybean cotyledonary cells. This uptake kinetic consists of a saturable component selective for sucrose and linear uptake component evident at higher substrate levels. The lack of a direct connection of the maternal and cotyledonary tissues (Thorne, 1981) has resulted in the necessity of the accumulation of solute into cotyledonary tissues from the apoplastic space. The apoplastic solute concentration and physiological conditions in cotyledonary tissues are therefore, determining the properties of the in vivo solute uptake into soybean seeds. Although the apoplastic solute concentration in soybean cotyledons has been reported differently with different measurement techniques (Hsu et al., 1984; Gifford and Thorne, 1985), reported values (i.e. higher than 15 mM) tend to suggest that the majority of solute is taken up via a linear uptake mechanism in soybean cotyledons in vivo.

Several recent reports have concentrated on the mechanism of the saturable sucrose uptake and suggesting that it is a proton-sucrose cotransport system (Giaquinta, 1983; Lin, 1985 a). In freshly excised soybean cotyledons, only a nonsaturable component was found for the amino-acid uptake (Bennett and Spanswick, 1983). The same authors have also reported that a high-affinity, saturable system can be detected upon the incubation of the cotyledon for 12-24 hours without a source of nitrogen (Bennett and Spanswick, 1983). Coincident with the appearance of the saturable uptake, the membrane potential becomes sensitive to added aminoacids, suggesting that the saturable component of uptake is via a proton cotransport system also. Recent studies have shown, at least for some solutes, a component of the linear uptake is sensitive to metabolic inhibitors, both in intact cotyledons (Bennett and Spanswick, 1984) and in protoplasts isolated from cotyledons (Lin, 1985 c).

However, the uptake mechanisms involved in the linear component is not clear. Further studies with a system in which the saturable component has been eliminated, as that used by Lin (1985 c), are needed to provide a better understanding of the mechanism involved in the linear component.

In this communication, I will examine published data plus our own unpublished ones which deal with the machanism of solute uptake into developing soybean cotyledons, with the emphasis on the role of tonoplast on the in vivo solute uptake.

DEVELOPMENT OF SOLUTE TRANSPORT IN SOYBEAN COTYLEDONS

Recently we have shown that the solute uptake capacity changes during soybean seed development (VerNooy et al., 1986 b). Throughout seed development apparent uptake of exogenous glucose is found to be by diffusion, with maximum rates on influxes in very young cotyledons (younger than 15 days after flowering of age) due to their high surface to volume ratio. Active uptake of dilute sucrose (0.5 mM) reached a maximal rate 20 to 25 days after flowering and declined as seed growth rate diminished. The influx of exogenous glutamine was maximal in very young cotyledons, declined progressively throughout maturity but increased again to nearly maximal as the cotyledons become yellow. About 50% of total amino-acid uptake is consistently energy dependent (sensitive to the uncoupler FCCP). The declining of solute uptake capacity in the saturable uptake component sucrose concentration range coincides with the declining of canopy photosynthesis and the rate of seed growth (VerNooy et al., 1986 a and b).

Studies of grain filling in several cereal crops have indicated that the end of photosynthate import and grain growth is related to the crushing of specific cells involved in photosynthate import and the formation of an impenetrable black layer in the hilum region (Daynard and Duncan, 1969; Eastin et al., 1973). In developing soybean seeds, the cessation of seed growth could be due to the yellowing of seed coat (Tekrony et al., 1979) and/or the loss of photosynthetic competency in the leaves (Benner and Nooden, 1984). Recent study (VerNooy et al., 1986 a) on the in vitro assimilate uptake and metabolism in embryos of soybean seeds isolated from mid-podfilling through physiological maturity, has led us to suggest that the drop of sucrose uptake coincides with the cessation of seed growth as well as rapid decline in leaf photosynthesis rate that preceded leaf and seed coat yellowing, while the sustained import and metabolism of amino-acids is resulted from the remobilization from senescing leaves which prolong seed growth beyond loss of photosynthetic competency and sucrose availability.

SOLUTE UPTAKE INTO EXCISED INTACT COTYLEDONS
ISOLATED FROM DEVELOPING SOYBEAN SEEDS

The transport of nutrients to the developing soybean seeds occurs through an apoplastic pathway. Photosynthetic assimilates are unloaded into the apoplast separating seed coat and embryo prior ot their accumulation into the developing seeds (Thorne, 1981). Recent studies with excised intact developing soybean cotyledons have shown that sucrose uptake into cotyledons occurred by both a substrate-saturable component observable at low exogenous sucrose concentrations and a linear component apparent at high sucrose concentration (Thorne, 1980; Litchner and Spanswick, 1981 b). Like in most source regions of plants, sucrose uptake into intact soybean cotyledons (Litchner and Spanswick, 1981 a and b; Thorne, 1982) is (a) markedly inhibited by alkaline pH; (b) inhibited by the non-penetratring sulfhydryl group modifier, PCMBS; and (c) stimulated by FC, a potent stimulator of active proton/potassium exchange in plant cells. These finding provided circumstantial evidence for the involvement of a sucrose-specific proton cotransport (Giaquinta, 1983) in the sucrose uptake in soybean cotyledons.

436

Three major obstacles in our understanding of the mechanism(s) of amino-acid uptake in higher plants are : (a) the rapid intercellular metabolism of most natural amino-acids; (b) the influence of the medium pH on the charge species of the acid; and (c) the possible binding of the acid to components of cell wall and membrane. Depending upon the species of amino-acid and the tissue which was used, a proton-cotransport, proton-antiport, or neutral-transport (transport of the charged acid molecule) mechanism has been proposed for amino-acid transport in plant tissues (Reinhold and Kaplan, 1984). Unlike that for sucrose, less research effort was made in studying amino-acid uptake in developing soybean seeds. In freshly excised soybean embryos, Bennett and Spanswick (1983) could observed a nonsaturable amino-acid uptake component only. However, a high-affinity, saturable system appears upon incubation of the cotyledons for 12-24 hours without a source of nitrogen and the membrane potential becomes more sensitive to added amino-acids (Bennett and Spanswick, 1983), suggesting that the saturable component of uptake via a proton cotransport system.

SOLUTE UPTAKE INTO PROTOPLASTS ISOLATED FROM DEVELOPING SOYBEAN COTYLEDONS

Freshly isolated protoplasts from developing soybean cotyledons have been proved to process normal solute, e.g., sugar, transport properties as that of the tissue they are derived from (Lin et al., 1984; Schmitt et al., 1984). These protoplasts take up sucrose preferentially over glucose, this difference in sugar uptake is mainly due to the presence of a saturable uptake component for sucrose but not for glucose (Lin et al., 1984). Sucrose uptake into isolated protoplasts showed a biphasic concentration dependent uptake kinetic (Lin et al., 1984). The removal of cell wall had facilitated the study of the energetic of solute transport (Schmitt et al., 1984; Lin, 1985 b). We have recently provided several strong conclusive evidences for the existence of a sucrose/proton cotransport system in soybean cotyledonary cells (Lin et al., 1984; Schmitt et al., 1984; Lin, 1985 b). With isolated protoplasts, we have shown that the addition of exogenous sucrose causes : (a) a transient alkalinization of the protoplast suspension medium; (b) a transient acidification of the internal pH of protoplasts; (c) a transient depolarization of the membrane potential of protoplasts; and (d) a close to one stoichiometry of proton to sucrose for the uptake of sucrose (Lin, 1985 b). All of these transient phenomena reflect the entrance of charged protons with sucrose molecules into protoplasts and the re-establishment of a proton gradient across cell membrane by plasmalemma ATPases within a short period of time.

Unlike sucrose, uptake of glucose or 3-O-methyl-glucose lacks a pH dependency and saturable uptake component (Lin et al., 1984). Glucose uptake into isolated soybean cotyledonary protoplasts has been shown to be insensitive to several membrane impermeable metabolic inhibitors (Lin et al., 1984). However, lipid soluble inhibitors did show a similar inhibitory effects on both sucrose and glucose uptake (Lin et al., 1984). One possible explanation of the lack of pH response and the discrepancy of the metabolic inhibitors' effect on glucose uptake could be due to that the active glucose uptake is localized on the tonoplast and the active sucrose uptake is localized on the plasmalemma. Change of the external pH does not cause a corresponding pH change of the cytoplasm, and consequently the tonoplast controlling process will not be affected by the changes of the extracellular pH.

Attempts of measuring active amino-acid uptake into isolated soybean cotyledonary protoplasts have been shown to be unsuccessfull (VerNooy and Lin, unpublished). The rapid metabolism of amino-acid in soybean cotyledonary cells and the possible lack of a direct involvement of plasmalemma on amino-acid uptake could be account for such failure. Use of isolated intact vacuoles or tightly sealed tonoplast vesicles from soybean cotyledons could provide some insights into our understanding to the mechanism of amino-acid uptake in soybean cotyledonary tissues.

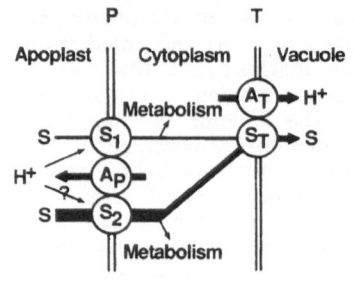

Figure 1

Proposed Mechanism of Solute Transport
in Soybean Cotyledonary Cells

A_p : Plasmalemma ATPase
A_T : Tonoplast ATPase
P   : Plasmalemma
S   : Solute (e.g. sucrose, glucose, amino-acid, etc.)
S_1 : Plasmalemma Carrier mediates saturable solute uptake
S_2 : Plasmalemma Carrier mediates linear solute uptake
S_T : Tonoplast Solute Uptake Carrier
T   : Tonoplast

POSSIBLE ROLE OF TONOPLAST IN SOLUTE TRANSPORT
IN SOYBEAN COTYLEDONS

    Under the physiological condition, the solute concentration of apoplastic space surrounding the majority of cotyledonary cells could be as high as 200 mM (Gifford and Thorne, 1985). At this solute concentration, the saturable uptake component become minimal and the majority of solute would be taken up through a diffusion process (Schmitt et al., 1984). Internal sucrose concentration in freshly isolated soybean cotyledonary protoplasts was estimated to be around 200 mM (Schmitt and Lin, unpublished result). Since the protoplast isolation has subjected cotyledonary cells under sucrose free condition for more than 3 hours (Lin et al., 1984), the majority of this intercellular sucrose must be reflecting that in the vacuoles. It is well known that in all storage plant tissue, carbohydrates, proteins, and other storage materials are restricted in vacuole or specialized vacuole-like structure (Giaquinta, 1983; Boller and Wiemken, 1986). Isolated vacuoles have been found to possess transport systems for a variety of substances. There are specific transport systems for the glucose analog 3-O-methylglucose and for sucrose (see Boller and Wiemken, 1986). It is suggested, as illustrated in Figure 1 that in soybean cotyledonary cells, in addition to a saturable-high-sucrose-affinity plasmalemma carrier protein, there is a common sugar transporter which allows majority of sucrose and glucose to enter the cells (linear uptake component) and the tonoplast is the one which regulate the accumulation of sugar as well as other solute in the vacuole. Further studies to use isolated intact vacuole or tightly-sealed tonoplast vesicles in solute transport is needed to elucidate the role of tonoplast in solute transport in soybean cotyledonary cells.

REFERENCES

Bennett, A. B, and Spanswick, R. M., 1983, Depression of amino-acid/proton co-transport in developing soybean embryos, Plant Physiol., 72:781.

Bennett, A.B., and Spanswick, R. M., 1984, Non-saturating mechanisms of solute uptake by developing soybean embryos, in : "Membrane Transport in Plants", Cram, W. J., Janacek, K., Rybova, R., and Sigler, K., ed., Academia, Prague.

Boller, T., and Wiemken, A., 1986, Dynamics of vacuolar compartmentation, Ann. Rev. Plant Physiol., 37:137.

Daynard, T. B., and Duncan, W. G., 1969, The black layer and grain maturity in corn, Crop Sci., 9:473.

Eastin, J. D., Hultquist, J. H., and Sullivan, C. Y., 1973, Physiologic maturity in grain sorghum, Crop Sci., 13:175.

Giaquinta, R. T., 1983, Phloem loading of sucrose, Ann. Rev. Plant Physiol., 34:347.

Gifford, R. M., and Thorne, J. H., 1985, Sucrose concentration at the apoplastic interface between seed coat and cotyledons of developing soybean seeds, Plant Physiol., 77:863.

Hsu, F. C., Bennett, A. B., and Spanswick, R. M., 1984, Concentrations of sucrose and nitrogeneous compounds in the apoplast of developing soybean seed coats and embryos, Plant Physiol., 75:181.

Lichtner, F. T., and Spanswick, R. M., 1981 a, Electogenic sucrose transport in developing soybean cotyledons, Plant Physiol., 67:869.

Lichtner, F. T., and Spanswick, R. M., 1981 b, Sucrose uptake by developing soybean cotyledons, Plant Physiol., 68:693.

Lin, W., 1985 a, Energetics of membrane transport in protoplasts, Physiol Plant., 65:102.

Lin, W., 1985 b, Energetics of sucrose transport in protoplasts isolated from deveping soybean cotyledons, Plant Physiol., 78:41.

Lin, W., 1985 c, Linear sucrose transport in protoplasts isolated from developing soybean cotyledons, Plant Physiol., 78:649.

Lin, W., Schmitt, M. R., Hitz, W. D., and Giaquinta, R. T., 1984, Sugar transport in soybean cotyledon protoplasts. I. Protoplast isolation and general characteristics of sugar transport, Plant Physiol., 75:936.

Reinhold, L., and Kaplan, A., 1984, Membrane transport of sugars and amino-acids, Ann. Rev. Plant Physiol., 35:45.

Schmitt, R. M., Hitz, W. D., Lin, W., and Giaquinta, R. T., 1984, Sugar transport in soybean cotyledon protoplasts. II. Sucrose transport kinetics, selectivity, and modeling studies, Plant Physiol., 75:941.

Thorne, J. H., 1980, Kinetics of ^{14}C-photosynthate uptake by developing soybean fruit, Plant Physiol., 65:975.

Thorne, J. H., 1981, Morphology and ultrastructure of maternal seed tissues of soybean in relation to the import of photosynthate, Plant Physiol., 67:1016.

Thorne, J. H., 1982, Characterization of the active sucrose transport system of immature soybean embryos, Plant Physiol., 70:953.

VerNooy, C. D., Thorne, J. H., Lin, W., and Rainbird, R. M., 1986 a, Cessation of assimilate uptake in maturing soybean seeds, Plant Physiol., in press.

VerNooy, C. D., Thorne, J. H., and Lin, W., 1986 b, Changes in solute transport in developing soybean cotyledons, Plant Physiol., in press.

THE VACUOLE IN RELATION TO PLANT GROWTH REGULATORS

Philip John

Department of Agricultural Botany
Plant Science Laboratories, University of Reading.
Reading RG6 2AS, United Kingdom

INTRODUCTION

They are five types of plant growth regulator : The auxins, gibberellins, cytokinins, abscisic acid and ethylene. In the present paper, I shall briefly review the role of the vacuole in their mode of action, compartmentation and biosynthesis, and then I shall discuss in more detail the evidence that indicates an involvement of the vacuole in ethylene biosynthesis. The terms "hormone" and "growth regulator" will be used interchangeably, in line with current usage (see, for example, Venis, 1985).

VACUOLES AND GROWTH REGULATORS

Considerable experimental effort and ingenuity have been invested in recent years into the identification in plant cells of macromolecular receptors which recognize specific growth regulators. The existence of these hormone-receptor sites is inferred from the stringent structural and stereochemical requirements shown by plant growth regulators for physiological activity. However, it is clear from recent reviews of this topic (Venis, 1985; Libbenga et al., 1986; Hall, 1986; Stoddart, 1986) that far less progress has been made in the identification and isolation of receptor sites from plant cells compared with the advances made with animal cells. Consequently almost no description can be given of how the hormonal signal is translated into biochemical events in plant cells. Nevertheless what limited knowledge is available does indicate that the vacuole is unlikely to be an important location of hormone receptor sites. Thus binding sites for auxins have been located on membranes derived from the plasma membrane and the endoplasmic reticulum (Venis, 1985; Libbenga et al., 1986). In addition there is evidence that the fungal toxin, fusicoccin, which can elicit many of the physiological responses of auxins, has a receptor site at the plasma membrane (Venis, 1985; Libbenga et al., 1986). Little is known with certainty of where binding occurs in the cell or where the primary action is either with gibberellins (Stoddart, 1986) or with cytokinins (Venis, 1985). By contrast there is good evidence that the abscisic acid receptor is located at the plasma membrane, at least in the case of guard cells of Vicia faba (Hornberg and Weiller, 1984). Sites showing many of the properties expected of ethylene receptors have been found in plant membranes (Hall, 1986). Activities of marker enzymes associated with the membranes indicate that they are derived mainly from the endoplasmic reticulum and are unlikely to be of vacuolar origin.

Despite the lack of evidence for a specific interaction of plant growth regulators with the vacuole, this organelle does not appear to contain significant amounts of some plant growth regulators. Thus indole-3-acetic acid (IAA) has been observed to enter the vacuole of the giant alga Hydrodictyon africanum (Raven, 1975) and abscisic acid has been identified in the vacuolar sap of ripe citrus fruit (Milborrow, 1984). Moreover when radioactive gibberellin (GA_1) was applied to leaves 30 - 100% of the radioactivity present in the protoplasts prepared from the leaves was recovered in the vacuole fraction (Ohlrogge et al., 1980). The gibberellin was probably in the vacuolar sap rather in the tonoplast since it was released on lysis of the vacuoles. It was estimated that about 50% of the gibberellin present in the leaves was lost during the preparation of the protoplasts (Ohlrogge et al., 1980). The vacuolar gibberellin is likely to have been more resistant to loss than extravacuolar gibberellin, and therefore the extent to which the vacuole accumulates gibberellin in vivo may have been overestimated in this study (Ohlrogge et al, 1980). Vacuoles isolated from leaves of barley and cowpea have also been shown (Knuth et al., 1983) to take up gibberellin (GA_1). In the case of the vacuoles from cowpea leaves uptake was stimulated by ATP, thus indicating the presence of an active uptake system.

Since ethylene is a gas it cannot readily be contained by any organelle of the plant cell. However, its immediate precursor, 1-aminocyclopropane-1-carboxylate (ACC) is known to be present in vacuoles isolated from pea mesophyll cells (Guy and Kende, 1984 b). It was reported that about 90% of the ACC present in protoplasts was recovered in vacuoles isolated from the protoplasts.

Generally the vacuole is by far the largest organelle of the differentiated plant cell. Thus an even concentration of growth regulators throughout the plant cell would ensure that the vacuole contained much of the cell content of growth regulator. However, in the absence of active transport systems indole-3-acetic acid, gibberellin, abscisic acid and ACC would be excluded from a vacuolar sap which is more acid than the cytosol (Raven, 1975; Milborrow, 1984), because they are all weak acids which, in the protonated form, have appreciable permeability coefficients for membranes. Transport systems must therefore exist at the tonoplast to regulate the vacuolar content of at least some of the plant growth regulators. If it is assumed that the receptor sites for these growth regulators are entirely extravacuolar, as discussed above, then the vacuolar sequestration of growth regulators would effectively reduce their active concentration in the plant cell.

Representatives of all the main types of plant growth regulator (except ethylene) are known to exist in a conjugated form, usually as a glycoside derivative. Thus inositol and inositol-arabinoside derivatives of IAA are, together with insoluble β-(1→4)-glucan derivatives, the main esters of IAA in kernels of Zea mays (Cohen and Bandurski, 1982). Glucosyl, n-propyl and methyl esters of gibberellins have been identified in a wide range of plant material (Bearder, 1980). The glucosyl ester of abscisic acid is also likely to be of widespread occurrence (Bearder, 1980; Neill et al., 1983). Zeatin and ribosyl-zeatin are both active as cytokinins and both can be converted by plants to the glucosyl derivative (Horgan, 1984). Although analogous derivatives of ethylene are unknow, its immediate precursor ACC is readily converted to the malonyl derivative by plant tissues (Amrhein et al., 1981). The location of these derivatives in the plant cell is as yet unknown, but they are generally viewed as inactive forms of the growth regulator and are probably stored in the vacuole. In some cases, exemplified by the soluble IAA derivatives, the active form of the growth regulator is readily released under physiological conditions. Where, as in this case, the derivative acts as a temporary store, derivatisation and release can be viewed as components of a homeostatic mechanism helping to maintain appropriate tissue levels of the active growth regulator. In other cases, for example the glucose ester of abscisic acid (Neill et al., 1983), and the malonyl derivative of ACC (Hoffman et al., 1983), release of the abscisic acid and ACC respectively appear to be too slow to be of physiological significance, and derivatisation is more appropriately viewed as a process of detoxification. The formation of derivatives of the kinds observed with the growth regulators described here is a common response of plant tissue to exogenously supplied compounds, and therefore it is difficult to assess the significance of this process for the compartmentation of plant growth regulators per se.

BIOSYNTHESIS OF ETHYLENE

The biosynthesis of auxins, gibberellins, cytokinins and abscisic acid does not appear to involve the vacuole directly. However, there is some evidence that the tonoplast is the location of the ultimate step in the biosynthetic pathway to ethylene. This pathway is : methionine → S-adenosylmethionine → 1-aminocyclopropane-1-carboxylate (ACC) → ethylene. The enzyme responsible for the synthesis of ACC, the ACC synthase, is active in vitro, readily purified by conventional procedures (Yang and Hoffman, 1984), and is presumably cytoplasmic. By contrast it has been generally found that activity of the enzyme responsible for the generation of ethylene from ACC, the so-called ethylen-forming enzyme (EFE) disappears when cell integrity is lost, so that despite the fact that EFE activity in vivo is high compared with ACC synthase activity it has proved difficult to demonstrate EFE activity in vitro (Yang and Hoffman, 1984). Cell-free preparations derived from a variety of plant tissues can convert ACC to ethylene, but these EFE-like activities (McKeon and Yang, 1984; Venis, 1984) do not show the specificity toward stereoisomers of the higher analogue of ACC that is shown by EFE in vivo (Hoffman et al., 1982); and for this and other reasons these in vitro activities have been attributed (Yang and Hoffman, 1984; Pirrung, 1986; Diolez et al., 1986) to a non-enzymatic, chemical reaction between ACC and products of the in vitro enzyme activities, rather than to a direct action by the EFE on ACC.

An important advance was made therefore when Guy and Kende (1984 a and b) demonstrated that the EFE activity of vacuoles isolated from pea and bean leaves retained, among other features, the stereodiscrimination showed by EFE activity in vivo. Moreover the vacuoles accounted for some 80 % of the EFE activity of the protoplasts from which they were derived. Thus it was concluded (Guy and Kende, 1984 b) that the tonoplast is a subcellular location of the EFE.

Previous failures to demonstrate a true EFE activity in vitro had been attributed by John (1983) to a mechanistic requirement by the EFE for a transmembrane electrical potential. However, subsequent determinations of EFE activity in vivo at different cell potentials failed to provide a clear indication of a potential-dependance for EFE activity, at least at the plasma membrane of plant cells (John et al., 1985). Thus, it was of particular interest that the vacuolar EFE activity described by Guy and Kende (1984 b) seemed to depend, although in an unspecified way, on a transmembrane ion gradient; first, because activity was lost when the vacuoles were lysed (Guy and Kende, 1984 b; Mayne and Kende, 1986), secondly, because protonophores and other ionophores inhibited EFE activity (Mayne and Kende, 1986).

Recently Porter et al. (1986) have extended the work of Guy and Kende (1984 a and b) to determine the extent to which the EFE activity observed with isolated vacuoles, protoplasts and cells represents the EFE activity of the intact parent tissue. Porter et al. (1986) confirmed the finding of Guy and Kende (1984 b) that the EFE activity observed with protoplasts prepared from pea and bean leaf tissue is largely vacuolar. However, in extending the comparison to leaf tissue it was discovered that the EFE activity of protoplasts is < 4% that of intact mesophyll tissue (Porter et al., 1986). These new findings underline the view already expressed by Guy and Kende (1984 b) that "it is premature to conclude that the vacuole is the sole site of ethylene synthesis in the cell".

Table 1 shows that the most dramatic loss of EFE activity occurred in the separation of cells from the intact tissue. The isolated cells and protoplasts used in the experiments of Guy and Kende (1984 a and b) and Porter et al. (1986) were obtained from mesophyll tissue which had been incubated for extended periods

Table 1

Ethylene Production by Different Systems
Derived from Pea and Bean Leaves

System	Ethylene produced (nl mg chlorophyll^{-1})[a]	
	Pea	Bean
Leaf discs	1080 $\pm$ 26	473 $\pm$ 35
Isolated cells	30.7 $\pm$ 1.4	24.0 $\pm$ 1.5
Protoplasts	25.0 $\pm$ 1.5	11.7 $\pm$ 1.4
Vacuoles[b]	16.2 $\pm$ 0.8	4.6 $\pm$ 0.2
Lysed vacuoles	< 0.01	< 0.01

[a]Assays were performed over a period of 3 h in the dark, with the addition of 1 mM ACC.
[b]For the vacuole suspension the chlorophyll content used in the calculations is that of the protoplasts from which the vacuoles were derived (data of Porter et al., 1986).

of time (> 3 h) in the presence of cell-wall degrading enzymes. This treatment may have been responsible for the loss of activity of an EFE located at the tonoplast. Thus we have recently determined the EFE activity of tonoplast membranes isolated without lengthy incubation of the original tissue.

As described elsewhere in this volume, Hevea latex has proved to be a useful source of readily isolated vacuoles for biochemical study. However we were unable to detect appreciable EFE activity in crude latex tapped from a variety of succulent plants.

The advantages of the juice of the Kiwi fruit (Actinidia chinensis) as a source of tonoplast membranes had been noted by Chedhomme and Rona (1986). We discovered that washed membranes from this juice have a high EFE activity (Table 2). Moreover, when these membranes are presented with racemic mixtures of 2-ethyl-ACC enantiomers only the mixture containing the (1R,2S)-enantiomer proved to be an effective substrate for 1-butene formation (Table 2). Thus the membranes from the Kiwi fruit juice show the same stereospecificity as intact tissues (Hoffman et al., 1982; Baldwin et al., 1985) and they differ from the lipoxygenase system which, like many other in vitro systems (McKeon and Yang, 1984; Venis, 1984), shows no specificity towards stereoisomers of 2-ethyl-ACC (Table 2). From these observations it is concluded that the membranes of the Kiwi fruit juice resemble the vacuoles isolated from pea and bean mesophyll cells (Guy and Kende, 1984 a and b) in possessing an authentic EFE activity. However, like the vacuoles, the EFE activity of the membranes of the Kiwi fruit juice is very much less than that of the tissue from which they were obtained (A. J. R. Porter, unpublished data). The juice expressed from Kiwi fruit contains three main types of membrane vesicle. These arise from the revesicularization of the tonoplast or plasma membrane of cell of the outer pericarp (Chedhomme and Rona, 1986). Whatever the relative contributions of these different vesicles to the observed EFE activity, membranes of the Kiwi fruit juice should improve a useful starting material for the isolation of EFE.

Table 2

Ethylene and 1-Butene Production
by a Variety of Systems

System	Substrate	Ethylene (nl)	1-Butene
Mung bean hypocotyls[a]	ACC	1200 ± 23	–
	(1R,2R)- and (1S,2S)- 2-ethyl-ACC	–	0
	(1R,2S)- and (1S,2R)- 2-ethyl-ACC	–	194 ± 27
Lipoxygenase system[b]	ACC	87 ± 3.0	–
	(1R,2R)- and (1S,2S)- 2-ethyl-ACC	–	12 ± 0.7
	(1R,2S)- and (1S,2R)- 2-ethyl-ACC	–	10 ± 2.0
Membranes from Kiwi fruit juice[c]	ACC	47 ± 3.0	–
	(1R,2R)- and (1S,2S)- 2-ethyl-ACC	–	0
	(1R,2S)- and (1S,2R)- 2-ethyl-ACC	–	41 ± 2.9

Temperature, 25°C; substrates added at 0.5 mM.

[a]Segments of hypocotyl (0.5 g) incubated in 20 mM Tris-Mes (pH 6.2) for 16 h as in John et al. (1985).

[b]Lipoxygenase system (Diolez et al., 1986) consists of lipoxygenase (400 units.ml^{-1}), linoleic acid (5 mM) and $MnCl_2$ (0.1 mM); incubated in 20 mM Epps buffer (pH 7.5) for 6 h.

[c]Membranes from the juice of about 10 g of ripe Kiwi fruit were sedimented at 2000 g for 10 min, washed in 25 mM Tris-Hepes (pH 7.5) containing 0.7 M mannitol, and incubated in this medium for 16 h.

AKNOWLEDGEMENTS

I am grateful for support from the S.E.R.C. (United Kingdom) and the generous gift of ethyl-ACC from Robert Adlington.

REFERENCES

Amrhein, N., Scheebeck, D., Skorupka, H., Tophof, S., and Stockigt, J., 1981, Identication of a major metabolite of the ethylene precursor 1-amino-cyclopropane-1-carboxylic acid in higher plants, Naturwiss., 68:619.

Baldwin, J. E., Adlington, R. M., Lajoie, G. A., and Rawlings, B. J., 1985, On the biosynthesis of ethylene. Determination of the stereochemical course using modified substrates, J. Chem. Soc. Chem. Commun., 21:1496.

Bearder, J. R., 1980, Plant hormones and other growth substances their background, structure and occurrence, in : "Encyclopaedia of Plant Physiology", Vol. 9, Hormonal Regulation of Development, J. McMillan, ed., Springer-Verlag, Berlin.

Chedhomme, F., and Rona, J. P., 1986, Isolation and electrical characterization of tonoplast vesicles from Kiwi fruit (Actinidia chinensis), Physiol. Plant., 67:29.

Cohen, J. D., and Bandurski, R. S., Chemistry and physiology of the bound auxins, Annu. Rev. Plant Physiol., 33:403.

Diolez, P., Davy de Virville, J., Latché, A., Moreau, F., Pech, J. C., and Reid, M., 1986, Role of the mitochondria in the conversion of 1-amino-cyclopropane-1-carboxylic acid to ethylene in plant tissues, Plant Science Letters, 43:13.

Guy, M., and Kende, H., 1984 a, Ethylene formation in Pisum sativum and Vicia Faba protoplasts, Planta, 160:276.

Guy, M., and Kende, H., 1984 b, Conversion of 1-amino-cyclopropane-1-carboxylic acid to ethylene by isolated vacuoles of Pisum sativum L., Planta, 160:281.

Hall, M. A., 1986, Ethylene receptors, in : "Hormones, Receptors and Cellular Interactions in Plants", C. M. Chadwick and D. R. Garrod, eds., Cambridge University Press, Cambridge.

Hoffman, N. E., Fu, J., and Yang, S. F., 1983, Identification and metabolism of 1-(malonylamino)-cyclopropane-1-carboxylate in germinating peanut seeds, Plant Physiol., 71:197.

Hoffman, N. E., Yang, S. F., Ichihara, A., and Sakamura, S., 1982, Stereospecific conversion of 1-amino-cyclopropane-carboxylic acid to ethylene by plant tissue, Plant Physiol., 70:195.

Horgan, R., 1984, Cytokinins, in : "Advanced Plant Physiology", M. B. Wilkins, ed., Pitman, London.

Hornberg, C., and Weiler, E. W., 1984, High affinity binding sites for abscisic acid on the plasmalemma of Vicia faba guard cells, Nature, 310:321.

John, P., 1983, The coupling of ethylene biosynthesis to a transmembrane electrogenic proton flux, FEBS Letters, 152:141.

John, P., Porter, A. J. R., and Miller, A. J., 1985, Activity of the ethylene forming enzyme measured in vivo at different cell potentials, J. Plant Physiol., 121: 397.

Knuth, M. E., Keith, B., Clark, C., Garcia-Martinez, J. L., and Rappaport, L., 1983, Stabilization and transport capacity of cowpea and barley vacuoles, Plant Cell Physiol., 24:423.

Libbenga, M. E., Mann, A. C., Van Den Linde, P. C. G., and Mennes, A. M., 1986, Auxin receptors, in : "Hormones, Receptors and Cellular Interactions in Plants", C. M. Chadwick and D. R. Garrod, eds., Cambridge University Press, Cambridge.

McKeon, T. A., and Yang, S. F., 1984, A comparison of the conversion of 1-amino-2-ethyl-cyclopropane-1-carboxylic acid stereoisomers to 1-butene by pea epicotyls and by a cell-free system, Planta, 160:84.

Mayne, R. G., and Kende, H., 1986, Ethylene biosynthesis in isolated vacuoles of Vicia faba L. - requirement for membrane integrity, Planta, 167:159.

Milborrow, B. V., 1984, Inhibitors, in : "Advanced Plant Physiology", M. B. Wilkins, ed., Pitman, London.

Neill, S. J., Horgan, P., and Heald, J. K., 1983, Determination of the levels of abscisic acid glucose ester in plants, Planta, 157:371.

Ohlrogge, J. B., Garcia-Martinez, J. L., Adams, D., and Rappaport, L., 1980, Uptake and subcellular compartmentation of gibberellin A_1 applied to leaves of barley and cowpea, Plant Physiol., 66:422.

Pirrung, M. C., 1986, Mechanism of a lipoxygenase model for ethylene biosynthesis, Biochemistry, 25:114.

Porter, A. J. R., Borlakoglu, J. T., and John, P., 1986, Activity of the ethylene-forming enzyme in relation to plant cell structure and organization, J. Plant Physiol, in press.

Raven, J. A., 1975, Transport of indole-acetic acid in plant cells in relation to pH and electrical potential gradients and its significance for polar IAA transport, New Phytol., 74:163.

Stoddart, J. L., 1986, Gibberellin receptors, in : "Hormones, Receptors and Cellular Interactions in Plants", C. M. Chadwick and D. R. Garrod, eds., Cambridge University Press, Cambridge.

Venis, M. A., 1984, Cell-free ethylene-forming systems lack stereochemical fidelity, Planta, 162:85.

Venis, M. A., 1985, "Hormone Binding Sites in Plants", Longman, London.

Yang, S. F., and Hoffman, N. E., 1984, Ethylene biosynthesis and its regulation in higher plants, Annu. Rev. Plant Physiol., 35:155.

INVOLVEMENT OF VACUOLES IN ETHYLENE METABOLISM

IN PLANT CELLS

Moudher Bouzayen[*], André Latché[*], Jean-Claude Pech[*]
and Gilbert Alibert[**]

[*]Ecole Nationale Supérieure Agronomique
145, Avenue de Muret, and [**]Centre de Physiologie Végétale
Université Paul Sabatier, Unité Associée C.N.R.S. No. 241
118, Route de Narbonne, F-31.062-Toulouse, France

INTRODUCTION

The previous demonstration that isolated plant protoplasts were able to produce ethylene (Anderson et al., 1979) has opened new perspectives in the knowledge of the compartmentation of ethylene metabolism. More recently, new technologies of vacuole isolation led to the demonstration that 1-amino-cyclopropane1-carboxylic acid (ACC), the immediate precursor of the plant hormone (Adams et al., 1979) is located both in the vacuole and the cytoplasm (Guy and Kende, 1984 a; Bouzayen et al., 1986). ACC can also be found as a malonyl conjugate (Arheim et al., 1981; Hoffman et al., 1982) which is sequestered in the vacuole, its unique site of storage (Bouzayen et al., 1986).

The synthesis of ACC and its metabolization into ethylene probably occur in two separate compartments of the cell (Guy and Kende, 1984 b). Moreover, several sites for ethylene production have been suggested : The plasma membrane (Anderson et al., 1979; Forney and Arteca, 1982), the tonoplast (Guy and Kende, 1984 a; Mayne and Kende, 1986) and the mitochondria (Vinkler and Apelbum, 1983). The role of the mitochondria in ethylene production is highly questionable (Diolez et al., 1986). The tonoplast is so far the only membrane system which resembles the in vivo system by being able to discriminate between stereo-analogs of ACC (Guy and Kende, 1984 a).

In this work, we use protoplasts and vacuoles isolated from Acer pseudoplatanus cells in order to first examine the site of synthesis of MACC and its translocation into the vacuole and second to search for the site(s) of conversion of ACC into ethylene.

MATERIAL AND METHODS

Cell Suspension Cultures of Acer Pseudoplatanus

Acer pseudoplatanus cells were grown as previously described (Alibert et al., 1982). Cells were always sampled during the exponential phase, i.e. between days 6 and 9 after transfer into new medium.

Chemicals and Radiochemicals

Caylase was from Cayla (Toulouse, France) and Pectolyase Y23 from Seishim Pharmaceutical Co. (Chibaken, Japan). (2,3 ^{14}C) 1-aminocyclopropane-1-carboxylic acid (2.96 GBq mmol^{-1}) was kindly prepared by CEA (France). The other chemicals were the same as those used by Alibert et al. (1982).

Preparation of Protoplasts and Vacuoles

The original method of Alibert et al. (1982) was used with the modification described elsewhere (Bouzayen et al., 1986). Protoplasts were numbered using a Fuschs-Rosenthal haemacytometer. The volume of 1.10^6 protoplasts and the corresponding vacuoles were estimated to be 9 μl and 7.35 μl respectively by measurement of the space accessible to ^3H_2O but inaccessible to ^{14}C-dextran (Komor et al., 1982).

Ethylene Determination

Ethylene was determined by gas chromatography and its radioactivity measured by liquid scintillation spectrometry after absorption in 0.25 M Hg(ClO$_4$)$_2$.

Extraction, Assay and Estimation of Free and Conjugated ACC in the Various Cell Compartments

Protoplasts or vacuoles were lysed by thawing and ACC and conjugated ACC extracted as described elsewhere (Lizada and Yang, 1979). The ACC content of the extract was determined by the method described elsewhere (Mansour et al., 1986). Conjugated ACC was purified by anion exchange chromatography and hydrolyzed as described elsewhere (Bouzayen et al., 1986). The free ACC liberated upon hydrolysis was then assayed according to the method described previously (Lizada and Yang, 1979) except that HgCl$_2$ and NaOCl concentrations were increased to 6 mM and 67 mM respectively. Radioactivities were determined by scintillation counting (LKB 1215).

Vacuole to protoplast content in ACC and MACC was calculated on the basis of α-mannosidase activity, a vacuolar marker (Alibert and Boudet, 1982).

Feeding protoplasts with (2,3 ^{14}C) ACC

2 ml of protoplast suspension (20 10^6 protoplasts ml^{-1} in 25 mM Tris-Mes buffer pH 6.5, 0.7 M mannitol) were introduced into 27 ml glass vials supplemented with 0.125 mM ACC (2,960 MBq mmol^{-1}) and incubated in a shaker batch at 25°C. At selected times vacuoles were isolated, ACC and MACC content and ethylene production determined. More detailed experimental conditions are given in the legends of Tables and Figures.

RESULTS

Intracellular Distribution of ACC and MACC

Table 1 shows 67% of free ACC is located in the vacuole. In contrast to ACC, all the conjugated form of ACC appears to be localized in the vacuole of Acer.

Kinetics of MACC Formation and Accumulation

The kinetics of ^{14}C-MACC formation has been followed after feeding proto-plasts with ^{14}C-ACC. Fig. 1 shows that the rate of ^{14}C-MACC formation is rapidly increasing in the protoplasts during the first 2 hours and then reaches a constant

Table 1

Endogenous ACC and MACC Distribution Between Protoplasts and Vacuoles
in <u>Acer</u> cells cultured <u>in vitro</u>

	Total nmol per		Percent in vacuoles	nmolar concentrations	
	10^6 protoplast	Vacuoles corresponding to 10^6 protoplasts		Cytosol	Vacuoles
ACC	$9.60 \pm 0.12^*$	$6.43 \pm 0.45^*$	67	1.92	0.87
MACC	$1.34 \pm 0.26^{**}$	$1.49 \pm 0.06^{**}$	111	0	0.20

* mean $\pm$ SE of two replicates.
** mean $\pm$ SE of three replicates.

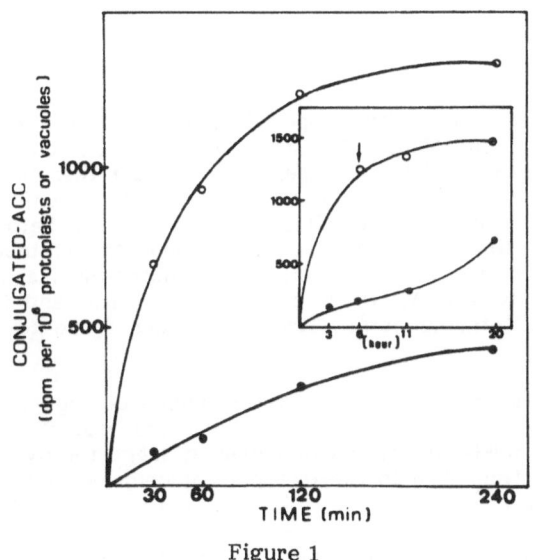

Figure 1

Kinetics of MACC formation
in protoplasts (O) and corresponding vacuoles (●)
after feeding with exogenous 0.125 mM ^{14}C-ACC (2960 MBq mmol^{-1})

Insert : In a similar experiment a chase was performed after 6 hours by adding
12.5 mM cold ACC (arrow) to the protoplast suspension. Changes of the radiolabeled
MACC were followed during the next 14 h.

level. Most of the MACC synthesized during the first hours was present in the cytosol, the vacuolar content being very low but increasing steadily with time. When a chase was performed by adding cold ACC (12.5 mM) after 6 h incubation in the presence of labelled ACC, 50 % of the synthesized MACC was then recovered in the vacuole after 14 h incubation.

Localization of the Sites of Ethylene Synthesis

Since ACC is converted into equimolar concentrations of ethylene, it could be anticipated that, in the compartment where ACC is converted into ethylene, the specific radioactivity of ACC should be similar to that of ethylene evolved from the protoplast. The specific radioactivity of ACC in the various compartments was therefore followed during a 4 hours incubation period of the protoplasts in the presence of exogenous (2,3 ^{14}C) ACC and was compared to the specific radioactivity of ethylene. Results presented in Fig. 2 show that, in our experimental conditions and at any incubation time, the specific radioactivity of the external ACC is at least 38 to 380 times higher than that of cytoplasmic or vacuolar ACC. The specific radioactivity of ethylene evolved varied from 1764 to 888 MBq mmol^{-1} which is closed to specific radioactivity in the external medium (2530 to 1864 MBq mmol^{-1}). Moreover the kinetics of formation of ethylene follow the same pattern as external ACC.

As expected, the specific radioactivity of ACC was decreasing with time in the external compartment and increasing in the cytosol and vacuole as a consequence of internal ACC efflux and ^{14}C-ACC uptake by the protoplasts. The specific radioactivity of ethylene evolved was also decreasing as well as the estimated percentage of ethylene synthesized from the outer side of the plasma membrane (76% after 60 mn to 49% after 240 mn).

DISCUSSION

The results presented in this paper demonstrate that the vacuole represents the main site of ACC and MACC deposition in the cell. Even though the chemical nature of the conjugated ACC has not been determined in these materials, it probably corresponds to malonyl-ACC which is the major or unique conjugated form of ACC encountered in various plant species (Hoffman et al., 1983; Satoh and Esashi, 1984; Amrhein et al., 1982; Knee, 1985). The sequestration of this metabolite in the vacuole which is reported here for the first time, is not surprising inasmuch as conjugated forms of hormones (Garcia-Martinez et al., 1981; Bray and Zeevaart, 1985) or phenolic glycosides (Rataboul et al., 1985) have been found to accumulate in this compartment.

Kinetic studies suggest that the ACC-malonyl conjugate is first synthesized in the cytosol and thereafter transfered into the vacuole. They exclude the localization of the transferase at the tonoplast as suggested by Matern et al. (1983) and therefore its participation to the transport of conjugated ACC into the vacuole (Matern et al., 1984).

Our results also demonstrate that the specific radioactivity of the ethylene evolved by the protoplasts is much higher than that of the cytosolic or vacuolar ACC and lower than that of external ACC. It can be concluded that ethylene is at least partly formed by an enzymatic complex located on the plasmalemma as suggested earlier by Anderson et al. (1979) and Forney and Arteca (1982). However since the variations of the specific radioactivities of external ACC and ethylene are not exactly superposable, we can assume that another site of ethylene synthesis is present within the cell. The rate of dilution may be used as an index for calculating the relative importance of both sources of ethylene : 76% at the beginning of the experiment (60 mn), only 49% after 240 mn. These results demonstrate for the first time that the plasmalemma is indeed involved in the conversion of ACC into

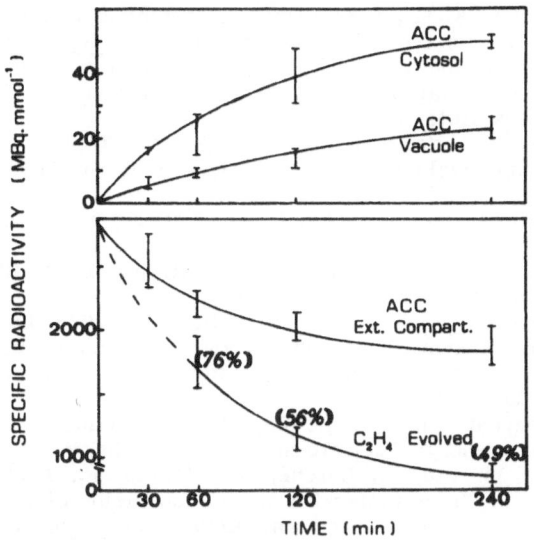

Figure 2

Changes in the specific radioactivity of ethylene and ACC
in the various cell compartments
and percentage of ethylene synthesized in the external compartment

Acer pseudoplatanus protoplasts were incubated in the presence of 0.125 μM (2,3 ^{14}C) ACC (2960 MBq mmol^{-1}). Vacuoles were isolated at various times intervals as described in Materials and Methods. Each value represents the mean $\pm$ SE of three replicates except for ACC in the supernatant (two replicates). The percentage of external ethylene production X (values in parentheses) was calculated according the following formula :

$$X = \frac{(a + b)/2 + c}{(a + b)/2 + d} \cdot 100$$

  a = Specific radioactivity of ACC in the cytosol

  b = Specific radioactivity of ACC in the vacuole

  c = Specific radioactivity of ACC in the extracellular medium

  d = Specific radioactivity of ethylene evolved

ethylene, but that other internal sites exist inside the cell. The vacuole may be one of these sites as claimed by Guy and Kende (Guy and Kende, 1984 a and b; Mayne and Kende, 1986).

ACKNOWLEDGEMENTS

This research was partially supported by the Centre National de la Recherche Scientifique (R.C.P. No. 725 and U.A. No. 241). We acknowledge the contribution of the Commissariat à l'Energie Atomique (Service des Molécules Marquées, L. Pichat et coll., CEA, France) in synthesizing ^{14}C ACC and the help of Sylvie Del Col in typing the manuscript.

REFERENCES

Adams, D. O., and Yang, S. F., 1979, Ethylene biosynthesis : Identification of a 1-aminocyclopropane-1-carboxylic acid as an intermediate in the conversion of methionine to ethylene, Proc. Natl. Acad. Sci. U.S.A., 76:170.

Alibert, G., and Boudet, A. M., 1982, Progrès, problèmes et perspectives dans l'obtention et l'utilisation de vacuoles isolées, Physiol. Vég., 20:289.

Alibert, G., Carrasco, A., and Boudet, A. M., 1982, Changes in biochemical composition of vacuoles isolated from Acer pseudoplatanus L. during cell culture, Biochim. Biophys. Acta, 721:2229.

Amrhein, N., Breuing, F., Eberle, J., Skorupka, H., and Tophof, S., 1982, The metabolism of 1-aminocyclopropane-1-carboxylic acid, in: "Plant Growth Substances", P. F. Wareing, ed., Academic Press, London.

Amrhein, N., Schneebeck, D., Skorupka, H., Tophof, S., and Stockigt, J., 1981, Naturwissenschaften, 68:619.

Anderson, J. D., Lieberman, M., and Stewart, R. N., 1979, Ethylene production by apple protoplasts, Plant Physiol., 63:931.

Bouzayen, M., Latché, A., Pech, J.-C., and Alibert, G., 1986, Localisation subcellulaire des lieux de synthèse et de stockage de l'acide 1-amino-cyclopropane-1-carboxylique conjugué dans des cellules d'Acer pseudoplatanus L., C. R. Acad Sci., Sér. D, 303:425.

Bray, E. A., and Zeevaart, J. A. D., 1985, The compartmentation of abscisic acid and β-D-glucopyranosyl abscisate in mesophyll cells, Plant Physiol., 79:719.

Diolez, P., Davy de Virville, J., Latché, A., Moreau, F., Pech, J.-C., and Reid, M., 1986, Role of the mitochondria in the conversion of 1-amino-cyclopropane 1-carboxylic acid to ethylene in plant tissues, Plant Science, 43:13.

Forney, C. F., and Arteca, R. N., 1982, Effects of amino- and sulfhydryl reactive agents on respiration and ethylene production in tomato and apple fruit discs, Physiol. Plant., 54:329.

Garcia-Martinez, J. L., Ohlrogge, J. B., and Rappaport, L., 1981, Differential compartmentation of gibberellin A_1 and its metabolites in vacuoles of cowpea and barley leaves, Plant Physiol., 68:865.

Guy, M., and Kende, H., 1984 a, Conversion of 1-amino-cyclopropane-1-carboxylic acid to ethylene by isolated vacuoles of Pisum sativum L., Planta, 160:281.

Guy, M., and Kende, H., 1984 b, Ethylene formation in Pisum sativum and Vicia faba protoplasts, Planta, 160:276.

Hoffman, N. E., Fu, J. R., and Yang, S. F., 1983, Identification and metabolism of 1-(malonylamino)-cyclopropane-1-carboxylic acid in germinating peanuts seeds, Plant Physiol., 71:197.

Hoffman, N. E., Yang, S. F., and McKeon, T., 1982, Identification of 1-(malonylamino)-cyclopropane-1-carboxylic acid, an ethylene precursor in higher plants, Biochem. Biophys. Res. Commun., 104:765.

Knee, M., 1985, Metabolism of 1-amino-cyclopropane-1-carboxylic acid during apple fruit development, J. Exp. Bot., 36:670.

Komor, E., Thom, M., and Maretzki, A., 1982, Vacuoles from sugarcane suspensions cultures. III. Protonmotive potential difference, Plant Physiol., 69:1326.

Lizada, M. C. C., and Yang, S. F., 1979, A simple and sensitive assay for 1-aminocyclopropane-1-carboxylic acid, Anal. Biochem., 100:140.

Mansour, R., Latché, A., Vaillant, V., Pech, J.-C., and Reid, M. S., 1986, Metabolism of 1-amino-cyclopropane-1-carboxylic acid in ripening apple fruits, Physiol. Plant., 66:495.

Matern, U., Feser, C., and Heller, W., 1984, N-malonyl-transferases from peanut, Arch. Biochem. Biophys., 235:218.

Matern, U., Heller, W., and Himmelspach, K., 1983, Conformational changes of apigenin, 7-O-(6-O-malonylglucoside), a vacuolar pigment from parsley, with solvent composition and proton concentration, Eur. J. Biochem., 133:439.

Mayne, R. G., and Kende, H., 1986, Ethylene biosynthesis in isolated vacuoles of Vicia faba L. Requirement for membrane integrity, Planta, 167:159.

Rataboul, P., Alibert, G., Boller, T., and Boudet, A., 1985, Intracellular transport and vacuolar accumulation of o-coumaric acid glucoside in Melilotus alba mesophyll cell protoplasts, Biochim. Biophys. Acta, 816:2536.

Satoh, S., and Esashi, Y., 1984, Identification and content of MACC in germinating cocklebur seeds, Plant Cell Physiol., 25:1277.

Vinkler, C., and Apelbum, A., 1983, Ethylene formation from 1-amino-cyclopropane-1-carboxylic acid in plant mitochondria, FEBS Letters, 162:

THE VACUOLE : POSSIBLE ROLE IN SIGNAL TRANSDUCTION

VERSUS CYTOPLASMIC HOMEOSTASIS ?

Alain M. Boudet, Gilbert Alibert, Gérard Marigo
and Raoul Ranjeva

Centre de Physiologie Végétale, Université Paul Sabatier
Unité Associée au C.N.R.S. n° 241
118, Route de Narbonne, F-31.062-Toulouse-Cedex, France

INTRODUCTION

Experimental evidence demonstrates that plant vacuoles play an important role in homeostasis. They may store temporarily excesses of solutes which are released at a later stage when cytoplasmic concentrations decrease. Such a dynamic role in the accumulation and mobilization of organic compounds and mineral ions has been deduced from the measurement of solute concentrations in different compartments in vivo using non destructive methods or in vitro on isolated protoplasts and vacuoles. Yeast cells with selectively permeabilized plasma membrane have also been used for such a purpose. One of the most impressive examples of vacuole/cytoplasm reversible exchanges concern the organic acids in crassulacean metabolism. They accumulate into the vacuole during the night and are progressively released during the day depending on cytoplasmic needs. Other clear evidences support the role of vacuoles in cytoplasmic pH homeostasis (Torimitsu et al., 1984) or in the maintenance of cytoplasmic phosphate (Rebeille et al., 1983) and potassium concentrations (Pitman et al., 1981; Storey and Leigh, 1987). As in yeast, it is most likely that vacuoles ensure the stability of aminoacid pools in the cytoplasm.

The mechanisms by which the related transport systems operating in both directions at the tonoplast are controlled remain to be elucidated. These mechanisms are probably quite complex and must involve an interplay of different regulatory parameters. This can be deduced from the interesting work of Perry et al. (1987) on beet storage root. In this material, the carbohydrates stored in the vacuole play a role in the maintenance of turgor and they cannot be released if there no exogenous supply of alternative vacuolar osmotica. In this way, it is suggested that the maintenance of turgor would take priority over cytoplasmic needs that may exist for sugars and that the export of vacuolar solutes is under the control of cell water relations.

At the present time, only relatively large solute concentration changes occurring over a long term period (hours) can be appreciated by the available methods. Apart from these modifications, it can be assumed that more subtle and transient exchanges between the vacuolar and cytoplasmic compartments are likely to occur. In this paper, we would like to propose as a working hypothesis that vacuoles play the role of an internal store for second messengers or more generally metabolic effectors that may be released in response to endogenous or extracellular stimuli.

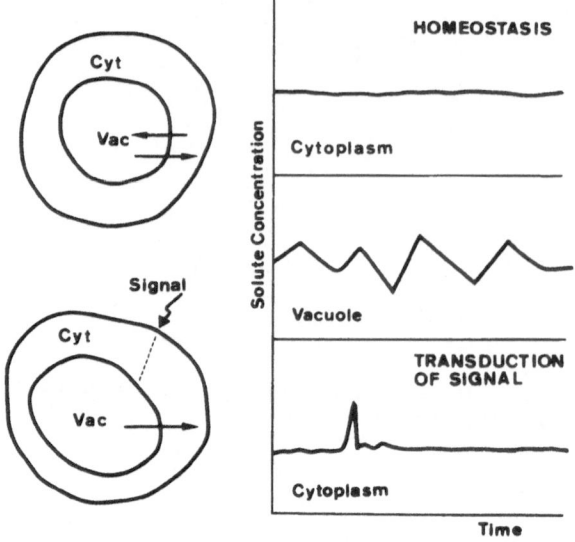

Figure 1

Vacuoles : Possible role in signal transduction
versus cytoplasmic homeostasis

In this way and in addition to the maintenance of homeostasis, vacuoles could play an important role in signal transduction (Fig. 1).

In the first part of this article, we will present two sets of experimental results illustrating the cytoplasmic control of vacuolar nitrate mobilization on one hand and the reversible phosphorylation of proteins located at the tonoplast on the other hand. In the second part of the paper, we will provide more speculative views based partly on the functioning of animal systems, on the possible release from the vacuole of messengers involved in signal transduction.

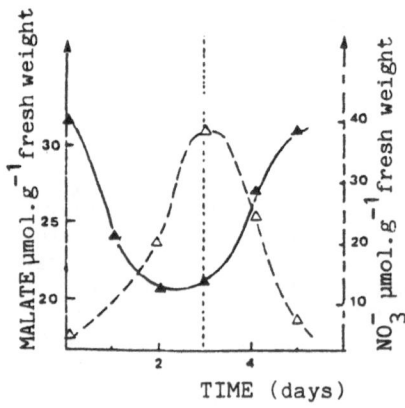

Figure 2

Time-course changes in malate (▲) and nitrate (△) concentrations
in Catharanthus roseus cells
after transfer on a fresh medium

When <u>Catharanthus</u> <u>roseus</u> cell suspension cultures are transferred to a fresh medium containing nitrate, the malate concentration of the cells drops dramatically and conversely, nitrate concentration increases. After three days, the malate level starts again to increase while the nitrate level falls (Fig. 2) (Marigo et al., 1986). The two anions which may accumulate at high concentrations (around 35 mM) are as expected essentially located in the vacuole (Marigo et al., 1985). Their opposite concentration changes support the idea, already presented by other authors, of the necessity of a balance between the vacuolar accumulation of different solutes in order to maintain the turgor pressure of the cell.

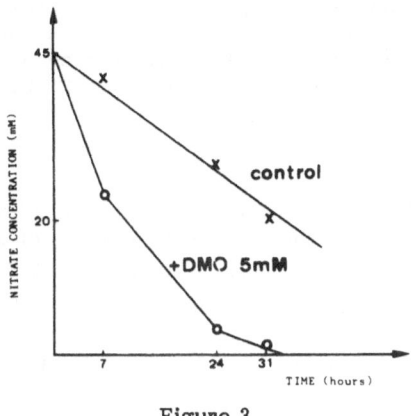

Figure 3

Effect of DMO on nitrate mobilization in <u>Catharanthus</u> <u>roseus</u> cells

DMO was added at the third day of the culture

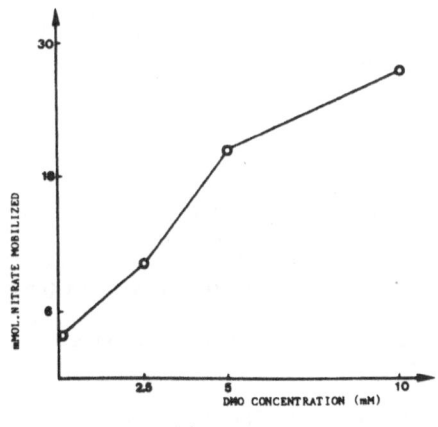

Figure 4

Effect of DMO concentration on the quantity of nitrate mobilized
within 7 hours after the addition of DMO

As the changes in cellular nitrate concentration can be roughly assimilated to changes in vacuolar nitrate concentration, this biological system appears particularly suitable to study the extent of nitrate mobilization in intact cells submitted to different experimental conditions. This involves the release of the anion from the vacuole and its subsequent reduction in the cytoplasm, a step which consumes a large amount of reducing power. We have thus applied to the cells different treatments susceptible to modify the proton cytoplasmic concentration and we have checked their possible effects on the mobilization of vacuolar nitrates. In this way, 5,5-dimethyl-oxazolidine-2,4-dione (DMO) was provided to the cells at different concentrations. This weak acid crosses the plasma membrane in a unionized form and releases protons in the cytoplasm. In presence of DMO, the nitrate mobilization occurs more rapidly than in the control (Fig. 3) and the phenomenon is directly correlated to the DMO concentration used (Fig. 4).

Additional experiments were performed with calmidazolium (R 24571 Janssen Pharmaceuticals), a calmodulin inhibitor which has been shown to inhibit a redox pump located on the plasmalemma of Catharanthus roseus cells (Belkoura, 1986). When calmidazolium was provided to the cells, the pH values of both the culture medium and the vacuole which normally decrease in the control, are stabilized during twenty hours (Table 1).

Table 1

Effects of a Treatment by Calmidazolium
on the pH values of the culture medium and of the vacuoles

	External pH		Vacuolar pH	
	Control	Assay	Control	Assay
5 h	6.73	6.83	6.13	6.22
10 h	6.31	6.94	6.04	6.31
22 h	6.00	6.83	5.95	6.15

Determinations were performed on the cell sap (Marigo et al., 1985) at different times after the addition of calmidazolium at the 3rd day of culture.

This compound could inhibit in some ways proton extrusion at the plasmalemma and the tonoplast and consequently increase proton availability in the cytoplasm. In presence of calmidazolium, the rate of nitrate mobilization is increased (Fig. 5) but cytoplasmic pH values measured by ^{14}C-DMO distribution remain constant all along the duration of the treatment (results not shown).

These preliminary results have to be confirmed by further approaches but they strongly suggest that :

- the reducing power level in the cytoplasm could in part control the rate of nitrate reduction and secondarily vacuolar nitrate efflux which appears to be a passive phenomenon,
- the vacuolar mobilization of nitrates could play a part in the maintenance of cytoplasmic pH homeostasis.

It is widely accepted that nitrates inhibit the tonoplast H^+-ATPase and this property is quite consistent with a possible autoregulation mechanism of nitrate cytoplasmic concentration. The reduction of proton efflux at the tonoplast would indeed increase nitrate assimilation.

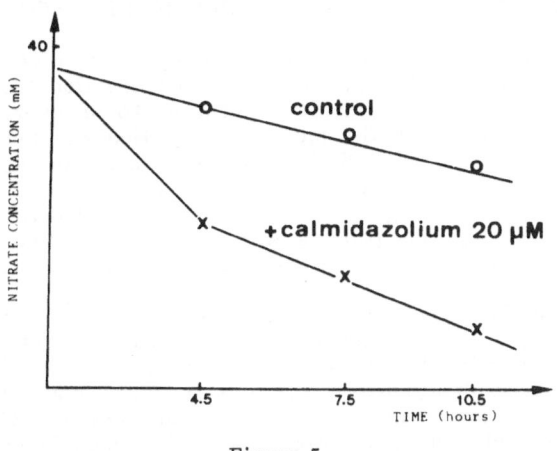

Figure 5

Effect of calmidazolium on nitrate mobilization
in Catharanthus roseus cells

Calmidazolium was added at the third day of the culture.

These overall observations performed on intact cells where the vacuoles are maintained in their normal environment, underline the complexity of the interactions occurring between the cytoplasm and the vacuole in order to maintain the delicate equilibria required for cytoplasmic homeostasis.

Other results showing the dynamic of vacuolar nitrate concentration have also been obtained in spinach leaves by Steingrover et al. (1986). These authors have observed that vacuolar nitrate concentration showed a diurnal rhythm in the leaves and they conclude that nitrate is taken up for osmotic purposes when light conditions are poor due to a lack of organic solutes.

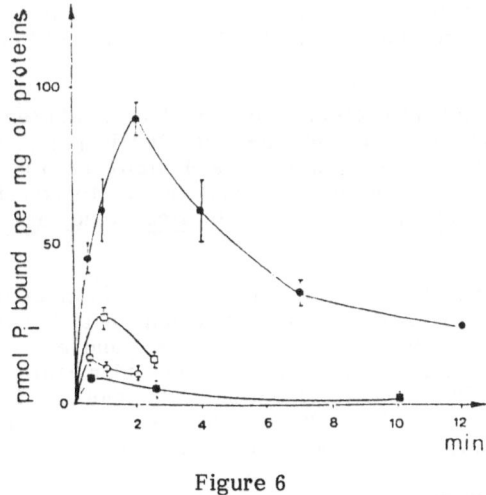

Figure 6

Time-dependent phosphorylation of Acer tonoplast proteins

$\circ$, without Ca^{2+}-CaM; $\square$, with Ca^{2+}; $\bullet$, with Ca^{2+}-CaM; $\blacksquare$, minus Ca^{2+}-CaM, plus exogeneous protein kinase. Mean $\pm$ SD of at least three determinations.

REVERSIBLE PHOSPHORYLATION OF TONOPLAST PROTEINS

Starting from <u>Acer</u> <u>pseudoplatanus</u> vacuoles (Alibert et al., 1982) we have been able to prepare a highly purified tonoplast fraction (Alibert et al., 1986) on which we have studied the capability of proteins to be reversibly phosphorylated (Teulières et al., 1985). When the membrane fraction was incubated in presence of $\gamma^{32}P$ ATP, there was a very limited phosphorylation of proteins (Fig. 6). The addition of Ca^{2+} slightly stimulated the rate of phosphorylation. However, the addition of calmodulin along with Ca^{2+} raised considerably the extent of phosphorylation (Fig. 6). This process was rapid since it took place within two minutes, after which time a spontaneous dephosphorylation occurred.

Fig. 7 shows the dependence on calmodulin of the phosphorylation process. SDS-PAGE electrophoresis of tonoplast bound proteins revealed the presence of at least eight ^{32}P-labelled polypeptides ranging from 10 to 60 KDa (results not shown). These results indicate that the tonoplast contains both endogenous protein kinases and phosphoprotein phosphatases and that several of its constitutive proteins can be reversibly phosphorylated in a Ca^{2+}/calmodulin dependent manner.

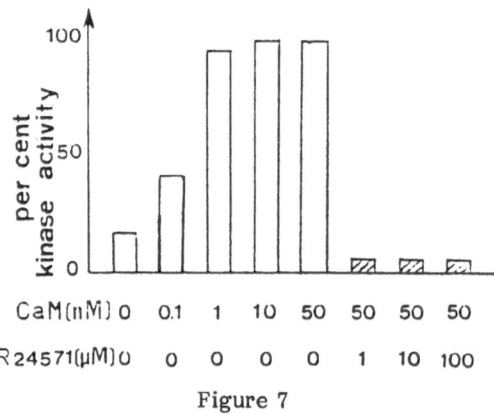

Figure 7

Effect of increasing concentration of CaM or CaM antagonist (R 24571) on the phosphorylation of the tonoplast-bound proteins

The phosphorylation of proteins in crude microsomal preparations or in purified plasmalemma fractions has already been reported in plants in the last few years (see Ranjeva and Boudet, 1987, for a review). However, this is the first evidence for the phosphorylation of proteins for a well defined tonoplast fraction. In contrast to other plant membranes, this process in <u>Acer</u> tonoplast is strictly controlled by the calcium calmodulin complex.

In relation with this CaM dependence, it is of interest to mention the results of Dauwalder et al. (1986) on the immunocytochemical localization of calmodulin in pea seedlings. In epidermal cells of the stem and all along the underside of the leaf, there was an intense staining of the vacuolar contents. Guard cells lacked this vacuolar stain. This specific location of calmodulin is rather intriguing. It seems unlikely that calmodulin itself could be reversibly released from the vacuoles but it can be hypothetized that vacuolar calmodulin could facilitate the formation of an easily exchangeable calcium pool inside the vacuole. As epidermal cells are located at the interface between the plant and its environment, this would be consistent with a possible role of vacuolar calcium in signal transduction. In the same context, it has to be stressed that Ca^{2+} concentration estimated by X-ray microanalysis is much more important in epidermal vacuoles than in mesophyll vacuoles from barley leaves (Storey and Leigh, 1987).

The functional role of the phosphorylated proteins on tonoplast needs to be elucidated. However, the exciting work of Zocchi (1985) has provided experimental evidence suggesting the control of the H^+-ATPase activity through Ca^{2+}/calmodulin dependent reversible phosphorylation. Interestingly, the inhibitory effect of phosphorylation is greater on the NO_3^--sensitive H^+-ATPase activity than the vanadate-sensitive activity. Thus, the H^+-ATPase on the tonoplast seems mainly concerned by this mechanism and would be a possible substrate for the tonoplast located protein kinases.

More generally, the phosphorylation of membrane proteins might induce dramatic changes in electrostatic charges and subsequently in the permeability or in the functions of transport systems or membrane-bound enzymes. In animal cells, one can observe that reversible phosphorylation seems to be a molecular mechanism regulating the functioning of ion channels (Levitan, 1985) and it may be anticipated that such modifications of tonoplast proteins are involved in the opening or closing of channels or pores on this membrane.

VACUOLES AS POTENTIAL INTERNAL STORES FOR SECOND MESSENGERS OR REGULATORY EFFECTORS OF METABOLISM

In animal systems, the coupling of a primary stimulus to a cellular response may involve the following sequence of discrete steps :

1 - Perception of the stimulus at the membrane level;

2 - Activation of membrane activities including opening of channels, production of "second messengers" by hydrolysis of membrane components, namely phosphoinositides or stimulation of enzymes such as adenylate kinase;

3 - Increase in the cytoplasmic concentration of second messengers (inositol-phosphate and diacylglycerol, cAMP, calcium);

4 - Activation of second messenger dependent enzymes, especially protein kinases which may phosphorylate different protein substrates.

Steps 2 and 3 are involved in what is called the transduction of the stimulus or the signal that is to say its conversion into a biochemical information easy to use by the cell.

In addition to calcium which is now recognized as a second messenger in plants (Hepler, 1985) different vacuolar located solutes are susceptible of modifying cellular metabolism when released from the vacuoles. These include for example ions such as H^+ and P_i or hormone precursors such as ACC. Conversely, a stimulated vacuolar uptake of these cytoplasmic effectors might also induce metabolic changes.

In following paragraphs we will discuss such a potential involvement of vacuoles in relation to phosphate, proton and calcium concentration changes.

PHOSPHATE

Most of the results obtained by ^{31}P-NMR spectroscopy which allows simultaneous measurement of cytoplasmic and vacuolar phosphate pools agree to suggest that a controlled flux of P_i to and from the vacuole (which contains much of the cellular P_i) maintains constant the P_i cytoplasmic concentrations. This has been for example elegantly demonstrated in sycamore cells (Rebeille et al., 1983). However, the results obtained by Strasser et al. (1983) show some modifications of cytoplasmic phosphate concentration in response to an external stimulus. Addition to parsley cell suspension cultures of an elicitor preparation which is not toxic to the cells led to the temporary increase of vacuolar phosphate at the expense of cytoplasmic phosphate. The elicitor-induced synthesis of coumarinic phytoalexins in the parsley cells was ascribed at least in part to the temporarily decreased cytoplasmic phosphate concentration. Even though the mechanisms responsible for the phosphate redistribution are unknown, these results provide a very interesting example of the potential role of cytoplasm/vacuole solutes exchange in stimulus response coupling. Orthophosphate plays a central role in cell functioning and regulates the activity of many reactions and metabolic sequences. For example, it is clear that even very modest changes in cytosolic P_i can rapidly and deeply affect some of the earliest events in photosynthesis (Walker and Sivak, 1986). Alterations of P_i movements between the vacuole and the cytosol that normally ensure that cytosolic P_i is maintained within carefully defined limits may thus considerably modify cell metabolism and be involved in the transduction of signals.

PROTONS

The value of cytoplasmic pH is carefully controlled in plants through different mechanisms which have been extensively reviewed (Smith and Raven, 1979). Evidence for a role of the vacuole in buffering cytoplasmic pH changes was provided by ^{31}P-NMR studies of cells incubated in buffers of different pH values (Torimitsu et al., 1984). However, despite the interest of this non invasive method for pH measurement in intact cells, the estimation of pH changes over short-term period (interior to half an hour) is difficult.

Recently, transient cytoplasmic pH changes (over 2 to 3 minutes) were reported by Steigner et al. (1986) on the green alga Eremosphaera viridis. Using pH-sensitive microelectrodes, they have shown that small and reversible changes in cytoplasmic pH (0.2 units) may occur in response to light/dark transitions. Light-off induces a rapid acidification and light-on a rapid alkalinisation. Such a result emphasizes the relative flexibility of cytoplasmic pH in plants. Animal cells may utilise cytoplasmic pH as a regulator of cellular functions, particularly those that are essential for the onset of cell proliferation and development (Swann and Whitaker, 1985; Moolenaar, 1986). For example, in human fibroblasts the epidermal growth factor induces a small (0.2 units) and rapid alkalinisation of cytoplasmic pH by stimulation of the Na^+/H^+ exchanges at the plasmalemma. This small but persistent alkaline pH shift may accelerate such diverse pH sensitive processes as glycolysis, protein synthesis and cytoskeletal organization. It is then proposed that the Na^+/H^+ exchanger may function as a transmembrane signal transducer involved in different biological processes such as sea urchin egg activation, increased metabolism and cell proliferation in response to growth factors.

In plant tissues, similar involvement of cytoplasmic pH changes in signal transduction have been reported by Chrestin et al. (1984) in laticiferous cells of Hevea. Treatment of Hevea bark with ethrel increased the activity of lutoid H^+-ATPase (Chrestin et al., 1987). This activation of ATPase induced a marked cytosolic alkalinisation resulting in the activation of metabolism in laticifers. Thus, in addition to modifications at the plasmalemma, one can observe changes in the activity of tonoplast located transport systems in response to external stimuli which might be more generally involved in the transduction of external signals through cytoplasmic pH shifts.

CALCIUM

As far as signal transduction is concerned, it seems that plasma membrane is the most likely candidate for the perception of external stimuli on membrane receptors and the subsequent release of second messengers. However, in the case of calcium it clearly appears at least in animal cells that internal compartments such as endoplasmic reticulum play an important part in the supply of second messengers to the cytoplasm. It is now widely admitted in plants as in animal cells that cytoplasmic Ca^{2+} concentration is maintained at a lower level (around 10^{-7} M) by active extrusion mechanisms occurring at the plasmalemma and on the membranes of Ca^{2+}-sequestering organelles. In plants, in addition to the plasma membrane transport system that actively pumps Ca^{2+} out of the cytosol, large quantities of Ca^{2+} are sequestered into the vacuole. A Ca^{2+}/H^+ antiport system has been characterized on the tonoplast by different groups (Schumaker and Sze, 1985; Blumwald and Poole, 1986). This transport system is very active and has been assumed to be one very important way in which vacuoles function in Ca^{2+} homeostasis in the cytoplasm of plant cells. However Ca^{2+} is not only stored in vacuoles but can be released (Schumaker and Sze, 1985). Evidence has been provided supporting the view that the Ca^{2+}/H^+ antiporter is reversible and Ca^{2+} channels are likely to occur on the tonoplast. Thus the possibility is open that vacuoles may release Ca^{2+} ions to the cytoplasm in response to environmental or endogenous stimuli. In animal cells, inositol-triphosphate released from the plasma membrane in response to external signals reacts with a calcium channel on the endoplasmic reticulum and induces the release of Ca^{2+} into the cytosol. As we have characterized inositol-triphosphate in plants (unpublished) a similar mechanism could occur at the tonoplast and vacuole is a possible candidate in signal transduction events involving Ca^{2+} as a second messenger (Fig. 8). Such a proposition provides a background in agreement with the hypothesis of Pickard (1984) who postulated that the elicitation of ethylene synthesis by auxin involves the opening of Ca^{2+} channels on the tonoplast by the hormone. The subsequent release of Ca^{2+} into the cytosol would promote a sequence of events leading to the stimulation of ACC synthesis.

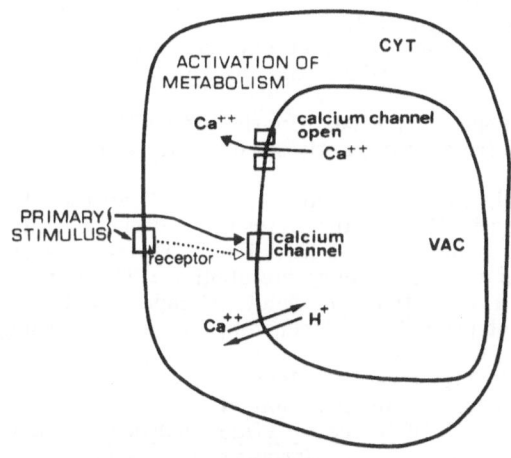

Figure 8

Hypothetical scheme on the role of vacuoles
in cytoplasmic Ca^{2+} homeostasis and signal transduction
through the release of Ca^{2+}

CONCLUSION

There is certainly a long way before generalizing the idea that vacuoles release messages involved in signal transduction. The main problem in this area remains the simultaneous in vivo measurement of effector concentrations both in cytoplasmic and vacuolar compartment during short term experiments. The experimental tools available remain limited to solve this complex problem (Thellier, 1984). NMR has been used for the estimation of ^{31}P and this method should be improved and extended to other ions in the future. Nevertheless indirect evidence based on the study of transport systems at the tonoplast and individual changes of solutes in specific compartments would extend our knowledge of these putative interrelations.

The perception at the tonoplast of a signal or a messenger followed by a vacuolar efflux of potential second messengers would represent a convenient means for regulating cytoplasmic events in plant cells. Such a cascade of events would be consistent with the observed pleiotropic effects caused by different stimuli such as hormones, light (phytochrome), and elicitors.

ACKNOWLEDGEMENTS

The authors wish to thank the students which have participated in some of the experimental work reported in this article and particularly Chantal Teulières and Véronique Atteia. The help of Annie Boudet in preparing the drawings was very much appreciated.

REFERENCES

Alibert, G., Carrasco, A., and Boudet, A. M., 1982, Changes in biochemical compotion of vacuoles isolated from Acer pseudoplatanus L. during cell culture, Biochim. Biophys. Acta, 721:22.

Alibert, G., Carrasco, A., and Citharel, B., 1986, Biochemical characteristics of tonoplast preparation isolated from Acer pseudoplatanus cell suspension cultures, Physiol. Vég., 24:85.

Belkoura, M., 1986, Système redox et extrusion de protons au niveau du plasmalemme des cellules de Catharanthus roseus L. G. Don., Ph. D. Thesis, University of Toulouse,

Blumwald, E., and Poole, R., 1986, Kinetics of Ca^{2+}/H^{+} antiport in isolated tonoplast vesicles from storage tissue of Beta vulgaris L., Plant Physiol., 80: 727.

Chrestin, H., Gidrol, X., and Marin, B., 1987, Transtonoplast ΔpH changes as a signal for activation of the cytosolic metabolism in Hevea latex cells in response to treatments with Ethrel (an ethylene releaser). Implications for the production of a secondary metabolite : The natural rubber, in : "Plant Vacuoles. Their Importance in Solute Compartmentation and Their Applications in Biotechnology", B. Marin, ed., Plenum Publishing Corporation, New-York.

Chrestin, H., Gidrol, X., Marin, B., Jacob, J. L., and D'Auzac, J., 1984, Role of the lutoidic tonoplast in the control of the cytosolic homeostasis within the laticiferous cells of Hevea, Z. Pflanzenphysiol., 114:269.

Dauwalder, M., Roux, S. J., and Hardison, L., 1986, Distribution of calmodulin in pea seedlings : Immunocytological localization in plumules and root spices, Planta, in press.

Hepler, P. K., and Wayne, R. O., 1985, Calcium and plant development, Annu. Rev. Plant Physiol., 36:397.

Levitan, I. B., 1985, Phosphorylation of ion channels, J. Membrane Biol., 87:117.

Marigo, G., Bouyssou, H., and Belkoura, M., 1985, Vacuolar efflux of malate and its influence on nitrate accumulation in Catharanthus roseus cells, Plant Science, 39:97.

Marigo, G., Bouyssou, H., and Boudet, A. M., 1986, Accumulation des ions nitrate et malate dans les cellules de Catharanthus roseus et incidence sur le pH vacuolaire, Physiol. Vég., 24:15.

Moolenaar, W. H., 1986, Regulation of cytoplasmic pH by Na^{+}/H^{+} exchange, Trends Biochem. Sci., 11:141.

Perry, C. A., Leigh, R. A., Deri Tomos, A., and Hall, J. L., 1987, Osmotic factors affecting the mobilization of sucrose from vacuoles in red beet storage tissues, in : "Plant Vacuoles. Their Importance in Solute Compartmentation and Their Applications in Biotechnology", Plenum Publishing Corporation, New-York.

Pickard, B. G., 1984, Voltage transients ellicited by sudden step-up of auxin, Plant Cell Environ., 7:771.

Pitman, M. G., Läuchli, A., and Stelzer, R., 1981, Ion distribution in roots of barley seedlings measured by electron probe X-ray microanalysis, Plant Physiol., 68:673.

Ranjeva, R., and Boudet, A. M., 1987, Phosphorylation of proteins in plants : From metabolic regulation to potential involvement in stimulus/response coupling, Annu. Rev. Plant Physiol., 38, in press.

Rebeille, F., Bligny, R., Martin, J. B., and Douce, R., 1983, Relationship between the cytoplasm and the vacuole phosphate pool in Acer pseudoplatanus cells, Arch. Biochem. Biophys., 225:143.

Schumaker, K. S., and Sze, H., 1985, A Ca^{2+}/H^+ antiport system driven by the proton electrochemical gradient of a tonoplast H^+-ATPase from oat roots, Plant Physiol., 79:1111.

Smith, F. A., and Raven, J. A., 1979, Intracellular pH and its regulation, Annu. Rev. Plant Physiol., 30:289.

Steigner, W., Kohler, K., Urbach, W., and Wilhelm, S., 1986, Transient cytoplasmic pH changes after light-off and light-on in Eremosphaera viridis, Plant Physiol. Suppl., 80:144.

Steingröver, E., Ratering, P., and Siesling, J., 1986, Daily changes in uptake, reduction and storage of nitrate in spinach grown at low light intensity, Physiol. Plant., 66:550.

Storey, R., and Leigh, R. A., 1986, Distribution of potassium between vacuole and cytoplasm in response to potassium deficiency, in : "Plant Vacuoles. Their Importance in Solute Compartmentation and Their Applications in Biotechnology", B. Marin, ed., Plenum Publishing Corporation, New-York.

Strasser, H., Tietjen, K. G., Himmelspach, K., and Mattern, U., 1983, Rapid effect of an elicitor on uptake and intracellular distribution of phosphate in cultured parsley cells, Plant Cell Reports, 2:140.

Swann, K., and Whitaker, M., 1985, Stimulation of the Na^+/H^+ exchanger of sea urchin eggs by phorbol ester, Nature, 314:274.

Teulières, C., Alibert, G., and Ranjeva, R., 1985, Reversible phosphorylation of tonoplast proteins involves tonoplast bound calcium-calmodulin-dependent protein kinase(s) and protein phosphatase(s), Plant Cell Reports, 4:199.

Thellier, M., 1984, From classical to modern methods for element localization in plant systems, Physiol. Vég., 22:867.

Torimitsu, K., Yazaki, Y., Nagasuka, K., Ohta, E., and Sakata, M., 1984, Effect of external pH on the cytoplasmic and vacuolar pHs in mung bean root tip cells : A ^{31}P nuclear magnetic resonance study, Plant Cell Physiol., 25:1403.

Walker, D. A., and Sirvak, M. N., 1986, Photosynthesis and phosphate : A cellular affair ? Trends Biochem Sci., 11:176.

Zocchi, G., 1985, Phosphorylation/dephosphorylation of membrane proteins controls the microsomal proton-translocating ATPase activity of corn Zea mays roots, Plant Science Letters, 40:153.

TRANSTONOPLAST   pH CHANGES AS A SIGNAL FOR ACTIVATION OF THE
CYTOSOLIC METABOLISM IN HEVEA LATEX CELLS IN RESPONSE TO
TREATMENTS WITH ETHREL (AN ETHYLENE RELEASER). IMPLICATIONS FOR
THE PRODUCTION OF A SECONDARY METABOLITE : NATURAL RUBBER

Hervé Chréstin, Xavier Gidrol and Bernard Marin

Laboratoire de Physiologie Végétale, Centre ORSTOM
d'Adiopodoumé, Boite Postale N° V-51, Abidjan, Ivory Coast

The latex of Hevea brasiliensis is a specialized true fluid cytoplasm which
is expelled from wounded latex vessels (D'Auzac et al., 1982). It is collected indus-
trially for its high content of a secondary metabolite of high economic interest
: Natural rubber.

The application of Ethrel (an ethylene releaser) or auxin-like substances
to the bark of rubber trees induces a sharp, though transient, increase in latex
latex production (D'Auzac and Ribaillier, 1969). The main known effect of ethylene
was shown to consist of a significant alkalinization of latex cytosol. This increase
in the cytosolic pH occurs some 24 hours after treatment and is parallel with increase
in the production of latex and hence of rubber (Fig. 1) (Tupy, 1973; Coupé et al.,
1976; Coupé, 1977; Brzozowska-Hanover et al., 1979).

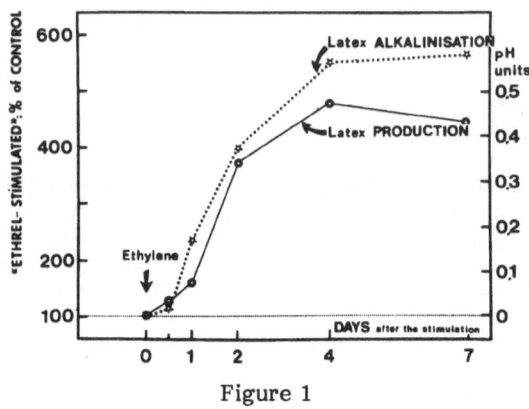

Figure 1

Kinetics of effect of treatment of Hevea bark with Ethrel
on the production of natural rubber
and on the pH changes (alkalinization) of the latex cytosol

467

Laticiferous cells contain a vacuolar compartment : the so-called lutoids which can easily so isolated intact and purified by simple centrifugation (D'Auzac et al., 1982).

We reported (see Chrestin et al., 1987) that highly significant inverse relationships have been demonstrated between the pH of the cytosol and of the vacuolar compartment, as well as between the vacuolar pH and the production of rubber. Accordingly direct correlation was demonstrated linking the transtonoplastic pH gradient and the production of rubber. Moreover, significant acidification of the intravacuolar fluid was shown to be a response to treatment of Hevea with Ethrel (Coupé, 1977; Coupé and Lambert, 1977; Brzozowska-Hanower et al., 1979; Chrestin, 1985).

These findings and correlations could be fully explained by demonstration of the functioning of two opposing H$^+$ pumps located on the lutoidic tonoplast :

- one is an Mg-dependent ATPase which catalyses the electrogenic influx of protons in the vacuolar compartment (vacuolar acidification) and the alkalinization of the cytosol (Marin et al., 1981; Marin, 1982; Crétin, 1982; Chréstin, 1985; Marin et al., 1986; Chréstin et al., this issue).

- the second is a tonoplast redox system which consumes NADH and cytochrome c (used as artificial acceptor). When assayed with very freshly isolated intact lutoids, its functioning induced an electrogenic efflux of protons from the vacuolar space into the cytosol, causing vacuolar alkalinization and concomitant acidification of the cytosol (Crétin, 1983; Chréstin, 1985; Chréstin et al., 1987).

The present paper demonstrates the functioning of the two opposing proton-pumping systems as a true tonoplastic biophysical pH-stat, and shows some conditions of the maintenance of a high transtonoplastic $\Delta\overline{\mu}_H$+ in quasi in vivo conditions.
The pyrophosphatase activity found in Hevea latex was demonstrated to be soluble and exclusively located in the cytoplasmic compartment (Jacob and Marin, unpublished data). This activity do not participate at all in any transfer of protons across the tonoplast. Such results contrast with those described in the literature concerning the tonoplast of the different higher plants studied until today.

It especially focuses on the activation of the ATPase moiety of this tonoplastic pH-stat through increase of the ATP content and of potential tonoplastic ATPase activity in the latex after treatment of rubber trees with Ethrel (an ethylene releaser).

THE TWO TONOPLASTIC OPPOSING H$^+$-PUMPS WORK AS A BIOPHYSICAL pH-STAT IN QUASI IN VIVO CONDITIONS

Plotting the activity of the two H$^+$-pumps as a function of the pH of the medium (buffered deproteinized latex cytosol obtained by ultrafiltration on PM-10 membrane) clearly shows that the ATPase remains at its maximal potential activity over the physiological pH range (6.5 to 7.3: Brzozowska-Hanower et al., 1979) while the tonoplast e$^-$ transport system, being more pH sensitive in the same pH range, becomes more efficient at a slightly alkaline pH (Fig. 2). This suggests that an excess alkalinization of latex cell cytosol, possibly caused by excessive working of the H$^+$-pumping ATPase, will be efficiently counteracted by the activation of the redox chain-dependent efflux of protons from the lutoids. Cytosolic pH will then theoretically tend to stabilize itself in the pH range comprised between the optimum pH of each proton pump, insofar as their respective substrate availability is not limiting in the latex (cf. different papers referenced in Marin, 1985; Chrestin, 1985).

We therefore conclude that this system consists of a typical biophysical (bio–osmotic) pH–stat, based on controlled (energy–dependent) transtonoplastic fluxes of free H^+. Its operation will actively participate in the fine regulation of cytosolic pH, and be closely involved in the control of the cytosolic metabolism of latex (cf. Marin, 1985).

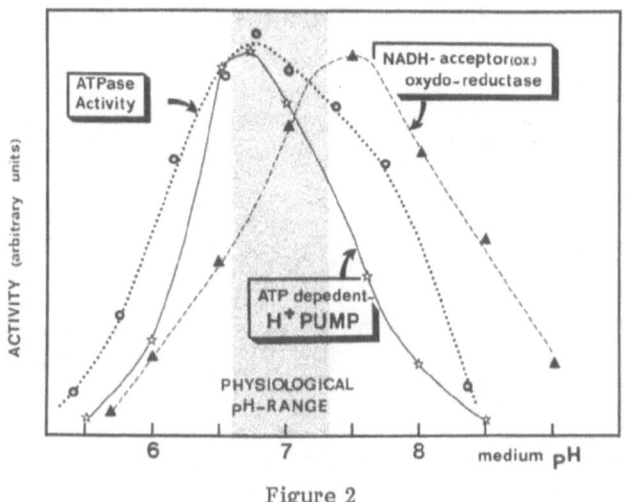

Figure 2

Dependence on pH of tonoplastic ATPase
and NADH–cytochrome c oxido–reductase activities,
and of their proton pumping efficiency,
measured in buffered ultrafiltered cytosol from Hevea latex

OBLIGATORY CONTINUOUS SUPPLY OF ATP TO MAINTAIN
HIGH TRANSTONOPLASTIC ELECTROCHEMICAL PROTON GRADIENT

When fresh lutoids were preincubated with ^{14}C–methylamine in the presence of glucose (10 mM) they also accumulated the pH probe (as in the absence of glucose Crétin, 1982), according to their initial transtonoplastic proton gradient (Fig. 3–A). The addition of MgATP (2.5 mM) also induced rapid further influx of methylamine indicating a marked vacuolar acidification caused by the well–demonstrated functioning of the tonoplastic H^+–pumping ATPase. When, after 30 min incubation, the same lutoid suspension was provided once more with an extra but limited (1.5 mM) amount of MgATP (without hexogenous hexokinase : circles and dotted lines in Fig. 3–A), it could be seen that the lutoids maintained, and even slightly amplified, their ATP–dependent transtonoplastic pH gradient. Fig. 3–A (triangles and dotted line) reports a similar experiment, where the second addition of MgATP was accompanied by a addition of a saturating amount of exogenous hexokinase (in presence of glucose 10 mM) in order to use and then eliminate as rapidly as possible all the ATP present in the medium.

It was observed that under these conditions (lack of ATP), the lutoids progressively lost the pool of free protons they had previously accumulated during the functioning of the tonoplastic H_+–pumping ATPase when ATP was still available in the medium (i.e. before addition of hexokinase).

As addition of the hexokinase reaction product, glucose-6-P, did not lead to efflux of protons (not shown), the pool of free electrogenic protons accumulated in the vacuolar compartment (against the thermodynamic equilibrium) during the functioning of the H^+-pumping ATPase, clearly tends to dissipate with time if insufficient ATP is available.

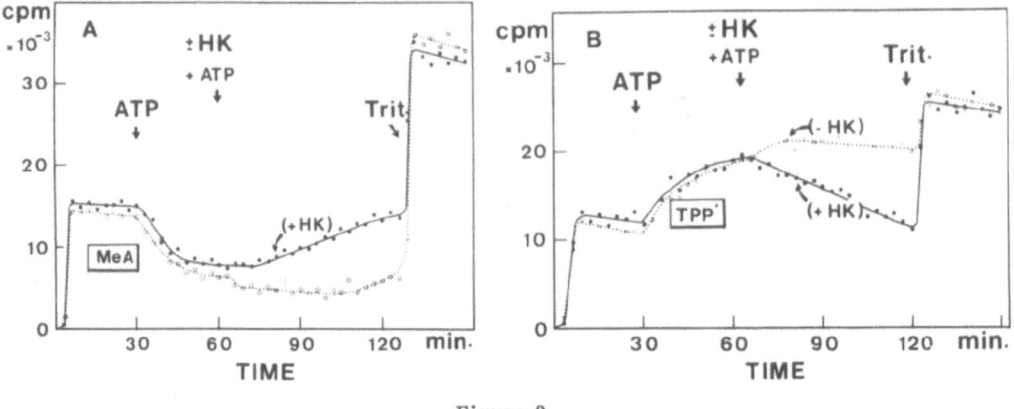

Figure 3

Efflux of protons and tonoplast depolarization of intact lutoids
in a medium without ATP
after its complete consumption by hexokinase plus glucose

The assays were performed with a fresh lutoid suspension using flow dialysis (Crétin, 1982) at pH 7.1, in the presence of glucose (10 mM). In A: Fluxes of ^{14}C-methylamine (using as pH probe) in the presence of MgATP alone (2.5 mM), then MgATP (1.5 mM) in the presence (full symboles and lines) or not (open symboles and dotted lines) of saturating amounts of hexokinase. In B : as in A but fluxes of ^{14}C-TPP$^+$ were measured to monitor the transmembrane potential changes.

Fig. 3-B demonstrates that the dissipation of the transtonoplastic gradient of free protons in the absence of ATP is accompanied by polarization of the membrane (interior more negative), as shown by the concomitant influx of the potential probe TPP$^+$ in the vacuoles when hexokinase was added.

We conclude that high transtonoplastic electrochemical proton gradient, then intravacuolar acidity and hence cytosolic (external medium) pH are actively controlled by a continuous, energy-consuming H^+-pump. This leads to proposing "kinetic control" of the H^+-pumping ATPase, essentially by the availability of its own substrate.

INCREASE OF THE TRANSTONOPLASTIC ΔpH
AND ACTIVATION OF THE LUTOIDIC H^+-PUMPING ATPase IN HEVEA
BY TREATMENT OF BARK WITH ETHREL

Compartmental pH Changes Induced by Treatment With Ethrel

The "stimulation" of rubber production by treatment of Hevea bark with Ethrel (ethylene releaser), is known to induce a rapid alkalinization of the latex cytosol, and a more or less delayed acidification of the vacuolar compartment (Coupé, 1977; Coupé and Lambert, 1977; Brzozowska-Hanower et al., 1979; Chréstin, 1985). We attempted to see if the functioning of the tonoplastic proton pumps were involved in these compartmental pH changes.

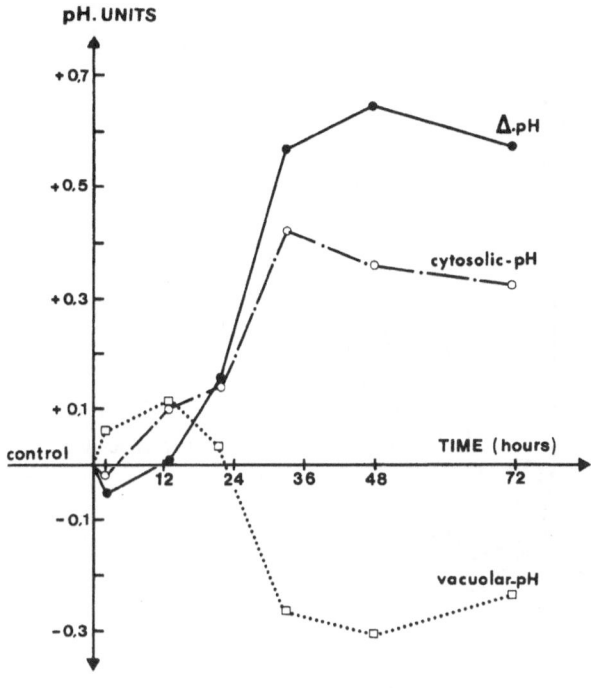

Figure 4

Kinetics of the effects of Ethrel
on the cytosolic and vacuolar pH,
and on the resulting transtonoplastic ΔpH in Hevea latex

The results are expressed as mean differences in latex characteristics between
Ethrel-treated trees and control (basal line).

Fig. 4 shows some kinetic aspects of the effects of Ethrel on the cytosolic
and lutoidic (vacuolar) pH changes in the latex vessels. It can be seen that the
response to Ehtrel treatment could be subdivided into two distinct phases :

– an initial stage (lasting about 21 hours), characterized by slow, slight
alkalinization of the cytosol and a transient, slight rise in the intravacuolar pH
(maximum after 13 hours). During this early stage, responses of the pH of the two
compartments varied in the same way and with similar amplitude. The transtonoplastic pH gradient (ΔpH) then remained constant for at least 13 hours.

– a delayed wide response with simultaneous alkalinization of the cytosol
and acidification of the lutoids (vacuoles). The pH of the cytosol increased by 0.42
unit compared with control (maximum after 33 hours) and then remained 0.3 unit
higher for more than 2 days. At the same time, the intravacuolar pH decreased
to 0.2 to 0.3 unit below control (Fig. 4 and 5).

The resulting transtonoplastic ΔpH increased by 0.57 pH unit (33 h) and then
by 0.55 pH unit against the control. Although the pH changes of the two cellular
compartments moved in symmetrically opposite ways, the cytosolic pH was shown
to be more affected than the intravacuolar pH. This could be explained by the greater
buffering capacity of the intravacuolar fluid (Chréstin, 1985).

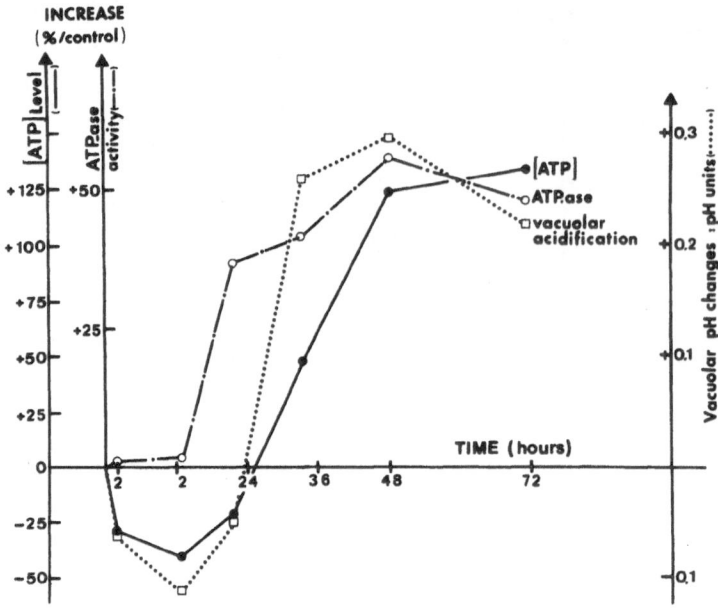

Figure 5

Kinetics of the effect of Ethrel
on the potential tonoplastic ATPase activity
and cytosolic ATP level, expressed in % variation compared with control
and on vacuolar acidification in latex from Ethrel-treated trees

Base line = Control

Increase in the Tonoplastic ATPase Activity

It can be seen in Fig. 5 that Ethrel induced a sudden increase in the tonoplastic ATPase activity in vitro 21 hours after treatment. The specific activity of the H^+-pumping ATPase measured on intact lutoids after 21 hours was shown to increase by 37% compared with the control, and then by 55 to 60% after 48 hours.

The specific ATPase activity measured on purified tonoplast, and in optimal conditions of pH and substrates (Fig. 6-A) was shown to be four times higher in membranes from Ethrel-treated trees than in membranes of the control.

Kinetic studies (Fig. 6-B) indicate that the K_m of the enzyme for MgATP remained unchanged whereas increase in V_{max} could be interpreted as an increase in the number of catalytic sites on the tonoplast, probably owing to activation of the de novo synthesis of the enzyme (Gidrol, 1984; Chréstin, 1985).

Increase in ATP and Total Adenine Nucleotides Pools in the Latex Cytosol from Ethrel-treated Trees

Fig. 5 shows that as early as 2 hours after treatment with Ethrel, the ATP content of latex cytosol from "stimulated" trees fell by 30% below control, then fell again by 47% after 13 hours. Following this transient decrease, the ATP content increased rapidly and exceeded control level about 24 hours after treatment. Later, the ATP content in the latex from "stimulated" trees reached values 225% (after 33 h) and then 240% (after 71 h), of that of control.

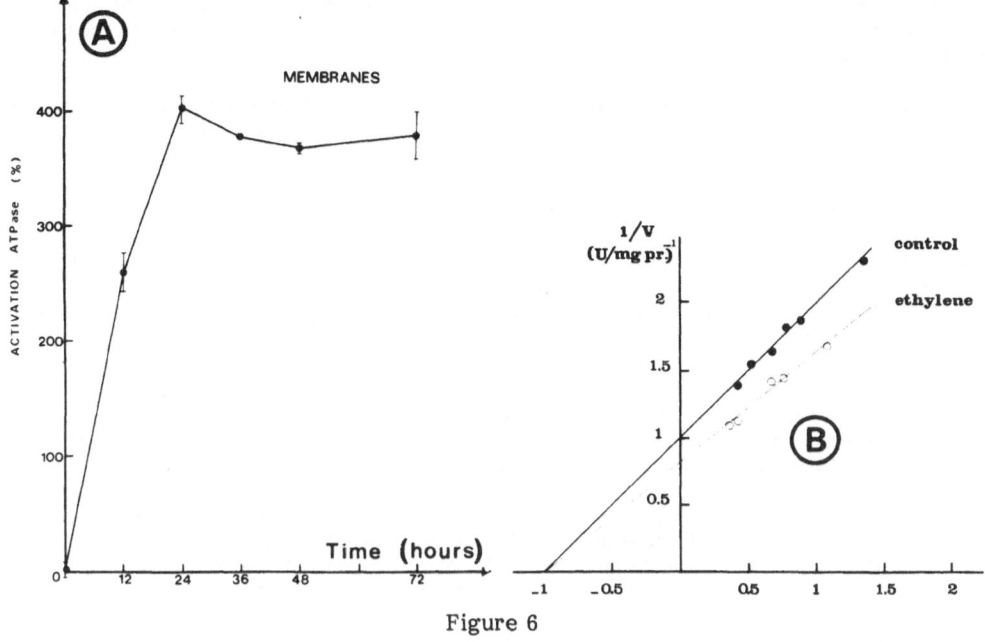

Figure 6

Effects of Ethrel treatment
on the potential activity and on the kinetic parameters
of the tonoplastic ATPase as measured on the purified tonoplast

In A: Time-course of the Ethrel treatment on the potential ATPase activity measured on purified tonoplast; in B : Effect of the Ethrel treatment on the K_m and V_{max} of tonoplastic ATPase (Ethrel : circles; control : black dots).

In addition, it was found that the total adenine nucleotides pool (ATP + ADP + AMP) in fact closely followed the kinetic variations of the ATP content alone, resulting in scarcely detectable changes in the energy charge and ATP:ADP ratio (Chréstin, 1985) in the latex cytosol. This led us to conclude that de novo synthesis of adenine nucleotides was induced by ethylene (at least 24 h after the treatment), with rebalancing of the energy charge, possibly through adenylate kinase and metabolic activities.

It can be seen from Fig. 5 that the intravacuolar pH changes closely followed the variations in ATP content in the latex. Lower concentrations of ATP were associated with a less acid vacuolar pH, and higher ATP concentrations were associated with considerable acidification of the vacuolar compartment. These data clearly show that the early, marked activation of the tonoplastic proton-pumping ATPase (between 13 and 21 h) did not immediately result in any expected vacuolar acidification. This might be attributed to the transient relative lack in availability of the substrate for the H^+-pumping ATPase (i.e. : ATP) as an early "side-effect" of treatment with ethylene releaser.

DISCUSSION AND CONCLUSIONS

The variations in the total adenine nucleotide pool, and especially the decrease in the ATP content are the earliest known biochemical characteristics of latex to be affected after treatment with Ethrel. Early disturbance in the adenine nucleotide pool might result from the enhancement of the utilization of triphosphate-nucleotides in the synthesis of nucleic acids and proteins. In support of such an hypothesis, Coupé et al. (1976) and Coupé (1977) reported that treatment of Hevea bark with Ethrel (or with auxin-like substances) induced a increase in the total RNA and polyribosome content in latex during the first 12 hours following treatment. It is then possible that ethylene induces a true de novo adenine nucleotide synthesis within less than 24 hours, and that the total adenine nucleotide pool will be conti-nuously rebalanced.

The striking biochemical event which follows treatment with Ethrel is the marked sharp increase in tonoplastic ATPase activity after a time-lag of 13 to 21 hours. This observed in vitro in intact washed lutoids and also in purified lutoidic tonoplast (Crétin et al., 1983; Gidrol, 1984), hence in the absence of any possible soluble cytosolic activator. This increase in specific ATPase activity was shown elsewhere to be concomitant with significant stimulation of protein synthesis at the lutoidic tonoplast (Gidrol, 1984). Thus the lutoids from "Stimulated" trees keep – at tonoplast level – an indelible print of ethylene treatment ! The fact that only V_{max} was affected, and not K_m for MgATP, supports the hypothesis that there is an increase in the number of ATPase catalytic sites on the tonoplast, ascribable to de novo ATPase synthesis, as an early biochemical event induced by ethylene.

The simultaneous opposite changes in cytosolic and vacuolar pH, together with the satisfactory stoicheiometry of the estimated transmembrane H^+ fluxes (according to the respective buffer capacity of each compartment : Chrestin, 1985), suggest some stimulation of tonoplastic proton pumps by ethylene.

The fact that intravacuolar acidification was shown to run parallel with the increase in the cytosolic ATP content leads to concluding that transtonoplastic H^+ fluxes might depend on the lutoidic ATPase activity, as controlled by the availability of ATP. The K_m of ATPase for MgATP was shown to be about 0.5 to 0.7 mM measured in physiological conditions (with ultrafiltered cytosol at pH 6.8 to 7.2) (Gidrol et al., 1985; Gidrol, 1984) while the mean cytosolic ATP content was shown to remain less than 0.25 mM in the control (i.e. : not Ethrel treated) (Chréstin, 1985).

Thus, the tonoplastic H^+-pumping ATPase always operates at far less than its maximum potential in vivo, and its real activity depends in a linear manner on the ATP content of the cytosol. The initial decrease in the ATP content in the latex (30 to 40% less than control) is then assumed to result in a decrease of the ATPase activity and cause some vacuolar alkalinization. The data showing the progressive discharge of the transtonoplastic ΔpH under ATP depletion in vitro (Fig. 3) agree with such an hypothesis.

In contrast, when increase in potential ATPase activity was followed by increase in cytosolic ATP, efficient activation of the H^+-pumping ATPase occurred in vivo. It is suggested that this accounts for the rise in the transtonoplastic ΔpH and for the alkalinization of the latex cytosol which were demonstrated as favouring active metabolism in the latex.

As a conclusion, it is established that the two opposing proton pumps located on the lutoidic tonoplast can work as a true biophysical pH-stat under in vivo conditions. It is suggested that the differential functioning of the two moïties is strictly controlled by the availability of their respective substrates. This was demonstrated in particular for the proton-pumping ATPase in vitro and in vivo. The ATPase moiety of the tonoplastic pH-stat was shown to be activated by exogenous ethylene (Ethrel) because of at least 2 early biochemical events :

- increase in the potential ATPase activity probably through de novo synthesis of the enzyme,

- and, increase in the real ATPase activity through a marked rise in the ATP content in the cytosol.

The combination of these two biochemical events induced by ethylene brings about efficient activation of the tonoplastic proton-pumping ATPase, and results in an increase in the transtonoplastic ΔpH (favouring detoxification of the cytosol) and in the concomitant alkalinization of the cytosol; both these phenomena favour the latex metabolism and hence rubber production (Chrestin et al., 1987).

These results should be regarded as an interesting model for the participation of vacuoles in the transduction of signals regulating the cytoplasmic metabolism and the production of secondary metabolites (natural rubber in this case) in higher plants.

REFERENCES

D'Auzac, J., Crétin, H., Marin, B., and Lioret, C., 1982, A plant vacuolar system : The lutoids from Hevea brasiliensis, Physiol. Vég., 20:311.

D'Auzac, J., and Ribaillier, D., 1969, L éthylène, nouvel agent stiumulant de la production de latex chez Hevea brasiliensis, C. R. Acad. Sci., Paris, Série D, 268:3046.

Brzozowska-Hanover, J., Crétin, H., Hanower, P., and Michel, M., 1979, Variations de pH entre les compartiments vacuolaire et cytoplasmique au sein du latex latex d'Hevea brasiliensis. Influence saisonnière et action du traitement par l'Ethrel, générateur d'éthylène. Répercussion sur la production et l'apparition d'encoches sèches, Physiol. Vég., 17:851.

Coupé, M., 1977, Etudes physiologiques sur le renouvellement du latex d'Hevea brasiliensis : Action de l'éthylène. Importance des polyribosomes, Thèse Doctorat Etat Sci. Nat., Montpellier.

Coupé, M., and Lambert, C., 1977, Absorption of citrate by the lutoids from the latex vessels, and rubber production by Hevea, Phytochemistry, 16:45.

Coupé, M., Lambert, C., and D'Auzac, J., 1976, Etude comparative des polyribosomes du latex d'Hevea sous l'action de l'Ethrel et d'autres produits augmentant l'écoulement du latex, Physiol. Vég., 14:391.

Chréstin, H., 1985, "La Vacuole dans l'Homéostasie et la Sénescence des Cellules Laticifères d'Hevea", Etudes et Thèses, ORSTOM, Paris.

Crétin, H., 1982, The proton gradient across the vacuo-lysosomal membrane of lutoids from the latex of Hevea brasiliensis. I. Further evidence for a proton-translocating ATPase on the vacuo-lysosomal membrane of intact lutoids, J. Membrane Biol., 65:175.

Crétin, H., 1983, Efflux transtonoplastique de protons lors du fonctionnement d'un système transporteur d'électrons (la NADH-cytochrome c réductase) membranaire des vacuo-lysosomes du latex d'Hevea brasiliensis, C. R. Acad. Sci., Paris, Série III, 296:101.

Crétin, H., Gidrol, X., and Monoury, G., 1983, Ethylene stimulation of the ATP-dependent transtonoplastic proton fluxes in Hevea cells, in : "Membrane Transport in Plants", W. J. Cram, K. Janacek, R. Rybova, and K. Sigler, eds., Academia, Publishing House of the Czechoslovak Academy of Sciences, Prague.

Gidrol, X., 1984, Caractérisation de l'ATPase tonoplastique de la cellule laticifère d'Hevea brasiliensis, Thèse Troisième Cycle, Université d'Aix-Marseille II.

Gidrol, X., Marin, B., Chréstin, H., and D'Auzac, J., 1985, The functioning of the tonoplast H^+-translocating ATPase from Hevea latex in physiological conditions, in : "Biochemistry and Function of Vacuolar Adenosine-tri-phosphatase in Fungi and Plants", B. P. Marin, ed., Springer-Verlag, Berlin, Heidelberg, New-York and Tokyo.

Marin, B., 1982, "Le Fonctionnement du Transporteur Tonoplastique du Citrate du Latex d'Hevea brasiliensis", Trav. Doc. ORSTOM, Vol. 144, ORSTOM, Paris.

Marin, B., 1985, "Biochemistry and Function of Vacuolar Adenosine-triphosphatase in Fungi and Plants", Springer-Verlag, Berlin, Heidelberg, New-York and Tokyo.

Marin, B., and Chrestin, H, |984, Citrate compartmentation in the latex from Hevea brasiliensis. Relationship with rubber production. Compte-Rendu du Colloque Hevea 84 : Exploitation. Physiologie et Amélioration de l'Hevea, I.R.C.A-G.E.R.D.A.T., Paris and Montpellier.

Marin, B., Gidrol, X., Chréstin, H., and D'Auzac, J., 1986, The tonoplast proton-translocating ATPase of higher plants as third class of proton pumps, Biochimie, 68:1263.

Marin, B., Marin-Lanza, M., and Komor, E., 1981, The proton-motive potential difference across the vacuo-lysosomal membrane of Hevea brasiliensis (rubber tree) and its modification by a membrane-bound adenosine-triphosphatase Biochem. J., 198:365.

Tupy, J., 1973, The activity of latex invertase and latex production in Hevea brasiliensis, Physiol. Vég., 11:13.

PRODUCTION OF PLANT SECONDARY METABOLITES

BY PLANT CELL CULTURES

IN RELATION TO THE SITE AND MECHANISM OF THEIR ACCUMULATION

Michael Wink

Genzentrum der Universität München
Pharmazeutische Biologie
D-8000-München 2, Federal Republic of Germany

INTRODUCTION

Higher plants are characterized by their capacity to synthesize a wide variety of organic compounds, the so-called secondary metabolites. Due to the powerful means of modern analytical chemistry, many thousands of secondary products have been described (Table 1). Since only 5 to 15% of all higher plants have been studied so far, it is safe to estimate that the real number of existing natural products is substantially larger.

The main biological function of plant secondary metabolites is that of chemical defense (Whitacker and Feeny, 1971; Levin, 1976; Swain, 1977; Harborne, 1982). They are directed against microorganisms (viruses, bacteria, fungi), other plants (allelopathy), and phytophageous animals (insects, molluscs, vertebrates). Plants also use some of these compounds to attract pollinating and seed-dispersing animals.

The biological and evolutionary background seems to be unrelated to biotechnology on the first glance. However, the structures of plant secondary metabolites have not been formed just at random, but they constitute structures that have been shaped and optimized during evolution. As a result they are biologically active and can be exploited by Man for a variety of purposes. Important applications are found as pharmaceuticals, spices, fragrances, natural colours, stimulants, halucinogens, pesticides, poisons, etc. These compounds form a reservoir for lead structures which can be further optimized by the synthetic chemist.

Since many natural products are economically important (Table 2), ways to obtain these compounds have been explored. Traditionally, the compounds are extracted from plants grown in the field. If the chemical synthesis is possible, it is usually the cheaper alternative. However, due to the complex stereochemistry of most natural products, the chemical synthesis is often not economic for many products. A third way would be biotechnology : it is possible today to establish cell cultures of nearly every plant species. The idea is to use these cell cultures for the production of valuable compounds (Zenk, 1978 and 1982), similar to the situation of fungi and bacteria employed for the production of antibiotics.

Table 1

Number of Known Plant Secondary Structures

Compounds	Number of known structures
Monoterpenes	1000
Sesquiterpenes	1500
Diterpenes	1000
Triterpenes/Steroids	800
Tetraterpenes	350
Polyketides	700
Polyacetylenes	750
Flavonoids	1200
Phenylpropanoids	500
Amines	100
Alkaloids	7000
Nonprotein amino-acids	400
Cyanogenic glycosides	50
Glucosinolates	100

Table 2

Market For Some Secondary Products
(after Curtin, 1983)

Compound	Use	Whole sale price ($)	Estimated retail market
Vinblastine/ vincristine	Leucemia	5000 / g	18-20 Mio (US)
Ajmalicine	Heart disorder	1500 / kg	5 Mio (World)
Digitalis glycosides	Heart disorder	3000 / kg	20-55 Mio (US)
Quinine	Flavour	100 / kg	25 Mio (World)
Codeine	Sedative	650 / kg	50 Mio (US)
Shikonine	Dye	4500 / kg	0.6 Mio (Japan)
Jasmine	Fragrance	5000 / kg	0.5 Mio (World)
Pyrethroids	Insecticide	300 / kg	20 Mio (US)
Spearmint	Flavour	30 / kg	90 Mio (US)
Plant drugs	Pharmaceuticals		8000 Mio (US)
	Fragrances/ Flavours		1000 Mio (World)

After nearly 30 years of research efforts into the production of secondary metabolites by plant cell cultures, some notable success has been achieved. Over 30 cell culture systems are known which produce a higher amount of a natural product as compared to the intact plant (Table 3).

Table 3

Example for Successfull Cell Culture Systems

Product	Species	Yield		
		Cell culture		Intact plant
		% DW	g/l	% DW
Shikonine	Lithospermum	23	1.5	12
Rosmarinic acid	Coleus	25	4	3
Ginsengosides	Panax	27		5
Anthraquinones	Rubia	20		1
	Morinda	18	3	2
	Cassia	6		0.6
Berberine	Coptis	13	1.4	
Jatrorrhizine	Berberis	10	2	
Alkannine	Echium	12		0.04
Shikimic acid	Galium	10		
Verbascoside	Syringa	16	1.4	
Cinnamoyl-putrescine	Nicotiana	10	1	
Nicotine	Nicotiana	5	0.9	
Glycyrrhizine	Glycyrrhiza	4		
Diosgenine	Dioscorea	4	0.05	2
Biscoclaurine	Stephania	3		0.8
Ajmalicine	Catharanthus	1	0.5	0.3
Serpentine	Catharanthus	0.8	0.2	0.5
Ubiquinone-10	Nicotiana	0.5	0.05	
Benzo-phenanthridines	Eschscholtzia	0.5	0.05	
Caffeine	Coffea	1.6		1.6

[a]Data from Zenk (1978), Berlin (1984), Constabel et al. (1981), Wink (1986)

However, probably more than 90% of all cell culture systems studied so far, failed to produce a desired compound. Unfortunately, many of the economically important products (Table 2) fall into this category. To understand the underlying difficulties and problems we have to know the basic biochemical and physiological parameters which lead to product formation in the intact plant. The production of a compound reflects an equilibrium between the rate of biosynthesis and that of accumulation and turnover. These processes are usually interrelated by spatial (compartmentation) and temporal (developmental) gene expression. In this paper emphasis is laid on the process of product accumulation and the underlying mechanisms.

SITE OF BIOSYNTHESIS AND OF ACCUMULATION

Although all cells of a plant species are theoretically competent to express all the genes of their genome, only a fraction is usually expressed in a tissue- and development-specific manner. The genes for the biosynthesis of secondary metabolites are no exception in this sense, and we find in many cases that their biosynthesis is restricted to a specific organ and/or developmental stage. For example, the roots or rhizomes are the main site of synthesis for anthraquinones, rotenoids, glycyrrhizine, berberine, jatrorrhizine, shikonine, ajmalicine, tropane alkaloids, nicotine, ginsengosides, sanguinarine, rutacridone alkaloids, betalaines, etc . (Wink, 1986). Shoots are the site for quinine, cinchonine, quinolizidine alkaloids, vindoline, cardenolides, terpenes, phylloquinones, essential oils, cyanogenic glycosides, glucosinolates, polyacetylenes, flavonoids (e.g. in flowers), and anthocyanines. Within these tissues often a further specialization can be observed in that particular cell layers or cells synthesize a secondary metabolite; e.g. the epidermis with its hairs and trichomes is often the site for anthocyanine, and terpenoid biosynthesis (for review see Molisch, 1923; Mothes, 1955; Wiermann, 1981; Wink, 1986).

If a metabolite has been found in a special plant part this does not necessarily mean, that it had also seen synthesized there. We can distinguish three possibilities respectively : i) a compound is stored in the same cell where it had been formed; ii) a metabolite is accumulated in cells adjacent to the cells of synthesis; and iii) product is stored in organs which differ totally from the plant part of its synthesis. In the latter case, long-distance transport has to be taken into account.

For a number of secondary metabolites situation "1" and "2" is certainly valid. Interestingly, many of the compounds that are formed successfully by plant cell cultures (Table 3) fall in this category (see Wink 1986). For tropane alkaloids and nicotine, case "3" is relevant (Mothes, 1955) : These alkaloids are synthesized in the roots and are then transported by the xylem all over the plant. For quinolizidine alkaloids and the indolizidine alkaloid swainsonine phloem transport from leaves to other plant parts has been described (Wink and Witte, 1984; Wink, 1986; Dreyer et al., 1985). These compounds are usually formed by undifferentiated cells, i.e. cell cultures, in rather low yields, if at all (Wink, 1986).

MECHANISMS OF PRODUCT ACCUMULATION

To fulfil the biological and ecological functions which were mentioned in the introduction, plant secondary metabolites have to be present at the right time and place and usually in rather large quantities. Since many metabolites can be toxic for the producing cell, special mechanisms must have been evolved to store these compounds in a high concentration but also at a safe place.

According to the experimental evidence, we can assume that most hydrophilic metabolites are stored in the central vacuole (Matile, 1978 and 1984). There, the toxic compounds are separated from the cytoplasm by the tonoplast. In the vacuole concentrations of natural products of more than 200 to 500 mmol/l have been recorded. Since most compounds have been synthesized in the cytoplasm or cytoplasmic organelles, a transport or diffusion across the tonoplast against a concentration gradient has to take place. What are the mechanisms involved ?

The vacuole also stores primary metabolites, i.e. amino-acids, sugars, and inorganic ions. For these compounds it was established during the last decade that they cross the tonoplast by specific carrier proteins. The transport is driven by

480

an H^+-substrate antiport system and the necessary proton gradients are built up by a tonoplast ATPase and pyrophosphatase (Sze, 1984; see different communications in this book).

There is good evidence, that a similar mechanism also works for secondary metabolites. Deus-Neumann and Zenk (1984 and 1986) showed that indole and isoquinoline alkaloids are taken up by isolated vacuoles from suspension-cultured cells in an ATP-, pH-, time-, and concentration-dependent fashion (Table 4). The uptake was highly specific : Vacuoles from Fumaria discriminated between the S- and R-enantiomers of reticuline and scoulerine and vacuoles from alkaloid-free plants did not accumulate alkaloids at all. However, Renaudin et al. (1985 and 1986) assume that lipophilic indole alkaloids of a low pK-value enter the vacuole by diffusion. Since the vacuolar pH is acidic, the alkaloids become charged when they enter the vacuole and thus cannot diffuse back into the cytoplasm ("ion trap mechanism"). Similar views have been expressed by Matile (1978) and Neumann et al. (1985).

Table 4

Characteristics of Alkaloid Uptake into Vacuoles Isolated
from Suspension-cultured Cells
Study I = Deus-Neumann and Zenk (1984); II = Deus-Neumann
and Zenk (1986); III = Mende and Wink (1987)

Parameter	Study		
	I	II	III
Species	Catharanthus	Fumaria	Lupinus
Alkaloid	Vindoline	S-reticuline	Lupanine
pH-optimum	6.5	7.5	7.5
Activation energy	n.d.	n.d.	54 kJ/mol
ATP-stimulation	-	5-fold	30-fold
KCl-stimulation	7-fold	n.d.	20-fold
K_m (μMol/l)	0.3	0.3	94
Specificity	high	high	high
Inhibition			
vanadate	n.d.	0	0
DCCD	100 %	100 %	96 %
CCCP	n.d.	100 %	100 %
PHMB	n.d.	91 %	64 %
valinomycine	n.d.	n.d.	61 %
NaN_3	n.d.	n.d.	5 %

n.d. = not determined; DCCD = N,N'-dicyclohexylcarbodiimide; CCCP = Carbonylcyanid-m-chlorophenylhydrazone; PHMB = p-Hydroxymercuribenzoate.

But we could provide further evidence recently for a carrier-mediated uptake of alkaloids under physiological conditions (Mende and Wink, 1987). The quinolizidine alkaloid lupanine is a rather strong base and is present as the charged species at pH 7 by nearly 100%. Isolated vacuoles from Lupinus polyphyllus cell suspension cultures (but not vacuoles from other species) take up lupanine. Because lupanine is a charged molecule under physiological conditions, a simple diffusion is rather unlikely. Since the uptake is activated by ATP (up to 30-fold) and temperature-time-, pH-, and concentration-dependent and highly specific (Table 4), we conclude that lupanine is transported into the vacuole by a carrier protein and an H^+-lupanine antiport mechanism.

There is also evidence that coumaroyl glucosides, cardiac glycosides, and flavonoids are transported across the tonoplast by specific carriers (Werner and Matile, 1985; Kreiss et al., 1987; Matern et al., 1986).

Assuming the presence of specific carrier proteins for secondary metabolites has interesting consequences. Since these carriers are proteins, they must be coded by genes. Because of the large number of secondary metabolites (Table 1), which would all require their own carrier protein, an enormous number of carrier genes should exist. To obtain the production of a natural product it would be necessary to express the genes concomitantly for both, biosynthesis and accumulation. Biosynthesis alone will not lead to accumulation, since most products can and will be degraded by the plant cells (Barz, 1977; Barz and Köster, 1981). Only when the storage capacity is given, i.e. when the respective genes for carrier proteins are expressed, then the accumulation of a metabolite can take place.

This is probably also true for cell cultures. We can assume that it is easier to select producing cells in those instances where the biosynthesis and accumulation take place within the same cell (i.e. as is the case for species listed in Table 3). Whereas it might be difficult to obtain the concomitant expression of storage and biosynthesis in those cultures, where these processes are located at different sites. Interestingly, many of the products which were difficult to produce by plant cell culture, fall in this groupe (e.g. morphine, tropane and quinolizidine alkaloids, cardenolides, monoterpenes, etc, .).

Plant genes are switched on or off by tissue-, organ-, or development-specific factors, which activate the gene promotors. This would mean, that to switch on the genes for a biosynthesis which takes place in root cells, we require a root-specific signal, that is probably coded by a tissue-specific master gene. We also need to activate the genes for accumulation, which could be triggered by the same signal. If two or more tissues are involved we certainly require the participation of different master genes. Unfortunately, we have no detailed information on the regulation of genes involved in secondary metabolism of plant cell cultures. We need to under-stand the basic principles of the biochemistry and molecular genetics of secondary metabolite formation to devise strategies to overcome the problems envisaged in this paper. Then the biotechnological exploitation of natural products may become also feasible in those systems, which fail at the present time.

ACKNOWLEDGEMENTS

I would like to thank the Deutsche Forschungsgemeinschaft for grants and a Heisenberg Fellowship.

REFERENCES

Barz, W., 1977, Catabolism of endogenous and exogenous compounds by plant cell cultures, in : "Plant Tissue Culture and its Biotechnological Application", W. Barz, E. Reinhard, and M. H. Zenk, ed., Springer-Verlag, Berlin.

Barz, W., and Köster, J., 1981, Turnover and degradation of secondary (natural) products, in : "The Biochemistry of Plants", E. E. Conn, ed., Vol. 7, Academic Press, New-York.

Berlin, J., 1984, Plant cell cultures - a future source of natural products ?, Endeavour, New Ser., 8:5.

Constabel, F., Kurz, W. G. M., and Kutney, J. P., 1982, Variation in cell cultures of periwinkle, Catharanthus roseus, in : "Plant Tissue Culture 1982", A. Fujiwara, ed., IAPTC, Tokyo.

Curtin, B., 1983, Biotechnology, 1:649.

Deus-Neumann, B., and Zenk, M. H., 1984, A highly selective alkaloid uptake system in vacuoles of higher plants, Planta, 162:250

Deus-Neumann, B., and Zenk, M. H., 1986, Accumulation of alkaloids in plant vacuoles does not involve an ion-trap mechanism, Planta, 167:44

Dreyer, D., Jones, K. C., and Molyneux, R. J., 1985, Feeding deterrency of some pyrrolizidine, indolizidine, and quinolizidine alkaloids towards pea aphid (Acyrthosiphon pisum) and evidence for phloem transport of the indolizidine alkaloid swainsonine, J. Chem. Ecol., 11:1045

Harborne, J. B., 1982, "Introduction to Ecological Biochemistry", Academic Press, London.

Kreis, W., May, U., and Reinhard, E., 1987, Subcellular localization of primary and second cardiac glycosides, and the related 16'-O-glucosyl-transferase in cell suspension cultures of Digitalis lanata, in : "Plant Vacuoles.
Their Importance in Solute Compartmentation and Their Applications in Biotechnology", B. Marin, ed., Plenum Publishing Corporation, New-York.

Levin, D. A., 1976, The chemical defenses of plants to pathogens and herbivores, Annu. Rev. Ecol. Syst., 7:121

Matile, P., 1976, Localization of alkaloids and mechanisms of their accumulation in vacuoles of Chelidonium majus lacticifers, Nova Acta Leopoldina, Suppl. 1976: 139

Matile, P., 1978, Biochemistry and function of vacuoles in plants, Annu. Rev. Plant Physiol., 29:193

Matile, P., 1984, Das toxische Kompartiment der Pflanzenzelle, Naturwissenschaften, 71:18

Mende, P., and Wink, M., 1987, Uptake of the quinolizidine alkaloid lupanine by protoplasts and isolated vacuoles of suspension-cultured Lupinus polyphyllus cells. Diffusion or carrier-mediated transport ?, submitted for publication.

Molisch, H., 1923, "Mikrochemie der Pflanze", G. Fischer, Jena.

Mothes, K., 1955, Physiology of alkaloids, Annu. Rev. Plant Physiol., 6:393

Neumann, D., Krauss, G., Hieke, M., and Gröger, D., 1983, Indole alkaloid formation and storage in cell suspension cultures of Catharanthus roseus, Planta Med., 48:48

Renaudin, J. P., Brown, S. C., Barbier-Brygoo, H., and Guern, J., 1985, Compartmentation of alkaloids in a cell suspension culture of Catharanthus roseus : A reappraisal of the role of pH gradients, in : "Primary and Secondary Metabolism of Plant Cell Cultures", K. H. Neumann, W. Barz, and E. Reinhard, ed., Springer-Verlag, Heidelberg, Berlin.

Renaudin, J. P., Brown, S. C., Barbier-Brygoo, H., and Guern, J., 1986, Quantitative characterization of protoplasts and vacuoles from suspension cultured cells of Catharanthus roseus, Physiol. Plant., in press.

Swain, T. L., 1977, Secondary compounds as protective agents, Annu. Rev. Plant Physiol., 28:479

Wiermann, R., 1985, Secondary products and cell and tissue differentiation, in : "The Biochemistry of Plants", E. E. Conn, ed., Vol. 7, Academic Press, New-York.

Wink, M., 1986, Physiology of the accumulation of secondary metabolism with special reference to alkaloids, in : "Cell Culture in Phytochemistry", Vol. 4, in press.

Wink, M., 1987, Why do cell suspension cultures of lupins fail to produce alkaloids in large quantities ? Plant Cell Tissue Organ Cult., in press.

Wink, M., and Witte, L., 1984, Turnover and transport of quinolizidine alkaloids : Diurnal variation of lupanine in the phloem sap, leaves and fruits of Lupinus albus L., Planta, 161:519

Witacker, R. H., and Feeny, R. P., 1971, Allelochemics : Chemical interaction species, Science, 171:757

Zenk, M. H., 1978, The impact of plant cell culture on industry, in : "Frontiers of Plant Tissue Culture 1978", T. A. Thorpe, ed., IAPTC, Calgary.

Zenk, M. H., 1982, Pflanzliche Zellkulturen in der Arzneimittel Forschung, Naturwissenschaften, 69:534.

PRODUCTION OF SECONDARY METABOLITES IN CELL CULTURES

OF SOME TERPENOID-INDOLE ALKALOIDS PRODUCING PLANTS

R. Verpoorte[*], R. Wijnsma[*], P. A. A. Harkes[**],
H. J. G. ten Hoopen[***], J. J. Meijer[***] and W. M. Van Gulik[***]

[*]Center for Bio-Pharmaceutical Sciences
State University of Leiden, P.O. Box No. 9502, 2300-RA-Leiden
[**]Department of Plant Molecular Biology
State University of Leiden, Nonnensteeg 2, 2311-VJ-Leiden
and [***]Department of Biochemical Engineering
Delft University of Technology, Julianalaan 67, 2628-BC-Delft
The Netherlands

INTRODUCTION

Within the group of terpenoid-indole alkaloids and related compounds quite a few have pharmaceutical interest. As most of them have quite high prices on the market, the production by means of biotechnological processes seems attractive. Thus in the past years a number of studies have been dealing with plant cell, tissue and organ cultures of terpenoid-indole alkaloids producing plants from genera like Rauwolfia, Catharanthus, Vinca, Tabernaemontana, Voacanga, Amsonia and Cinchona. Three of these are at present studied in our Biotechnology Delft Leiden (BDL) project group, viz. Catharanthus, Tabernaemontana and Cinchona with the aim to improve yields of alkaloids in the cell cultures. Recently we have reviewed the work on Cinchona cell cultures (Verpoorte et al., 1985; Wijnsma and Verpoorte, 1986 e). Here we shall confine ourselves to a brief review of some of the results recently in our laboratories.

Cinchona proved to be difficult to grow as a cell suspension culture. When this problem was finally solved, it was found that such cultures only produce low amounts of alkaloids; the production capacity even decreased in the course of time. Several methods have been tried to improve the production of alkaloids in these cultures. Epigenetic manipulation, i.e. varying medium constituents (Harkes et al., 1985; Wijnsma et al., 1986 f), and feeding of various precursors were the two major strategies followed so far.

FEEDING OF PRECURSORS TO CINCHONA CELL CULTURES

The influence of various precursors from the terpenoid-indole pathway, e.g. tryptophan, tryptamine and secologanine (scheme 1), on the alkaloid production in the cell cultures was studied. Feeding L-tryptophan resulted already at a concentration of 0.05 mM to severe growth inhibition; at 0.1 mM all cells were dead after four days. At none of the concentrations tested an increase of quinoline alkaloid content could be observed in the cell cultures. However, in the HPLC

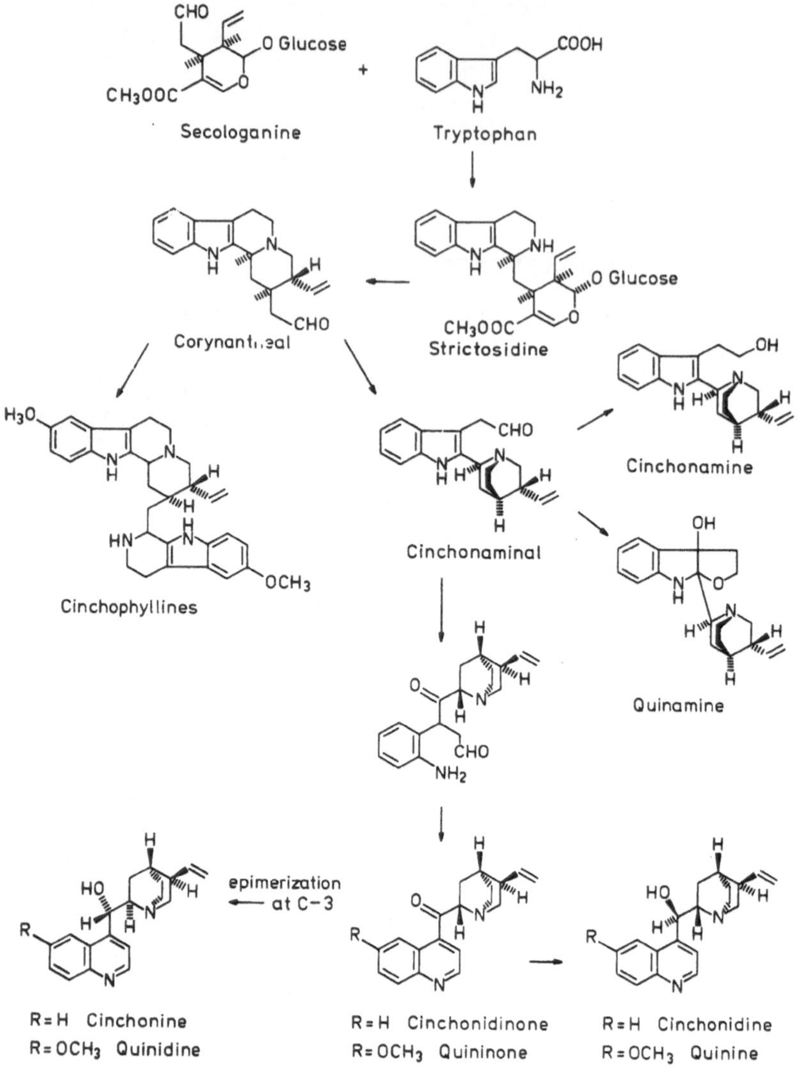

Scheme 1

Biosynthesis of <u>Cinchona</u> Alkaloids

chromatograms other blue fluorescent compounds were found to be present in the alkaloid extracts (Harkes et al., 1986; Wijnsma et al., 1986 b). These compounds proved to be harman and norharman. Further studies showed that these compounds are formed spontaneously from tryptophan, particularly during sterilizing of the medium. The presence of cell material (dead or alive) resulted in higher levels of these two compounds. The time course of the norharman production is shown in Fig. 1. From these results it becomes clear that harman derivatives as products of biotransformation from tryptophan and tryptamine have to be considered with caution. Feeding with tryptamine showed that this compound is not as toxic as tryptophan; still no influence on the alkaloid production could be observed, however, except for the harman derivatives. Also with secologanine feeding no increase of alkaloid production could be observed.

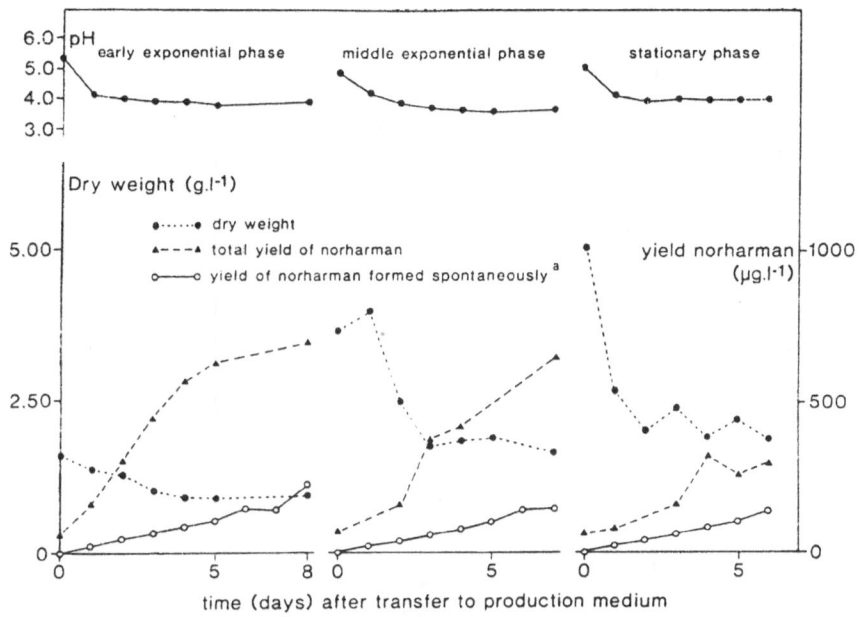

Figure 1

Time-course of norharman yield[a]
in Cinchona ledgeriana cell suspension cultures
after transfer to Zenk's alkaloid production medium,
containing tryptophan (Zenk et al., 1977)

[a] spontaneously : Norharman found in the medium without cells.

By means of a newly developed assay for the tryptophan decarboxylase activity, using HPLC in combination with fluorescence monitoring for the analysis of tryptamine (BDL, unpublished results), it was found that in fact the Cinchona cell culture is devoid of any tryptophan decarboxylase activity, even after transferring to media which induce this enzyme in Catharanthus roseus and Tabernaemontana cell cultures.

Further feeding experiments were done with corynantheal, the putative intermediate in the pathway from strictosidine towards the quinoline alkaloids (scheme 1) (Battersby and Parry, 1971). This compound was fed to Cinchona cell cultures. Sampling at regular time intervals in a seven-day period showed that corynantheal was rapidly absorbed by the cells; within a few hours most of the alkaloid is found in the cells (Fig. 2). Subsequently the corynantheal concentration decreases, resulting in the formation of one major compound, with in its turn decreases in concentration, whereas some fluorescent compounds increase in concentration. The first product formed could be identified as corynantheol by means of UV and mass spectrometry and comparison with a reference compound. The fluorescent compounds were not identical with the quinoline alkaloids. Based

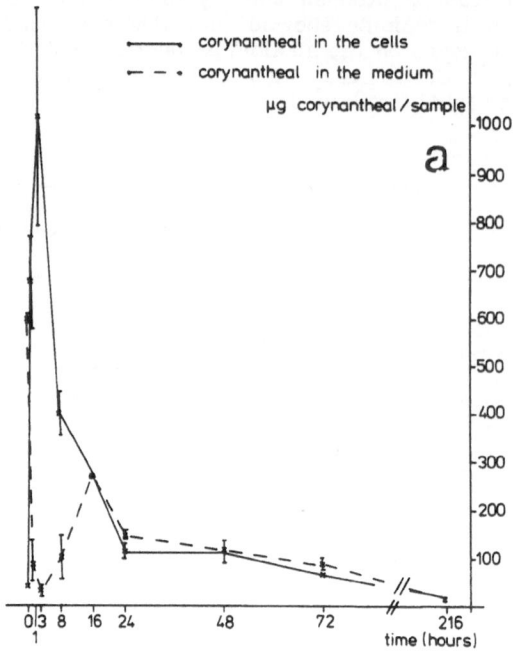

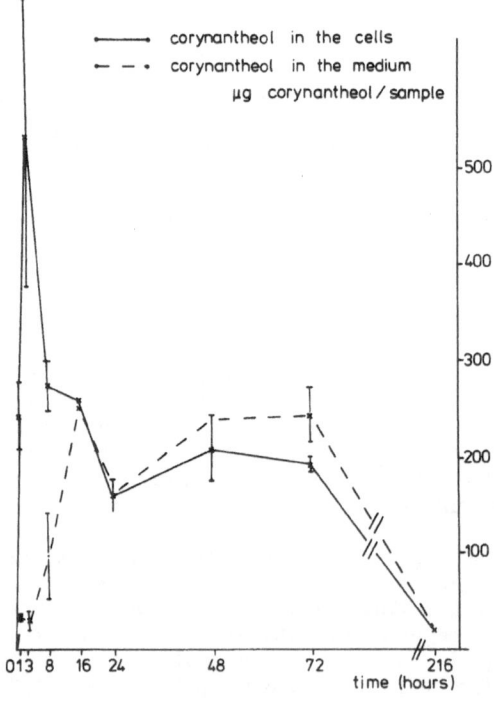

Figure 2

Time-course of corynantheal levels (a) and corynantheol formation (b)
in cells and medium in Cinchona ledgeriana cell suspension culture

on their UV and mass spectra they are believed to be the products of oxidation of corynantheal and corynantheol in the C-ring, yielding the 3,4-dehydro and 3,4,5,6-tetradehydro derivatives, the latter having the typical harman chromophore. The bioconversion of corynantheal is thus as summarized in scheme 2. The major route is the reduction of the 17-aldehyde function, followed by the oxidation of the C-ring. This experiment does not allow any conclusion abour the possible role of corynantheal as an intermediate; neither does it allow any conclusion about the possibility of the absence of a corynantheal converting enzyme. Both options are still possible; further experiments with an alkaloid producing cell line are necessary to solve this problem.

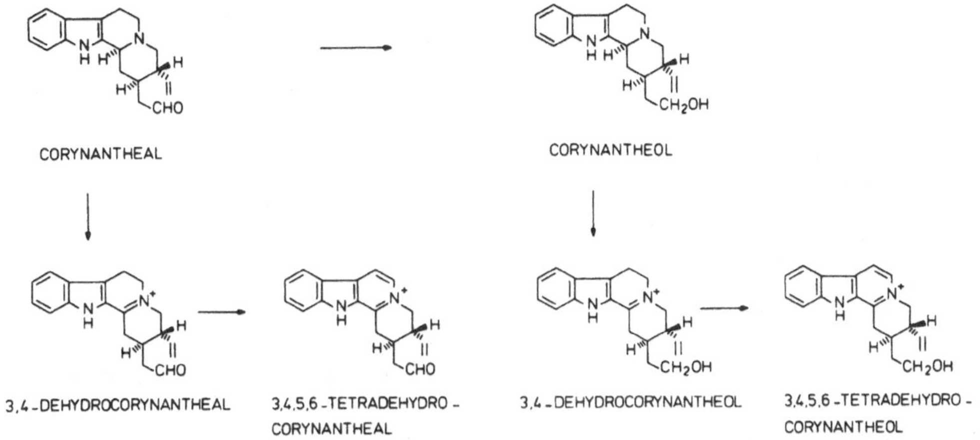

CORYNANTHEAL   CORYNANTHEOL

3,4-DEHYDROCORYNANTHEAL   3,4,5,6-TETRADEHYDRO-CORYNANTHEAL   3,4-DEHYDROCORYNANTHEOL   3,4,5,6-TETRADEHYDRO-CORYNANTHEOL

Scheme 2

Bioconversion of Corynantheal
by Cinchona ledgeriana Cell Suspension Cultures

ANTHRAQUINONES IN CHICHONA CELL CULTURES

During the studies of Cinchona cell cultures it has been noted that often large amounts of yellow-orange compounds are formed. By using a two-phase system in the culture flasks, these compounds can be captured in the organic phase (paraffine). In fermentors even deposits of these compounds are found on the bottom and on the walls in quite large amounts. Isolation of these compounds from callus cultures of Cinchona ledgeriana and Cinchona pubescens allowed us to identify them as anthraquinones. From each plant about 15 compounds were identified from the complex mixture of more than 30 different anthraquinones present (Wijnsma et al., 1984 and 1986 a; Mulder-Krieger et al., 1984). There were all of the type common for Rubiaceae (Wijnsma and Verpoorte, 1986 d). Subsequent studies have shown that the production of the anthraquinones is induced by biotic elicitors like sterilized mycelia of Phytophtora cinnamomi and Aspergillus niger and enzyme preparations like cellulase and pectinase (Table 1). An example of this is presented in Fig. 3 (Wijnsma et al., 1985). The anthroquinones also exhibit strong antimicrobial activity against gram(+), gram(-) bacteria and yeasts. These facts lead us to the following conclusion : the anthroquinones act as phytoalexines in Cinchona. This could further confirmed by studies with whole plants.

In Cinchona trees infected with Phytophtora cinnamomi, anthraquinones were detected in the infected parts, whereas in healthy plants of the same tree no anthraquinones could be found (Wijnsma et al., 1986 c). Infection of 1-year-old plantlets with Agrobacterium tumefaciens or Phytophtora cinnamomi resulted in induction of the anthraquinone biosynthesis.

Table 1

Some Elicitors of Anthraquinone Biosynthesis
in Cinchona ledgeriana Cell Suspension Cultures
(BDL, unpublished results)

Elicitor	Total production anthraquinones as % of control	Days after addition of elicitor
sterilized Aspergillus niger preparation	390 %	11 days
sterilized Agrobacterium tumefaciens preparation	525 %	11 days
cellulase	230 %	3 days
pectinase	180 %	3 days

As Agrobacterium tumefaciens is very sensitive to the antimicrobial activity of the anthraquinones, this could cause problems in transforming Cinchona with Agrobacterium tumefaciens. The use of glyphosine or glyphosate to suppress the anthraquinone biosynthesis could be considered to solve this problem, as in cell cultures of Cinchona the biosynthesis of the anthraquinone is inhibited by these agents (BDL, unpublished results).

Apparently Cinchona fine cell suspension cultures are better producers of anthraquinones than of alkaloids, the former being present at levels of about 1 – 2% on dry-weight basis, the latter only at levels of 0.01%, although under certain conditions alkaloid levels can be increased to a 0.1 - 1% level (BDL, unpublished results).

GROWTH CONDITIONS OF CELL CULTURES

As growth of cell cultures is an important factor in determining the economy of a biotechnological production process, and because of the changes in secondary metabolism with various medium compositions, studies have been made of the influence of some media constituents on the growth and alkaloid production in both Catharanthus, Tabernaemontana and Cinchona (Harkes et al., 1985; Wijnsma et al., 1986 f; BDL, unpublished results). Among others, the carbon source and oxygen were studied as possible growth limiting factors.

The utilization of various carbon sources was studied by means of cell cultures of the mentioned species. Glucose and sucrose proved to be suitable, both resulting in similar yields of biomass. Fructose on the other hand severely inhibited growth of the cells; in the case of Cinchona a clear increase of anthraquinone levels could be noted with this sugar. Sucrose is already to a variable degree hydrolyzed in glucose and fructose during sterilization. For reproducible results, therefore, glucose should be preferred as carbon source, or, instead of heat sterilization, filter sterilization should be preferred. The course of sugar utilization in a cell culture is presented in Fig. 4. In fact, the cells of all three above-mentioned types of plants prefer the use of sucrose and glucose; fructose is the last sugar to be utilized.

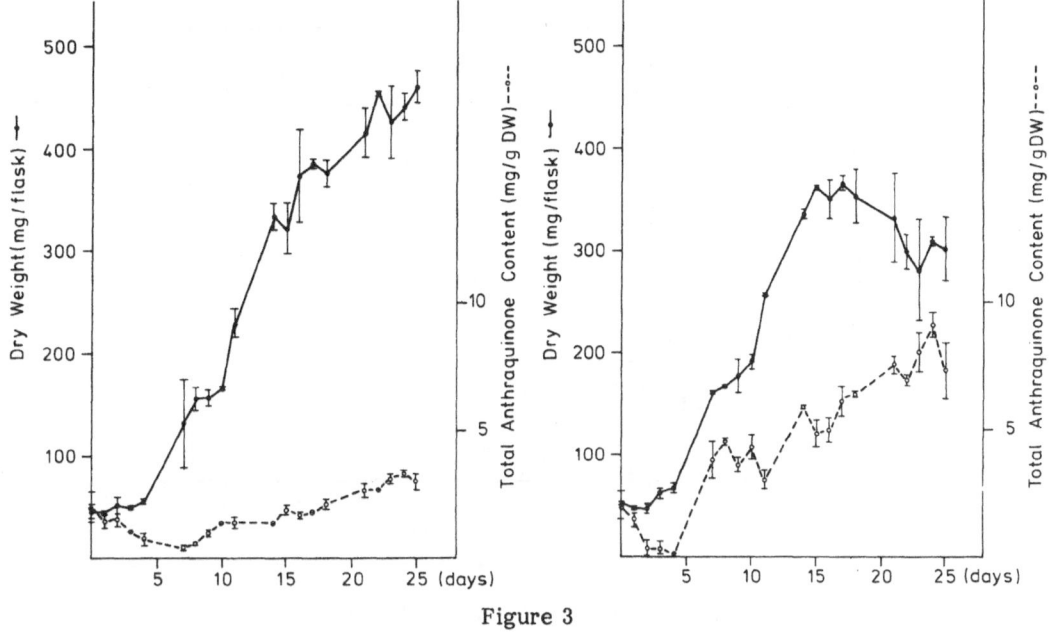

Figure 3

Growth and anthraquinone production
in a cell suspension culture of <u>Cinchona</u> <u>ledgeriana</u>

Values ± S.D., n = 2. a. Control. b. After addition of a sterilized mycelium preparation of <u>Phytophthora</u> <u>cinnamomi</u>.

Another factor which may influence growth is the oxygen supply. Particularly in stirred fermentors the oxygen supply may be a problem in the thick biomass at the end of the growth phase. Therefore, studies have been performed on the oxygen consumption of the cell cultures. Preliminary studies of the oxygen supply in batch-type cultures in flasks showed that the way of closing the flasks at the top is of great importance for the oxygen supply. A cotton plug allowed the highest oxygen transport rates (K_W = 3 10^{-7} m^3 s^{-1}), a silicon cap being nearly as good. Silicon stoppers had a K_W about 10 times lower. Aluminium foil showed 4 - 50 times lower transport rates for oxygen, the rate being very much dependable on individual differences in applying the foil. In fact, oxygen limitation may occur if the foil is tightened too much.

Also in fermentors oxygen supply may cause problems. Plant cells are very friable, so only low stirring rates can be used infermentors. This results in poor oxygen transport from the gas phase to the liquid phase. In spite of the low oxygen demand of plant cells, oxygen limitation is likely to occur, especially in the thick biomass at the end of the growth phase. To solve these and other problems preventing the large-scale culturing of plant cells in fermentors, kinetic data such as maximum specific growth rate, oxygen demand at different growth rates, maintenance energy requirements and the effects of shear forces on the cells are needed. Lack of such data prevents the design of large-scale fermentation equipment of plant cells.

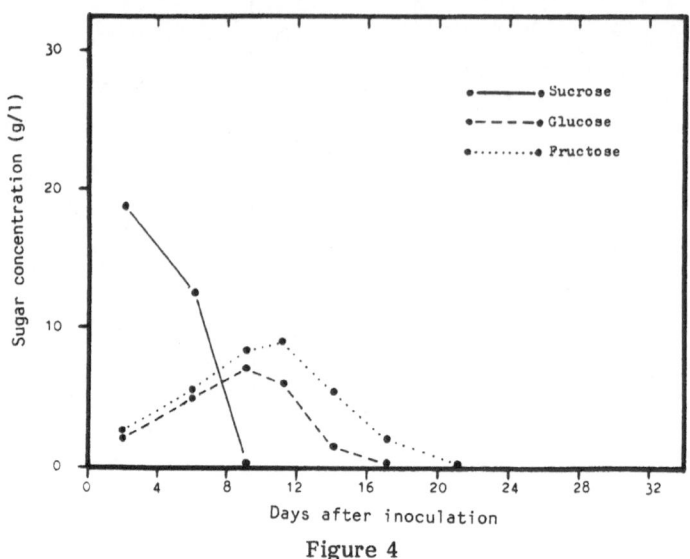

Figure 4

Time-course of sugar metabolism
in a <u>Catharanthus</u> <u>roseus</u> cell suspension culture

Initial sugar concentration was 3%.

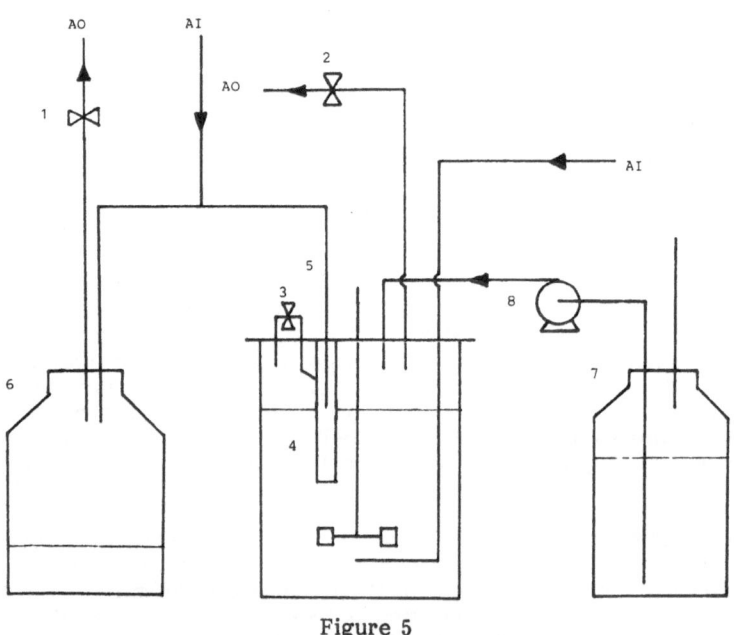

Figure 5

Scheme continuous culture system

Key : 1, 2 and 3 pneumatic valves; 4 glass tube; 5 suspension/air outlet; 6 effluent vessel; 7 influent vessel; 8 medium pump; AI air inlet; A0 air outlet.

To answer the above-mentioned questions, a continuous culture system was set up for plant cells. A 3 l stirred fermentor was used as a basis. A special device was constructed to provide a homogenous effluent from the culture at low dilution rates (Fig. 5). Effluent was driven out of the fermentor by overpressure at certain time intervals. Medium was pumped in continuously. The length of the time interval was chosen in such a way that volume changes of the culture were ca. 1% (see for the principle of the system, Fig. 5). Normally valves 1 and 3 are closed; used air leaves the fermentor through valve 2, while suspension is driven out of pipe 4 by a continuous air stream. At the end of each time interval, valve 2 was closed, while valves 1 and 3 were opened. This resulted in a rapid rise of the suspension inside pipe 4. At that moment, suspension and used air can only leave the fermentor through pipe 5. Suspension was collected in vessel 6. When the suspension level is below the end of pipe 5, only air leaves the fermentor. After 30 seconds, valves 1 and 3 were closed and valve 2 was opened again. This system proved to be satisfactory in a series of preliminary experiments.

Continuous culture experiments are at present being performed with Catharanthus roseus in the above-mentioned set-up.

REFERENCES

Battersby, A. R., and Parry, R. J., 1971, Biosynthesis of Cinchona alkaloids : Middle stages of the pathway, Chem. Commun., 30.

Harkes, P. A. A., Krijbolder, L., Libbenga, K. R., Wijnsma, R., and Verpoorte, R., 1985, Influence of various media constituents on the growth of Cinchona ledgeriana tissue culture and the production of alkaloids and anthraquinones therein, Plant Cell Tissue Organ Cult., 4:199.

Harkes, P. A. A., De Jong, P. J., Wijnsma, R., Verpoorte, R., and Van Der Leer, T., 1986, Influence of production media on Cinchona cell cultures. Spontaneous formation of –carbolines from L-tryptophan, Plant Science Letters, submitted.

Mulder-Krieger, Th., Verpoorte, R., Van Der Kreek, M., and Baerheim-Svendsen, A., 1984, Identification of alkaloids and anthraquinones in Cinchona pubescens callus cultures; the effect of plant growth regulators and light on the alkaloid content, Planta Medica, 50:17.

Verpoorte, R., Wijnsma, R., Mulder-Krieger, Th., Harkes, P. A. A., and Baerheim-Svendsen, A., 1985, Plant cell and tissue culture of Cinchona species, in : "Primary and Secondary Metabolism in Plant Cell Cultures", K. H. Neumann, W. Hüsemann and E. Reinhard, eds., Springer-Verlag, Heidelberg.

Wijnsma, R., Verpoorte, R., Mulder-Krieger, Th., and Baerheim-Svendsen, A., 1984, Anthroquinones from callus cultures of Cinchona ledgeriana, Phytochemistry, 23:2307.

Wijnsma, R., Go, J. T. K. A., Van Weerden, I. N., Harkes, P. A. A., Verpoorte, R., and Baerheim-Svendsen, A., 1985, Anthraquinones as phytoalexins in cell and tissue cultures of Cinchona species, Plant Cell Reports, 4:241.

Wijnsma, R., Go, J. T. K. A., Harkes, P. A. A., Verpoorte, R., and Baerheim-Svendsen, A., 1986 a, Anthraquinones in callus of Cinchona pubescens, Phytochemistry, 25:1123.

Wijnsma, R., Van Der Leer, T., Verpoorte, R., Harkes, P. A. A., De Jong, P. J., and Baerheim-Svendsen, A., 1986 b, Feeding of indole alkaloid precursors, L-tryptophan and tryptamine to cell suspension cultures of Cinchona ledgeriana, Plant Cell Reports, submitted.

Wijnsma, R., Van Weerden, I. N., Verpoorte, R., Harkes, P. A. A., Lugt, Ch. B., Scheffer, J. J. C., and Baerheim-Svendsen, A., 1986 c, Anthroquinones in Cinchona ledgeriana bark infected with Phytophtora cinnamomi, Planta, Medica, 211.

Wijnsma, R., and Verpoorte, R., 1986 d, Anthraquinones in Rubiaceae, Fortsch. Chem. Organisch. Natur., in press.

Wijnsma, R., and Verpoorte, R., 1986 e, Quinoline alkaloids in cell and tissue cultures of Cinchona species, in : "Cell Culture and Somatic Genetics of Plants", Vol. 5, I. K. Vasil and F. Constabel, eds., Academic Press, in press.

Wijnsma, R., Verpoorte, R., Harkes, P. A. A., Van Vliet, T. B., Ten Hoopen, H. J. G., and Baerheim-Svendsen, A., 1986 f, The influence of initial sucrose and nitrate concentrations on the growth of Cinchona ledgeriana cell suspension cultures and the production of alkaloids and anthraquinones therein, Plant Cell Tissue Organ Cult., submitted.

Zenk, M. H., El-Shagi, H., Arens, H., Stöckigt, J., Weiler, E. W., and Deus, B., 1977, in : "Plant Tissue Culture and its Biotechnological Application", W. Barz, and M. H. Zenk, eds., Springer-Verlag, Berlin.

CONTINUOUS PRODUCTION OF INDOLE ALKALOIDS

BY GEL-ENTRAPPED CELLS OF CATHARANTHUS ROSEUS

Alain Pareilleux and Florence Majerus

Département de Génie Biochimique et Alimentaire
Unité Associée C.N.R.S. No. 544
Institut National des Sciences Appliquées
Avenue de Rangueil, 31.077-Toulouse-Cedex, France

INTRODUCTION

Plant cell cultures present a great potential for secondary metabolite production (Staba, 1980). Recently, plant cell immobilization techniques have been reported including entrapment in Ca-alginate (Brodelius et al., 1979) and others matrices (Brodelius and Nilsson, 1980; Jirku et al., 1981; Schuler, 1981; Galun et al., 1983; Lindsey et al., 1983), allowing continuous production of useful substances to be obtained. It is obviously important that the working period should be maintained as long is possible with a minimum loss of cell viability and a sustained biosynthetic activity. Besides there is evidence that limitation of growth is essential for the mechanical stability of the Ca-alginate matrix and often results in enhanced production of metabolites (Lindsey and Yeoman, 1983).

Herein we report investigations relating to growth limiting conditions consistent with the entrapment of Catharanthus roseus cells in Ca-alginate beads (Majerus and Pareilleux, 1986 a) and results concerning the production of indole alkaloids using immobilized cells in a continuous flow reactor (Majerus and Pareilleux, 1986).

MATERIALS AND METHODS

Cell Suspensions Cultures

Cell suspensions cultures of Catharanthus roseus (L.) G. Don were subcultured in the medium described by Gamborg et al. (1968) supplemented with 4.5 μM 2,4-D and 0.28 μM kinetin and addition of sucrose as the carbon source. The suspensions were cultivated in the dark on a rotary shaker (110 rpm) at 27°C in Erlenmeyer flasks (100 ml per 250 ml flask). When other media were tested, the cells were harvested by filtration, washed and resuspended in the appropriate medium. The composition of the limiting medium was the same as the basal medium except the concentrations of 2,4-D, phosphate and $CaCl_2$ which were changed (45 nM, 100 μM and 5 mM respectively).

Immobilization Procedure

Cells of a 7-day old supension were retained on a 40 μm net and 250 g wet weight of cells were suspended in 300 ml of 3% alginate dissolved in the limiting medium; the suspension was then added dropwise to the medium with the addition of 50 mM CaCl$_2$. After 1 hour the beads were washed and collected. In batch experiments 10 g of beads per flask were used. The bioreactor (working volume 1 l) was inoculated with 200 g of beads (2.7 g cell dry weight); after a 7 day batch culture it was continuously fed with limiting medium buffered with 50 mM Mes at pH 6 and changed to non-buffered medium at pH 4.3 at day 52.

Analytical Procedures

They were described elsewhere (Majerus and Pareilleux, 1986 a). Extraction of the alkaloids and quantitative determination of tryptamine, ajmalicine and serpentine were performed using an HPLC method previously described (Vinas and Pareilleux, 1982).

RESULTS AND DISCUSSION

Growth Limitation of Free Cell Suspensions

In an attempt to improve the stability of the beads, growth limitations by hormonal or nutrient supply were investigated. When cell suspensions were subcultured into medium devoid of 2,4-D a lowered growth was observed during the third subculture, with a proportional increase of cell death (Fig. 1).

Compared with normal medium grown, cells of a second subculture in 2,4-D free medium showed enhanced ajmalicine and serpentine levels (Table 1).

Decreasing the phosphate concentration of the medium resulted in a progressive limitation of the growth with a concurrent increase of the ajmalicine content (Table 2). Similar results were reported by Knobloch and coworkers (1981

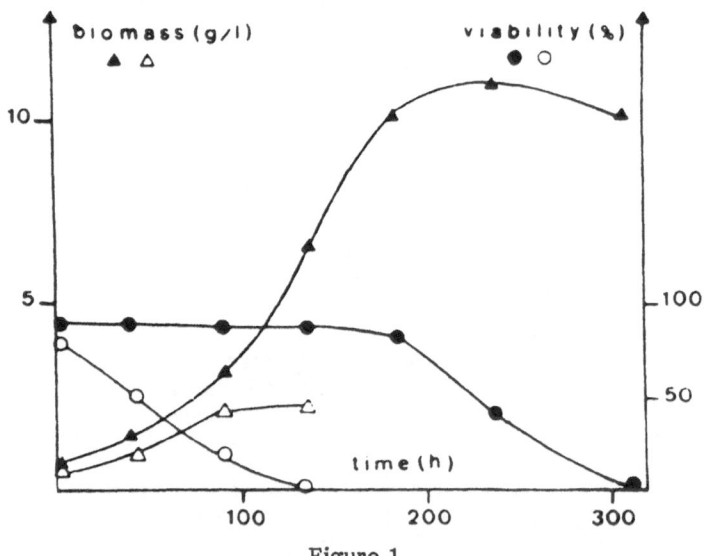

Figure 1

Growth and variability of <u>Catharanthus</u> <u>roseus</u> cells
in 2,4-D free medium after successive subcultures from normal medium

Closed symbols : second transfer (first transfer results were analogous) ; open symbols : third transfer.

Table 1

Maximum Levels of Tryptamine, Ajmalicine and Serpentine
in Normal or 2,4-D Free Medium Cultured Cells

Product	Alkaloid content(μg/g. DW)	
	Normal medium	2,4-D free medium (2nd subculture)
Tryptamine	540	345
Ajmalicine	20	72
Serpentine	11	120

and 1983) concerning the effect of phosphate level on the growth and the accumulation of secondary metabolites. From these observations a growth limiting medium was formulated (45 mM 2,4-D, 100 μM phosphate) in which product accumulation was stimulated; in addition, it contained 5 mM $CaCl_2$ for the stabilization of the alginate beads.

Table 2

Levels of Tryptamine, Ajmalicine and Serpentine
at day 7 of culture for various phosphate concentrations

Phosphate (μM)	Alkaloid content(μg/g. DW)		
	Tryptamine	Ajmalicine	Serpentine
0	135.5	30.2	2.6
25	175.2	23.3	4.3
50	210.0	21.5	3.2
100	247.3	17	3.7
250	387.75	12.2	6.0
500	420	7.6	5.0
1100	543	2.1	3.5

Metabolic Activity of Free or Entrapped Cells

Cell growth and respiration rates for both free and entrapped cells were examined using the above described medium and presented in Fig. 2. Immobilized cells showed lower initial rates of growth and respiration compared to free cells, possibly due to a diffusion barrier. Whereas a decrease in the respiration activity with an increasing cell lysis occurred for free cell suspensions, the oxygen uptake of the entrapped cells remained constant between day 8 - 26 of the experiment. As a consequence of the entrapment, a more prolonged period of ajmalicine accumulation within the cells and higher amounts of the product were found (Fig. 3). The tryptamine and serpentine contents were not strongly affected.

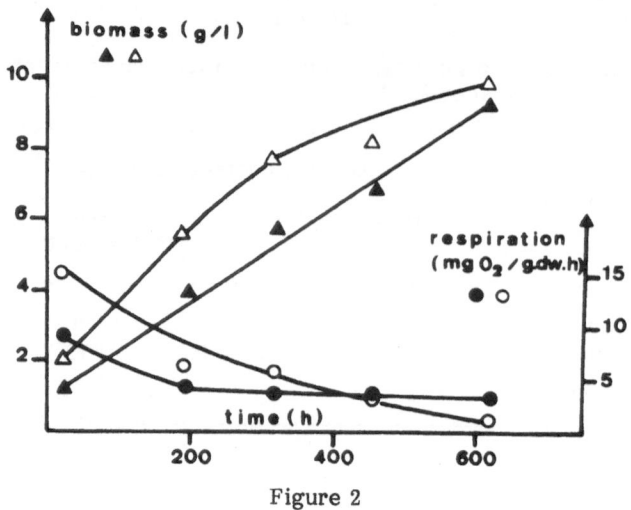

Figure 2

Time-course of biomass formation
and respiratory activity for free (open symbols) and entrapped
(closed symbols) cells of Catharanthus roseus
grown in the limiting medium

Continuous Flow Reactor Using Entrapped Cells

Fig. 4 shows the recovery of the products in the effluent medium using various operating conditions. With regard to the ajmalicine concentrations steady states were observed for a variety of dilution rates when the reactor was supplied with buffered medium. The product excretion could be characterized from these results. According to the balance equation :

$$V (dC/dt)_a = V (dC/dt)_p - Q.C$$

with V = reactor volume, C = product concentration in the medium, Q = flow rate and where a and p represent the accumulation and the production in the reactor, the excretion rate r_e and the production rate P were equivalent during steady state periods; then, it can be written

$$r_e = P = D.C$$

with D = dilution rate. As shown in Fig. 5 plotting the production rate against the concentration in the outlet medium, a linear relationship was obtained,

$$P = \alpha(C^* - C)$$

accounting for a simple diffusion mechanism and where C^* is given as the concentration at the thermodynamic equilibrium. From the empirical equation,

$$P = 0.288 (350 - C)$$

a maximum production rate of 100 μg/l day might be expected. Assuming a low C^* value compared to the product level within the cells observed during a batch experiment (0.5 to 50 mg/l cell soap) it can be concluded that only a little part of the metabolites was able to be excreted by passive diffusion. This model is consistent with a vacuolar storage of the protonated form of the alkaloids resulting

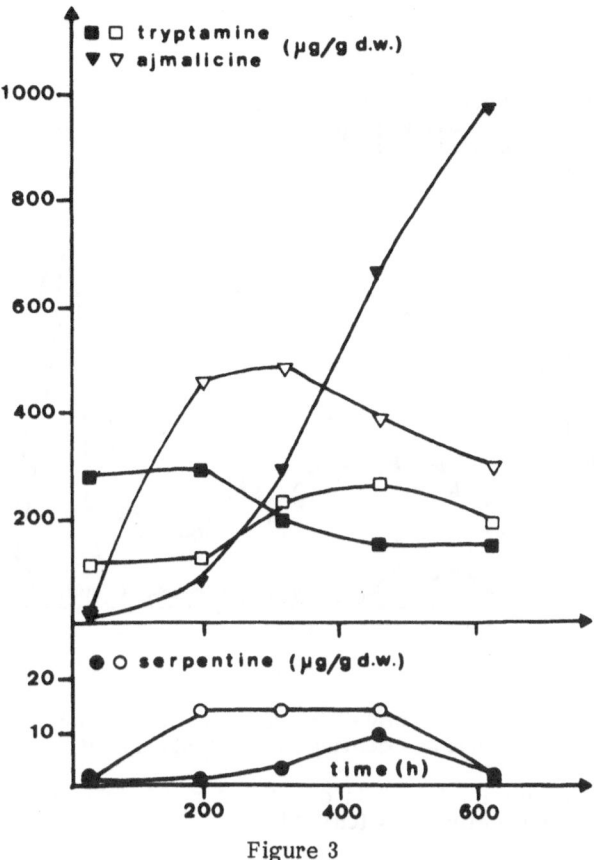

Figure 3

Kinetics of product accumulation
within free (open symbols) and entrapped (closed symbols) cells
of Catharanthus roseus grown in the limiting medium

of an ion-trap mechanism, as suggested by Renaudin and Guern (1982) and Neumann et al. (1983). Consequently, the product diffusion would be dependent on the pH gradient between the vacuole and the culture medium.

As expected, enhanced product concentrations were obtained when a pH 4.3 non buffered medium was used as the feeding medium at day 52 of the experiment (see Fig. 4). In consequence of the pH modification (pH 6 to pH 5.5) the maximum production rate of ajmalicine increased from 70 to 340 μg/l d, probably due to the change of the C^* value. Afterwards, an enhancement of tryptamine concentration up to 490 μg/l was observed. However, no steady states were obtained, and after the maxima were reached, a rapid decrease in product concentrations occurred due to either their wash out from the cells and/or a detrimental effect of the pH decrease.

CONCLUSION

In a growth limiting medium, Ca-alginate entrapped cells of Catharanthus roseus sustained their biosynthetic activity over long periods of time with a maintained stability of the polymeric matrix. Thus, a continuous flow reactor using

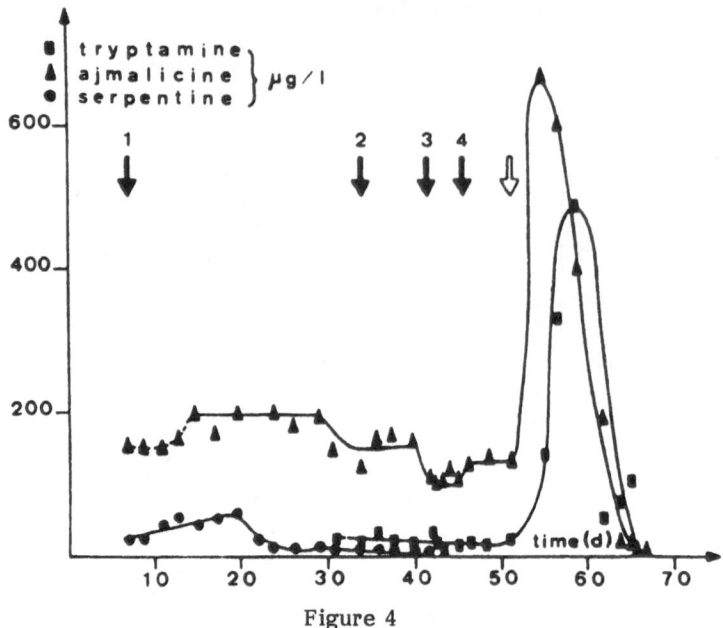

Figure 4

Time-course of metabolite concentrations in the effluent
during a continuous flow experiment

↓, dilution rate change, 1 : D = 0.24 d^{-1}, 2 : D = 0.72 d^{-1}, 3 : D = 0.72 d^{-1}, 4 : D = 0.51 d^{-1}; ⇓, pH modification, D = 0.51 d^{-1}.

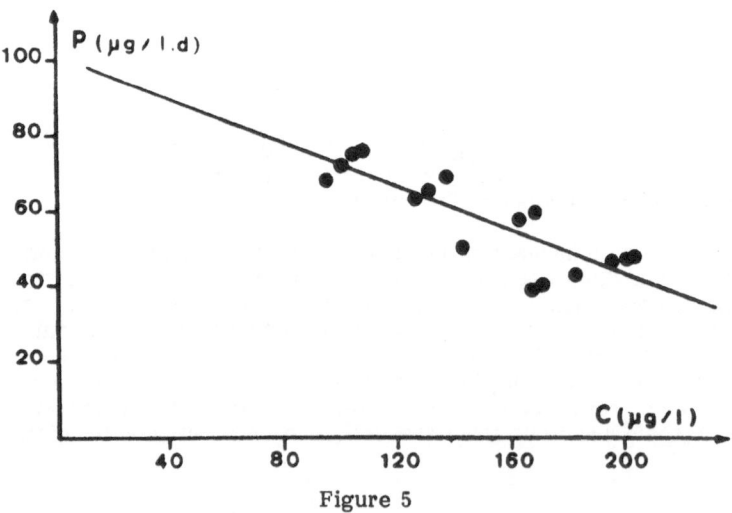

Figure 5

Relation between the production rate
and the concentration of ajmalicine in the effluent

such immobilized cells was functional for more than two months. Although a number of problems including the selection of high producing cell lines with improved product excretion have to be overcome, continuous flow processes using immobilized plant cells shows high potential as a means of producing secondary metabolites.

REFERENCES

Brodelius, P., Deus, B., Mosbach, K., and Zenk, M. H., 1979, Immobilized plant cells for the production and transformation of natural products, FEBS Letters, 103:93.

Brodelius, P., and Nilsson, K., 1980, Entrapment of plant cells in different matrices. A comparative study, FEBS Letters, 122:312.

Galun, E., Aviv, D., Dantes, A., and Freeman, A., 1983, Planta Medica, 49:9.

Gamborg, O. L., Miller, R. A., and Ojima, K., 1968, Nutrient requirements of suspension cultures of soybean root cells, Exptl. Cell Res., 50:151.

Jirku, V., Mocek, T., Vanek, T., Krumphanzl, V., and Kubanek, V., 1981, Continuous production of steroid glycoalkaloids by immobilized plant cells, Biotechnol. Letters, 3:447.

Knobloch, K. H., Beutnagel, G., and Berlin, J., 1981, Influence of accumulated phosphate on culture growth and formation of cinnamoyl-putrescines in medium-induced cell suspension cultures of Nicotiana tabacum, Planta, 153:582.

Knobloch, K. H., Hansen, B., and Berlin, J., 1981, Medium-induced formation of indole-alkaloids and concomitant changes of interrelated enzyme activities in cell suspension cultures of Catharanthus roseus, Z. Naturforsch., 36c:40.

Knobloch, K. H., and Berlin, J., 1983, Influence of phosphate on the formation of indole-alkaloids and phenolic compounds in cell suspension cultures of Catharanthus roseus : 1. Comparison of enzyme activities and product accumulation, Plant Cell Tissue Organ. Cult., 2:333.

Lindsey, K., and Yeoman, M. M., 1983, Novel experimental systems for studying the production of secondary metabolites by plant tissue cultures, in : "Soc. Exptl. Biol. Sem. Ser.", Vol. 18 : Plant Biotechnology, pp. 39-66, S. H. Mantell and H. Smith, eds., Cambridge University Press, Cambridge.

Lindsey, K., and Yeoman, M. M., 1983, The relationship between growth rate, differentiation and alkaloid accumulation in cell cultures, J. Exptl. Bot., 34:1055.

Lindsey, K., Yeoman, M. M., Black, G. M., and Mavituna, F., 1983, A novel method for the immobilization and culture of plant cells, FEBS Letters, 155:143.

Majerus, F., and Pareilleux, A., 1986 a, Plant Cell Reports, accepted.

Majerus, F., and Pareilleux, A., 1986 b, Biotechnol. Letters, submitted.

Neumann, D., Krauss, G., Hieke, M., and Groger, D., 1983, Indole alkaloid formation and storage in cell suspension cultures of Catharanthus roseus, Planta Med., 48:20.

Renaudin, J. P., and Guern, J., 1982, Compartmentation mechanisms of indole-alkaloids in cell suspension cultures of Catharanthus roseus, Physiol. Vég., 20:533.

Schuler, M. L., 1981, Production of secondary metabolites from plant cell cultures. Problems and prospects, Ann. N. Y. Acad. Sci., 369:65.

Staba, E. J., 1980, Secondary metabolism and biotransformation, in : "Plant Tissue Culture as a Source of Biochemicals", E. J. Staba, ed., CRC Press Inc., Boca-Raton, pp. 59-98.

Vinas, R., and Pareilleux, A., 1982, Production d'alcaloïdes par des suspensions cellulaires de Catharanthus roseus cultivées in vitro, Physiol. Vég., 20:219.

FACTORS AFFECTING ALKALOID PRODUCTION

IN PLANT CELL CULTURES

Margaret F. Roberts

The School of Pharmacy, University of London
29 - 39, Brunswick Square, London WC1N 1AX, United Kingdom

INTRODUCTION

Higher plants are the indispensable producers of medicinal substances such as steroids, cardiac glycosides and alkaloids. Over the past decade, plant cell cultures have been investigated as an alternative means of producing commercialy important secondary products. The potential advantage over traditional field methods of cultivation includes independence from geographical, climatic and political problems (Ellis, 1984; Heinstein, 1985; Staba, 1985; Zenk et al., 1985).

By 1982 at least 30 secondary products were known to accumulate in plant cell culture systems in concentrations which were either equal to or higher than those in the parent plant. As Deus and Zenk (1982) succinctly pointed out, industrial application of plant cell culture can occur only if the price of the field produced product exceeds that of the cell culture produced product; thus, high priced products with a limited demand are more likely to justify production by cell culture. Despite considerable effort, however, only shikonin is currently produced commercially by cell culture techniques (Staba, 1985).

A wide range of alkaloids have been successfully detected in cell cultures, covering all the main alkaloids types (Anderson et al., 1985 and 1987). Major research interest on alkaloid production in plant cell cultures has focused on the production of indoles (Zenk, 1980; Neumann et al., 1983) and isoquinolines (Zenk, 1985). Among the indoles, prime effort has gone into attempts to produce the highly priced Catharanthus alkaloids vinblastine and vincristine utilized in cancer therapy; in the isoquinolines, interest has focused on the production of berberine and the morphinan alkaloids used as analgesic and antitussive agents.

The problems and difficulties found in working with these systems are typical of research involving plant cell cultures. There have been some successes and it is now possible to achieve relatively high yields of the indole alkaloids ajmalicine, serpentine, and vindoline (Zenk et al., 1977; Kutney et al., 1986) and the isoquinolines berberine and jatrorrhizine, with yields maintained when cells are grown on a large scale in bioreactors (Zenk, 1985; Breuling et al., 1986).

Improvements in low yield or nonoccurrence of secondary metabolites in plant cell cultures have been attempted in five general areas :

1 - The selection of high yielding cells with a stable genome.

2 - The use of various media, hormones, precursors, and elicitors.

3 - Knowledge of the occurence and regulation of the enzymes of a biosynthetic sequence.

4 - Understanding the methods of product accumulation.

5 - Investigation into product metabolism.

SELECTION OF HIGH YIELDING CELLS FOR CULTURE

The variability in any population for any trait has long been recognized. The production of a secondary metabolite by a plant is no exception; so the selection of a specimen or a variety with a high yield of a desired natural product is a prerequisite to the development of a cell culture with high yields.

The development of radioimmunoassay (RIA), a highly sensitive technique for the quantitative analysis of constituents in plant extracts has revolutionized the selection of high yielding cell clones. An admirable example has been the selection of Catharanthus roseus plants with high yields of vindoline, ajmalicine and serpentine, 300 mg l^{-1} (Zenk et al., 1977) and more recently 2g l^{-1} (Kutney et al., 1986). The use of cell fluorescence and RIA made it possible to select individual cells with high yields of a given natural product for growth into a producing culture. Deus and Zenk (1984) found, however, that to maintain high yielding cell cultures, continuous selection was necessary. In a similar manner, it was also found possible to develop cells high yielding for berberine, 2.25 g l^{-1} (Hinz and Zenk, 1981; Sato and Yamada, 1984; Breuling et al., 1986).

Our laboratory is currently investigating the three Ailanthus altissima the leaves and stems of which were used in traditional medicine. In our cell cultures of Ailanthus we found that the quassinoids present in the whole plant were absent but the cell suspensions produced high yields of the indole alkaloids canthin-6-one and 1-methoxy–canthin-6-one. The levels of the alkaloids varied depending on the source of seed and hence alkaloid yield was improved by plant screening. In this particular case, culture yields have remained stable for several years (Anderson et al., 1986).

MEDIA, HORMONES, PRECURSORS AND ELICITORS

Alkaloid yield may be improved by optimisation of cultural conditions (Mantell and Smith, 1983). In particular, variations in medium produced significantly different levels of 1-methoxy–canthin-6-one in Ailanthus cell suspension cultures (Anderson et al., 1986).

Changes in hormones can be useful, a fact which was used to good effect in the development of production media for alkaloid production from Catharanthus roseus (Deus and Zenk, 1983) and Thalictrum minus (Nakagawa et al., 1986). Kutney et al. (1986) have also improved yields of ajmalicine in Catharanthus roseus by using bioregulators.

Other factors may also become critical, for example, increasing the dissolved oxygen content of the medium to 50% of air saturation considerably improves the production of berberine and jatrorrhizine by cell culture. But increasing the oxygen tension by increasing the aeration rate leads to growth reduction due, probably, to shear stress (Breuling et al., 1986).

With Papaver in cell culture, these techniques for improving alkaloid yield have been found to have less clear cut advantages, although conditions which promote cell differentiation appear in many instances to improve both production and yield of morphinan alkaloids (Constabel, 1985; Roberts, 1986).

It has been found that in some instances supplementing the culture medium with one of the biosynthetic precursors increases alkaloid yield (Mantell and Smith, 1983). Deus and Zenk (1983) used L-tryptophan and L- tryptamine to good effect with Catharanthus roseus cultures producing indole alkaloids and we have shown increases of 50 - 100% in the production of 1-methoxy-canthin-6-one in cell suspensions fed L-tryptophan at 250-500 mg l^{-1} (Anderson et al., 1986).

The use of fungal elicitors of secondary metabolites is a recent approach (Di Cosmo and Tallevi, 1985). Using autoclaved conidia of pathogens, Heinstein (1985) showed that increases in the morphinan alkaloids codeine and morphine in Papaver somniferum cell suspensions could be obtained. More commonly, fungal elicitors in Papaver somniferum cell cultures cause increased yields of the benzophenanthridine alkaloid sanguinarine which is far in excess of levels of this alkaloid found in the whole plant (Eilert et al., 1985). A 50% increase in 1-methoxy-canthin-6-one in Ailanthus cell suspensions was obtained using autoclaved mycelium of Colletotrichum lindemuthanium (Anderson et al., 1986). Halbrock's group treated parsley cells with fungal elicitors to show coordinated changes in transcription rates, m-RNA amounts, and translation activities which effect activities of two enzymes crucial to phenylpropanoid metabolism (Schmelzer et al., 1985). The use of elicitors merits further exploration to determine its potential.

ENZYME ISOLATION AND LOCALISATION

The characterisation and localisation of the enzymes involved in the synthesis of a secondary metabolite is essential to understanding the production both in the whole plant and in cell cultures. This area has been slow to develop despite the elucidation of many metabolic pathways in the 1950's and 60's. In the last ten years, however, two areas have been looked at in detail.

The isolation of the enzymes required for the production of alkaloids in Catharanthus roseus has been a major focus for a number of research groups (Anderson et al., 1985). Enzymes required for the synthesis of strictosidine and cathenamine from tryptamine and secologanin have been isolated from cell culture (Stockigt, 1979). The major steps in the biosynthetic sequence leading to the heteroyohimbine alkaloids, via strictosidine, were known by 1980 (Zenk, 1980), the NADPH enzyme responsible for the formation of tetrahydroalstonine has been recently characterised (Hemschidt and Zenk, 1985). Studies by other workers has resulted in the isolation of enzymes responsible for the final steps in the biosynthesis of vindoline, the monomeric half of vinblastine (Fahn et al., 1985).

The other major interest has been in the isolation of the enzymes associated with the production of the isoquinoline alkaloids (Zenk, 1985; Zenk et al., 1985). Berberis and Coptis cell cultures have proved rich sources of both alkaloids and the enzymes associated with alkaloid biosynthesis. (S)-reticuline is the precursor of a number of isoquinoline alkaloids and the sequence from dopa to (S)-reticuline is now largely understood (Zenk, 1985; Zenk et al., 1985). The key enzyme required for the biosynthesis of berberine, (S)-tetrahydroberberine oxidase (STOX) (Amann et al., 1984), the berberine bridge enzyme (Steffens et al., 1985), and the methylene-dioxy group forming enzyme (Rueffer and Zenk, 1985) have all been isolated (Fig. 1).

The key to the problem of morphinan production in tissue culture appears to be the lack of ability in most cells to convert (S)-reticuline to (R)-reticuline, for whilst a large proportion of the non-morphinan alkaloids are derived from (S)-laudanosoline via (S)-reticuline, morphinan production has an absolute requirement for the (R)-isomer. Since conversion of (S)-reticuline to 1,2-dehydroreticuline appears

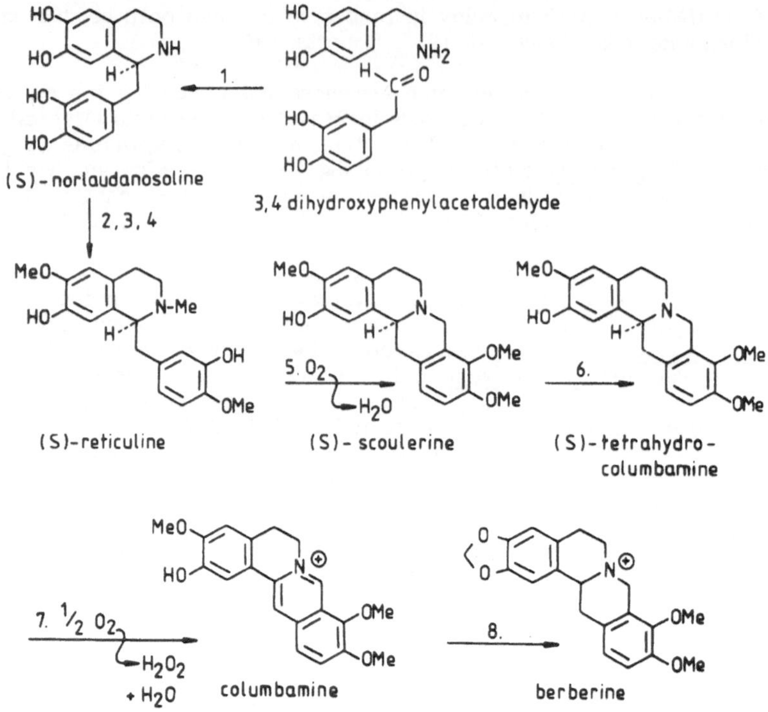

(S)-norlaudanosoline

3,4 dihydroxyphenylacetaldehyde

(S)-reticuline

(S)-scoulerine

(S)-tetrahydro-columbamine

columbamine

berberine

Enzyme	Source
1. (S)-norlaudanosoline synthase	Eschscholtzia
2. 6-O-methyltransferase	Argemone
3. 4'-O-methyltransferase	Glaucium
4. N-methyltransferase	Berberis
5. Berberine bridge enzyme (BBE)	Berberis
6. (S)-Scoulerine 9-O-methyltransferase	Berberis
7. (S)-Tetrahydroprotoberberine oxidase (STOX)	Berberis
8. Berberine synthase	Berberis

Figure 1

Biosynthesis of berberine

to require the ubiquitous STOX enzyme of the berberine pathway, it may be that the enzyme(s) required for the conversion of the 1,2–dehydroreticuline to (R)–reticuline are either absent or are not active in many cell cultures (Fig. 2).

Hodges and Rapoport (1982) reported the enzymic conversion of (R)–reticuline to salutaridine in cell free systems from the whole plant. Furuya et al. (1984) characterized the conversion of codeinone to codeine using cell free systems and,

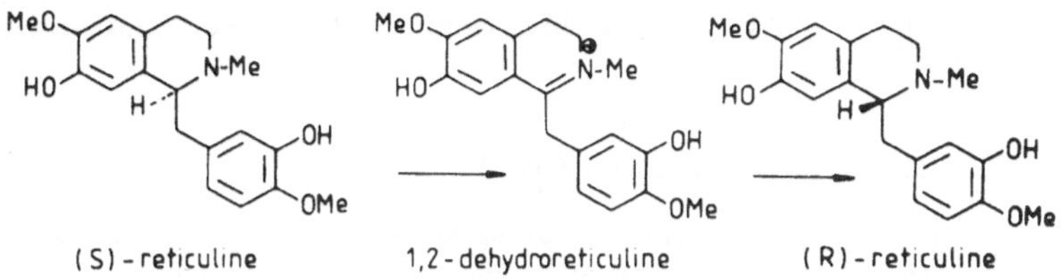

Figure 2

The conversion of (S)-reticuline To (R)-reticuline

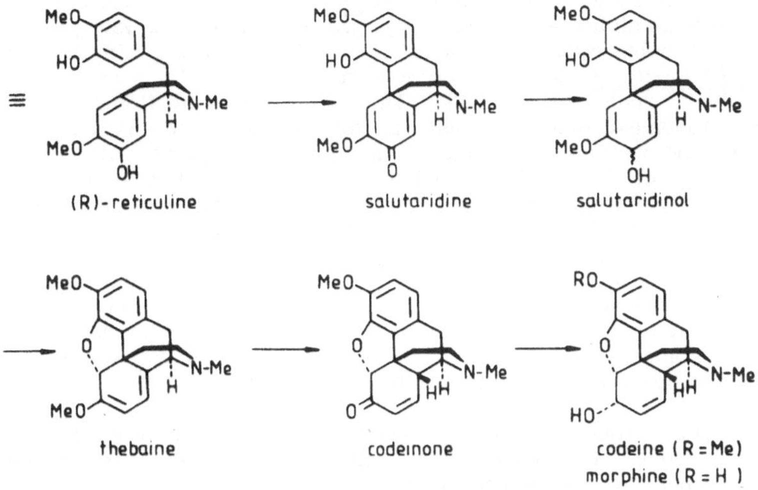

Figure 3

The conversion of (R)-reticuline to codeine and morphine

also, immobilized cells. Further studies should help to elucidate the requirements and components of the fully functioning pathway to the morphinans (Fig. 3).

The location of the enzymes for alkaloid production can be very critical. It is possible that the alkaloid could be produced in one cell and stored in another type of cell or produced in one compartment of a cell and stored in another. Since cell selection has been on the basis of cells with high alkaloid content, it is important to know whether these cells also synthesize alkaloids. Amann et al. (1986) have successfully localized two of the membrane-bound enzymes of berberine synthesis which were found in the same vesicle. These vesicles aggregate with the cell vacuole, releasing their alkaloidal contents into the newly formed vacuolar compartment where enzyme activity appears to be lost. The involvement of the cell vesicles both in alkaloid synthesis and storage has implications concerning the lack of formation of (R)-reticuline in many cell cultures and suggests that specialist vesicles associated with differentiated tissue are required (Homeyer and Roberts, 1984; Kutchen et al., 1985).

PRODUCT ACCUMULATION OR EXCRETION

The ability of a cell culture to store high levels of alkaloid or to excrete the alkaloid into the medium is critical to the development of high yielding culture. From a commercial point of view the fact that plants tend to store their secondary products in vacuoles can be a disadvantage and with immobilized cells, excretion into the medium is advantageous.

While Matile (1978) suggested that alkaloids were taken up into vacuoles of Chelidonium majus latex by a process of simple diffusion, Renaudin and Guern (1983) using isolated Catharanthus roseus cells have postulated that the non-protonated alkaloids are trapped by protonation in the low pH vacuoles (pH values of 3 - 5 being given by various groups). Recent work using isolated vacuoles of Catharanthus roseus suggests active transport of alkaloids across the tonoplast into the vacuolar space utilizing ATP. The movement of alkaloids in both directions through the tonoplast indicated that the alkaloid was not modified subsequent to passage into the vacuole and that in this particular plant cell culture, alkaloid sequestration is not the result of an ion trap mechanism (Deus-Neumann and Zenk, 1986). Papaver somniferum plants sequester alkaloids in vesicles in the latex while in cell culture, vacuoles containing the morphinan alkaloid thebaine were found only associated with laticifer-like cells (Kutchan et al., 1985). In Papaver somniferum as with Catharanthus roseus specificity of uptake for alkaloids indigenous to the plant was observed using latex vacuoles. These vacuoles took up large amounts of morphine which remained within the vacuole and could not be displaced, suggesting that in this instance an ion-trap mechanism might be involved. No clear requirement for MgATP was observed (Homeyer and Roberts, 1984).

It has been found by Furuya's group (1984) investigating the conversion of codeinone to codeine that immobilized cells of Papaver excrete alkaloids into the medium. Our non-morphinan producing cell cultures of Papaver somniferum also readily excrete the alkaloid cryptopine into the medium (Anderson et al., 1983) and we have experienced similar results with other members of the Papaveraceae. In some cases, cells may be made to excrete material into the medium by making the cell tonoplast more permeable. The extent to which this is a practical possibility has been investigated (Lundberg et al., 1986). Wink (1984) suggested, as a result of experiments with Lupin cell cultures that the medium may be used as a lytic compartment by the cells.

ALKALOID METABOLISM

One under-researched area is that of alkaloid metabolism in plants. The extent to which alkaloids accumulate either in a plant or in a cell culture may depend not only on an ability to sequester alkaloids but on the rate at which they are further metabolised. While Fairbairn showed metabolism of morphine (16 %) by Papaver somniferum latex (Fairbairn and Steele, 1981), Furuya's group (1978) has not observed metabolism of morphine in cell cultures. On the other hand, Ailanthus cultures apparently show rapid turnover of labelled 1-methoxy-canthin--6-one (Anderson et al., 1986).

CONCLUSIONS

The development of cell suspension cultures with high levels of secondary natural products depends on successfully manipulating all of these factors. Research has reached a stage where gene induction and activation and the regulation of enzymes is a real possibility. Success in these areas would enhance our ability to manipulate plant cell cultures for the commercial production of natural products.

REFERENCES

Amann, M., Nagakura, N., and Zenk, M. H., 1984, (S)-tetrahydroprotoberberine oxidase the final enzyme in protoberberine biosynthesis, Tetrahedron Letters, 25:953.

Amann, M., Wanner, G., and Zenk, M. H., 1986, Intracellular compartmentation of two enzymes of berberine synthesis in plant cell cultures, Planta, 161: 310.

Anderson, L. A., Homeyer, B. C., Phillipson, J. D., and Roberts, M.F., 1983, Dopamine and cryptopine production by cell suspension cultures of Papaver somniferum, J. Pharm. Pharmac., 35:21P.

Anderson, L. A., Phillipson, J. D., and Roberts, M. F., 1985, Biosynthesis of secondary products by cell cultures of higher plants, in : "Advances in Biochemical Engineering and Biotechnology", Vol. 31 : Plant Cell Culture, A. Fiechter, ed., Springer-Verlag, Berlin.

Anderson, L. A., Hay, C. A., Roberts, M. F., and Phillipson, J. D., 1986, Studies on Ailanthus altissima cell suspension cultures : Precursor feeding of L--(methylene ^{14}C)-tryptophan and L-tryptophan, Plant Cell Reports, in press.

Anderson, L. A., Phillipson, J. D., and Roberts, M. F., 1987, Alkaloid production by plant cells, in : "Plant and Animal Cell Cultures : Process possibilities", C. Webb and F. Mavituna, eds., Ellis Horwood, Ltd., Chichester.

Breuling, A. W., Alfermann, A. W., and Reinhard, E., 1986, Cultivation of cell cultures of Berberis wilsonae in 20 l air lift bioreactors, Plant Cell Reports, 4:220.

Constable, F., 1985, Morphinan alkaloids from plant cell cultures, in : "The Chemistry and Biology of Isoquinoline alkaloids", J. D. Phillipson, M. F. Roberts and M. H. Zenk, eds., Springer-Verlag, Heidelberg.

Deus-Neumann, B., and Zenk, M. H., 1984, Instability of indole alkaloid production in Catharanthus roseus cell suspension cultures, Planta Medica, 50:427.

Deus-Neumann, B., and Zenk, M. H., 1986, Accumulation of alkaloids in plant vacuoles does not involve an ion-trap mechanism, Planta, 167:44.

Di Cosmo, F., and Tallevi, S. G., 1985, Plant cell cultures and microbial insult : Interactions with biotechnological potential, Trends Biotechnol., 3:110.

Ellis, B. E., 1984, Probing secondary metabolism in plant cell cultures, Can. J. Bot., 62:2912.

Eilert, U., Kurz, W. G. W., and Constabel, F., 1985, Stimulation of sanguinarine accumulation in Papaver somniferum cell cultures by fungal elicitors, J. Plant. Physiol., 119:65.

Fahn, W., Laussermair, E., Deus-Neumann, B., and Stockigt, J., 1985, Late enzymes of vindoline biosynthesis. S-adenosyl-L-methionine:11-0-demethyl-17--O-deacetylvindoline 11-O-methyl-transferase and unspecific acetyl-esterase, Plant Cell Reports, 4:337.

Fairnbairn, J. W., and Steele, M. F., 1981, Biosynthetic and metabolic activities in some organelles in Papaverum somniferum latex, Phytochemistry,20:1031.

Furuya, T., Nakano, M., and Voshikawa, T., 1978, Biotransformations of (RS)-reticuline and morphinan alkaloids by cell cultures of Papaver somniferum, Phytochemistry, 17:891.

Furuya, T., Yoshikawa, T., and Taira, M., 1984, Biotransformations of codeinone to codeine by immobilized cells of Papaver somniferum, Phytochemistry, 23:999.

Heinstein, P. F., 1985, Future approaches to the formation of secondary natural products in plant cell suspension cultures, J. Nat. Prod., 48:1.

Hemscheidt, T., and Zenk, M. H., 1985, Partial purification and characterisation of a NADPH dependent tetrahydroalstonine synthase from Catharanthus roseus cell suspension cultures, Plant Cell Reports, 4:216.

Hinz, H., and Zenk, M. H., 1981, Production of protoberberine alkaloids by cell suspension cultures of Berberis species, Naturwissenschaften, 68:620.

Hodges, C. C., and Rapoport, H., 1982, Enzymic conversion of reticuline to salutaridine by cell-free systems from Papaver somniferum, Biochemistry, 21:3729.

Homeyer, B. C., and Roberts, M. F., 1984, Alkaloid sequestration by Papaver somniferum latex, Z. Naturforsch., 39c:876.

Kutchan, T. M., Ayabe, S., and Coscia, C. J., 1985, Cytodifferentiation and Papaver alkaloid accumulation, in : "The Chemistry and Biology of the Isoquinoline Alkaloids", J. D. Phillipson, M. F. Roberts, and M. H. Zenk, eds., Springer-Verlag, Berlin.

Kutney, J. P., Aweryn, B., Chatson, K. B., Choi, L. S. L., and Kurz, W. G. W., 1986, Alkaloid production in Catharanthus roseus (L.) G. Don cell cultures. XIII. Effects of bioregulators on indole alkaloid biosynthesis, Plant Cell Reports, 4:259.

Lundberg, P., Linsefors, L., Vogel, H. J., and Brodelius, P., 1986, Permeabilisation of plant cells : ^{31}P-NMR studies on the permeability of the tonoplast, Plant Cell Reports, 5:13.

Mantell, S. H., and Smith, H., 1983, Cultural factors that influence secondary metabolite accumulations in plant cell and tissue cultures, in : "Plant Biotechnology", S. H. Mantell and H. Smith, eds., Cambridge University Press, Cambridge.

Matile, P., 1978, Biochemistry and function of vacuoles, Ann. Rev. Plant Physiol., 29:193.

Nakagawa, K., Fukui, H., and Tabata, M., 1986, Hormonal regulation of berberine production in cell suspension cultures of Thalictrum minus, Plant Cell Reports, 5:69.

Neumann, D., Krauss, G., Heike, M., and Groger, D., 1983, Indole alkaloid formation in storage suspension cultures of Catharanthus roseus, Planta Medica, 48:187.

Renaudin, J.-P., and Guern, J., 1983, Compartmentalisation mechanisms of indole alkaloids in cell suspension cultures of Catharanthus roseus, Physiol. Vég., 20:533.

Roberts, M. F., 1986, Papaver, in : "Cell Culture and Somatic Cell Genetics of Plants", Vol. 5 : Phytochemicals in Cultured Cells, I. K. Vasil and F. Constabel, eds., Academic Press, Orlando.

Rueffer, M., and Zenk, M. H., 1985, Berberine synthase, the methylene dioxy group forming enzyme in berberine synthesis, Tetrahedron Letters, 26:201.

Sato, F., and Yamada, Y., 1984, High berberine producing cultures of Coptis japonica cells, Phytochemistry, 23:281.

Schmetzer, E., Somissich, I., and Hahlbrock, K., 1985, Coordinated changes in transcription and translation rates of phenylalanine-ammonialyase and 4-coumarate:CoA ligase mRNA's in elicitor-treated Petroselinum crispum cells, Plant Cell Reports, 4:293.

Staba, E. J., 1985, Milestone in plant tissue culture systems for production of secondary products, J. Nat. Prod., 48:203.

Steffens, P., Nagakura, N., and Zenk, M. H., 1985, Purification and characterization of the berberine bridge enzyme from Berberis beaniana cell cultures, Phytochemistry, 24:2577.

Stockigt, J., 1979, Enzymic formation of intermediates in the biosynthesis of ajmalicine, strictosidine and cathenamine, Phytochemistry, 18:965.

Wink, M., 1984, Evidence for an extracellular lytic compartment of plant cell sussion cultures : The cell culture medium, Naturwissenschaften, 71:635.

Zenk, M. H., 1980, Enzymic synthesis of ajmalicine and related indole alkaloids, J. Nat. Prod., 43:438.

Zenk, M. H., 1985, Enzymology of benzylisoquinoline alkaloid formation, in : "The Chemistry and Biology of Isoquinoline Alkaloids", J. D. Phillipson, M. F. Roberts, and M. H. Zenk, eds., Springer-Verlag, Heidelberg.

Zenk, M. H., El-Shagi, H., Arens, H., Stockigt, J., Weiler, E. W., and Deus, B., 1977, Formation of the indole alkaloids, serpentine and ajmalicine in cell suspension cultures of Catharanthus roseus, in : "Plant Tissue Culture and Its

Biotechnological Application", W. Barz, E. Reinhard, and M. H. Zenk, eds., Springer-Verlag, Berlin.

Zenk, M. H., Rueffer, M., Amann, M., and Deus-Neumann, B., 1985, Benzylisoquinoline biosynthesis by cultivated plant cells and isolated enzymes, J. Nat. Prod., 48:725.

PAPAVER LATEX AND ALKALOID STORAGE VACUOLES

Margaret F. Roberts

School of Pharmacy, University of London
29/39, Brunswick Square
London, England, WC1N 1AX, U. K.

Papaver somniferum L. - the opium poppy - is an annual herb 50 - 150 cm in height. It is grown commercially under licence as the major source of the opiates codeine and morphine. The creamy colored latex oozes from the cut, unripe capsule and provides a readily available source of the laticifer contents. Opium, the dried exuded latex of the poppy, normally contains at least 25 alkaloids, which probably occur as salts of meconic acid or sulphate. The morphinan alkaloids are the predominant alkaloids in opium with morphine (up to 52% of the total alkaloids), codeine, and thebaine normally present. Papaverine, noscopine, and narceine are also commonly found in significant amounts. Microscopic study by Thureson-Klein (1970) has shown latex to be a multitude of particles suspended in a large central vacuole. This work, with the electron microscopy reported by Dickenson and Fairbairn (1975) and Nessler and Marlberg (1977), has established the presence of fragments of the endoplasmic reticulum, nuclei, mitochondria, Frey-Wyssling particles, and spherical bodies referred to as the 1000xg vacuoles. The bulk of these vacuoles contain alkaloids (Fairnbairn and Djote, 1980; Roberts, 1971) and form a distinct pellet when the latex is centrifuged at 1000 x g for 30 minutes. The formation of these vacuoles within the laticifers results from localised dilatation of elongated stacks of endoplasmic reticulum (Thureson-Klein, 1970; Nessler and Marlberg, 1977). Membrane staining with zinc iodine osmium tetroxide suggests that they are analogous to the central vacuole of other cells. However, this has not been conclusively demonstrated in view of the similar staining of dictyosome derived vacuoles reported by Danwalder and Whaley (1973) and Marty (1973 a and b).

METABOLIC ACTIVITY

Meissner (1966 a and b) and Meissner and Mothes (1964) studied the metabolic activity of the expelled latex of 29 species of plants. Evidence of ribosomes in poppy latex was found and a high rate of gasous exchange not due to microorganisms was reported. Fairbairn et al. (1968) showed the presence of a Papaver somniferum latex fraction with a P/O ratio characteristic of mitochondria. The inhibition of the oxygen uptake by malonate, an inhibitor of succinate dehydrogenase, indicated that at least part of the respiration observed in the latex resulted from Krebs cycle activity. The polyphenolase activity observed by Meissner (1966 a) was found to be entirely confined to the smaller particles of the 1000xg fraction (Roberts, 1971). The polyphenolase complex isolated from the 1000xg vacuoles oxidized a variety of phenolic substrates, including p-cresol, catechol, p-coumaric acid, hydroquinone, and tyrosine but not observed to oxidize (R)-reticuline to salutaridine, intermediates in morphinan alkaloid biosynthesis from tyrosine. The polyphenolase complex was found to be composed of both soluble and membrane-bound fractions only one of which, the soluble form, utilized tyrosine, a probable precursor of the morphinan alkaloids (Roberts, 1974). Using sucrose gradients it was possible to show that the catecholase (polyphenolase) activity of the 1000xg fraction was localised in two distinct types of vacuoles. The lighter vacuoles which sedimented at the 30% sucrose level contained a soluble enzyme which was readily released on vacuole plasmolysis, whereas the catecholase found within the heavier vacuoles, sedimenting at 55-60% sucrose, was membrane-bound and showed significant activity only in the presence of Triton X-100 (Roberts et al., 1983).

In efforts to elucidate its role in alkaloid biosynthesis, a number of enzymes involved in general metabolism have been identified in Papaver somniferum latex. Enzymes involved in the glyoxylic acid and tricarboxylic acid cycles have been found, namely, aconitase, isocitrate dehydrogenase, succinate dehydrogenase, fumarase, malate dehydrogenase and isocitrate lyase. Enzymes associated with lysosomes (arylesterase and acid phosphatase) have also been isolated. Finally, the occurrence of some enzymes previously reported as occurring in poppy seedlings (Jindra et al., 1966) has been investigated : peroxydase, glutamate-oxaloacetate and glutamatepyruvate transaminases; and phenylalanine, tyrosine, dopa and glutamic acid decarboxylases (Antoun and Roberts, 1975 a and b). Almost all these enzymes were found in the latex supernatant, i.e., the latex less the 1000xg fraction. Acid phosphatase activity was found principally in a fraction which sedimented at 25,000 x g (Antoun and Roberts, 1975 b). That acid phosphatase and arylesterase are absent from the 1000xg fraction suggests that the 1000xg vacuoles are not lysosomes, which contrasts with the findings for particles of similar density in the latex of Hevea brasiliensis (Pujarniscle, 1968) and Chelidonium majus (Matile et al., 1970; Matile, 1976). Another difference between the latex of Papaver somniferum and that of Hevea is the apparent absence of peroxidase and catalase in poppy latex, which is the more surprising since both enzymes are present in poppy seedlings.

Two of the key enzymes involved in the initial phases of alkaloid biosynthesis, a decarboxylase and a transaminase, have been shown to occur in acetone powder preparations of Papaver somniferum root (Jindra et al., 1966). The presence of L-dopa decarboxylase has been demonstrated in Papaver somniferum latex through the use of L-dopa-1- ^{14}C and L-dopa-3- ^{14}C as substrates (Roberts and Antoun, 1978). Recent work (Zenk, 1985; Zenk et al., 1985) gives a clear indication of the involvement, in isoquinoline alkaloid biosynthesis, of an aromatic decarboxylase in the conversion of L-Dopa during the formation of the condensation product (S)-norlaudanosoline.

The probable involvement of tyramine in the early steps of alkaloid biosynthesis is interesting. In feeding experiments with 7-^{14}C-tyramine, 1-2% incorporation into the morphinan alkaloids is obtained, with 97% of the label found in the tetrahydroisoquinoline (TIQ) segment of the molecule (C-15, Fig. 1) (Roberts et al., 1987). Similar results were found in tissue culture experiments with Berberis but on that occasion, the label was split between the TIQ and the benzyl derived segments of the berberine molecule on a 75-25% basis (Zenk et al., 1985). Clearly tyramine appears to have an important role in alkaloid formation and its formation from L-tyrosine would also require an aromatic amino-acid decarboxylase (Roberts et al., 1987).

Other enzymes established as occurring in the latex are the methyltransferases responsible for the conversion of (S)-norlaudanosoline to (S)-reticuline. Again, these enzymes were found to occur in the latex supernatant (Antoun and Roberts, 1975 c), and have been examined in some detail in tissue cultures of Argemone, Berberis, Eschscholtzia and Glaucium by Zenk's group who have now isolated all the enzymes of the pathway to berberine using plant cell cultures (Zenk, 1985). Recently, Amann et al. (1986) have shown the occurrence of vacuoles in Berberis wilsoniae containing two of the eight enzymes necessary for the synthesis from (S)-reticuline to berberine.

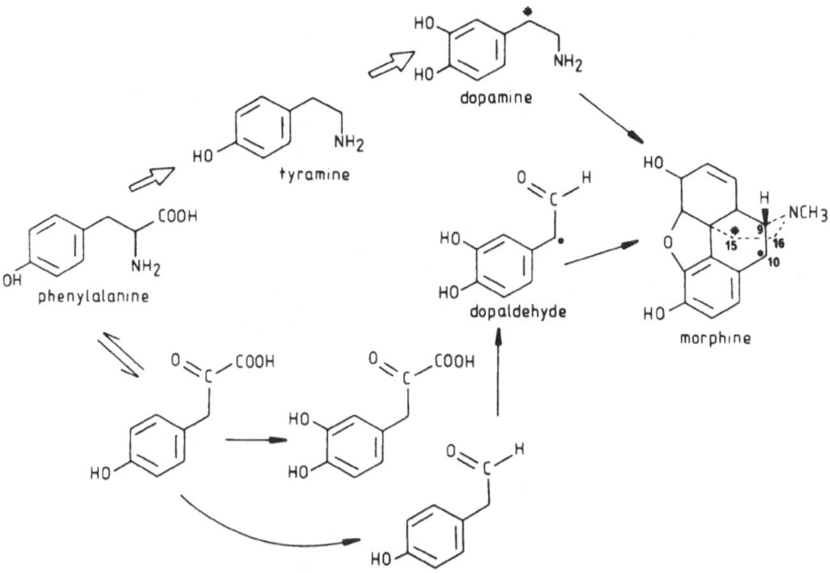

Figure 1

Biosynthesis of morphine :
Pathways through dopamine and dopaldehyde

While the two enzymes do not act sequentially, the end result is that the product of the second enzyme is trapped in the vacuole, which merges with other vesicles to form vacuoles which release their contents (protoberberine alkaloids) into a central vacuole.

LOCALIZATION OF STORED ALKALOIDS AND DOPAMINE

The number of enzymes associated with alkaloid biosynthesis and found outside the 1000xg fraction is apparently at odds with the reported biosynthesis and metabolism of the alkaloids by the 1000xg fraction (Meisner and Mothes, 1964; Bohm et al., 1972; Fairbairn et al., 1974; Fairbairn and Steele, 1981). Since different centrifugation programs have been used (Fairbairn and Steele, 1981; Roberts and Antoun, 1978) it is possible that some of the enzymes responsible for alkaloid bio-synthesis are located in particles somewhat lighter than the 1000xg fraction, but not always clearly separated from it. It is suggested that more than one type of vacuole may occur within the 1000xg fraction, and, therefore, variable sedimentation of the lighter vacuoles of this non-homogeneous fraction may account for the difference in reported results. Separating the 1000xg fraction on sucrose gradients has shown vacuoles which differ in morphology and density as well as in enzyme and metabolite content. The most dense population contains the bulk of the alkaloids and should be designated "alkaloid containing vacuoles" (ACV). The gradient work (Roberts et al., 1983) established that the group of vacuoles which banded at 30% sucrose contained the soluble catecholase as well as some dopamine, while the very dense vacuoles - which banded at 55% sucrose - contained membrane-bound catecholase, together with most of the dopamine and alkaloids (Fig. 2).

Alkaloid levels in the ACV showed a rapid (4-fold) increase from bud stage to 7 days after petal opening. ($^{14}CH_3$)-morphine is rapidly taken up by the latex vacuoles, which adds support to the view that they are storage vacuoles and the apparent location of ($^{14}CH_3$)-morphine, as observed in Fig. 2-A, suggests that variation exists in the efficiency of morphine uptake into the different vacuoles.

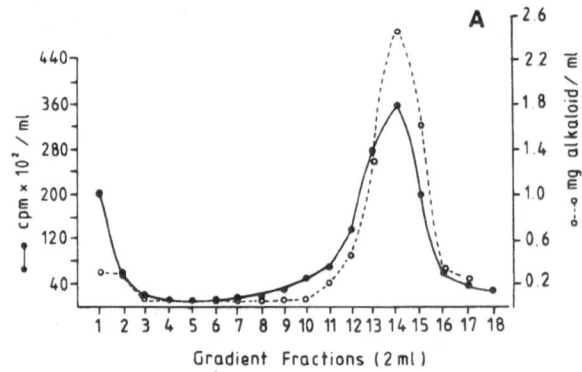

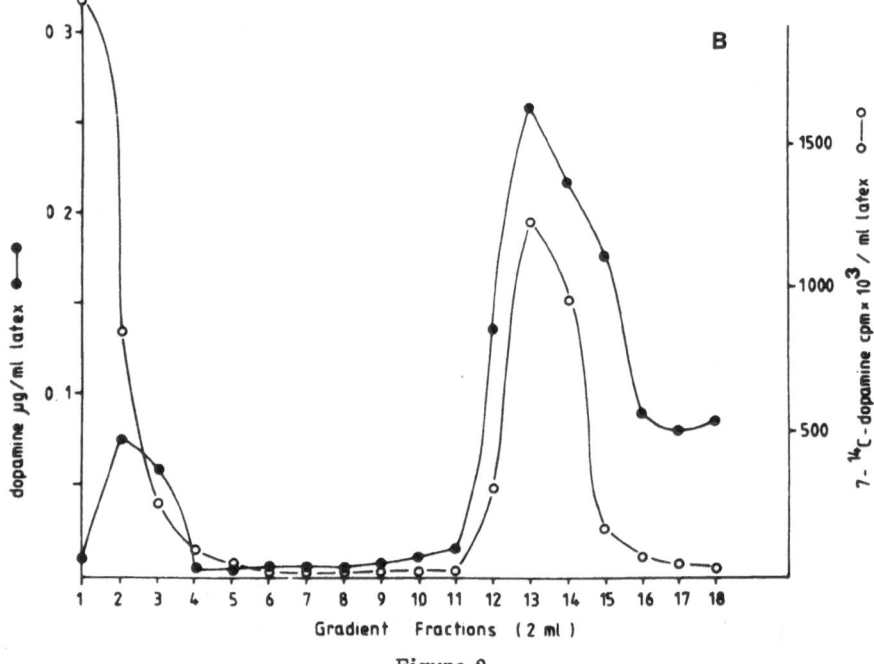

Figure 2

Sucrose gradient separation of <u>Papaver</u> <u>somniferum</u> latex
1000xg vacuoles

A. Distribution and uptake of morphine.
B. Distribution and uptake of dopamine.

DOPAMINE ACCUMULATION

The high level of dopamine, a principle precursor of morphinan alkaloids, present and apparently stored in the 1000xg fraction of <u>Papaver</u> <u>somniferum</u>, is interesting from a biosynthetic point of view and may well account for the low incorporation of added L-dopa and dopamine, since the labelled precursor would be taken up by a system which is certainly not deficient, and probably near to saturation with native dopamine. The extent to which the dopamine inside the vacuoles is involved in alkaloid biosynthesis is conjectural, since no further metabolism of dopamine within the vacuole has been observed. It is clear that the polyphenolases in the vacuole, upon exposure to atmospheric oxygen subsequent to wounding, form a melanin-like polymer from the vacuolar dopamine. Consequently, regardless of metabolic purpose, the two exist as component of an effective defensive system.

Previous work on the occurrence of dopamine, which is required for isoquinoline alkaloid synthesis in Papaver, is sparse and suitable models for the production and vacuolar storage of dopamine in plants are not available. In animals, dopamine and other catecholamines are important in neurotransmission and are found stored in the adrenal medula in neurotransmitter storage vacuoles or chromaffin granules (Slotkin et al., 1978; Scherman et al., 1983). The uptake of dopamine by these storage vacuoles is highly dependent on temperature and has an absolute requirement for ATP and Mg^{2+}. Catecholamine uptake may be directly correlated to an inside positive membrane potential ($\Delta\Psi$) and a proton gradient (ΔpH) rather than to ATPase activity. Further, it has been found that a specific monamine carrier mediates in the coupling between the generated electrochemical gradient and substrate accumulation. Although there is some evidence for unique vacuolar storage of dopamine in plant cell tissue cultures of Papaver bracteatum (Kutchan et al., 1983 and 1986), it has not been possible to separate the latex 1000xg vacuoles of Papaver somniferum into those which accumulate alkaloids and those which accumulate dopamine. Most of the dopamine was to be found with the alkaloids in the vacuoles which banded at the 50-60% sucrose interphase using discontinuous gradients. More dopamine was found at the surface of the gradient than alkaloid and this may reflect the rupture of 1000xg vacuoles during handling – which, if correct suggests that dopamine is accumulating in separate vacuoles from the alkaloids since the ratio of alkaloid to dopamine in the supernatant is much lower than in the 1000xg pellet.

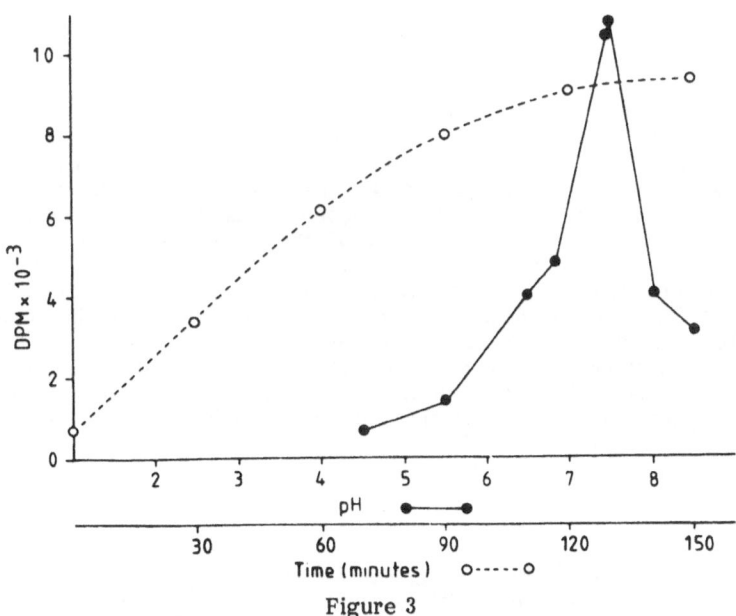

Figure 3

Dopamine accumulation vs. time and pH
of Papaver somniferum latex 1000xg vacuoles

Dopamine accumulation in the 1000xg fraction is a relatively slow process in that the accumulation with time remains linear over the first 60 minutes at 25°C (Fig. 3) (Homeyer and Roberts, 1984 a). This contrasts sharply with the rapid accumulation of dopamine by the neurotransmitter storage vacuoles in mammals, where accumulation is linear with time for only the first 4 minutes of exposure (Slotkin et al., 1978), as well as with the rapid accumulation of alkaloids by the latex 1000xg vacuoles.

In Papaver somniferum latex, dopamine uptake is markedly sensitive to temperature and has maximal uptake into the vacuoles at 30°C. Accumulation is also sharply affected by pH with maximal activity at pH 7.5 (Fig. 3). The use of ascorbate (20 mM) and DIECA (50 mM) demonstrated that the loss of activity at alkaline pH was not attributable to the oxidation of dopamine. The accumulation of dopamine showed saturation kinetics with an apparent K_m of 4.37 mM. Exogenous ATP and Mg^{2+} caused no significant stimulation.

Table 1

Catecholamine Uptake by 1000xg Vacuoles

Catecholamine assayed	Uptake (as % of increase in dopamine concentration)
Adrenaline	17
Hordenine	0
Noradrenaline	8
Tyramine	22

Mammalian chromaffin granules demonstrate the ability to take up dopamine, adrenalin, serotonin or noradrenaline with each of these compounds able to competitively inhibit the uptake of the others, thus providing the support for the concept that a single carrier is likely responsible for the transport of the several amines (Kanner et al., 1979). Uptake of catecholamines by the 1000xg vacuoles demonstrated considerable specificity for dopamine as shown in Table 1. The uptake of noradrenaline was less than 10% that of the weight of dopamine taken up by a vacuole control sample in these experiments; in contrast, dopamine and noradrenaline are taken up to about the same extent by chromaffin granules (Pletscher, 1977). The formation of noradrenaline from dopamine could not be demonstrated in 1000xg vacuoles, and therefore the existence of a membrane dopamine-β-hydroxylase as found in the mammalian adrenal membrane (Pletscher et al., 1977) is unlikely. The accumulation of dopamine in the 1000xg vacuole fraction of Papaver somniferum latex is of particular interest in view of its role as a prime precursor of the morphinan alkaloids.

ALKALOID ACCUMULATION

While the isolated latex of Papaver has been shown to carry out limited biosynthesis, the previous work gives no clear evidence that the latex is the primary site of alkaloid biosynthesis; latex, however, and in particular the latex vacuoles sedimenting at 1000xg, has been shown to be the major site of alkaloid accumulation in the plant (Fairbairn and Steele, 1981; Roberts et al., 1983; Bohm et al., 1972). Previous work on alkaloid storage in latex has investigated the uptake of sanguinarine and chelerythrine by the latex vacuoles of Chelidonium majus. Sanguinarine and chelerythrine have a greater affinity than do berberine, coptisine, and dihydrocoptisine for the contents of the vacuolar space since these latter alkaloids can be displaced from the vacuoles through the addition of sanguinarine. It is suggested that the alkaloids are only bound in the vacuoles by phenolic compounds and chelidonic acid, with chelidonic acid responsible for the differential affinity of the alkaloids. It is further suggested that complexes are formed with phenols and chelidonic acid which cannot move through the tonoplast (Matile et al., 1970; Jans, 1973; Matile, 1978). In this case, the process of accumulation in the vacuoles did not appear to require energy, however the uptake of alkaloids into cell vacuoles isolated from tissue cultures of Catharanthus roseus was stimulated by additions of ATP and Mg^{2+} (Deus-Neumann and Zenk, 1986).

Table 2

Levels of Morphinan Alkaloids in Bands A and B
of Vacuoles obtained with Discontinuous Gradients

Alkaloid	μg/mg protein	
	Band A	Band B
Narcotine	49.9	52.6
Papaverine	51.7	100.0
Codeine	96.5	118.3
Thebaine	89.6	316.9
Morphine	600.0	1326.8
Dopamine	nil	178.6

Investigation using the light microscope with a phase contrast objective has shown the 1000xg vacuoles of Papaver somniferum latex to be variable in size, ranging from 100 to 770 nm in diameter. On discontinuous sucrose gradients, the larger vacuoles were trapped at the 50-60% interphase (Band B) while the lighter vacuoles banded at the 40-50% interphase (Band A). the two groups of vacuoles were somewhat different in the composition of their alkaloid content (Table 2). The lighter vacuoles (Band A), while containing alkaloids, did not contain dopamine whereas the more dense vacuoles (Band B) contained both alkaloids and dopamine. Band B has not yet been separated into alkaloid and dopamine containing populations.

Uptake of thebaine and morphine by 1000xg vacuoles from the latex of either Papaver somniferum or Papaver bracteatum was not markedly temperature dependent (Table 3) (Roberts and Homeyer, 1985). Uptake was rapid and complete in 6 minutes and precluded estimation of an apparent K_m for morphine as has been obtained for dopamine. Sequential additions of morphine (up to 320 μg) were made to 300 μl samples of 1000xg vacuoles and in each case, ca. 73% of the added sample was taken up within 4 minutes. Uptake was dependent on pH and despite a native pH of 6.8 for latex, maximal activity occurred at a pH of 7.3-7.5 (Fig. 4) (Homeyer and Roberts, 1984 b). Vacuoles preloaded with ($^{14}CH_3$)-morphine did not leak significant levels of alkaloid into the medium over the pH range tested and were capable of high levels of morphine uptake even at pH 8.5.

Table 3

Temperature Dependence of Alkaloid Uptake

	Q_{20} of labelled morphine uptake by latex vacuoles	
	pH 6.2	pH 7.0
P. somniferum	3.95	1.59
P. bracteatum	4.12	3.18

Morphine uptake was measured at 10°, 30°, 0° and 20°
Q_{20} = the average ratio of high to low temperature sample 1000xg vacuoles separated from the latex, and suspended in 200 mM Hepes, 500 mM mannitol; resuspended, adjusted for pH and samples maintained at the designated temperature; $^{14}CH_3$-morphine uptake was measured after 4 mins and the reaction stopped by centrifugation through a 20% sucrose gradient.

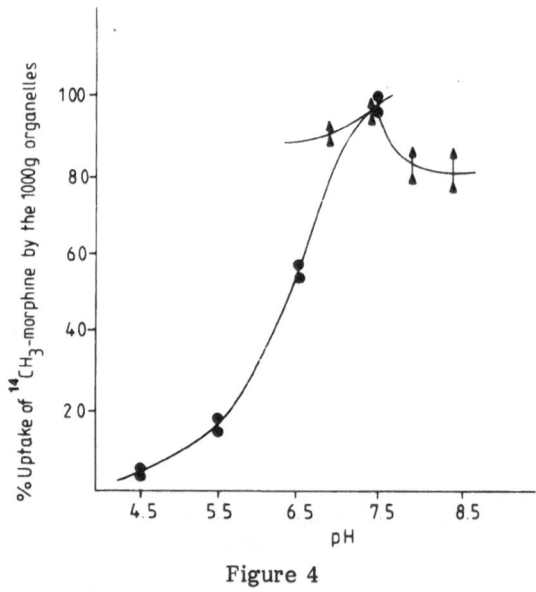

Figure 4

Morphins accumulation vs. pH
by Papaver somniferum latex 1000xg vacuoles

Essentially all the alkaloids in the latex are found in the 1000xg vacuoles. Recent experiments investigating alkaloid uptake have concentrated on morphine, partly because of the ready availability of $(^{14}CH_3)$-morphine but also because it is the major alkaloid accumulated in Papaver somniferum. Since codeine and thebaine are similar in structure to morphine, the uptake of these alkaloids was investigated as was the uptake of papaverine, sanguinarine, and a series of non-papaver alkaloids

(Table 4). The morphinan alkaloids and the benzylisoquinoline alkaloid papaverine are taken up to a similar extent; however, neither sanguinarine nor the non-papaver alkaloids were taken up by the vacuoles, clearly indicating that the vacuoles have a degree of specificity for the morphinans and papaverine. Uptake of morphine was unaffected by the presence of codeine or thebaine and consequently, a possible difference in ports of entry is suggested. The poor sensitivity of uptake to temperature, the rapid uptake of morphine, and the specificity for certain alkaloids suggests a channel protein mechanism rather than a protein carrier mechanism for passage through the vacuole membrane. The channel protein mechanism, in general, does not involve coupling to an energy source and permits much greater fluxes across the membrane than do carrier protein mechanisms (Alberts et al., 1983).

Table 4

Uptake of Alkaloids
by the 1000xg Vacuoles of Papaver somniferum

Alkaloid	% Uptake of alkaloids administered
Morphine	68
Codeine	63
Thebaine	64
Papaverine	50
β-coniceine	nil
Reserpine	nil
Atropine	nil
Quinine	nil
Cytisine	nil
Sanguinarine	nil

For the concentration of morphine and other alkaloids to build up within the vacuole, it seems apparent that some trapping mechanism must be involved. Matile et al. (1970) suggest that in the case of sanguinarine accumulation by Chelidonium majus latex vacuoles, the ion trap mechanism is limited by the available anions, in this instance, phenolics. In the case of the Papaver alkaloids, the association of meconic acid with the 1000xg vacuoles (Fairbairn and Williamson, 1978) may be an important factor in alkaloid accumulation. Renaudin and Guern (1982) and Neumann et al. (1983) suggested that alkaloids could accumulate by an ion trap mechanism in acidic vacuoles as a result of protonation, the vacuolar membrane being only slightly permeable to the protonated form. Salt formation or binding to organic substances might also occur (Jans, 1973). The sequestration of alkaloids operates against a large concentration gradient, shows a high degree of specificity for the morphinan alkaloids and is both pH and temperature sensitive (Homeyer and Roberts, 1984 a; Roberts and Homeyer, 1985).

The use of neutral red (Strugger, 1969) indicated that a proton gradient was present between vacuole and supernatant in Papaver latex which was also confirmed by the effect of changes in the pH of the bathing medium on the vacuolar uptake of morphine. At pH 4.5 uptake was negligable, increasing rapidly to a plateau at pH 7.5-8.5 (Fig. 4). In many instances (Schuldiner et al., 1978) proton gradients result from coupling of a membrane ATPase to a proton pump; however, preliminary experiments showed no stimulation of morphine uptake on addition of ATP and Mg^{2+} (1 mM) nor was any significant effect observed with known ATPase inhibitors chlormadinone acetate (2.5×10^{-5} M), orthovanadate (10^{-4} M) and ouabain (10^{-4}

M). Subsequent work was therefore designed to investigate the development and maintenance of the vacuolar acidity and to determine the mechanism by which morphine was taken up into the vacuoles and stored, since in the case of Papaver no displacement of alkaloid from the vacuoles was observable once uptake had taken place (Homeyer and Roberts, 1984 a; Roberts and Homeyer, 1987).

Almost all other examples of accumulation of organic substances, specifically : arginine in yeast (Boller et al., 1975), citrate in Hevea lutoids (D'Auzac and Lioret, 1974; Crétin, 1982), sucrose in sugar beet (Doll et al., 1979) and sugarcane (Komor et al., 1982), and alkaloids in Catharanthus roseus (Deus-Neumann and Zenk, 1984) are likely to be by active transport and there is considerable evidence to support the involvement of a membrane-bound ATPase pump (Leigh and Walker, 1980; Marin et al., 1981; Chrestin et al., 1984; Deus-Neumann and Zenk, 1986).

Although initial experiments with poppy latex vacuoles gave no evidence to support the direct involvement of an ATPase proton pump in the accumulation of morphinan alkaloids, FCCP caused considerable reduction in alkaloid uptake, suggesting that a proton gradient across the membrane is essential for alkaloid accumulation, which, in turn, implies the existence of a proton pump.

As a preliminary to investigating the mechanism of morphine accumulation by Papaver somniferum alkaloid vacuoles, we investigated methods for measuring the pH gradient (ΔpH) across the membrane of isolated vacuoles and for estimating transmembrane potentials ($\Delta\Psi$). (^{14}C)-methylamine was used to probe ΔpH changes and tetra-(^{3}H)-phenylphosphonium bromide (TPP) was used as a probe for $\Delta\Psi$. The required measurements of vacuolar volume were made using (^{14}COOH)-dextran and tritiated water. The method using methylamine was essentially that of Rottenberg (1975) which was based on the equilibrium of uncharged methylamine across cell membranes and the impermeability of membranes to charged methylamine. The distribution of TPP across the membrane was used to calculate the membrane potential using the Nernst equation (Sorgato et al., 1978; Mitchell, 1966). Adsorption of methylamine and TPP to membranes was measured, found to be negligible and disregarded. The characteristics of Papaver latex have been found to vary significantly with the age of the plant, the developmental level of the capsule, the weather conditions prior to harvest, and the hour of the day at which harvest was accomplished. In light of this variation, the measurement of ΔpH and $\Delta\Psi$ can at best be regarded only as indicative of absolute values. More important in determining the relevance of the components of the proton-motive force in effecting alkaloid accumulation in poppy latex vacuoles are the small changes in ΔpH and $\Delta\Psi$ observed as a result of the addition of various reagents such as FCCP, valinomycin, and MgATP.

The proton-motive force component contribution to morphine uptake by Papaver somniferum latex 1000xg vacuoles was investigated using ammonium chloride (1 - 50 mM) and thiocyanate (1 - 50 mM), substances which produced dose dependent decreases in ΔpH, and $\Delta\Psi$, respectively (Johnson and Scarpa, 1979). Additions of morphine at the concentrations used in the experiments (0.27 mg ml^{-1}) had no significant effect on ΔpH or $\Delta\Psi$; the uptake of morphine into the vacuoles was, however, much more radically affected by changes in ΔpH than by changes in $\Delta\Psi$ (Fig. 5-A and B).

The effect of ATP + Mg^{2+} (1 mM) on ΔpH was negligible, with higher levels of ATP/Mg^{2+} having a marked effect on ΔpH and $\Delta\Psi$ which was found to be inversely related to the concentration of ATP (Table 5), whereas less change in morphine uptake was observed. Since Thom and Komor (1984) have demonstrated that the inhibition of an ATPase driven uptake by excess ATP is unlikely, we conclude that there is an ATPase present, but the relation to alkaloid accumulation is not clear. The possibility that vacuolar uptake of phosphate, which could result in depolarisation of the membrane, was refuted when no significant incorporation

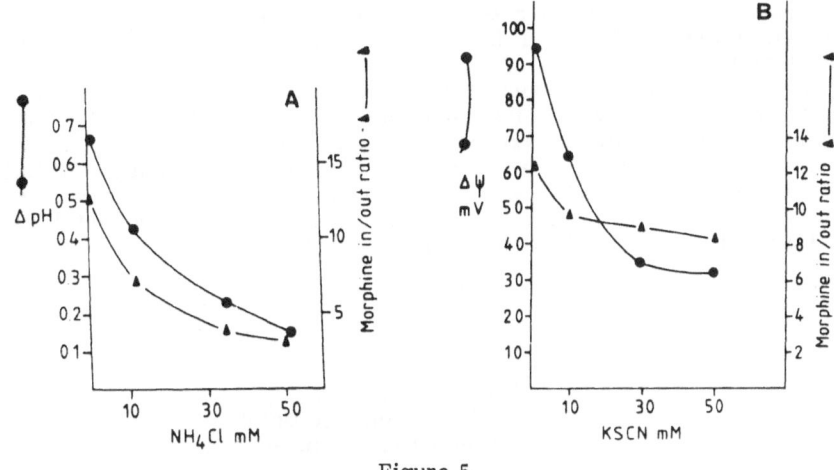

Figure 5

Morphine accumulation vs. proton-motive force effectors
in Papaver somniferum latex 1000xg vacuoles

A. Ammonium ion, with effect on transmembrane proton gradient; B. Thiocyanate,
with effect on transmembrane potential.

of ^{32}P into the vacuoles was observed using AT^{32}P at the same concentrations
(1 – 5 mM) as in morphine uptake experiments. Subsequent experiments with ATP
+ Mg^{2+} (1 mM) gave variable results in that some 1000xg vacuoles showed marginal
stimulation of uptake of morphine while others showed slight inhibition.

Table 5

The Effect of ATP and Mg^{2+}
on ΔpH and $\Delta\Psi$ and Morphine Uptake
of 1000xg Latex Vacuoles

Conditions of incubation	ΔpH	$\Delta\Psi$ (mV)	(^{14}CH$_3$)-morphine ratio[a]
Control	0.69	– 101	12.9
ATP + Mg^{2+}			
2.5 mM	0.66	– 87	12.6
5.0 mM	0.62	– 75	12.3
10.0 mM	0.58	– 74	11.2

[a]ratio of intravacuolar to extravacuolar concentration.
Latex vacuole samples and probes as for Fig. 5; ATP and Mg^{2+}
concentrations as shown; pre-incubations with ATP/Mg^{2+} 30 mins;
incubations with probes 30 mins with morphine 4 min.; temperature,
25°.

A number of uncouplers and KCl exerted effects on ΔpH and $\Delta\Psi$. Since Marin et al. (1981) have shown a dose related response of membrane potential with the vacuo-lysosomal membrane of <u>Hevea</u> latex using KCl, an investigation of the effects of K^+ on ΔpH was conducted. The experiments were designed to demonstrate effects on the proton concentration within the vacuole due to H^+/K^+ exchange or the transport of K^+ into the vacuole. The results, shown in Table 6, demonstrate no direct requirement for K^+ in the formation of membrane potential. Perturbations in ΔpH and $\Delta\Psi$ were only observed when KCl was used in the presence of nigericin or valinomycin; it seems unlikely that the K^+ ion, under the experimental conditions used, is responsible for any diffusion potential observed.

Table 6

The Effect of (K^+)
on ΔpH, $\Delta\Psi$ and Morphine Uptake
Using Latex 1000xg Vacuoles

Conditions of incubation	ΔpH	$\Delta\Psi$	$(^{14}CH_3)$-morphine ratio[a]
Control	0.97	− 96	39.0
+ KCl 30 mM	0.99	− 95	38.0
+ KCl 120 mM	0.92	− 85	33.1
+ KCl 30 mM + nigericin	0.75	− 93.5	39.0
+ KCL 30 mM + valinomycin	0.97	− 55.6	37.0

[a]ratio of intravacuolar to extravacuolar concentrations.
Latex vacuole samples and probes as for Fig. 5; KCl, nigericin and valinomycin at concentrations shown with 30 min preincubation followed by addition of probes or morphine; incubation times : 30 mins with probes, 4 min with morphine; temperature : 25°.

The variable nature of the results with ATP and Mg^{2+} suggested that more than one type of vacuole might exist in the 1000xg fraction. The hypothesis was that morphine uptake was only significantly stimulated in the lighter vacuoles of this fraction and that these were variably sedimenting through the 20% sucrose gradient used to separate the 1000xg vacuoles from substrate in short time morphine accumulation experiments. Latex samples taken over the maturing period of <u>Papaver somniferum</u> plants were separated into a 900xg fraction and an 1100xg fraction. The effect of MgATP on morphine accumulation by the vacuoles of these fractions was measured and found to be variable, despite the repeated washing (two centrifugations and resuspensions in buffer) of the 900xg fraction (Table 7). The 1100xg fraction consistently showed stimulation of uptake with MgATP amounting to over 50%. Experiments with compounds which inhibited or stimulated tonoplast ATPase showed that ΔpH and morphine accumulation were similarly affected by these reagents. The difference in the centrifuge speed at which the two groups of vacuoles sedimented could be due to either density or size, or indeed, to both. The presence of both 700 μm and 100 μm vacuoles in the 1000xg fraction suggests that the 900xg fraction is probably composed of the larger vacuoles, and the 1100xg fraction of the smaller. The difference in volume of the two types of vacuoles would result in a more readily detectable proton pump in the smaller population, since the addition of protons via an ATPase proton pump would have over 300 times the effect in the 100 μm vacuoles as in the 700 μm, all other factors remaining

Table 7
The Effect of Exogenous MgATP
on Papaver somniferum Latex Vacuoles
Sedimenting at 900xg and 1,100xg

Latex collection	on ΔpH % control		on $^{14}CH_3$-morphine uptake % control	
	900xg	1,100xg	900xg	1,100xg
10 Sept. 86	121.1	127.5	105.51	107.08
13 Oct. 86	n.d.	n.d.	n.d.	186.57
21 Oct. 86	n.d.	n.d.	106.42	302.14
26 Oct. 86	n.d.	161.8	n.d.	n.d.

Samples taken from first flowering through the period of rapid natural morphine accumulation (10-day from petal opening). Vacuoles incubated with 2 mM ATP/Mg^{2+} for 30 mins at 20°. Assays used $^{14}CH_3$-methylamine, 1 mM (ΔpH) and $^{14}CH_3$morphine 0.27 mM with incubation times of 30 and 4 mins respectively; vacuolar void volume 900xg = 40 μl and 1,100xg = 3 μl.

Table 8

Effect of Inhibitors
on ΔpH and $^{14}CH_3$-morphine Accumulation
Using Papaver somniferum Latex Vacuoles
Sedimenting at 900xg and 1,100xg

Inhibitor	on ΔpH % control		on Morphine Uptake % control	
	900xg	1,100xg	900xg	1,100xg
FCCP	58	86.1	96.8	86.0
DCCD	95.8	82.2	96.7	79.6
pCMB	81.1	94.2	69.8	n.d.
KNO_3	n.d.	82.4	51.2	82.8
orthovanadate	100	100	96.8	99.4

n.d. = not determined
Vacuoles incubated with 2 mM ATP for 30 min at room temperature (20°). Inhibitor added (FCCP, 5 10^{-4} M; DCCD, 5 10^{-4} M; KNO_3, 5 10^{-3}; orthovanadate, 5 10^3 M) for 30 min incubation. Morphine added for 4-6 min incubation. Control is 2 mM ATP + Mg^{2+}. Assay used 1 mM methylamine and 0.27 mM morphine. Vacuolar void volumes : 900xg = 40 μl; 1,000xg = 3 μl.

constant. Although exogenous MgATP stimulates morphine accumulation, it is not essential. Earlier experiments showed that the 1000xg vacuoles had an excellent capacity for morphine uptake over repeated additions of morphine to a vacuole population. The proton gradient was not readily dissipated in the 900xg fraction, although it was more sensitive to pCMB than the 1100xg fraction, and the effect of DCCD, a recognized inhibitor of membrane-bound ATPase, is much more pronounced in the 1100xg than in the 900xg fractions (Table 8).

A number of questions remain : Is an ATP/Mg^{2+} proton pump solely responsible for the observed proton gradient of the alkaloid accumulating vacuoles ? What, precisely is the role of meconic acid, and is it formed within the vacuole or accumulated as is morphine ? Does meconic acid "trap" the morphinan alkaloids within the vacuoles through the formation of a salt ? Is alkaloid biosynthesis primarily associated with the laticiferous system ?

REFERENCES

Alberts, B., Bray, D., Lewis, J., Raff, M., Roberts, K., and Watson, J. D., 1983, "Molecular Biology of the Cell", Garland, New-York.
Amann, M., Wanner, G., and Zenk, M. H., 1986, Intracellular compartmentation of two enzymes of berberine synthesis in plant cell culture, Planta, 167:310.
Antoun, M. D., and Roberts, M. F., 1975 a, Some enzymes of general metabolism in the latex of Papaver somniferum, Phytochemistry, 14:909.
Antoun, M. D., and Roberts, M. F., 1975 b, Phosphatases in the latex of Papaver somniferum, Phytochemistry, 14:1275.
Antoun, M. D., and Roberts, M. F., 1975 c, Enzymic studies with Papaver somniferum L. : 5. The occurrence of methyl-transferase enzymes in poppy latex, Planta Medica, 28:6.
d'Auzac, J., and Lioret, C., 1974, Mise en évidence d'un mécanisme d'accumulation du citrate dans les lutoides du latex d'Hevea brasiliensis (Kunth Mull. Arg.), Physiol. Vég., 12:617.
Bohm, H., Olesch, B., and Schulze, C. H., 1972, Weitere untersuchungen uber die biosynthese von alkaloiden in isoliertem milchsaft des schlafmohns Papaver somniferum L., Biochem. Physiol. Pflanzen, 163:126.
Boller, T., Durr, M., and Wiemken, A., 1975, Characterization of a specific transport for arginine in isolated yeast vacuoles, Eur. J. Biochem., 54:81.
Chréstin, H., Gidrol, X., Marin, B., Jacob, J. L., and D'Auzac, J., 1984, Role of lutoidic tonoplast in the control of the cytosolic homeostasis within the laticiferous cells of Hevea, Z. Pflanzenphysiol., 114:269.
Crétin, H., 1982, The proton gradient across the vacuo-lysosomal membrane of lutoids from the latex of Hevea brasiliensis, J. Membrane Biol., 65:175.
Danwalder, M., and Whaley, W., 1973, Staining of cells of Zea mays root apices with the osmium-zinc iodide and osmium impregnation techniques, J. Ultrastruct. Res., 45:279.
Deus-Neumann, B., and Zenk, M. H., 1984, A highly selective alkaloid uptake system in vacuoles of higher plants, Planta, 162:250.
Deus-Neumann, B., and Zenk, M. H., 1986, Accumulation of alkaloids in plant vacuoles does not involve an ion-trap mechanism, Planta, 167:44.
Dickenson, P. B., and Fairbairn, J. W., 1975, The ultrastructure of the alkaloid vesicles of Papaver somniferum, Ann. Bot., 39:707.
Doll, S., Rodier, F., and Willenbrink, J., 1979, Accumulation of sucrose in vacuoles isolated from red beet tissue, Planta, 144:407.
Fairbairn, J. W., and Djote, M., 1970, Alkaloid biosynthesis and metabolism in an organelle fraction of Papaver somniferum, Phytochemistry, 9:739.
Fairbairn, J. W., and Steele, M. J., 1981, Biosynthetic and metabolic activities of some organelles in Papaver somniferum latex, Phytochemistry, 20:1031.
Fairbairn, J. W., and Williamson, E. M., 1978, Meconic acid as a chemotaxonomic marker in the Papaveraceae, Phytochemistry, 17:2087.
Fairbairn, J. W., Hakim, F., and El Kheir, Y., 1974, Alkaloidal storage metabolism

and translocation in the vesicles of Papaver somniferum latex, Phyto-chemistry, 13:1133.

Fairbairn, J. W., Palmer, J. M., and Paterson, A., 1968, The alkaloids of Papaver somniferum L. : 8. Organelle activity of the isolated latex, Phytochemistry, 7:2117.

Homeyer, B. C., and Roberts, M. F., 1984 a, Dopamine accumulation in Papaver somniferum L. latex, Z. Naturforsch., 39c:1034.

Homeyer, B. C., and Roberts, M. F., 1984 b, Alkaloid sequestration by Papaver somniferum L. latex, Z. Naturforsch., 39c:876.

Jans, B., 1973, Untersuchungen am milchsaft des schollkrautes (Chelidonium majus L.), Ber. Schweiz. Bot. Ges., 83:306.

Jindra, A., Kovacs, P., Pittnerova, Z., and Psenak, M., 1966, Biochemical aspects of the biosynthesis of opium alkaloids, Phytochemistry, 5:1303.

Johnson, R. G., and Scarpa, A., 1979, Proton-motive force and catecholamine trans-port in isolated chromaffin granules, J. Biol. Chem., 254:3750.

Kanner, B. I., Fishkes, H., Maron, R., Sharon, I., and Schuldiner, S., 1979, Reserpine as a competitive and reversible inhibitor of the catecholamine transporter of bovine chromaffin granules, FEBS Letters, 100:1.

Komor, E., Thom, M., and Maretzki, A., 1982, Vacuoles from sugarcane. III. Proton-motive potential difference, Plant Physiol, 69:1326.

Kutchan, T. M., Ayabe, S., Krueger, R. J., Coscia, E. M., and Coscia, C. J., 1983, Cytodifferentiation and alkaloid accumulation in cultured cells of Papaver bracteatum, Plant Cell Reports, 2:281.

Kutchan, T. M., Rusch, M. D., and Coscia, C. J., 1986, Subcellular localization of alkaloids and dopamine in different vacuolar compartments of Papaver bracteatum, Plant Physiol., 81:161.

Leigh, R. A., and Walker, R. R., 1980, ATPase and acid phosphatase activities asso-ciated with vacuoles isolated from storage roots of red beet (Beta vulgaris L.), Planta, 150:222.

Marin, B., Marin-Lanza, M., and Komor, E., 1981, The proton-motive potential difference across the vacuo-lysosomal membrane of Hevea brasiliensis (rubber tree) and its modification by a membrane-bound adenosine-tri-phosphatase, Biochem. J., 198:365.

Marty, F., 1973 a, Sites réactifs à l'iodure de zinc tetroxide d'osmium dans les cel-lules de la racine d'Euphorbia characias L., C. R. Acad. Sci., Sér. D, 277: 1317.

Marty, F., 1973 b, Dissemblance des faces golgiennes et activité des dictyosomes dans les cellules en cours de vacuolisation de la racine d'Euphorbia cha-racias L., C. R. Acad. Sci., Sér. D, 277:1749.

Matile, P., 1978, Biochemistry and function of vacuoles, Annu. Rev. Plant Physiol., 29:193.

Matile, P., 1976, Localization of alkaloids and mechanisms of their accumulation in vacuoles of Chelidonium majus laticifers, in : "Secondary Metabolism and Coevolution", M. Luckner, K. Mothes and L. Nover, eds., Nova Acta, Leopold., Suppl. 7.

Matile, P., Jans, B., and Rickenbacher, R., 1970, Vacuoles of Chelidonium latex : Lysosomal property and accumulation of alkaloids, Biochem. Physiol. Pflanzen., 161:447.

Meissner, L., 1966 a, Uber den RNS- und proteingehalt isolierter milchsafte und den einbau radioaktiv markierter aminosauren in die latexproteine, Flora, 156:634.

Meissner, L., 1966 b, Uber den gasstoffwechsel isolierter milchsafte, Flora, 157:1.

Meissner, L., and Mothes, K., 1964, Uber stoffwechselaktivitat im latex von Papaver somniferum L., Phytochemistry, 3:1.

Mitchell, P., 1966, Chemiosmotic coupling in oxidative and photosynthetic phospho-rylation, Biol. Ref. Camb. Philos. Soc., 41:445.

Nessler, C. L., and Mahlberg, P. G., 1977, Ontogeny and cytochemistry of alkaloid vesicles in laticifers of Papaver somniferum, Am. J. Bot., 64:541.

Neumann, D., Krauss, G., Heike, M., and Groger, D., 1983, Indole alkaloid formation in storage suspension cultures of Catharanthus roseus, Planta Medica, 48:187.

Pletscher, A., 1977, Effect of neuroleptics and other drugs on monoamine uptake by membranes of adrenal chromaffin granules, Br. J. Pharmacol., 59:419.

Pujarniscle, S., 1968, Caractère lysosomal des lutoides du latex d'Hevea brasiliensis Mull. Arg., Physiol. Vég., 6:27.

Renaudin, J. P., and Guern, J., 1982, Compartmentation mechanisms of indole alkaloids in cell suspension cultures of Catharanthus roseus, Physiol. Vég., 20:533.

Roberts, M. F., 1971, Polyphenolases in the 1000xg fraction of Papaver somniferum latex, Phytochemistry, 10:3021.

Roberts, M. F., 1974, Oxidation of tyrosine by Papaver somniferum latex, Phytochemistry, 13:119.

Roberts, M. F., and Antoun, M. D., 1978, The relationship between L-dopa-decarboxylase in the latex of Papaver somniferum and alkaloid formation, Phytochemistry, 17:1083.

Roberts, M. F., and Homeyer, B. C., 1985, Effect of pH on temperature-dependent morphine uptake by Papaver somniferum and Papaver bracteatum latex, J. Pharm. Pharmacol., 37:140P.

Roberts, M. F., McCarthy, D., Kutchan, T. M., and Coscia, C. J., 1983, Localisation of enzymes and alkaloidal metabolites in Papaver latex, Arch. Biochem. Biophys., 222:599.

Roberts, M. F., Kutchan, T. M., Coscia, C. J., and Brown, J., 1987, to be published.

Robert, M. F., and Homeyer, B. C., 1987, to be published.

Rottenberg, H., 1975, The measurement of transmembrane electrochemical proton gradients, J. Bioenerg., 7:61.

Scherman, D., Jaudon, P., and Henry, J. P., 1983, Characterization of the monoamine carrier of chromaffin granule membrane by binding of $(2-^3H)$-dihydrotetra-abenazine, Proc. Natl. Acad. Sci. U.S.A., 80:584.

Schuldiner, S., Fishkes, H., and Kanner, B. J., 1978, Role of a transmembrane pH in epinephrine transport by chromaffin granule membrane vesicles, Proc. Natl. Acad. Sci. U.S.A., 75:3713.

Slotkin, T. A., Salvaggio, M., Lau, C., and Kirksey, D. F., 1978, ^{3}H-dopamine uptake by synaptic storage vesicles of rat whole brain and brain regions, Life Sciences, 22:823.

Sorgato, M. C., Ferguson, S. J., Kell, D. B., and John, P., 1978, The proton-motive force in bovine heart submitochondrial particles. Magnitude, site of generation, and comparison with the phosphorylation potential, Biochem. J., 174:237.

Strugger, S., 1969, "Prakticum der Zell - und Gewebe - Physiologie der Pfflanzen", Springer-Verlag, Berlin, Gottingen and Heidelberg.

Thom, M., and Komor, E., 1984, Effect of magnesium and ATP on ATPases of sugarcane vacuoles, Planta, 161:361.

Thureson-Klein, H., 1970, Observations on the development and fine structure of the articulated laticifers of Papaver somniferum, Ann. Bot., 34:751.

Zenk, M. H., 1985, Enzymology of benzylisoquinoline alkaloid formation, in : "The Chemistry and Biology of Isoquinoline Alkaloids", J. D. Phillipson, M. F. Roberts and M. H. Zenk, eds., Springer-Verlag, Heidelberg.

Zenk, M. H., Rueffer, M., Amann, M., and Deus-Neumann, B., 1985, Benzylisoquinoline biosynthesis by cultivated plant cells and isolated enzymes, J. Nat. Prod., 48:725.

INFLUENCE OF EXTERNAL pH ON ALKALOID PRODUCTION AND EXCRETION BY CATHARANTHUS ROSEUS RESTING CELL SUSPENSIONS

Claudine Nef[*], Christian Ambid[**] and Jean Fallot[**]

[*]Laboratoire de Physiologie Végétale, ORSTOM
B. P. No. V-51, Abidjan, Côte-d'Ivoire
and [**]Laboratoire de Biotechnologie Végétale, E. N. S. A. T.
145, Avenue de Muret, 31076-Toulouse-Cedex, France

INTRODUCTION

Catharanthus roseus (L.) G. Don, cell line C20, cultivated in a batch culture system with a modified Gamborg's B5 medium (Gamborg et al., 1968) deprived in auxin (2,4-dichlorophenoxy-acetic acid) and enriched in mannitol (200 mM) were maintained in survival conditions during 10 days after the growth phase (Ambid et al., 1982). The production of two alkaloids : ajmalicine and serpentine, was then largely emphazised (Roustan et al., 1982) and these compounds were essentially located within the cells.

In the present study, the same medium was used in a closed continuous culture system, and the alkaloid production and excretion were then examined, in relation with the external culture pH.

MATERIAL AND METHODS

The culture was achieved in a bioreactor (2 liters capacity) by inoculating 250 ml of a 7 days old batch culture grown in standard B5 medium (4.5 μM 2,4-D, 0.28 μM kinetin) in the vessel containing modified B5 medium at pH 5.5.

Cells were mechanically stirred (180 rpm) and continuously aerated (0.2 v.v.m. at the end of the growth phase); the culture was regularly supplied (0.8 ml/h) with B5 modified medium at pH 5.5 (or pH 4.3 for the experiment at acidic pH values).

Cells were maintained in the reactor and separated from the effluent by a decantation system.

Measurements of suspension growth and alkaloids extraction were carried out according to the procedure previously described (Ambid et al., 1982). For each experiment, alkaloids content was measured in the cells (internal production), in the reactor medium, and in the effluent (external production) using HPLC technique (Roustan et al., 1982); it was expressed in μg 10^{-6} cells.

RESULTS

Growth Characteristics in the Culture System

In this system, where medium is continuously supplied, we observed a growth phase of 10 days (with maximal growth rate : 0.158 j^{-1}) followed by a stationary phase of 20 days characterized by a constant cell number (Fig. 1). The rate of cellular mortality increased slowly, reaching 20% after 27 days of culture. In the reactor, the pH of the medium stayed near 6.0 value when the resting cells suspension was obtained.

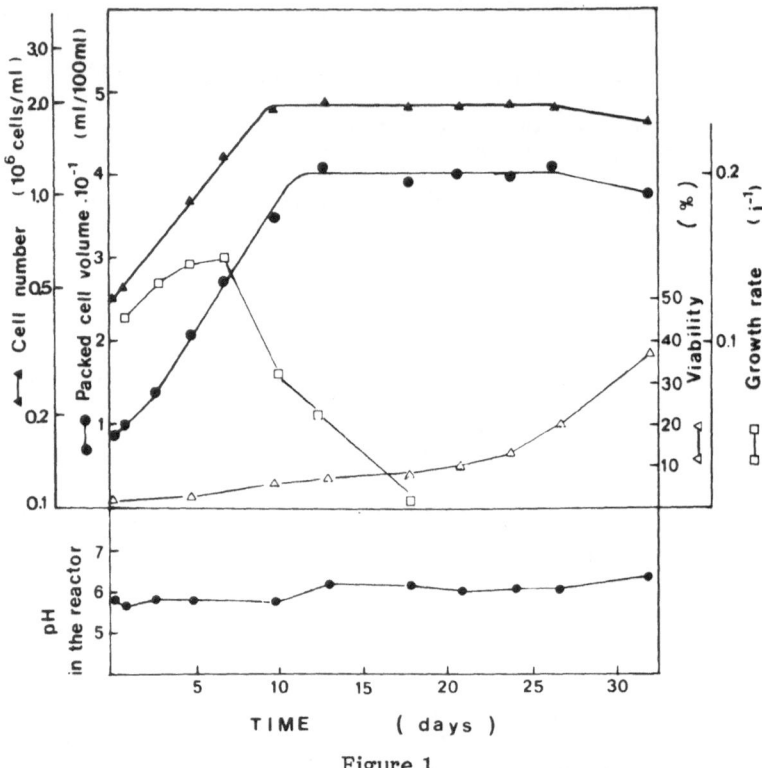

Figure 1

Growth characteristics of a <u>Catharanthus roseus</u> cells suspension
cultivated in a bioreactor with a renewed medium
deprived in 2,4–D and enriched in mannitol

Production and Distribution of the Alkaloids

The production of the two alkaloids was important only during the stationary phase, after 10 days of growth. At the end of the culture, this production reached respectively 0.3 and 0.7 μg 10^{-6} cells for serpentine and ajmalicine (Fig. 2-A and 2-B), with a constant amount for the ajmalicine.

These two compounds were present within the cells and, also, in the culture medium. Serpentine was largely located in the cells (Fig. 2-A) while the external amount of ajmalicine was important (Fig. 2-B).

530

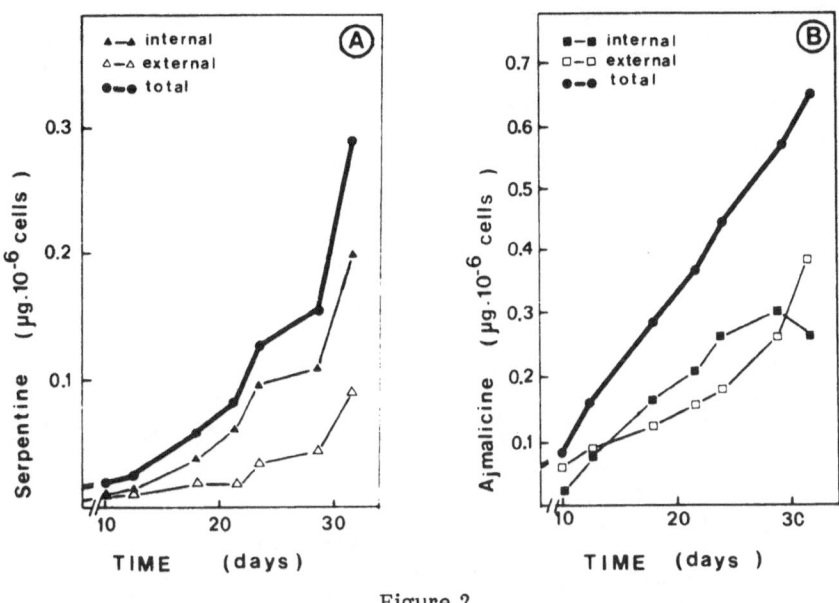

Figure 2

Production and distribution of serpentine (A) and ajmalicine (B)
between cells and medium
for Catharanthus roseus resting-cells suspension

Table 1

Evolution of the Ratio : External Level of Alkaloid/Total Level
during the Stationary Phase of Catharanthus roseus Cell Culture (in per-cent)

Alkaloid	Days of culture					
	10	13	18	21	24	28
Serpentine	50	43	32	30	28	30
Ajmalicine	63	47	48	46	46	43

If we examined the distribution between cells and medium, calculating the
ratio : external amount on total amount of each alkaloid (Table 1), we noticed that
these ratios were different for the two compounds, and relatively constant for
each of them : 30% of the total serpentine and about 45% of ajmalicine were present
in the medium.

Acidification of the Renewed Medium

Eight days after the beginning of the stationary phase, when the medium
of the resting cells suspension was continuously supplied with the acidic medium
(pH 4.3), no alteration was observed in the parameters of growth, but the distribution and the production of the two major alkaloids were largely modified.

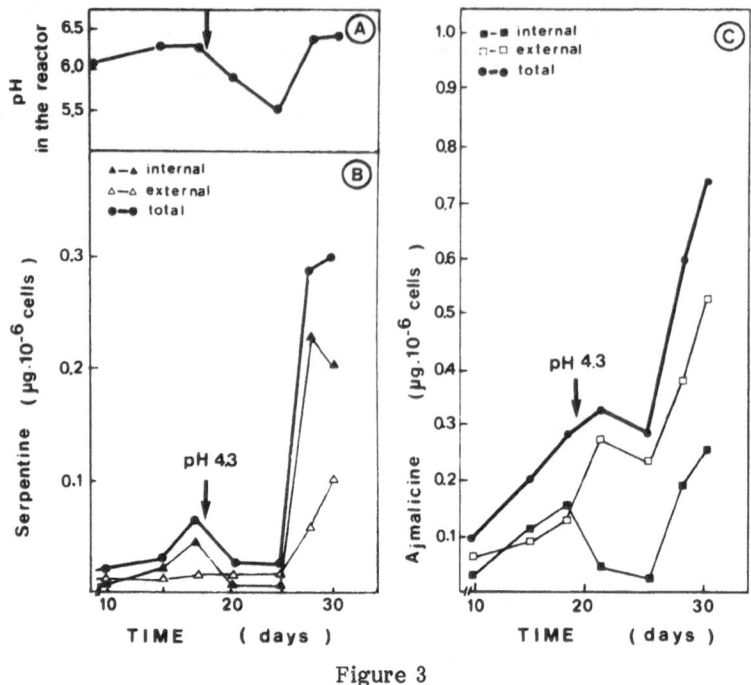

Figure 3

Influence of the acidification of the renewed medium
on the pH in the reactor (A), on the production and distribution
of serpentine (B) and ajmalicine (C)

In the first 6 days of renewal at pH 4.3, the pH in the reactor was slowly acidified (Fig. 3-A). The total production of serpentine decreased (65% of the compound disappeared in the first 3 days). Its internal level decreased in an important way, but the external one remained constant, showing that there were poor production and no excretion during this period (Fig. 3-B). The total production of ajmalicine also decreased and it could be noticed that 85% of this alkaloid was found in the external medium because of the increase of its excretion (Fig. 3-C).

After 6 days of renewal at pH 4.3, the pH of the reactor rised (autoregulation of the pH by the cells). This rise was accompanied by a new synthesis of serpentine and ajmalicine (Fig. 3-B and 3-C). Serpentine was largely located in the cells, while ajmalicine was in majority present in the medium : respectively 30% and 64% of the total production of serpentine and ajmalicine were extracellular.

DISCUSSION

Maintaining a <u>Catharanthus roseus</u> cells suspension in survival conditions allowed us to increase the production of the two major alkaloids : serpentine and ajmalicine, as well in batch conditions as in a continuous culture system. The latter allowed the continuous excretion, then the collection, of these compounds in the medium.

The constant supply of the renewed medium (pH 5.5) maintained the pH of the cell suspension about 6.0 during the stationary phase, while in batch conditions, it was shown to be subject to variations (slight alkalinisation) (Roustan et al., 1982). Continuous culture system allowed to control modifications on the pH in the reactor and, by the way, to modified considerably the distribution and the rate of production of the two alkaloids. These compounds are weakly basic and some models of transport mechanisms across cell membranes have been proposed : diffusion and accumulation in the vacuoles according to the ion-trapping model (Renaudin et al., 1982) or active and specific transport (Deus-Neumann et al., 1986).

Nevertheless, whatever the transport and accumulation mechanisms involved, non-growing cells suspensions obtained in a continuous culture system allowed important excretion of the alkaloids. This system could be an attractive biotechnological process for producing and collecting secondary metabolites of medicinal interest.

REFERENCES

Ambid, C., Roustan, J. P., Nef, C., and Fallot, J., 1982, Influence of 2,4-dichloro-phenoxyacetic on the production of ajmalicine and serpentine by cell suspension cultures of Catharanthus roseus, in "Proceedings of the Fifth International Congress of Plant Tissue and Cell Culture",

Deus-Neumann, B., and Zenk, M. H., 1986, Accumulation of alkaloids in plant vacuoles does not involve an ion-trap mechanism, Planta, 167:44.
Gamborg, O. L., Miller, R. A., and Ojima, K., 1968, Nutrient requirement of suspension cultures of soybean root cells, Exp. Cell Res., 50:151.
Renaudin, J. P., and Guern, J., 1982, Compartmentation mechanisms of indole alkaloids in cell suspension cultures of Catharanthus roseus (L.) G. Don, Physiol. Vég., 20:533.
Roustan, J. P., Ambid, C., and Fallot, J., 1982, Influence de l'acide 2,4-dichloro-phénoxyacétique sur l'accumulation de certains alcaloides indoliques dans les cellules quiescentes de Catharanthus roseus cultivées in vitro, Physiol. Vég., 20:523.

STRESS METABOLISM IN PLANT CELL CULTURES : ENHANCEMENT OF

PRIMARY AND SECONDARY METABOLITES IN OSMOTICALLY AND SALT

ADAPTED CELL CULTURES OF CATHARANTHUS ROSEUS

Kathryn Rudge and Philip Morris[*]

Wolson Institute of Biotechnology, University of Sheffield
Sheffield S10 2TN, England, and [*]Welsh Plant Breeding
Station, Plas Gogerddan, Aberystwyth, Wales, United Kingdom

INTRODUCTION

The work described is part of a study concerned with the effects of environmental stress on secondary metabolite production and identification of common responses to varying stress forms. Of additional interest were the mechanisms of stress adaptation induced in the cultures. Previous work in the area of stress research in plant cell cultures has focussed upon development of tolerant lines for agricultural and horticultural use (Katz and Tal, 1980), or as a tool for studying cellular events such as differentiation (Brown et al., 1979) or vacuolar acidification (Marigo et al., 1983).

The effects of two types of stress are reported here : (a) osmotic stress, with mannitol as a non-penetrating solute, and (b) salt stress (NaCl), which has both osmotic and ion toxicity components. Stress responses were measured as changes in cell growth, solute levels and alkaloid accumulation.

RESULTS

The results described below will deal primarily with the effects of mannitol induced osmotic stress, using the salt stress results to highlight differences and similarities between the two types of stress. It should also be noted that the two stresses were applied differently. The salt stress experiments used cells conditionned to growth on 150 mM NaCl ("+" line), and a parallel culture on salt free medium as a control ("-" line). Previous growth experiments had shown the NaCl (-) line to be inhibited when grown on 150 mM NaCl. This was taken as being indicative of tolerance in the NaCl (+) line. The osmotic stress was imposed as a shock treatment on cells previously grown on a standard medium. Mannitol at 0.6 Molar was chosen for osmotic stress work, after initial experiments using mannitol in the range 0.05 to 1.54 M (Rudge et al., 1986).

Growth

The effect of osmotic stress on growth, measured as dry weight accumulation, was to decrease growth rates and give a later, slightly reduced, dry weight peak. The reduction in growth rate was approximately 50% (Table 1). Similar patterns

Table 1

Effects of Osmotic and Ionic Stress on Growth
in C87 and NaCl(-) and (+) Lines of <u>Catharanthus roseus</u>

Cells	Max. dry wt[a] (g/l)	Max. fresh wt[a] (g/l)	μ[b] (d^{-1})	Max. $\dfrac{\text{F.wt.}}{\text{D.wt.}}$
C87	11.2 ± 0.21	205.6 ± 7.4	0.633	32.1
C87 + 0.6 M mannitol	7.8 ± 0.07	43.3 ± 0.5	0.325	6.4
NaCl (−)	9.1 ± 0.6	158.1 ± 10.8	0.550	30.4
NaCl (+)	7.9 ± 0.2	113.9 ± 0.3	0.330	19.7

[a]Results are the mean +/- SEM of three 3ml samples.
[b]Specific growth rate calculated from dry weight increases.

were seen in the salt stressed cells (Table 1), although the growth rate was inhibited was less than in osmotically stressed cells.

Changes in fresh weight were more pronounced. The summary of growth data given in Table 1 shows a marked reduction both in fresh weight and in the fresh weight to dry weight ratio for the osmotically stressed cells, implying that only a limited osmotic adjustment had occurred. These effects were also found in salt stressed cells but to a lesser degree, the f.w.:d.w. ratio falling by 30% compared to 80% for osmotically stressed cells.

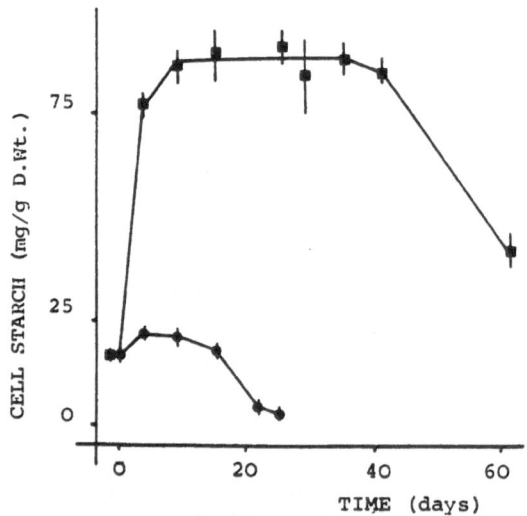

Figure 1

Starch accumulation in cells
grown with (■) or without (●) mannitol

Points represent the average and s.e.m. of three samples. Starch was estimated using an enzymatic assay (Kerr et al., 1984).

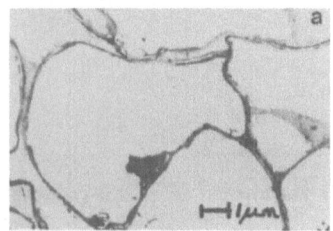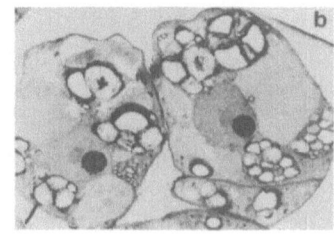

Figure 2

Transmission electron micrographs
of <u>Catharanthus</u> <u>roseus</u> cells

(a) C87 control and (b) C87 cells on 0.6 M mannitol.

Cell viability and morphology were affected by both types of stress. Mannitol stressed cells accumulated large amounts of starch, levels reaching 90 mg/g d.w. compared to 30 mg/g d.w. for unstressed cells (Fig. 1). Additionally, the rate of turnover of starch was reduced and this combination of effects was sufficient to extend the period of cell viability from 20 - 23 days for untreated cells to 70 days for stressed cells. The abundance of starch can also be clearly seen in transmission electron micrographs of the stressed cells (Fig. 2-b). As well as a prominence of starch grains, a further notable feature of the stressed cells was the absence of a central vacuole. The unstressed cells (Fig. 2-a) show a large central vacuole occupying over 95% of the total cell volume. However in the stressed cells the central vacuole was replaced by several smaller vacuoles which together occupy a reduced proportion of the total cell volume. The cells also contain an electron dense material not seen in the control cells. The salt stressed cells showed increased starch levels and extended viability although these effects have diminished with repeated subculturing of the cells. This can be interpreted as a continuing adaptation to stress by the salt treated cells.

Measurements of cell numbers and volumes highlight further differences between stressed and unstressed cells (Table 2). Cell division was inhibited by osmotic stress, with final cell numbers only one third those of the unstressed cells. This decrease was paralleled by a decrease in total cell volume so that the volumes per cell were quite similar at 43.5 and 39.4 $cm^3/10^9$ cells for control and stressed cells respectively. A similar pattern was found with the salt stressed cells (Table 2).

In summary the growth responses of stressed cells were as follows. Both stresses produced a smaller population of cells, with a similar volume and fresh weight per cell, but with a higher dry weight. These effects are less pronounced in the salt stressed cells, a probable reflection of the lower level of stress imposed and also the continuing adaptation of stressed cells to a high salt environment.

Alkaloid Accumulation

Unlike the effects of the two forms of stress on growth, where responses were similar, the effects on alkaloid accumulation were different for osmotic and salt stressed cells.

In osmotically stressed cells the final level of serpentine accumulated was four times greater than in controls (Fig. 3). This was due to a combination of factors

Table 2

Cell Number and Volumes for <u>Catharanthus</u> <u>roseus</u>
Line C87 with or without Mannitol
and NaCl(–) and (+) Lines

Cells	Cell number[a] (x 10^9/l)	Cell volume[b] (cm^3/l)	Cell volume[c] (cm^3/10^9 cells)
C87	4.28	186.3	43.5
C87 + 0.6 M mannitol	4.16	45.7	39.4
NaCl (–)	2.70	195.7	72.4
NaCl (+)	2.18	146.3	67.1

[a]Maximum values, counted after incubation in chromic acid.
[b]Measured using a dual isoltope dilution technique (Parr et al., 1984).
All points represent the average of three samples.
[c]Calculated from a + b.

: firstly, alkaloid accumulation occurred over the whole of the viable period which was three times longer in stressed cells, and secondly, the rate of accumulation was higher for stressed cells. When rates of production were calculated, values were obtained of 0.13 and 1.43 mg/10^9 cells/d for control and stressed cells respectively, a ten-fold increase. Similarly the calculated concentrations of alkaloids show higher levels in stressed cells. Serpentine, the major alkaloid, reached a level of 1.6 mM compared to 0.1 mM for the control line whilst ajmalicine levels were 0.05 and 0.14 mM for control and stressed cells respectively. As a vacuolar solute, the concentration of alkaloid in the unstressed cells represents a close estimate

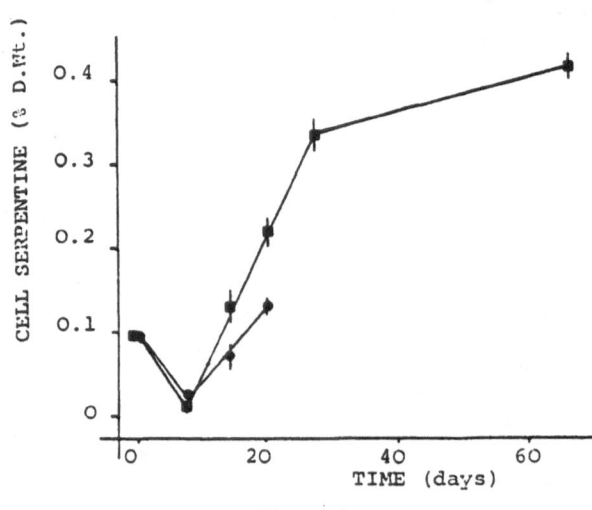

Figure 3

Cellular serpentine levels
in 0.6 M mannitol (■) and control (●) cells

Alkaloids were measured by HPLC (Morris et al., 1985). Points represent the average and bar, the s.e.m. of three samples.

of the actual cellular concentration. However for stressed cells the vacuolar space occupies a smaller proportion of the total cell volume and as such the expressed concentration of serpentine is probably an underestimate of the concentration in vacuoles of stressed cells. Further quantification would require measurement of alkaloids from isolated vacuoles of the two cell types.

Alkaloid levels in salt stressed cells were measured after transferring the cells from a 2,4-D containing medium into one in which the 2,4-D was replaced by IAA, as 2,4-D inhibits alkaloid accumulation (Morris et al., 1985). Alkaloid accumulation was followed over a single passage. Levels of alkaloids were low in both stressed and unstressed cultures, with serpentine levels of 0.045 and 0.03 mM and ajmalicine levels of 0.004 and 0.009 mM for control and stressed cells respectively. These low levels could be a consequence of a carry over of 2,4-D.

Primary Solutes

In order to determine how the cells were adopting to stress, measurements were made of the following solutes : Free amino-acids, proline, malate and, for the salt stressed cells, the inorganic ions Na^+, K^+ and Cl^-.

Proline in particular has been widely reported as accumulating under stress conditions both in whole plants (Rajagobal, 1983; Fitter and Hay, 1981) and in cell suspension cultures (Katz and Tal, 1980; Handa et al., 1983). As a compatible solute, it can be tolerated at high levels without being inhibitory to enzyme function (Fitter and Hay, 1981). Levels of several other amino-acids such as serine, glycine and alanine have also been found to increase with stress (Handa et al., 1983).

In osmotically stressed cells, proline was not a major amino-acid for either stressed or control cells and maximum accumulated levels were similar at 9.2 and accumulation was affected with a decrease in the rate of proline turnover so that

Table 3

Solutes Contributing to Osmotic Adjustement
in Mannitol and Salt Stressed Cells of Catharanthus roseus

Cells	Age (days)	Total AA[a] (mM)	Proline[b] (mM)	Malate[c] (mM)	Na+	K+ (mM)	Cl-[d]
C87	8	61.9	9.02	2.9	-	-	-
	20	12.0	0.05	3.5	-	-	-
C87+0.6M	6	55.7	10.9	1.4	-	-	-
mannitol	20	78.0	3.1	5.8	-	-	-
NaCl (-)	5	12.2	3.7	6.9	0	124	4.7
	22	13.2	1.1	7.8	-	-	-
NaCl (+)	5	7.1	2.8	9.2	237	41	21
	15	9.2	2.4	14.3	-	-	-

[a]Total amino-acid levels were measured with an amino-acid analyzer (Spakman et al., 1958).
[b]Independent analysis of proline using an acid-ninhydrin method (Bates et al, 1972).
[c]Malate was measured enzymatically (Gutman et al., 1974).
[d]Ion levels were measured with a flame spectrophotometer and chloridometer.
[ad]All values represent the average of three samples.

proline was maintained at higher levels in stressed cells than in controls (Table 3). This was also true for total free amino-acids so that at day 20, free aminoacid levels were 6 times greater in stressed cells than in controls (Table 3), indicating a possible osmoregulatory role for these molecules. The free amino-acid composition has been given elsewhere (Rudge et al., 1986). In salt stressed cells concentrations of free amino-acids were lower in both (+) and (-) lines but with proline making a greater contribution to the total in both cases. No differences of note were found between the salt stressed and control cell lines.

The organic acids, malate in particular, have also been reported as increasing under stress conditions (Lance and Rustin, 1984). Measurements of osmotically and salt stressed cells showed increases in both stress types (Table 3). However, all values lay approximately within the normal physiological range for plant cells (1 to 10 mM) (Agochukwu and Anosike, 1979), and did not make any major contributions to osmotic adjustment in either type of stressed cell.

In salt stressed cells, Na^+, K^+ and Cl^- measurements showed an increased accumulation of Na^+ in the salt stressed cells of 237 mM (Table 3). No equivalent uptake of Cl^- was found, and the level of K^+ decreased in the NaCl (+) line compared to the NaCl (-) line (Table 3).

DISCUSSION

The major feature of stress imposition is the alteration in growth patterns. The smaller cell number means a higher carbon resource per cell, leading to extended viability and adjustments in intracellular solute levels. These solute changes can partially offset induced water deficits, although changes in major osmotica, such as soluble sugars, have still to be characterised. Responses in the salt stressed cells were similar, but less pronounced, when compared to mannitol stressed cells.

Alkaloid accumulation was affected differently by the two stress forms. Under osmotic stress, cells produce larger amounts of alkaloids at a faster rate. It is therefore possible that this level of osmotic stress is an effector for increased alkaloid production; a phenomenon also reported for other forms of stress (Pate, 1983; Heinstein, 1985; Briske and Camp, 1982). It has not been possible to show increased alkaloid formation with ionic stress, but the comparatively lower level of stress imposed combined with the possible inhibitory effect of 2,4-D do not preclude this as a possibility worthy of further investigation.

REFERENCES

Agochukwu, E. N., and Anosike, E. O., 1979, Effect of storage under nitrogen on methanol, lactate, malate and their dehydrogenases in yam tubers, Phytochemistry, 18:1621.
Bates, L. S., Waldren, R. P., and Teare, I. D., 1972, Rapid determination of free proline for water stress studies, Plant and Soil, 39:205.
Briske, D. D., and Camp, B. J., 1982, Water stress increases alkaloid concentration in Threadleaf Groundsel (Senecio longilobus), Weed Sci., 30:106.
Brown, D. C. W., Leung, D. W. M., and Thorpe, T. A., 1979, Osmotic requirement for shoot formation in tobacco callus, Physiol. Plant., 46:36.
Fitter, A. H., and Hay, R. K. M., "Environmental Physiology of Plants", Academic Press, London.
Gutmann, I., and Wahlefeld, A. W., 1974), L(-)-malate. Determination with malate dehydrogenase and NAD, in : "Methods in Enzymatic Analysis", H. U. Bergmeyer, ed., Academic Press, New-York.
Handa, S., Bressan, R. A., Handa, A. K., Carpita, C., and Hasegawa, P. M., 1983, Solutes contributing to osmotic adjustement in cultured plant cells adapted to water stress, Plant Physiol., 73:834.

Heinstein, P. F., 1985, Future approaches to the formation of secondary natural products in plant cell suspension cultures, J. Nat. Prod., 48:1.

Katz, A., and Tal, M., 1980, Salt tolerance in the wild relatives of the cultivated tomato. Proline accumulation in callus tissue of Lycopersicon esculentum and L. peruvicanum, Z. Pflanzenphysiol., 98:429.

Kerr, P. S., Huber, S. C., and Israel, D. W., 1984, Effect of N-source on soybean leaf sucrose phosphate synthase, starch formation and whole plant growth, Plant Physiol., 75:483.

Lance, C., and Rustin, P., 1984, The central role of malate in plant metabolism, Physiol. Vég., 22:625.

Marigo, G., Delorme, Y. M., Luttge, U., and Boudet, A. M., 1983, Rôle de l'acide malique dans la régulation du pH vacuolaire dans les cellules de Catharanthus cultivées in vitro, Physiol. Vég., 21:1135.

Morris, P., Scragg, A. H., Smart, N. J., and Stafford, A., 1985, Secondary product formation by cell suspension cultures, in : "Plant Cell Cultures - A Practical Approach", R. A. Dixon, ed., IRL Press, Oxford.

Parr, A. J., Smith, J. I., Robbins, R. J., and Rhodes, M. J. C., 1984, Apparent free space volume estimation : A nondestructive method for assessing the growth and membrane integrity/viability of immobilised plant cells, Plant Cell Reports, 3:161.

Pate, D. W., 1983, Possible role of ultraviolet radiation in evolution of cannabis chemotypes, Econ. Bot., 37:396.

Rajagopal, V., 1983, Effect of irradiance of free proline accumulation in stressed barley leaves, Z. Pflanzenphysiol., 11:277.

Rudge, K., and Morris, P., 1986, The effect of osmotic stress on growth and alkaloid accumulation in Catharanthus roseus, in : "Secondary Metabolism in Plant Cell Cultures", P. Morris, A. H. Scragg, A. Strafford, and M. W. Fowler, eds., Cambridge University Press, Cambridge (1986).

Spakman, D. H., Stein, W. H., and Moore, S., 1958, Recording apparatus for use in the chromatography of amino-acids, Anal. Chem., 30:1190.

SIGNIFICANT INCREASE IN PYRIMIDINIC ALKALOID (TRIGONELLINE)

CAUSED BY IONIZING RADIATION

Michel Riedel[*]

Radiobiology Group, C. E. R. N.
Geneva, Switzerland

INTRODUCTION

The main theme of this colloquy is the importance of cellular compartmentalization allowed by vacuoles and the biotechnological applications which stem from this. The session during which this work was presented covered, in particular, large-scale production of compounds by plant cells.

We therefore describe here an example, which appeared to be interesting for several reasons, of an increase in the production of a pyrimidinic alkaloid caused by the radiation technology which can have several results : modification of gene expression, blockage of certain metabolic pathways, and also the occurrence of free radical-induced reactions leading either to the non-enzymatic formation of various compounds or to modifications of membrane permeability causing cell decompartmentalisation which may be temporary or definitive, depending on the dose and kind of radiation. Variation measured in the trigonelline (T) or N-methyl nicotinic acid contents provoked by ionizing radiation led us firstly to put forward an hypothesis concerning regulation of the cell cycle (phase G_2 M) and secondly to attempt to make all those concerned with modern biotechnology aware of the advantages of the use of ionizing radiation; one such advantage is that of causing at a precise, selected moment significant variations in the concentration of a given metabolite produced by plant tissue or a culture of microorganisms.

MATERIAL

Biological Material

Broad beans (Vicia faba L. var. Fillbasket, Graines d'Elite Clause) were supplied by Suttons Seeds Ltd. They are soaked in fresh, aerated filtered water at 19°C for three days until germination and then placed in sterile (170°C, 3 h) expanded vermiculite which has been humidified (approx. 155 ml/dm^3) and kept at 19°C. The seeds were numbered at the end of this period and cultivated at 19°C; the roots were kept in filtered running water (active carbon to absorb Cl) where they grew for a further three days. They were then selected in function of the

Currently at : Laboratoire d'Histophysiologie et de Radiobiologie Végétales, Université des Sciences et Techniques du Languedoc, F-34.060-Montpellier-Cedex, France size of the main root and its rate of growth (specimens whose size did not fall within the range of values calculated were excluded (x $\pm$ s and v $\pm$ s' (in which x = average size of roots, s = standard deviation of size values, v = average rate of growth of the roots, s' = standard deviation of growth rate)).

The beans were stored in circulating water at 4°C for 24 hours before irradiation; the tips of main roots were irradiated in water at 4°C in special tanks enabling irradiation in a steady position. Given the delivery rate (approx. 1.7 Gy/h) and the doses applied (0.54 and 2.48 Gy), the temperature of 4°C slowed the cell metabolism and practically halted growth during the irradiation period in order to prevent the root tips from leaving the beam. The roots were then put back immediately in water at 19°C in the culture tanks after being measured (time T_0). For ten days, the size of the roots was measured daily, tigelles and secondary roots were removed and any contaminated specimens were removed. After the last measurement the beans were prepared for freezing (a 4-cm segment was cut from the root tip; the rest of the root and the unmarked cotyledon were discarded). The samples were weighed, frozen rapidly by dipping in liquid nitrogen, wrapped and labelled and stored at − 30°C. 3 to 4 minutes elapsed between the removal of the roots from water and their placing in liquid nitrogen. The following samples were analysed : No. 105 (2.48 Gy, position 2) and No. 253 (0.54 Gy, position 2).

Type and Conditions of Radiation

The beam was of secondary 250 GeV/c hadrons which were mainly protons with about 10% positive pions, and was produced from primary 400 GeV/c protons incident on a target; the radiation was pulsed, each pulse being about 1.4 second long and occurring at a frequency of 1 per 9.6 seconds. For a detailed description, see the technical memorandum by A. H. Sullivan, C. E. R. N., HS/RB/TM/80 − 05.

Chromatographic Equipment

We used an HPLC chromatograph Spectra-Physics SP 8000 controlled by microprocessors (1 pump, ternary gradient, automatic injection, stabilised column temperature, degassed with helium, digital integrator) and fitted with an SP 8000 detector (UV absorbance, double beam).

Table 1
Programme of elution giving the composition of the mobile phase
for chromatographic analysis

Time (min)	CH_3OH (%)	H_2O (%)	CH_3CN (%)
0.0	50.0	50.0	0.0
5.0	50.0	50.0	0.0
10.0	70.0	25.0	5.0
12.0	70.0	25.0	5.0
15.0	100.0	0.0	0.0
17.0	100.0	0.0	0.0
19.0	50.0	50.0	0.0
20.0	50.0	50.0	0.0

METHOD

Preparation of samples

Frozen root tips were crushed in a mixture of water and methanol (50:50 v/v) at a low temperature (0°C) in a sterile medium for three minutes. This crushed material was ultrafiltered (Millex 0.45 m), placed in the automatic injector (at ambient temperature) and analysed. Solvents were HPLC grade.

Chromatographic Analyses

Chromatographic characteristics were identical for the two samples. Injection: 10 microliters. Column : Zorbax ODS 10 microns - 250 mm x 4.6 mm ID (temperature 40°C). Mobile phase : mode : "constant gradient" (cf. Table 1); flow rate : 1 ml/mn; nature : methanol (CH_3OH) - water (H_2O) - acetonitrile (CH_3CN) (HPLC grade). Detector : UV absorbance at 254 nm –Sensitivity 0.16 full scale optical density unit.

RESULTS

The sample irradiated at 2.48 Gy (dose at which root growth is small : 27% of that observed for non-irradiated roots) contained approximately twice as much trigonelline as the equivalent which was subjected to only weak radiation (0.54 Gy) and whose growth was almost normal (91% of that recorded in non-irradiated controls). Twenty chromatographic analyses were carried out; average variation of retention time was only 0.56%.

The average migration time of trigonelline under the conditions used for chromatographic analyses was 187 seconds. The two values of areas integrated below trigonelline peaks of the curve are 3471 and 1757 units (3471/1757 = 1.97).

DISCUSSION AND CONCLUSION

Regulation of the Cell Cycle

The reason for this nearly 100 % increase in trigonelline is not known, but an increase in the amount of N-methyl-nicotinamide excreted has also been observed in man after exposure to ionizing radiation (Gerber, 1971). Radio-induced variations in membrane permeability could occur (Grosch, 1979; Harris, 1970), as could modifications in the metabolism of molecules such as nicotinic acid and nicotinic amide (NAD pathway, polyA and DNA repair) (Oleinick et al., 1985; Cleaver et al., 1985). These modifications could be linked with the variations in cytokinin content, and the latter compounds may be involved, for example, in the methylation processes due to the S-adenosyl-methionine (SAM) - adenosyl-homocystein (SAH) system, or might be more directly the result of any radical process caused by ionizing radiation.

Damage caused by this kind of radiation includes on the one hand a certain amount of accumulation of cells in G_2 and inhibition of mitosis and, on the other (Webster et al., 1970), the formation of chromosomic aberrations and inhibition of DNA synthesis. In many cases, and in particular in <u>Vicia faba</u>, the G_2 phase appears to be the most sensitive (Rost, 1977).

It is now accepted that the cytokinins affect any processes which occur in G_2 or during the G_2-mitosis transition stage in tissue in which there is active division and which requires this type of hormone. In fact, without cytokinin, cells accumulate in G_2 for a moment and then leave the cell cycle or enter an endoreplication cycle in which successive replications of DNA are observed without mitosis or cytokinesis. It has been put forward that radioinduced delay at transitions ($G_1 \rightarrow S$ and $G_2 \rightarrow M$) reflect the repair time at the "main control point" (Van't Hof, 1973). It is therefore possible that part of the radio–restoration activity of these hypermodified purines could be attributed to a modification of the $G_2 \rightarrow M$ blockage. This could occur by means of specific action on one or more indispensable proteins involved in mitosis (e.g. MAPs : microtubule associated proteins). Cytokinins may also have an effect on membranes and regulate the amounts of divalent cations Ca^{2+}/Mg^{2+} (Mn^{2+}) which, among other things, control the assembling of the microtubules (Helper and Palevitz, 1974). Finally, modifications of superoxide–dismutase (SOD) activities may also be considered (Cu^{2+}, Mn^{2+}) and cytokinin interactions with free radicals could occur. Variations of topoisomerase, DNase and ligase activities are also possible (Mn^{2+}).

However, whatever the precise mechanism involved, regulation of the cell cycle at G_2 may therefore be :

a) by cytokinins or by a molecule with cytokinic activity,

b) by means of another hormone which is "antagonistic" to cytokinins in this phenomenon,

c) the result of harmonious equilibrium between the molecules mentioned above.

We have just stated that cytokinins are necessary to allow the transition $G_2 \rightarrow M$. Hypothesis (a) cannot therefore be excluded. Hypothesis (b) is also possible. The existence of the "G_2 factor", a molecule which preconditions cells to stop at G_2, has been confirmed and it has been purified and identified (Evans et al., 1979). It is trigonelline or, more precisely, the N-methyl derivative of nicotinic acid. However, it has also been shown that although the G_2 factor is necessary for a maximum of cells to be stopped at G_2 it is not sufficient for the cells to be polyploid after irradiation (Evans, 1978). The G_2 factor is thus not the only feature responsible for modifications in the cell cycle after irradiation. The hypothesis that cell cycle regulation at G_2 may be the result of a balance between molecules possessing cytokinic activity and trigonelline is fully plausible for normal tissue (Van't Hof, 1980; Tramontano et al., 1986) and in our opinion the capacity of these two types of molecule for complexing and possibly exchanging manganese and for modifying polyA-polymerase activity, or even the cyclic nucleotide metabolism, may be responsible for this phenomenon.

In addition, the CK:T ratio at the time of irradiation may account for certain variations in radiosensitivity. It is also particularly interesting to note the possibility of the existence of pyrimidine–indole complexes of the 1-methyl-nicotine amid (1-MNA)-indolyl-acetic acid (IAA) type (Ishida et al., 1980), and the possible physiological and radiobiological implications.

Although it is known today that gamma radiation can modify cytokinin and trigonelline levels, that radio–induced free radicals can react with manganese and that complexes of this cation with hormones can be affected, it is nevertheless not possible to affirm that this phenomenon is a major factor in radio–induced modifications in organisms or partly contribute to radiorestoration. It is probable but not certain.

Radiobiology and Biotechnology

Ionizing radiation is thus likely, at a given moment, to be able to vary the level of an important pyrimidinic alkaloid, which might block the cell cycle. However, the quantities of other molecules of considerable physiological and possibly biotechnological importance can vary after even short exposure to ionizing radiation. The substances concerned are hormones and growth factors such as abscissic acid

(Degani and Itai, 1978), cytokinins (Pandey et al., 1978), auxins (Skoog, 1935; Miura et al., 1974), gibberellic acid (Machaiah et al., 1976) and associated molecules such as cyclic nucleotides, polyamines, etc.

Since this type of molecule controls most secondary metabolisms, radiation is of real interest for laboratory study of metabolic pathways. However, applications are immediate : it is known, for example, that gamma radiation causes considerable reduction in auxin activity and an increase in certain cytokinic activities; in a fermenter this results in a fall in cell division and accentuated differentiation, which may be interesting for example for the production and turnover of certain secondary metabolites (Barz, 1977). In addition and simultaneously, radiation causes variations in membrane permeability either directly (free radicals and peroxidation of lipids) or through modification of hormone contents (cytokinins, membrane phosphorylation and lipid methylation); this leads to envisaging, depending on the case, either release into the medium of sequestrated metabolites whose production may also have been modified or, on the other hand, making it easier to introduce and even incorporate selected genetic material. In fact, conditions are particularly suitable for the insertion of such material because of the rearrangements (recombinations ?) caused by radiation and the stimulation of the various DNA processes (i.e. nucleases, polymerases, ligases, topoisomerases, etc.).

Nevertheless, high doses of radiation cause a great increase in the death rate and lower doses cause an increase in variability and harmful chromosome aberrations. "Provoked radiorestoration" methods have therefore been perfected to remedy these disadvantages and therefore to enhance survival and/or to reduce variability by treatment after exposure to radiation and possibly to enable the incorporation of particular genetic material. Very encouraging results have thus been obtained to date on rice embryos in which it appears that it has perharps been possible to incorporate a kanamycin-resistant gene. However, much verification remains to be carried out in order to be sure of being able to generalise the method in the near future.

A major development in the history of genetics was the discovery of H. J. Muller in 1927 of the modification of mutation rates caused by ionizing radiation (Muller, 1927). Since then radiobiology has continued to contribute both to agronomy and medicine, and is today aiding the development of biotechnology (particularly in "extreme" environments) in order to solve a number of problems to which other complementary approaches have still no provided satisfactory solutions. Modern "radiobiotechnology" deserves to be better known and more profitably used in the future.

REFERENCES

Barz, W., 1977, Catabolism of endogenous and exogenous compounds by plant cell cultures, in : "Plant Tissue Culture and its Biotechnological Application", W. Barz, E. Reinhard and M. H. Zenk, eds., Springer-Verlag, Berlin.

Cleaver, J. E., Milam, K. M., and Morgan, W. F., 1985, Do inhibitor studies demonstrate a role for poly(ADP-ribose) in DNA repair ?, Radiation Research, 101: 166.

Degani, N., and Itai, C., 1978, The effect of radiation on growth and abscissic acid in wheat seedlings, Environ. Exptl. Bot., 18:113.

Evans, L. S., 1978, Cell cycle kinetics of endoduplication in gamma-irradiated root meristems of Pisum sativum, Amer. J. Bot., 65:1084.

Evans, L. S., Almeida, M. S., Lynn, D. G., and Nakanishi, K., 1979, Chemical characterization of a hormone that promotes cell arrest in G_2 in complex tissues, Science, 203:1122.

Gerber, G. B., 1971, Studies on the mechanism of excess excretion of nucleic acid and nicotinamide-adenine-dinucleotide (NAD) metabolites after irradiation, in : "Biochemical Indicators of Radiation Injury in Man", IAEA WHO, ed., IAEA, Vienna.

Grosch, D. S., and Hopwood, L. E., 1979, "Biological Effects of Radiations", Academic Press, New-York.

Harris, J. W., 1970, Effects of ionizing radiation on lysosomes and other intracellular membranes, in : "Advances in Biological and Medical Physics", Academic Press, New-York.

Helper, P. K., and Palevitz, B. A., 1974, Microtubules and microfilaments, Annu. Rev. Plant Physiol., 25:309.

Ishida, T., Tomita, K., and Inoue, M., 1980, An X-ray study on the interaction between indole ring and pyridine coenzymes : Crystal structure of 1-methyl-3-carbamoyl-pyridinium:indole-3-acetic acid(1:1) monohydrate charge transfer complex, Arch. Biochem. Biophys., 200:492.

Machaiah, J. P., Vakil, U. K., and Sreenivasan, A., 1976, The effect of gamma irradiation on biosynthesis of gibberellins in germinating wheat, Environ. Exptl. Bot., 16:131.

Miura, K., Hashimoto, T. and Yamaguchi, H., 1974, Effect of gamma irradiation on cell elongation and auxin level in Avena coleoptiles, Rad. Bot., 14:207.

Muller, H. J., 1927, Artificial transmutation of the gene, Science, 66:84.

Oleinick, N. L., and Evans, H. H., 1985, Poly(ADP-ribose) and the response of cells to ionizing radiation, Rad. Bot., 101:29.

Pandey, K. N., Sabharwal, P. S., and Kemp, T. R., 1978, Cell division factors (cytokinins) from irradiated plant tissue, Nature, 271:449.

Rost, T. L., 1977, Responses of the plant cell cycle to stress, in : "Mechanisms and Control of Cell Division", T. L. Rost and E. M. Gifford, eds., Dowden Hutchinson and Ross Inc., Stroudsburg, Pennsylvania.

Skoog, F., 1935, The effect of X-radiation on auxin and plant growth, J. Cell. Comp. Physiol., 7:227.

Tramontano, W. A., Evans, L. S., McGinley, P. A., and Ciancaglini, E., 1986, Effects of cytokinins on the metabolism of nicotinic acid in cultured roots of Pisum sativum and Glycine max in relation to cell arrest in G_2, Environ. Exptl. Bot., 25:393.

Van't Hof, 1973, Two principal points of control in the mitotic cycle of pea meristem cells : Energy considerations, characterization, and radiosensitivity, in : "Advances in Radiation Research, Biology and Medicine", Vol. 2, J. F. Duplan and A. Chapiro, eds., Goedon and Breach Science Publishers, New-York.

Van't Hof, J., 1980, Pea (Pisum sativum) cells arrested in G_2 have nascent DNA with breaks between replicons and replication clusters, Exptl. Cell Research, 129:231.

Webster, P. L., and Van't Hof, J., 1970, Recovery of G_2 cells in pea root meristems : Survival and mitotic delay following irradiation, Rad. Bot., 10:145.

PARTICIPANTS[1]

CANADA

- Pr. R.J. POOLE (CS, C, L), Department of Biology, McGill University, 1206, Doctor Penfield Avenue, MONTREAL, Quebec H3A-1B1,
- Dr. P.A. REA (C, L), Department of Biology, University of York, HESLINGTON, Yorkshire Y01-5DD, England (GREAT-BRITAIN),

FRANCE

- Dr. G. ALIBERT (L), Centre de Physiologie Végétale, Université Paul Sabatier, 118, Route de Narbonne, 31077-TOULOUSE-CEDEX,
- Dr. J. ALEXANDRE (L), Laboratoire sur les Echanges cellulaires chez les Végétaux, Université des Sciences et Techniques de Rouen, Boite Postale No. 67, 76130-MONT-SAINT-AIGNAN,
- Dr. H. BARBIER-BRYGOO (L), Laboratoire de Physiologie Cellulaire Végétale, C.N.R.S., Avenue de la Terrasse, 91190-GIF-SUR-YVETTE,
- Dr. P. BESSON (P), L'Air Liquide, Direction des Centres de Recherches et de Développement, Centre de Recherches Claude-Delorme, Boite Postale No. 126, Les Loges-en-Josas, 78350-JOUY-EN-JOSAS,
- Pr. A.M. BOUDET (CS, C, L), Centre de Physiologie Végétale, Université Paul Sabatier, 118, Route de Narbonne, 31077-TOULOUSE-CEDEX
- Dr. M. BOUZAYEN (L), Laboratoire de Biotechnologie Végétale, Ecole Nationale Supérieure des Sciences Agronomiques, 145, Avenue de Muret, 31076-TOULOUSE-CEDEX,
- Dr. H. CANUT (L), Centre de Physiologie Végétale, Université Paul Sabatier, 118, Route de Narbonne, 31077-TOULOUSE-CEDEX,
- Dr. H. CHRESTIN (CO, CS, L), Laboratoire de Physiologie et de Biotechnologie Végétales, Centre ORSTOM d'Adiopodoumé, Boite Postale No. V-51, ABIDJAN (IVORY COAST),
- Dr. I. CLARIS (P), Laboratoire de Physiologie et de Biotechnologie Végétales, Centre ORSTOM d'Adiopodoumé, Boite Postale No. V-51, ABIDJAN (IVORY COAST),
- Dr. L. COYAUD (P), Laboratoire de Neurobiologie Cellulaire et Moléculaire, C.N.R.S., Avenue de la Terrasse, 91190-GIF-SUR-YVETTE,
- Dr. A. DUPAIX (L), Institut de Biochimie, Faculté des Sciences de Paris XI, Centre Universitaire d'Orsay, 91405-ORSAY,
- Dr. D. GAYRARD (P), Pierre-Fabre-Médicament, 17, Avenue J. Moulin, 81106-CASTRES,

- Dr. X. GIDROL (L), I.N.R.A., Centre de Recherches de Bordeaux, Station de Physiologie Végétale, Boite Postale No. 131. 33140-PONT-DE-LA-MAYE,

- Dr. R. GIORDANI (L), Laboratoire de Biologie de la Différenciation Cellulaire, U.A. C.N.R.S. No. 179, Faculté des Sciences de Luminy, 13288-MARSEILLE-CEDEX 9,

- Pr. J. GUERN (CS, C, L), Laboratoire de Physiologie Cellulaire Végétale, C.N.R.S., Avenue de la Terrasse, 91190-GIF-SUR-YVETTE,

- Dr. T. HARDY (P), Laboratoire de Physiologie et de Biotechnologie Végétales, Centre ORSTOM d'Adiopodoumé, Boite Postale No. V-51, ABIDJAN (IVORY COAST),

- Dr. M.A. HARTMANN (P), Institut de Botanique, Université Louis Pasteur, 28, Rue Goethe, 67083-STRASBOURG-CEDEX,

- Dr. A. KURKDJIAN (P), Laboratoire de Physiologie Cellulaire Végétale, C.N.R.S., Avenue de la Terrasse, 91190-GIF-SUR-YVETTE,

- Dr. D. LE QUOC (P), Laboratoire de Biochimie, Université de Franche-Comté, Route de Gray, 25030-BESANCON,

- Dr. K. LE QUOC (P), Laboratoire de Biochimie, Université de Franche-Comté, Route de Gray, 25030-BESANCON,

- Dr. F. MAJERUS (P), Département de Génie Biochimique et Alimentaire, Institut National des Sciences Appliquées, Avenue de Rangueil, 31077-TOULOUSE-CEDEX,

- Dr. B. MARIN (CO, CS, C, L), Unité de Recherches 603 :"Mécanismes biochimiques et physiologiques de la Production végétale", Département F, ORSTOM, 213, Rue Lafayette, 75480-PARIS-CEDEX 10, with permanent address : Laboratoire de Biotechnologie de la Cellule Végétale, Unité Fonctionnelle : Physiologie et Métabolisme Cellulaires, Centre ORSTOM de Montpellier, 2051, Avenue du Val de Montferrand, Boite Postale No. 5045. 34032-MONTPEL-LIER-CEDEX

- Dr. M. MONESTIEZ (L), Laboratoire d'Electrophysiologie des Membranes, U.A. C.N.R.S. No. 578, Université de Paris-VII, 2, Place Jussieu, 75251-PARIS-CEDEX 5,

- Dr. C. NEF (L), Laboratoire de Physiologie et de Biotechnologie Végétales, Centre ORSTOM d'Adiopodoumé, Boite Postale No. V-51, ABIDJAN (IVORY COAST),

- Dr. D. PARDO (CO), Centre de Transfert pour le Développement des Bio-Industries, Campus de Luminy, Case Postale No. 908, 13288-MARSEILLE-CEDEX 9,

- Pr. A. PAREILLEUX (L), Département de Génie Biochimique et alimentaire, Institut National des Sciences Appliquées, Avenue de Rangueil, 31077-TOULOUSE-CEDEX,

- Dr. V. PETIARD (CO), Groupe de Génétique Végétale, Francereco, Zone Industrielle, 30, Avenue G. Eiffel, Boite Postale No. 0166, 37001 TOURS,

- Dr. A. PUGIN (L), Laboratoire de Biochimie, Université de Franche-Comté, Route de Gray, 25030-BESANCON,

- Dr. J.P. RENAUDIN (L), Laboratoire de Physiologie Cellulaire Végétale, C.N.R.S., Avenue de la Terrasse, 91190-GI-SUR-YVETTE,

- Dr. M. RIEDEL (L), Radiobiology Research Group, C.E.R.N., GENEVA (SWITZERLAND). Presently : Laboratoire d'Histophysiologie et de Radiobiologie Végétales, Université des Sciences et Techniques du Languedoc, Place Eugène Bataillon, 34060-MONTPEL-LIER-CEDEX,

- Dr. J.P. RONA (L), Laboratoire d'Electrophysiologie des Membranes, U.A. C.N.R.S. No. 578, Université de Paris-VII, 2, Place Jussieu, 75251-PARIS-CEDEX 5,

- Dr. F.L. SOUQ (P), Société G. DELBARD, Centre de Recherches de La Malicorne, 03600-COMMENTRY,

- Dr. P. STECK (P), Département de Phytotechnologie, Elf-Biorecherches, Centre de Recherches de Labège-Innopole, Boite Postale No. 137, 31328-CASTANET-TOLOSAN-CEDEX,

- Pr. M. THELLIER (L), Laboratoire sur les Echanges Cellulaires chez les Végétaux, Université des Sciences et Techniques de Rouen, Boite Postale No. 67, 76130-MONT-SAINT-AIGNAN,

- Dr. M.F. TROUSLOT (P), Laboratoire de Physiologie et de Biotechnologie Végétales, Centre ORSTOM d'Adiopodoumé, Boite Postale No. V-51, ABIDJAN (IVORY COAST),

- Dr. P. VOLFIN (L), Institut de Biochimie, Faculté des Sciences de Paris XI, Centre Universitaire d'Orsay, 91504-ORSAY,

GERMANY

- Dr. A. EHMKE (P), Institute für Pharmazeutische Biologie, Mendelsohn-strasse 1, D-3300-BRAUNSCHWEIG,

- Dr. U.I. FLUGGE (L), Institut für Biochemie, Universität Göttingen, Untere Karspüle 2, D-3400-GOTTINGEN,

- Pr. R. HAMPP (CS, L), Institut für Biologie 1, Universität Tübingen, Auf der Morgenstelle 1, D-7400-TUBINGEN 1,

- Dr. R. HEDRICH (L), Max-Planck Institut für Biophysikalische Chemie, Am Fassberg, D-3400-GOTTINGEN,

- Dr. B. HOFFMANN (L), Botanisches Institut 1 der Justus-Liebig-Uni-versität, Senckenbergstrasse 17-21, D-6300-GIESSEN,

- Dr. E. JOHANNES (P), Botanisches Institut der Justus-Liebig-Universität, Senckenbergstrasse 17-21, D-6300-GIESSEN,

- Dr. G. KAISER (C, L), Lehrstuhl 1 der Universität, Mittlerer Dallen-bergweg 64, D-8700-WURZBURG,

- Dr. W. KREIS (P), Pharmazeutische Institüt, Universität Tübingen, Auf der Morgenstelle 8, D-7400-TUBINGEN,

- Pr. Th. RAUSCH (L), J.W. Goethe Universität, Siesmayerstrasse 70, D-6000-FRANKFURT,

- Dr. M.J. SCHRAMM (P), Institut für Botanik 1 der Universität, Mittlerer Dallenbergstrasse 64, D-8700-WURZBURG,

- Dr. M. STEINGRABER (P), Institut für Biologie, Universität Tübingen, Auf der Morgenstelle 1, D-7400-TUBINGEN 1,

- Dr. M. WINK (C, L), Pharmazeutische Biologie Institut, Universität München, Karlstrasse 29, D-8000-MUNCHEN,

- Pr. U. ZIMMERMANN (L), Lehrstuhl für Biotechnologie der Universität, Universität Wurzburg, Röntgenring 11, D-8700-WURZBURG,

ISRAEL

- Dr. M. KEREN-ZUR (P), International Genetic Sciences Partnership, P.O. Box No. 4330, IL-JERUSALEM 91042,

ITALY

- Dr. R. CERANA (L), Dipartimento di Biologia, Università degli Studi di Milano, Sezione di Fisiologia e Biochimica della Piante, Via Celoria 26, I-20133-MILANO,

- Dr. R. COLOMBO (L), Dipartimento di Milano, Università degli Studi di Milano, Sezione di Fisiologia e Biochimica delle Piante, Via Celoria 26, I-20133-MILANO,
- Dr. P. LADO (L), Dipartimento di Biologia, Università degli Studi di Milano, Sezione di Fisiologia e Biochimica delle Piante, Via Celoria 26, I-20133-MILANO,
- Dr. F. MACRI (L), Consiglio Nazionale della Ricerche, Istituto Biosyntesi Vegetali, Sezione di Padova, Corso Stati Uniti 4, I-35020-PADOVA,
- Pr. E. MARRE (CS, L), Dipartimento di Biologia, Università degli Studi di Milano, Sezione di Fisiologia e Biochimica delle Piante, Via Celoria 26, I-20133-MILANO,
- Dr. A. PERES (P), Dipartimento di Fisiologia e Biochimica Generali, Università degli Studi di Milano, Via Celoria, 26, I-20133-MILANO,
- Dr. A. VIANELLO (L), Consiglio Nazionale delle Ricerche, Istituto Biosintesi Vegetali, Sezione di Padova, Corso Stati Uniti 4, I-35020-PADOVA,

JAPAN

- Pr. Y. ANRAKU (CS, C, L), Department of Biology, Faculty of Sciences, University of Tokyo, HONGO, Tokyo 113,
- Dr. K. KITAMOTO (L), National Research Institute of Brewing, 2-830, Takinogawa, KITA-KU, Tokyo 114,
- Dr. Y. MORIYASU (P), Department of Botany, Faculty of Sciences, University of Tokyo, HONGO, Tokyo 113,
- Pr. K. NISHIDA (L), Kanazawa University, Department of Biology, 11, Marunouchi, KANAZAWA, Ishikawa,
- Dr. Y. YOSHIHASA (P), Department of Biology, Facultv of Sciences, University of Tokyo, HONGO, Tokyo 113,

NETHERLANDS

- Dr. I.J.M. BLOM (P), Universiteit Leiden, Botanisch Laboratorium, Nonnensteeg 3, NL-2311-VJ-LEIDEN,
- Dr. H. VAN DER VALK (P), Afdeling Plantenfysiologie van de Landbouwwhogeschool, Arboretumlaan 4, NL-6703-BD-WAGENIGEN,
- Dr. W.M. VAN GULIK (L), Center for Bio-Pharmaceutical Sciences, Gorlaeus Laboratories, P.O. Box No. 9502, NL-2300-RA-LEIDEN,

SWITZERLAND

- Dr. T. BOLLER (CS, C, L), Botanisches Institut, Universität Basel, Schönbeinstrasse 6, CH-4056-BASEL,
- Dr. F. KELLER (L), Universität Zürich, Institut für Pflanzenbiologie, Abteilung Physiologie, Zollikerstrasse 107, CH-8008-ZURICH,
- Dr. E. MARTINOIA (L), Laboratorium für Pflanzenphysiologie, Eidgenössischen, Technischen Hochschule Zürich, Sonneggstrasse 5, CH-8092-ZURICH,
- Pr. P. MATILE (C, L), Universität Zürich, Institut für Pflanzenbiologie, Abteilung Physiologie, Zollikerstrasse 107, CH-8008-ZURICH,
- Dr. C.H.W. WERNER (P), Universität Zürich, Institut für Pflanzenbiologie, Abteilung Physiologie, Zollikerstrasse 107, CH-8008-ZURICH,

UNITED KINGDOM

- Dr. W.F. BOSS (L), University of Edinburgh, Botany Department, The King's Building, Mayfield Road, EDINBURGH EH9-3JH,
- Dr. C.J. GRIFFITH (P), Department of Biology, University of York, HESLINGTON, Yorkshire YO1-5DD,
- Pr. J.L. HALL (CS), Biology Department, Building 44, The University, SOUTHAMPTON SO9-5NH
- Dr. B.C. HOMEYER (P), Department of Pharmacognosy, The School of Pharmacy, University of London, 29-30 Brunswick Square, LONDON WC1N-1AX,
- Dr. P. JOHN (CS, C, L), Department of Agricultural Botany, Plant Science Laboratory, University of Reading, Whiteknights, READING R6G-2AS,
- Dr. R.A. LEIGH (CS, C, L), Soils and Plant Nutrition Department, The Rothamsted Experimental Station, HARPENDEN AL5-2JQ,
- Pr. J.A. RAVEN (C, L), Department of Biological Sciences, University of Dundee, DUNDEE DD1-4MN,
- Dr. M.F. ROBERTS (C, L), Department of Pharmacognosy, The School of Pharmacy, University of London, 29-30 Brunswick Square, LONDON WC1N-1AX,
- Dr. K. RUDGE (L), The Wolfson Institute of Biotechnology, University of Sheffield, SHEFFIELD S10-2TN,
- Dr. J.A.C. SMITH (C, L), University of Edinburgh, Department of Botany, The King's Building, Mayfield Road, EDINBURGH EH9 3JH,

U.S.A.

- Pr. H. KENDE (L), Michigan State University, MSU-DOE Plant Research Laboratory, EAST-LANSING, Michigan 48824,
- Dr. W. LIN (L), E.I. Du Pont de Nemours and Company, Central Research and Development Department, Experimental Station, WILMINGTON, Delaware 19898,
- Dr. A. MARETZKI (P), The Hawaiian Sugar Planters' Association, 99-193, Aiea Hights Drive, AEIA, Hawaii 96701,
- Pr. J.D. MORRE (L), Purdue University, Biological Sciences and Medicinal Chemistry Department, WEST-LAFAYETTE, Indiana 47907,
- Dr. P.E. PFEFFER (L), U.S. Department of Agriculture, Agricultural Research Service, North Atlantic Area Eastern Regional Center, Department of Plant and Soil Biophysics, 600, East Mermaid Lane, PHILADELPHIA, Pensylvania 19118,
- Dr. J.K.M. ROBERTS (L), University of California at Riverside, Department of Biochemistry, RIVERSIDE, California 92521,
- Pr. L. TAIZ (L), Biology Department, Thimann Laboratories, University of California at Santa-Cruz, SANTA-CRUZ, California 95064,
- Dr. M. THOM (L), The Hawaiian Sugar Planters' Association, 99-193, Aiea Hights Drive, AEIA, Hawaii 96701,
- Pr. G.J. WAGNER (CS, C, L), Agronomy Department, College of Agriculture, University of Kentucky, LEXINGTON, Kentucky 40546-0091,
- Pr. R.L. WEISS (L), Department of Chemistry and Biochemistry, University of California at Los-Angeles, LOS-ANGELES, California 90014.

[1] C = Convenor, CO = Organizer, CS = Scientific Committee Member, P = Participant, L = Speaker